GABA Signaling System and Abiotic Stress Tolerance in Plants

Syed Uzma Jalil • Shamim Akhtar Ansari
Mohammad Israil Ansari

GABA Signaling System and Abiotic Stress Tolerance in Plants

Springer

Syed Uzma Jalil
Department of Biotechnology
Central Institute of Medicinal &
Aromatic Plants
Lucknow, Uttar Pradesh, India

Mohammad Israil Ansari
Department of Botany
University of Lucknow
Lucknow, Uttar Pradesh, India

Shamim Akhtar Ansari
Institute of Forest Productivity
Ranchi, Jharkhand, India

ISBN 978-981-95-0436-7 ISBN 978-981-95-0437-4 (eBook)
https://doi.org/10.1007/978-981-95-0437-4

This Springer imprint is published by the registered company Springer Nature Singapore Pte Ltd.
The registered company address is: 152 Beach Road, #21-01/04 Gateway East, Singapore 189721, Singapore

Preface

The book highlights the regulatory aspects of GABA in the defense mechanism of plants by regulating ROS production under unfavorable environments with an association of the GABA shunt pathway.

Excessive production of reactive oxygen species (ROS) due to biotic and abiotic stresses is the most common type of environmental contaminant, and its toxicity causes detrimental effects in plants and has a significant impact on crop yield globally. Environmental conditions generate an excessive concentration of ROS, which can induce lipid peroxidation, protein oxidation, enzyme inactivation, DNA damage, and interactions with other essential constituents of plant cells. Higher plants have evolved an intricate defense system to scavenge ROS. Being a central molecule of the defense system, gamma-aminobutyric acid (GABA) is ubiquitous in all from prokaryote to eukaryote cells, and engaged with both direct and indirect mitigation of ROS in plants. GABA plays a key role as a metabolite or endogenous signaling molecule in several regulatory mechanisms and accumulates in plant cells under stress conditions by the GABA shunt pathway. The GABA transporter supports GABA in stress tolerance by transporting the high-content Ca^{2+} into the cytosol of cells that are highly generated by attachment of GABA to the cell surface binding sites after the exposure of environmental stress stimuli. Ca^{2+} combines with CAM that activates the GAD enzyme for the conversion of glutamate into GABA. This synchronized process regulates the expression of GABA shunt genes during stress and enhances plant stress tolerance. The book chapters will provide detailed information on the signaling behavior of GABA to mitigate adverse impacts of stresses on plant growth and productivity via redirecting signaling pathways for acclimatization to the stress stimuli. Overall, the book will present a comprehensive scenario of GABA regulation in plants under adverse environmental conditions and will include the following topics:

- Elucidation of synthesis, signaling, and transportation of GABA in plants
- GABA modulation of plant physiological responses conferring tolerance against abiotic stress conditions
- Conferring the counterresponse of GABA against biotic stress conditions
- Function of GABA transporters in improving stress tolerance

- Molecular and functional role of GABA during senescence
- ROS production mechanisms under multitude of stress
- Promising role of GABA in postharvest management

The book chapters are based on extensive review of literature with logical illustrations. Hence, the book would be a new up-to-date addition to plant stress physiology and food security, encouraging entrepreneurs and professional scientists to acquire innovative ideas for sustainable improvement of global crop production. The chapter contents would also serve an updated comprehensive textbook to graduate students as well.

Lucknow, Uttar Pradesh, India — Syed Uzma Jalil
Ranchi, Jharkhand, India — Shamim Akhtar Ansari
Lucknow, Uttar Pradesh, India — Mohammad Israil Ansari

Competing Interests The authors have no competing interests to declare that are relevant to the content of this manuscript.

Contents

About the Authors

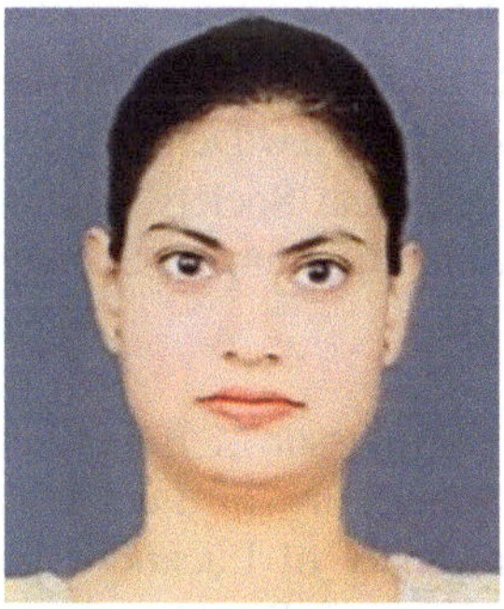

Syed Uzma Jalil is a distinguished researcher in Plant Biotechnology, with over eight years of experience in active research and innovation. She holds a PhD in Biotechnology from Amity University Uttar Pradesh, India, and an MSc in Biotechnology from Integral University, Lucknow, India. Her academic and professional pursuits center on gene transformation, plant tissue culture, metabolite engineering, molecular biology, biochemical assays, and nanoparticle synthesis and characterization. She has significantly advanced knowledge in these fields, contributing to the scientific development of horticultural and medicinal plants. She has successfully undertaken research projects at prestigious CSIR and ICAR institutes in India. Her scholarly contributions include over 35 publications and a comprehensive book, and she is a co-inventor of a granted US patent. Recognized internationally, she has presented her work at prominent conferences in Italy, Sri Lanka, and Singapore, earning multiple travel awards for her outstanding research contributions.

Shamim Akhtar Ansari Scientist G and Ex-Director, Institute of Forest Productivity, Ranchi, India, and an erudite academician and researcher with an experience of >38 years, is currently Visiting Professor at Integral University, Lucknow. He has specialized in Plant Physiology and Biochemistry during MS and doctoral work. Subsequently, he joined as Senior Scientist in Indian Council of Forestry Research and Education under Government of India Ministry of Environment, Forests and Climate Change. He has conducted extensive research, and developed expertise, in Clonal Forestry or Forest Biotechnology, including Tissue

Culture and Adventitious Rhizogenesis and Population (molecular) Genetics and Association Mapping of Tropical Deciduous Trees. He served as Assistant Editor of "TFRI Newsletter" for five years (1997–2002) and Editor of peer-reviewed international "Indian Journal of Tropical Biodiversity" for ten years (2002–2012). He has edited three books, co-authored two books, and published >100 research papers/review articles/popular science articles and received global citations. He has also guided 11 students for doctorate program. He has received several academic accolades, notable among them are University Gold Medal (Aligarh Muslim University, Aligarh) and R.D. Asana Gold Medal (Indian Society for Plant Physiology, New Delhi). He is an elected fellow of Indian Botanical Society, Indian Society for Plant Physiology, and the Linnean Society of London.

Mohammad Israil Ansari is Professor of Botany at the Department of Botany, University of Lucknow, India. Before joining the University of Lucknow, he was Professor at Amity Institute of Biotechnology, Amity University, Lucknow Campus, Lucknow (2007–2016). He graduated and post-graduated from the Department of Botany, Aligarh Muslim University, Aligarh, India, and doctoral degree from Dr. Ram Manohar Lohia Avadh University, Faizabad, India. He is an Academia Sinica Fellow from the Government of Taiwan and Fellow of Indian Botanical Society. He has more than 33 years of research and 22 years of teaching experience with 135 research papers, several books, book chapters, review articles, Post-doc, PhD, post-graduate summer projects as well as funded research projects to his credit. He has been associated with the Institute of Botany, Academia Sinica, Taiwan, Graduate Institute of Plant Biology, National Taiwan University, Taiwan and Institute of Plant and Microbial Biology, Academia Sinica, Taiwan (2000–2007). He had made significant contributions in the field of Nanotechnology, Molecular Biology of Leaf Senescence, Dehydration Stress and Red Rot Disease of Sugarcane. He is working on Bioengineered *Piper betle* GABA Functionalized Nanoparticles in Nutrient Enhancement of *Spinacia oleracia* and GABA-Transaminase gene that plays a very important role in leaf senescence process. He is editorial board member of several international journals as well as member of several Academic Board/Faculty Board/Board of Studies/Research and Development bodies. He has delivered several invited lectures at national and international level. He has completed several funded projects from the Department of Science and Technology, Govt. of India, New Delhi, Science and Engineering Research Board New Delhi, Department of Biotechnology, Govt. of India, Uttar Pradesh, Higher Education, Govt. of Uttar Pradesh, India.

Abbreviations[1]

$_1O^2$	Singlet oxygen
AAP3	Amino acid permease
ABA	Abscisic acid
ABTS	2,2′-azino-bis (3-ethylbenzothiazoline-6-sulfonic acid)
ACC	1-Aminocyclopropane-1-carboxylic acid
ACX	Acyl-CoA oxidase (EC 1.3.3.6)
ADC	Arginine decarboxylase (EC 4.1.1.19)
AG	Aminoguanidine
AKG	α-Ketoglutarate
Al	Aluminum
ALMT	Aluminum-activated malate transporter
ALMT9	Aluminum-activated malate transporter 9
AMADH	4-Aminobutyraldehyde dehydrogenase (EC 1.2.1.19)
AMT	Ammonium transporter
AO	Aldehyde oxidase (EC 1.2. 3.1)
APX	Ascorbate peroxidase (1.11.1.11)
AR	Adventitious root
As	Arsenic
AsA	Ascorbate
BAT1	Bidirectional transmembrane transporter
Ca^{2+}	Calcium ion
CAAT	Cationic amino acid transporter
Cad	Cadaverine
CaM	Calmodulin
CAT	Catalase (EC 1.11. 1.6)
Cd	Cadmium
CDPK	Calcium-dependent protein kinase (EC 2.7.1.37)
CML	Calmodulin-like protein
COR	Cold-responsive
Cr	Chromium

[1] Suffixing "s" to the abbreviation makes it plural, e.g., PA (Singular) and PAs (Plural).

CRISPR	Clustered regularly interspaced short palindromic repeats
CRISPR-Cas9	Specific molecular scissors
CuAO	Copper amine oxidase (EC 1.4.3.6)
DAO	Diamine oxidase (EC: 1.4.3.6)
DHAR	Dehydroascorbate reductase (1.8.5.1)
DPPH	2,2-Diphenyl-1-picrylhydrazyl
ETC	Electron transport chain
FAD	Flavin adenine dinucleotide
Fe	Iron
GA	Gibberellin
GABA	γ Aminobutyric acid
GABA-T	GABA transaminase (2.6.1.19)
GABA-TK	α-Ketoglutarate-dependent GABA transaminase (2.6.1.19)
GABA-TP	Pyruvate-dependent GABA transaminase (2.6.1.19)
GABP	GABA permease
GAD	Glutamate decarboxylase (EC: 4.1.1.15)
GAI	Gibberellin insensitive gene
GAT	GABA transporter
GDH	Glutamate dehydrogenase (EC 1.4.1.3)
GHB	γ-Hydroxybutyric acid
GHBDH	GHB dehydrogenase (EC 1.1.1.61)
GHS	Glutathione
GORK	Outward-rectifying Kv-like K(+) channel expressed in guard cells
GPX	Glutathione peroxidase (EC 1.11.1.9)
GR	Glutathione reductase (EC 1.8.1.7)
GS	Glutamine synthetase (EC 6.3.1.2)
GS/GOGAT	Glutamine synthetase/glutamate synthase (6.3.1.2)
GSA	Glutamic-γ-semialdehyde
GSH	Glutathione
GST	Glutathione S-transferase (EC 2.5.1.18)
H_2O_2	Hydrogen peroxide
Hg	Mercury
HR	Hypersensitive response
HSP	Heat shock protein
HSTF	Heat shock transcription factor
IAA	Indole-3-acetic acid
JA	Jasmonate/jasmonic acid
LOX	Lipoxygenase (EC 1.13.11.-)
MAPK	Protein kinase
MDA	Malonaldehyde
MDH	Malate dehydrogenase (EC 1.1.1.37)
MDHAR	Mon-dehydroascorbate reductase (EC 1.6.5.4)
MeJA	Methyl jasmonate
Mg	Magnesium

NADH-GOGAT	Glutamate synthase (EC 1.4.1.14)
NO	Nitric oxide
NR	Nitrate reductase (EC 1.6.6.1-3)
NRT	Nitrate transporter
O_2^-, $O_2^{\cdot -}$, $O_2\bullet-$	Superoxide radical
ODC	Ornithine decarboxylase (EC 4.1.1.17)
OH^-, $^{\cdot}OH$, •OH, OH•	Hydroxyl radical
P5C	Δ^1-Pyrroline-5-carboxylate
PA	Polyamine
PAO	Polyamine oxidase (EC 1.5.3.11)
Pb	Lead
PCD	Programmed cell death
POD	Peroxidase
ProT2	Proline transporter 2
PS	Photosystem
Put	Putrescine
PYRR-DH	Pyrroline-5-carboxylate reductase (EC 1.5.1.2)
RBOH	NADPH oxidase (EC 1.6.3.1)
Rboh	Respiratory burst oxidase homolog
RCC	Red chlorophyll catabolite
RNAi	RNA interference
RO•	Alkoxyl radicals
ROO•	Peroxy radicals
ROS	Reactive oxygen species
Rubisco	Ribulose 1,5 biphosphate carboxylase/oxygenase (EC 4.1.1.39)
SA	Salicylic acid
SAG	Senescence-associated gene
SAM	S-adenosyl methionine
SCoAL	Succinyl-CoA ligase (EC 6.2.1.4, EC 6.2.1.5)
SCOAS	Succinyl-CoA synthetase (EC 6.2.1.4, EC 6.2.1.5)
SOD	Superoxide dismutase (EC 1.15.1.1)
Spd	Spermidine
SSA	Succinate semialdehyde
SSADH	Succinate semialdehyde dehydrogenase (EC 1.2.1.24)
SSR	Succinate semialdehyde reductase (EC 1.2.1.76)
TCA	Tricarboxylic acid
UQ	Ubiquinone
XDH	Xanthine dehydrogenase (EC 1.17.1.4)
XOD	Xanthine oxidase (EC 1.17.3.2)
Zn	Zinc

1 GABA in Plants: An Introduction

Abstract

GABA is a nonprotein amino acid composed of a four-carbon backbone, which serves as a multifunctional signaling compound across diverse organisms, including both animals and plants. In plants, GABA plays an integral role in regulating a wide range of physiological and molecular processes. These include maintaining redox balance, stabilizing cytosolic pH, adjusting osmotic potential, and facilitating nitrate absorption and assimilation. It is also involved in carbon and nitrogen metabolism, directing pollen tube navigation, modifying cell walls, supporting adventitious root formation, and influencing plastid-related activities. In addition to its contributions to general plant growth and development, GABA is recognized for its role in enhancing tolerance to both biotic and abiotic stresses. It also has significant functions in regulating leaf senescence and maturation, as well as improving the quality and longevity of postharvest produce. Advancing our understanding of GABA's multifaceted roles is essential for optimizing plant performance across a range of environmental conditions, including both normal and stress-induced scenarios.

Keywords

GABA · Plant growth · ROS · Signaling molecule · Stress tolerance

Highlights GABA is a nonprotein amino acid with multifaceted signaling roles in plants, contributing to the regulation of redox homeostasis, cytosolic pH stabilization, and osmotic adjustment.

S. U. Jalil et al., *GABA Signaling System and Abiotic Stress Tolerance in Plants*,
https://doi.org/10.1007/978-981-95-0437-4_1

- It is integrally involved in coordinating Carbon and Nitrogen (C/N) metabolism, enhancing nitrate uptake, guiding pollen tube development, influencing cell wall restructuring, and promoting root growth.
- GABA supports plant tolerance to both biotic and abiotic stresses, aiding in growth maintenance and survival under challenging environmental conditions.
- It contributes significantly to the regulation of leaf senescence and is valuable in improving postharvest quality and storage, making it important for key developmental and postharvest phases.
- Exploring GABA's diverse functional roles offers valuable insights for enhancing plant health and productivity in both favorable and stressful environments.

1.1 Introduction

GABA is a crucial metabolite involved in stress responses across various biological systems, including prokaryotes, unicellular eukaryotes, and higher plants (Satyanarayan and Nair 1990; Shelp et al. 1999). Although GABA gained early attention due to its role in the vertebrate nervous system as a neurotransmitter (Ribak and Yan 2000), its significance in plant biology was recognized decades ago (Stewart 1949). Unlike several nonprotein amino acids restricted to specific taxa, GABA is nearly ubiquitous, having been initially discovered in potato tubers and later found across a wide range of organisms, including plants, where it accumulates in larger amounts compared to animals (Ramos-Ruiz et al. 2019; Al-Taher and Nemzer 2019).

In plants, GABA is found throughout various tissues and is also present in the root nodules of legumes as well as in vascular tissues such as xylem and phloem sap (Satyanarayan and Nair 1990). Predominantly existing in its zwitterionic form, GABA metabolism occurs in distinct cellular compartments, with biosynthesis mainly in the cytosol and degradation within the mitochondria. The compound serves as a key intermediate in the GABA shunt, where it is synthesized from L-glutamate and subsequently funneled into the tricarboxylic acid (TCA) cycle via succinate formation (Shi et al. 2010; Ansari et al. 2005, 2014).

Additionally, GABA can be generated through polyamine (PA) catabolism involving compounds such as putrescine (Put) and spermidine (Spd) (Bhatnagar et al. 2002; Shelp et al. 2012) or via nonenzymatic conversion of proline under oxidative stress (Signorelli et al. 2015; Michaeli and Fromm 2015).

The compound exerts multiple physiological roles in plants (Khan et al. 2021; Li et al. 2021a, b), including participation in central carbon metabolism, modulation of nitrogen assimilation, and functioning as a regulatory molecule during developmental and stress-related responses (Bouche and Fromm 2004; Fait et al. 2008; Bown and Shelp 2016; Ramesh et al. 2015). Its levels typically rise under stress conditions, such as drought, salinity, or aging (Ansari et al. 2005; Zarei et al. 2017; Oh et al. 2019; Jalil et al. 2017, 2019b).

Previous research has highlighted GABA's contributions to carbon/nitrogen balance, pH homeostasis, stress tolerance, and shelf-life extension in horticultural produce (Bouche and Fromm 2004; Ramos-Ruiz et al. 2019). Furthermore, its role in plant development encompasses pollen tube elongation, stem and root formation, and vascular differentiation (Palanivelu et al. 2003; Akama and Takaiwa 2007; Renault et al. 2011; Jalil et al. 2016).

1.2 GABA Characteristics in Plant Cells

GABA (Fig. 1.1) is an amino acid derivative and is structurally related to glutamate, another important amino acid. Chemically, it is composed of a straight carbon chain, terminating with an amino group (-NH_2) at one end and a carboxyl group (-COOH) at the opposite end. The carbon adjacent to the carboxyl group, known as the alpha carbon, is bonded to a hydrogen atom, contributing to its simple yet functionally significant structure.

GABA is highly water-soluble and exists in plants in both free and conjugated forms. In aqueous environments, it may adopt a flexible conformation, occasionally resembling the cyclic structure of proline. At physiological pH, GABA typically exists as a zwitterion—with the amino group carrying a positive charge (-NH_3^+) and the carboxyl group carrying a negative charge (-COO-)—enabling it to engage in various biochemical interactions (Shelp et al. 1999).

Its conformation is influenced by local conditions, such as pH, ionic strength, and interactions with enzymes, receptors, and transport proteins. GABA can form hydrogen bonds with surrounding water molecules or interact with solutes in the cytoplasm. The ability of GABA to undergo conformational flexibility enables it to participate effectively in metabolic and signaling networks.

GABA also interacts with specific membrane transporters and putative receptors in plant cells, potentially influencing its spatial configuration and activity. Such interactions are essential for its uptake, distribution, and functional responses. Environmental conditions, particularly those causing stress, can alter the protonation state of GABA and thereby its activity in cellular processes (Majumdar and Guha 1988; Sapse 2000).

GABA levels vary substantially depending on tissue type and environmental conditions, ranging from approximately 0.03 to 32.5 picomoles per gram fresh

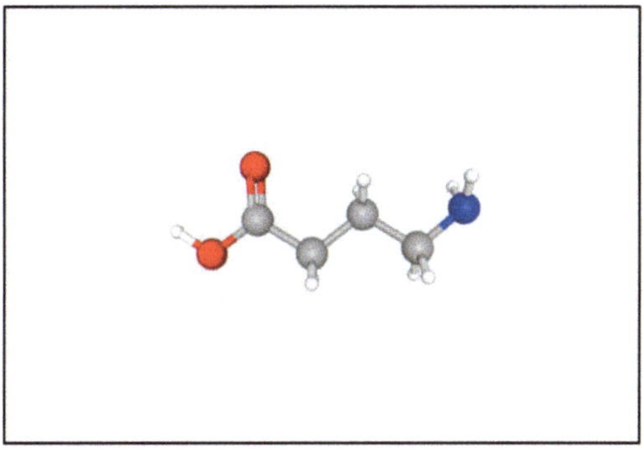

Fig. 1.1 GABA characteristics in plant cells

weight. Under normal conditions, the free GABA concentration in plant cells typically ranges between 0.03 and 2.0 micromoles per gram fresh weight (Shelp et al. 1999). However, during stress conditions, such as low oxygen availability, mechanical damage, temperature extremes, or excess water, its levels can increase rapidly (Wallace et al. 1984).

Additionally, in specific plant tissues like the root nodules of legumes, bound forms of GABA may represent a substantial portion—up to 6.6% of dry weight. The presence of different GABA-metabolizing enzymes and transport proteins across plant species and cellular compartments suggests that GABA may exhibit diverse conformational and functional characteristics depending on the context.

Further research is needed to elucidate the full range of GABA's structural dynamics and its molecular interactions in various plant systems. Nevertheless, its biochemical versatility and responsiveness to environmental cues make it a central player in plant adaptation and homeostasis.

1.2.1 GABA Biosynthesis in Plants

In plant cells, the synthesis of GABA is predominantly initiated in the cytoplasm through the decarboxylation of glutamate, a reaction catalyzed by the enzyme glutamate decarboxylase (GAD). This reaction is a key component of the GABA shunt pathway, which provides an alternative route to the TCA cycle.

Once synthesized, GABA is transported into the mitochondria, where it undergoes transamination by GABA transaminase (GABA-T). This enzyme uses either α-ketoglutarate (via GABA-TK) or pyruvate (via GABA-TP) as amino group acceptors to convert GABA into succinic semialdehyde (SSA). SSA is subsequently oxidized by succinic semialdehyde dehydrogenase (SSADH) to produce succinate, which feeds into the TCA cycle and supports both energy production and redox balance (Michaeli and Fromm 2015; Sita and Kumar 2020).

In addition to the canonical GABA shunt, other biosynthetic routes contribute to GABA accumulation, particularly under stress. For instance, polyamines (PAs) like spermidine (Spd) and putrescine (Put) can be catabolized to generate GABA, especially under oxidative stress conditions. These reactions are typically facilitated by enzymes such as diamine oxidase (DAO) and polyamine oxidase, which yield 4-aminobutanal—a compound that can spontaneously cyclize to pyrroline and be subsequently oxidized into GABA (Bouchereau et al. 1999; Shelp et al. 2012; Signorelli et al. 2015).

Proline degradation also contributes to GABA synthesis during stress. Glutamate is first converted to glutamate-5-semialdehyde by glutamate 5-kinase and dehydrogenase. This intermediate may cyclize to form 1-pyrroline-5-carboxylate, which is reduced to proline or diverted through ornithine into other pathways. Under certain stress conditions, proline can be converted nonenzymatically into GABA (Shelp et al. 2012).

The enzymatic breakdown of polyamines also plays a significant role in regulating GABA levels. For example, DAO catalyzes the oxidative deamination of Put, yielding hydrogen peroxide, ammonia, and 4-aminobutanal, which then enters the GABA biosynthetic route via pyrroline dehydrogenase activity.

Overall, GABA biosynthesis in plants is not limited to a single pathway but involves multiple interconnected metabolic routes that are often upregulated under stress. These alternative biosynthetic sources ensure GABA availability as both a metabolic intermediate and a signaling molecule, reinforcing its role in plant growth, stress tolerance, and metabolic regulation.

1.3 GABA: Roles and Functions

In plants, GABA is far more than a metabolic by-product; it plays a central role in coordinating physiological responses and cellular processes. Its function extends beyond stress mitigation to include regulation of growth, development, ion fluxes, carbon and nitrogen metabolism, and signal transduction. The multifunctional nature of GABA enables it to mediate responses to environmental cues and influence interactions with other organisms, including microbes and herbivores.

The diverse physiological roles of GABA are underpinned by its integration into key biochemical pathways, such as the GABA shunt, and its interaction with other signaling networks, including calcium and hormonal signals. GABA accumulation is typically induced under stress conditions, allowing plants to modulate internal processes in response to adverse environments.

Below are some of the major functional domains where GABA plays a vital role, beginning with its involvement in microbial stress responses.

1.3.1 GABA and Microbial Stress Response

GABA's involvement in stress physiology is not exclusive to plants; it also plays important roles in microbial systems. Many bacteria, including strains of *Pseudomonas aeruginosa*, *Pseudomonas fluorescens*, *Escherichia coli*, and *Lactobacillus* species, are capable of synthesizing and utilizing GABA. In these organisms, GABA can act as a signaling molecule that regulates various cellular activities, including stress adaptation and communication within microbial communities (Noe et al. 1958; Chou et al. 2008; Siragusa et al. 2007).

For instance, in *Pseudomonas fluorescens*, GABA has been shown to regulate biofilm formation and environmental stress responses through intercellular signaling pathways (Dagorn et al. 2013). Similarly, in *P. aeruginosa*, GABA affects the expression of genes involved in virulence and cyanide production under low oxygen availability, indicating a role in regulating pathogen–host interactions.

Mutants lacking functional genes in the GABA shunt often show increased sensitivity to acid stress and other environmental challenges. For example, *Listeria monocytogenes* mutants deficient in GAD activity exhibit reduced GABA accumulation and impaired acid resistance (Karatzas et al. 2012). Similarly, deletion of the *gabT* genes in *Pseudomonas syringae* strains impairs GABA utilization and reduces virulence during plant infection (Park et al. 2010).

In bacteria, GABA also serves as a source of carbon and nitrogen, particularly under nutrient-limited conditions. The tomato pathogen *P. syringae* has been observed to extract GABA from the plant apoplast and use it as a sole nutrient source (Rico and Preston 2008). In *E. coli*, strains with enhanced GABA transaminase activity can grow on GABA as their only carbon and nitrogen source (Dover and Halpern 1972).

GABA's influence extends to fungal and protozoan systems as well. In some fungi, it regulates hyphal development, sporulation, and cell wall remodeling (Kumar and Punekar 1997). Protozoan parasites such as *Plasmodium falciparum* and *Trypanosoma cruzi* show changes in GABA levels during various life stages, with evidence pointing to its role in energy metabolism and ion transport (MacRae et al. 2013; Galvez Rojas et al. 2015). In such parasites, GABA uptake and usage are often Adenosine Triphosphate (ATP)- and sodium-dependent, suggesting specific transport mechanisms.

These insights underscore the evolutionary conservation of GABA as a regulatory molecule across domains of life and point to its potential as a target for antimicrobial and antiparasitic strategies.

1.3.2 GABA in Yeast (*Saccharomyces cerevisiae*)

In *Saccharomyces cerevisiae*, the common baker's yeast, GABA is not merely a metabolic by-product but plays several critical physiological roles. As in plants and bacteria, GABA in yeast functions both as a metabolic intermediate and a stress-responsive signaling molecule.

One of the primary roles of GABA in yeast is in stress adaptation. It contributes to resistance against various environmental stressors, including oxidative and osmotic stress. Studies have shown that under such conditions, GABA levels rise, assisting in the stabilization of intracellular pH and redox balance (McKelvey et al. 1990; Cao et al. 2013).

In the context of nitrogen metabolism, GABA is generated through glutamate decarboxylation and can serve as a nitrogen reservoir, particularly when external nitrogen is scarce (Ramos et al. 1985). Yeast cells can store surplus nitrogen in the form of GABA and later utilize it during growth-limiting conditions, highlighting its role in nutrient homeostasis (Bach et al. 2009).

Furthermore, GABA influences intracellular pH regulation. The decarboxylation of glutamate into GABA consumes protons, which helps neutralize excess acidity within the cell. This buffering mechanism becomes especially critical under extreme pH conditions, contributing to cell survival and metabolic equilibrium (Dhakal et al. 2012; Reed 1950).

Emerging studies suggest that GABA also contributes to intercellular communication in yeast. Though not functioning as a neurotransmitter in the classical sense, GABA appears to influence cell signaling, potentially via GABA-like transporters and receptor-like proteins that coordinate metabolic responses and community behavior (Ekema et al. 2002).

Additionally, GABA metabolism affects yeast cell wall synthesis. The biosynthetic and regulatory roles of GABA extend to modulating the enzymes and structural components required for maintaining cell wall integrity, which is vital under changing environmental conditions (Lakhani et al. 2014).

While our understanding of GABA's function in yeast is still evolving, it is evident that this molecule contributes to the regulation of key cellular processes, from nitrogen assimilation and stress resistance to pH buffering and, possibly, even community signaling. These diverse functions suggest that GABA has a foundational role in yeast physiology, similar to its roles in plants and bacteria.

1.3.3 GABA Function in the Mammalian Brain

In mammals, GABA serves as the principal inhibitory neurotransmitter in the central nervous system. It plays a pivotal role in modulating neuronal excitability by dampening overactive nerve signals, thereby ensuring proper synaptic balance and preventing hyperexcitability that can lead to neurological disorders (Roberts 1986; DeFelipe 1993; Ribak and Yan 2000).

GABA exerts its inhibitory effects by binding to specialized receptors located on postsynaptic membranes. These receptors are primarily categorized into three types: GABA_A, GABA_B, and GABA_C. The GABA_A and GABA_C receptors are ionotropic, forming ligand-gated chloride channels that, when activated, allow chloride ions to flow into the neuron. This influx hyperpolarizes the membrane and suppresses the likelihood of action potential generation. In contrast, GABA_B receptors are metabotropic and function via G-protein-coupled mechanisms, influencing ion channel activity indirectly (Bowery et al. 1980; Owen and Kriegstein 2002).

The biosynthesis of GABA in mammals is primarily driven by GAD, which exists in two isoforms—GAD65 and GAD67. These isoforms are encoded by separate genes and differ in their expression patterns and regulatory mechanisms. GAD67 is especially important for maintaining baseline GABA levels in the brain and is essential for normal neural function. Mice lacking this gene exhibit severe deficits and typically die shortly after birth, underlining its critical role (Asada et al. 1997; Ji et al. 1999).

Once synthesized, GABA is metabolized via the GABA shunt pathway, involving enzymes such as GABA-T and SSADH. The proper functioning of this pathway is crucial for neuronal health. For instance, defects in SSADH can lead to γ-hydroxybutyric acid (GHB) accumulation, a condition associated with a rare metabolic disorder characterized by developmental delays and neurological symptoms (Jacobs et al. 1981; Hogma et al. 2001). Similarly, deficiencies in GABA-T, although uncommon, are linked with neuromotor and behavioral abnormalities, including hypotonia and psychomotor retardation (Medina Kauwe et al. 1999).

GABA transporters, such as GAT1, are also essential for regulating extracellular GABA levels by mediating its reuptake into neurons and glial cells. This recycling process ensures that GABA signaling is tightly controlled and that neurotransmitter balance is maintained (Roettger and Amara 1999; Martin and Tobin 2000).

GABA plays a fundamental role in the mammalian nervous system by providing inhibitory tone, regulating neural circuits, and maintaining synaptic homeostasis. Its disruption is associated with a range of neurological and psychiatric conditions, making it a key target for therapeutic interventions in disorders such as epilepsy, anxiety, and schizophrenia.

1.3.4 GABA in Plant Physiology and Stress Responses

In plants, GABA is emerging as a multifunctional compound that integrates signaling, metabolism, and stress responses. While originally recognized as a product of stress-related metabolic shifts, GABA is now acknowledged as a dynamic player in regulating development, photosynthesis, ion fluxes, and interactions with phytohormones (Khan et al. 2021, Li et al. 2021a, b).

1.3.4.1 Interplay of GABA in Carbon and Nitrogen Metabolism

GABA plays a central role in coordinating C and N metabolism in plants by serving as a metabolic link that facilitates communication between these pathways. Its involvement in regulating C/N balance, enhancing nutrient uptake, and modulating core metabolic fluxes highlights its importance in plant development and physiological stability. Acting as a critical intermediary, GABA connects the TCA cycle, which is primarily involved in carbon processing, with nitrogen assimilation pathways, significantly influencing the metabolic flow between these networks (Ansari et al. 2014).

GABA has been reported to promote nitrate uptake and increase nitrogen accumulation. Its application has been found to stimulate root development and enhance the activity of key nitrogen-metabolizing enzymes, including nitrate reductase, glutamate synthase, and glutamine synthetase. In *Arabidopsis* seedlings, treatment with exogenous GABA (50 mM) under controlled nitrate availability conditions resulted in increased root growth, nitrate absorption, and nitrogen content (Barbosa et al. 2010). This enhancement in nutrient acquisition is associated with elevated enzymatic activities involved in nitrogen metabolism, indicating GABA's role in fine-tuning the C/N ratio within plant cells (Barbosa et al. 2010).

Moreover, GABA contributes to maintaining an equilibrium between carbon and nitrogen pools within cells (Fait et al. 2008). Its regulatory function is responsive to the nitrate levels in plant tissues, thereby maintaining homeostasis between these two essential macronutrients. Maintaining this C/N balance is fundamental to optimal plant function and development (Coruzzi and Zhou 2001), and fluctuations in this ratio influence processes such as senescence and stress adaptation (Martin et al. 2002; Aoyama et al. 2014).

When nitrogen becomes limited, plants often accumulate excess sugars, leading to a carbon-rich state (Ono et al. 1996). To adapt to such imbalances, plants can sense and respond by adjusting metabolic partitioning of carbon and nitrogen sources (Aoyama et al. 2014). These adjustments help reestablish C/N homeostasis and are achieved through metabolic regulation and spatial allocation of available resources (Sato et al. 2011; Sulpice et al. 2013; Aoyama et al. 2014).

In terms of primary carbon metabolism, GABA influences the expression of genes related to the GABA shunt (such as *GAD2* and *SSADH*) and those linked to the central carbon metabolism (Fait et al. 2008). GABA transport from the cytosol into the mitochondria depends on carrier proteins. One such transporter, AtGABP from the APC transporter family, is crucial for shuttling GABA into the TCA cycle (Michaeli et al. 2011). Mutants deficient in this transporter (gabp lines) display disrupted mitochondrial metabolism, altered C/N balance, and elevated TCA cycle activity. Introducing a functional GABP gene into these mutants restores growth under carbon-limiting conditions, reinforcing the importance of GABA in maintaining cellular C/N equilibrium and preventing nitrogen loss (Michaeli et al. 2011).

Further evidence from *Xanthium strumarium* suggests that the GABA shunt operates as a central regulator of carbon and nitrogen metabolism in photosynthetic tissues (Tcherkez et al. 2009). Moreover, 14–3-3 proteins, which modulate the activity of key metabolic enzymes through phosphorylation-dependent interactions, also contribute to regulating C/N metabolism. GABA, in concert with calcium signaling, influences the expression of 14–3-3 proteins, which subsequently regulate critical enzymes, such as nitrate reductase, glutamine synthetase, starch synthase III, and glyceraldehyde-3-phosphate dehydrogenase (Comparot et al. 2003; Moorhead et al. 1999; Lancien and Roberts 2006; Carillo 2018).

GABA metabolism thus plays a key role in managing the dynamic interactions between carbon and nitrogen in plants. It helps fine-tune the metabolic activity and resource allocation to preserve internal balance. These interactions are governed by a complex network of signals, including those derived from nitrate, ammonium, and amino acids, such as glutamate, glutamine, and aspartate (Miller et al. 2008; Nunes-Nesi et al. 2010; Naliwajski and Skłodowska 2018). Signals from carbon metabolism, including CO_2 levels and carbon skeletons produced by the TCA cycle, also influence the coordination of nitrogen assimilation (Nunes-Nesi et al. 2010; Naliwajski and Skłodowska 2018).

1.3.4.2 GABA and Cytosolic pH Homeostasis

The GABA shunt plays an essential role in helping organisms, including bacteria such as *Escherichia coli*, withstand acidic stress. Under low pH conditions, GAD activity increases, catalyzing the conversion of glutamate into GABA. This reaction consumes protons, thereby contributing to intracellular pH stabilization. Once formed, GABA is exported out of the cell, further supporting cytoplasmic pH homeostasis (Carroll et al. 1994). This mechanism, first observed in microbial systems, has been proposed to function similarly in plants and other organisms (Crawford et al. 1994).

In plants, fluctuations in cytosolic pH, especially acidification, can negatively impact growth and cellular function. GABA production is closely tied to these pH changes. For instance, during hypoxic conditions such as flooding, cellular oxygen levels decline, leading to acid accumulation and a decrease in cytoplasmic pH. This acidification triggers GAD activity and enhances GABA synthesis (Roberts 1984; Kinnersley and Turano 2000). Under such stress, the degradation of membrane integrity results in the release of organic acids, exacerbating intracellular acidification and necessitating compensatory responses like GABA accumulation (Crawford et al. 1994).

Experimental data demonstrate the rapid responsiveness of GABA synthesis to pH shifts. One study showed that exposure to a weak acid caused a sharp drop in cytosolic pH by approximately 0.60 units within seconds. This was immediately followed by a notable increase in GABA levels within just 2 s of acid exposure (Crawford et al. 1994). Similarly, oxygen deprivation resulting in a sustained pH reduction of 0.5–0.6 units has been shown to induce GABA synthesis within minutes (Roberts et al. 1992).

Alongside the pH response, intracellular calcium levels also rise in response to acid stress (Gehring et al. 1990). This calcium surge is thought to activate GAD, possibly via a calcium/calmodulin-mediated pathway, further amplifying GABA production (Crawford et al. 1994). In carrot (*Daucus carota*) cell cultures, for example, lowered pH levels promote increased conversion of glutamate to GABA, a reaction that consumes excess protons and helps buffer cytosolic acidity (Carroll et al. 1994). The accumulated GABA thus acts as a pH modulator, reducing proton concentration and contributing to cytoplasmic pH stability.

1.3.4.3 GABA Orchestrating Chloroplast Function and Photosynthesis

GABA has emerged as a key modulator of chloroplast functionality and overall photosynthetic performance in plants. It influences various aspects of chloroplast activity by regulating the expression of genes linked to chlorophyll biosynthesis, photosystem assembly, and carbon fixation processes, thereby impacting the efficiency of photosynthesis (Xu et al. 2021).

External application of GABA has been shown to support photosynthetic function, particularly under abiotic stress conditions. It achieves this by promoting improved stomatal regulation, which minimizes excessive water loss, and optimizing transpiration rates to enhance water use efficiency. Additionally, GABA helps sustain enzymatic activity associated with photosynthesis, thereby counteracting the suppression of photosynthetic rates often observed during stress (Xu et al. 2021).

In the context of low-temperature stress, GABA alleviates growth inhibition in muskmelon seedlings by preserving photosynthetic integrity. This is attributed to GABA's ability to maintain and regulate components of the photosynthetic system, thereby improving light capture and carbon assimilation. GABA may also regulate pigment synthesis, particularly chlorophyll and carotenoids, which are vital for photosynthetic light absorption. Xiang et al. (2016) reported that GABA supplementation helps protect chlorophyll a and carotenoids from degradation during chilling conditions. It is thought to do so by enhancing the biosynthesis of chlorophyll a and its precursors while stabilizing chloroplast membranes.

Similarly, under saline–alkaline stress, GABA improves stress tolerance in *Cucumis melo* L. seedlings by enhancing antioxidant enzyme activity, such as the ascorbate–glutathione (AsA–GSH) cycle, and reducing malondialdehyde (MDA) accumulation. This action preserves cellular and chloroplast membrane integrity (Chen et al. 2018).

The architecture of mesophyll cells, particularly chloroplast ultrastructure, plays a defining role in determining their photosynthetic capacity (Araki et al. 2012). However, under environmental stress, the structure of chloroplasts can be severely impaired, resulting in compromised photosynthetic efficiency, reduced biomass accumulation, and lower crop productivity (Weston et al. 2000; Shao et al. 2014). GABA has been shown to protect tomato plants from chilling injury by maintaining thylakoid membrane structure and chloroplast envelope integrity. This reduces swelling and helps retain chlorophyll content, thereby supporting ongoing photosynthetic activity (Abd Elbar et al. 2021).

Photosystem II (PSII), a core component of the photosynthetic electron transport chain, is particularly vulnerable to stress-induced damage. It plays a central role in capturing light energy and facilitating electron transport (Li et al. 2018). Exposure to extremely low temperatures can damage PSII, as evidenced by reduced chlorophyll fluorescence parameters. However, treatment with GABA helps mitigate these effects by preventing excess energy loss, minimizing heat dissipation, and balancing excitation energy between PSII and PSI. These protective effects help sustain leaf-level photosynthesis and increase stress resilience, as observed in *Medicago ruthenica* (Ying et al. 2022).

1.3.4.4 GABA and Oxidative Stress Mitigation

GABA contributes significantly to plant defense mechanisms under oxidative stress, a condition resulting from the accumulation of reactive oxygen species (ROS) that surpasses the plant's detoxification capacity. These ROS, including superoxide radicals, hydrogen peroxide, and hydroxyl radicals, can severely damage cellular structures such as proteins, membranes, and nucleic acids. GABA functions both as a signaling molecule and as a direct participant in mitigating oxidative stress (Shelp et al. 2021).

One of GABA's protective roles involves its ability to scavenge ROS, thereby reducing their concentration and preventing oxidative damage. In *Arabidopsis* challenged with *Botrytis cinerea*, GABA has been shown to enhance the activity of antioxidant enzymes responsible for detoxifying ROS. Moreover, GABA accumulation in these conditions initiates metabolic shifts, including sugar accumulation and stimulation of the GABA shunt, which together promote cellular resilience and inhibit programmed cell death (Janse van Rensburg and Van den Ende 2020).

GABA-mediated signaling can also activate the expression of genes involved in antioxidant defense. It upregulates transcripts encoding ROS-detoxifying enzymes and other redox regulators, thereby bolstering the plant's defense capacity under oxidative conditions (Kumar et al. 2017; Liu et al. 2021; Qi et al. 2021). In tea plants exposed to heat stress, GABA has been reported to enhance polyphenol biosynthesis and activate antioxidant systems, highlighting its broader regulatory impact under thermal stress (Ren et al. 2021).

Additionally, GABA plays a role in preserving membrane integrity under oxidative stress. It minimizes lipid peroxidation, which helps maintain membrane structure and functionality (Shelp et al. 2021). GABA also contributes to the regulation of redox homeostasis by influencing the glutathione redox couple, the ratio of reduced glutathione (GSH) to oxidized glutathione glutathione disulfide (oxidized glutathione (GSSG), which is vital for redox balance in cells (Zhou et al. 2021).

Genetic studies in *Arabidopsis thaliana* have shown that mutants lacking SSADH are hypersensitive to oxidative stress, likely due to impaired hydrogen peroxide detoxification. Since the GABA shunt contributes succinate and Nicotinamide Adenine Dinucleotide (NADH) to the mitochondrial electron transport chain, defects in this pathway reduce the efficiency of ROS clearance. These effects are not restricted to plants; yeast mutants lacking GABA shunt enzymes also display increased sensitivity to hydrogen peroxide (Coleman et al. 2001; Bouché et al. 2003).

Overall, GABA mitigates oxidative stress through multiple interconnected mechanisms: acting directly as an antioxidant, enhancing enzymatic ROS detoxification, regulating defense-related gene expression, stabilizing cellular membranes, and maintaining redox equilibrium. These roles collectively strengthen the plant's ability to withstand oxidative stress and protect cellular structures under adverse environmental conditions.

1.3.4.5 GABA and Osmoregulation

Osmoregulation enables plants to maintain water and solute homeostasis under fluctuating environmental conditions, such as drought or high salinity (Wani et al. 2013). GABA plays a key role in this process by supporting various physiological and biochemical mechanisms that help plants withstand osmotic stress (Li et al. 2021a, b).

During stress-induced dehydration or salt exposure, plants must lower their osmotic potential to retain water. GABA contributes to osmotic adjustment by accumulating in plant cells, thereby increasing solute concentration and facilitating water retention (Nayyar et al. 2014; Michaeli and Fromm 2015). As an effective osmolyte, GABA helps maintain cell turgor and hydration under such conditions (Shelp et al. 2021; Singh et al. 2022).

Stomatal regulation is another vital component of plant water conservation strategies. Under osmotic stress, GABA modulates stomatal closure, reducing water loss through transpiration. This process is partially mediated by GABA-induced signaling pathways that influence ion fluxes, especially those involving calcium and potassium, which are central to stomatal function. As a result, water use efficiency improves, and drought tolerance is enhanced (Xu et al. 2021).

Osmotic stress often leads to overproduction of ROS, which can cause cellular damage. GABA contributes to oxidative stress mitigation during osmotic imbalance by regulating antioxidant enzyme activity and lowering ROS levels, thus protecting cells from oxidative injury (Barbosa et al. 2010; Jalil et al. 2017).

GABA levels are known to rise in response to salinity and flooding stress. In these conditions, GABA also influences transcriptional activity, particularly in roots, modulating gene expression to improve salt stress responses (Barbosa et al.

2010; Renault et al. 2010). GABA signaling regulates genes involved in the synthesis of key osmoprotectants like proline and glycine betaine, which aid in maintaining osmotic balance (Ghosh et al. 2021). This signaling extends to pathways associated with hormone biosynthesis, ROS modulation, and stress-induced protein turnover, as observed in *Caragana intermedia* (Shi et al. 2010).

In *Nigella sativa L.*, GABA supports osmotic adjustment under water-limited conditions by enhancing water balance (Rezaei-Chiyaneh et al. 2018). Similarly, in rice cultivars, GABA application promotes protein synthesis and enhances the activities of GAD and diamine oxidase (DAO), helping improve salinity–alkalinity stress tolerance (Kaur et al. 2020). These findings are consistent with reports showing that exogenous GABA enhances endogenous GABA levels, stimulates Pas biosynthesis, and reduces salt stress–induced damage in tomato plants (Hu et al. 2015; Wu et al. 2020).

GABA also supports stress protection in postharvest and horticultural contexts. For instance, its application helps reduce chilling injury in peach fruit by increasing endogenous GABA and proline levels (Shang et al. 2011). In perennial ryegrass and black pepper, GABA improves drought tolerance by maintaining membrane integrity (Krishnan et al. 2013; Vijayakumari and Puthur 2015).

Transport mechanisms for GABA also contribute to its role in osmoregulation. GABA can be transported by proline transporters, such as AtProT2 in *Arabidopsis thaliana* and LeProT1 in tomatoes. These transporters are particularly active under osmotic stress conditions, enabling the movement of GABA and other osmolytes to tissues where they help stabilize cellular environments and enhance stress resilience (Breitkreuz et al. 2003; Schwacke et al. 1999).

1.3.4.6 GABA as a Signaling Molecule

In vertebrates and mammals, GABA is well-known for its role as a neurotransmitter. Interestingly, emerging research indicates that GABA also acts as a signaling molecule in plants. Although the signaling mechanisms in plants differ from those in animals, there is growing evidence suggesting the existence of GABA perception systems in plants, possibly involving specific receptors or the interaction of GABA with plant glutamate receptor–like proteins.

Investigations using *Arabidopsis thaliana* mutants have offered insights into the functional relevance of the GABA shunt in plant signaling. The disruption of the gene encoding SSADH results in a range of stress-related phenotypes, including excessive accumulation of ROS, stunted growth, tissue necrosis, and enhanced sensitivity to environmental stressors (Bouche et al. 2003; Fait et al. 2005). These findings highlight a regulatory role for GABA metabolism in maintaining cellular homeostasis and responding to stress.

Despite these observations, the exact function of GABA in plant signaling remains under investigation. It is still unclear whether GABA acts purely as a metabolic intermediate en route to the TCA cycle or whether it fulfills a signaling role by interacting with specific receptors, being directed to subcellular compartments, or undergoing long-distance transport within the plant.

Support for GABA's role as a signaling molecule comes from studies on reproductive development. A concentration gradient of GABA has been observed along the path of the pollen tube from the stigma toward the ovule, which is believed to be essential for proper directional growth and successful fertilization (Palanivelu et al. 2003). In *Arabidopsis*, mutants deficient in GABA-T (POP2) lack this gradient, resulting in disrupted pollen tube navigation and consequent infertility. This strongly suggests that GABA functions as an intercellular signal in reproductive tissues.

The resemblance between GABA's proposed signaling role in plants and its well-established function in the animal nervous system is intriguing. If GABA operates as a signaling ligand in plants, it is likely that specific receptors and transporters are involved in perceiving and mediating its effects. This cell-to-cell communication mechanism would support the broader hypothesis that GABA regulates physiological processes not only through its metabolic function but also through active signaling pathways.

1.3.4.7 GABA and Pollen Tube Growth

Pollen tube growth is a vital process in plant reproduction, as it enables the delivery of sperm cells to the ovule for fertilization. This growth is tightly regulated by a variety of signals originating from both the pollen grain and the female reproductive tissues. Among these, GABA has emerged as an important signaling molecule involved in guiding and regulating pollen tube elongation.

Research in *Arabidopsis thaliana* has shown that mutations in the POP2 gene, which encodes a GABA-T enzyme, lead to the disruption of GABA metabolism and result in abnormal pollen tube guidance. These mutants exhibit elevated GABA levels in the pistil and a disrupted gradient, ultimately leading to infertility due to the inability of pollen tubes to properly locate and reach the ovules (Palanivelu et al. 2003; Renault et al. 2010; Ansari and Chen 2009).

Under normal conditions, the female reproductive tissues, especially the stigma, style, and transmitting tract that produce and secrete GABA, create a concentration gradient that increases toward the ovule. This gradient is crucial for orienting pollen tubes as they grow through the pistil. The breakdown of GABA by GABA-T is essential for maintaining this gradient and guiding pollen tubes with spatial accuracy.

The application of exogenous GABA has been found to promote pollen tube elongation, whereas inhibitors of GABA biosynthesis or metabolism impair growth, reinforcing its role in this process (Ling et al. 2013; Yu et al. 2014a, b). In *Picea wilsonii*, GABA signaling has been shown to regulate calcium dynamics, actin filament organization, and vesicle trafficking, mechanisms critical for the directional growth of the pollen tube.

These effects appear to be mediated by calcium-dependent signaling cascades, where GABA influences intracellular calcium levels, thereby affecting cytoskeletal organization and vesicle transport to the growing tip of the tube. Additionally, GABA may interact with specific transporters or receptor-like proteins on the pollen tube membrane to initiate downstream responses (Beuve et al. 2004).

GABA functions as a positive regulator of pollen tube development and navigation, integrating metabolic signals with guidance cues from the female tissues. This role is essential for ensuring successful fertilization and seed development in flowering plants.

1.3.4.8 GABA and Nitrate Acquisition and Assimilation

The efficient acquisition and assimilation of nitrogen, particularly in the form of nitrate, is crucial for plant growth and productivity. GABA has been found to influence these processes by modulating metabolic activity and nutrient transport under both normal and stress conditions.

Studies on *Arabidopsis* and muskmelon seedlings have demonstrated that exogenous GABA treatment enhances nitrate uptake by stimulating the expression of genes associated with nitrate transport and metabolism. These effects are accompanied by increased activity of nitrogen-assimilating enzymes, such as nitrate reductase (NR), glutamine synthetase (GS), and glutamate synthase (GOGAT), which are vital for incorporating nitrate into organic molecules (Barbosa et al. 2010; Zhen et al. 2018).

GABA also impacts ammonium assimilation by reducing deamination reactions mediated by glutamate dehydrogenase (GDH), thereby limiting the accumulation of toxic ammonium ions under stress conditions. This contributes to better nitrogen balance in plants experiencing calcium nitrate–induced stress (Zhen et al. 2018).

Further evidence suggests that GABA, as part of the GABA shunt, participates in maintaining the C/N balance by directing carbon from glutamate into the TCA cycle. This supports energy production while supplying carbon skeletons required for ammonium assimilation and amino acid synthesis (Michaeli and Fromm 2015).

Interestingly, while GABA application upregulates nitrate transporter gene expression, it does not always lead to a proportional increase in nitrate influx, suggesting that other layers of regulation, such as posttranslational modifications or transporter localization, may be involved in modulating nitrate uptake (Beuve et al. 2004).

Moreover, the effect of GABA on nitrogen acquisition appears to be context-dependent. For example, *Arabidopsis* seedlings show variable responses to GABA, depending on the external nitrate concentration in the medium. This indicates that GABA may function as a conditional signal that fine-tunes nitrogen metabolism according to environmental nutrient availability (Barbosa et al. 2010).

Overall, GABA plays a multifaceted role in nitrogen management by influencing nutrient uptake, modulating enzyme activity, and integrating metabolic responses through its link with carbon metabolism. These functions are particularly important under nutrient stress conditions and contribute to the plant's overall adaptability and efficiency in nitrogen use.

1.3.4.9 GABA and Adventitious Root Development

Adventitious roots (ARs) arise from non-root tissues and are essential for vegetative propagation, stress adaptation, and survival in many plant species. The development of these roots is regulated by a complex interplay of internal signals and environmental cues, including phytohormones, developmental stage, and various metabolites. Among these, GABA has recently emerged as a regulatory molecule influencing AR initiation and growth.

In *Arabidopsis*, mutants with disrupted GABA catabolism exhibit excessive accumulation of GABA, leading to inhibited cell elongation and stunted primary root growth (Renault et al. 2011). Similar inhibitory effects have been observed when exogenous GABA is applied, suggesting that GABA levels must be tightly regulated for normal root development.

Blocking GABA-related metabolic pathways also impacts AR formation. For instance, the inhibition of α-ketoglutarate dehydrogenase, an enzyme involved in the GABA shunt, suppresses AR elongation and alters the expression of genes linked to GABA metabolism (Yue et al. 2018). These findings suggest that GABA affects root growth not only via metabolic flux but also through transcriptional regulation.

In woody plants such as poplar, increased endogenous GABA delays the initiation of ARs and reduces their emergence. This inhibition is associated with changes in hormone signaling, particularly involving ethylene, abscisic acid (ABA), and auxin—three key regulators of root development (Xie et al. 2020).

Additional evidence comes from studies on *Malus xiaojinensis* and tobacco, where exogenous GABA application reduces both the number and length of adventitious roots. It was observed that GABA suppresses the auxin-induced expression of *PIN* genes, which encode the auxin transport proteins necessary for root patterning. This implies that GABA interferes with auxin transport, thereby modulating AR development (Li et al. 2021a, b).

Overall, GABA appears to function as a negative regulator of adventitious root formation under certain conditions. Its accumulation, whether through metabolic dysregulation or external application, can impair root initiation and elongation, likely by interacting with hormone signaling and auxin transport pathways. These findings suggest that while GABA contributes to stress resilience and developmental regulation, its role in root formation is context-specific and may involve antagonistic interactions with growth-promoting signals.

1.3.4.10 GABA and Senescence Regulation

Senescence in plants is a tightly controlled genetic process that involves the breakdown of functional cellular components, particularly proteins, to allow for the redistribution of nutrients to younger, actively growing tissues (Martinez et al. 2008; Buet et al. 2019). This reallocation of resources is critical for sustaining plant growth and supporting developmental processes in emerging organs, such as new leaves, flowers, and seeds (Himelblau and Amasino 2001; Ansari et al. 2014; Maillard et al. 2015).

Nitrogen, a vital macronutrient, plays a key role during senescence. As the aging leaf tissues degrade, nitrogen is recycled and redirected to developing parts of the plant, where it supports the synthesis of essential nitrogen-containing compounds (Ansari et al. 2014; Sakamoto and Takami 2014). Among the molecular pathways involved in this regulation, the GABA shunt is increasingly recognized for its involvement in nitrogen mobilization and metabolism during senescence (Ansari et al. 2005, 2014).

The GABA shunt operates as an auxiliary pathway of the TCA cycle and plays an important role in coordinating carbon and nitrogen flux. During senescence, this pathway becomes particularly active, facilitating the metabolic conversion of nitrogen from degraded macromolecules into reusable forms (Barbosa et al. 2010; Ansari et al. 2014). Proteins are broken down into amino acids like glutamine and asparagine, which are then translocated through the phloem to support sink tissues, such as seeds and fruits (Feller and Fischer 1994).

These amino acids can be further converted into glutamate and α-keto acids through transamination reactions, thereby feeding into the nitrogen and carbon metabolic networks (Ansari et al. 2014; Jalil et al. 2017, 2019b). Glutamate, in turn, can be decarboxylated by GAD to produce GABA, which enters the GABA shunt and contributes to both carbon recycling and energy production. This process not only supports respiration but also influences carbon–nitrogen balance in developing seeds, including the regulation of fatty acid biosynthesis and storage compound accumulation (Fait et al. 2011).

Metabolic profiling of *Arabidopsis thaliana* has shown that the GABA shunt serves as a central node, linking carbon and nitrogen metabolism during critical transitions such as seed imbibition and the early stages of germination (Fait et al. 2011; Jalil et al. 2017). These findings underscore the functional versatility of GABA during developmental shifts and highlight its regulatory influence during the senescence process.

1.3.4.11 GABA and Abiotic Stress Resilience

The involvement of GABA in enhancing plant tolerance to abiotic stress has become a prominent area of investigation. As a non-proteinogenic amino acid, GABA accumulates significantly under various stress conditions, such as drought, salinity, extreme temperatures, and oxidative challenges (Wallace et al. 1984; Zushi and Matsuzoe 2007; Renault et al. 2010). Its accumulation appears to contribute to cellular defense by functioning as an osmoprotectant, supporting osmotic homeostasis and reducing water loss when plants encounter dehydration.

Beyond osmotic adjustment, GABA stabilizes cellular membranes and proteins, thereby shielding plant cells from structural damage. Its compatibility with cellular systems allows it to accumulate without toxicity, even under salinity-induced stress. Moreover, it exhibits antioxidant activity, helping neutralize free radicals generated during stressful episodes (Carillo 2018; Jalil and Ansari 2020).

The GABA shunt, an alternative route of carbon flow within the TCA cycle, is believed to be upregulated in response to abiotic stresses, contributing to metabolic regulation under these conditions (Al-Quraan et al. 2016; Jalil et al. 2017, 2019a; Ansari et al. 2021). This pathway not only helps in maintaining cellular energy production but also supports stress-related signaling mechanisms.

ROS are commonly produced in excess during abiotic stress, leading to oxidative damage. GABA plays a key role in counteracting ROS toxicity, acting directly as a scavenger and indirectly by regulating antioxidant defenses (Jalil et al. 2019a;

Ansari et al. 2021). Exogenous GABA treatments have been reported to mitigate oxidative injury by downregulating ROS-producing genes and upregulating genes encoding protective enzymes like superoxide dismutase (SOD), catalase (CAT), and peroxidases (Shi et al. 2010; Jalil and Ansari 2020).

In addition to its antioxidative functions, GABA has shown beneficial effects on seed germination under salt stress, particularly in crops like tomatoes, where it enhances germination success and alleviates ion toxicity (Luo et al. 2011). Its ability to activate stress-responsive genes related to antioxidant enzyme production helps plants maintain redox equilibrium, thus improving survival during prolonged environmental stress (Shi et al. 2010; Vijayakumari and Puthur 2015; Li et al. 2017).

GABA also supports osmotic regulation through mechanisms that help sustain water uptake and retention, thereby strengthening resistance to drought and salinity (Yu et al. 2014; Vijayakumari and Puthur 2015).

Ion imbalance is another major consequence of abiotic stress. GABA contributes to restoring ion homeostasis by modulating the activity of specific ion channels and transporters, especially those involved in potassium (K^+) and calcium (Ca^{2+}) movement across membranes (Bouche et al. 2003; Gordeeva et al. 2003; Yuan et al. 2014; Demidchik and Shabala 2017; Wu et al. 2021). This regulation helps prevent cytotoxic ion accumulation and supports nutrient uptake.

Furthermore, GABA serves as a signaling molecule that modulates intracellular and systemic responses to stress. It interacts with hormonal signaling networks, including abscisic acid (ABA)–related pathways, and may influence kinase-mediated signaling events that activate broader stress adaptation responses (Podlešáková et al. 2019; Kaspal et al. 2021).

GABA accumulates in plants under adverse environmental conditions and performs multiple protective functions, ranging from osmotic regulation and antioxidant defense to ion homeostasis and stress signaling. Its multifunctional role highlights its potential as a target for biotechnological interventions aimed at improving crop resilience and productivity in the face of environmental challenges.

1.3.4.12 GABA and Biotic Stress Defense Mechanisms

GABA contributes to plant defense against biotic stress through multiple mechanisms, including the enhancement of antioxidant activity, modulation of signaling pathways, and interference with pathogen and pest development. Plants treated with GABA typically show elevated activity of enzymes involved in ROS detoxification, which helps minimize cell death associated with oxidative bursts triggered during pathogen invasion.

Beyond its protective role in plants, GABA also interacts with microbial communities. It serves as both a signaling molecule and a source of carbon and nitrogen for certain bacteria. For instance, elevated GABA levels have been shown to suppress the expression of hypersensitive response and pathogenicity (*hrp*) genes in bacterial pathogens, suggesting that GABA-mediated signaling can dampen bacterial virulence and improve plant resistance (Park et al. 2010). In *Escherichia coli* K12 mutants with enhanced GABA-T activity, GABA can serve as the sole nutrient source, supporting microbial proliferation (Dover and Halpern 1972).

During infection, GABA accumulation in the plant apoplast has also been observed. For example, in tomato plants infected with *Cladosporium fulvum*, GABA levels rise, which can be exploited by *Pseudomonas syringae pv. tomato* strain DC3000 in *Arabidopsis* using GABA as both a carbon and a nitrogen source (Rico and Preston 2008; Park et al. 2010).

GABA also contributes to defense against insect herbivores and fungal pathogens by activating immune responses and impairing pest physiology. Increased GABA accumulation has been associated with reduced insect feeding and altered insect development. Physical damage caused by chewing insects or mechanical wounding leads to localized GABA elevation, which can negatively affect herbivore development upon ingestion (Bown et al. 2002; Morse et al. 1979). Genetically engineered tobacco plants with higher GABA content display enhanced resistance to pests like root-knot nematodes and the tobacco budworm (Shelp et al. 2006).

GABA also acts as a direct anti-insect compound. Experiments have demonstrated its ability to slow larval development, reduce weight gain, and lower survival rates in insect species such as *Spodoptera littoralis* and *Choristoneura rosaceana* (Ramputh and Bown 1996; Bown et al. 2002, 2006; Scholz et al. 2015). Elevated GABA levels in insect-infested tissues of tobacco and soybean not only produce immediate toxic effects but also trigger secondary defense signaling (Bown et al. 2002).

As a signaling molecule, GABA orchestrates immune responses against fungal pathogens. In tomato, mutants challenged with *Botrytis cinerea* exhibit a two-phase defense response facilitated by GABA. The application of exogenous GABA restricts fungal spread by forming chlorotic rings around the infection site and boosting the activity of the GABA shunt, which mitigates ROS-induced cell death (Seifi et al. 2013). The GABA shunt also supports redox balance by producing NADH through SSADH and delivering carbon intermediates to the TCA cycle under oxidative stress.

This defense strategy extends to other crops as well. In rice, GABA shunt activation has been associated with reduced cell death following infection by *Magnaporthe oryzae*. Similarly, the application of GABA to tomato and pear plants has enhanced resistance to fungal pathogens, likely by fortifying the antioxidant defense system and reinforcing metabolic resilience (Fu et al. 2017; Yu et al. 2014; Yang et al. 2017).

1.3.4.13 GABA and Postharvest Value Addition

GABA contributes significantly to delaying senescence and preserving the postharvest quality of agricultural produce (Aghdam et al. 2022). It plays a regulatory role in controlling the synthesis and function of enzymes involved in cell wall degradation, which helps minimize the softening and structural breakdown of fruits and vegetables during storage (Ge et al. 2018; Han et al. 2018, 2021; Aghdam et al. 2019).

In several studies, GABA levels have been found to rise during the later stages of cold storage in various fruits. For example, cherimoya exhibits increased GABA content during prolonged chilling (Escribano and Merodio 2001), particularly after CO_2 treatment (Merodio et al. 1998). Similar trends have been reported in tomatoes maintained under modified atmospheres (Makino et al. 2008) and pears stored under controlled atmospheric conditions (Franck et al. 2007; Pedreschi et al. 2009).

In postharvest citrus, the GABA shunt plays a pivotal role in the metabolism of organic acids and contributes to amino acid synthesis. It supports nitrogen assimilation into amino acids such as aspartic acid, glutamic acid, glutamine, and asparagine, with the latter two being further integrated into extended metabolic functions essential for sustaining postharvest quality (Katz et al. 2011; Sun et al. 2013).

Moreover, GABA is known to participate in stress regulation and the ripening process of harvested produce. External application of GABA has been shown to alleviate oxidative stress and increase resilience to temperature fluctuations, water deficit, and salinity after harvest. In zucchini fruit, GABA treatment has improved cold storage tolerance by enhancing internal defense mechanisms (Palma et al. 2019).

Several investigations have revealed that the activation of the GABA shunt is crucial for reducing postharvest chilling injury, as it promotes ATP generation, reduces NADH accumulation, and limits ROS buildup. In the case of anthurium flowers, which are highly sensitive to low temperatures, postharvest treatment with salicylic acid via stem-end dipping has been found to activate the GABA shunt, allowing storage at temperatures below the optimal range of 12.5–20 °C (Aghdam et al. 2016).

GABA has also demonstrated antimicrobial activity, making it useful for managing postharvest diseases. Its application has been associated with the suppression of bacterial, fungal, and other microbial pathogens, which helps in reducing spoilage and prolonging shelf life (Yu et al. 2014; Fu et al. 2017; Sheng et al. 2017; Yang et al. 2017; Wang et al. 2022). For instance, hypotonic GABA treatment significantly decreased the occurrence and lesion size of gray mold in cherry tomatoes (Wang et al. 2022), while in button mushrooms, GABA treatments reduced cap browning and maintained sensory and nutritional quality during cold storage (Shekari et al. 2021).

Furthermore, the integration of GABA into food systems has drawn attention due to its potential health-promoting effects. Enrichment of food products with GABA, either through direct addition or via fermentation using GABA-producing microorganisms, has been explored to improve the nutritional value of processed foods (Ding et al. 2016, 2018; Heli et al. 2022).

While these findings highlight the promising role of GABA in postharvest preservation and food value enhancement, more research is needed to fully uncover its mechanisms and refine its practical applications in the agro-food industry.

1.4 Conclusion

GABA (Fig. 1.2) serves as a pivotal molecule in plant metabolism, acting as both a metabolic intermediate and a regulatory compound. It contributes to various metabolic networks by supplying carbon skeletons and energy and plays a role in stabilizing cytosolic pH and supporting nitrogen assimilation. Moreover, GABA links

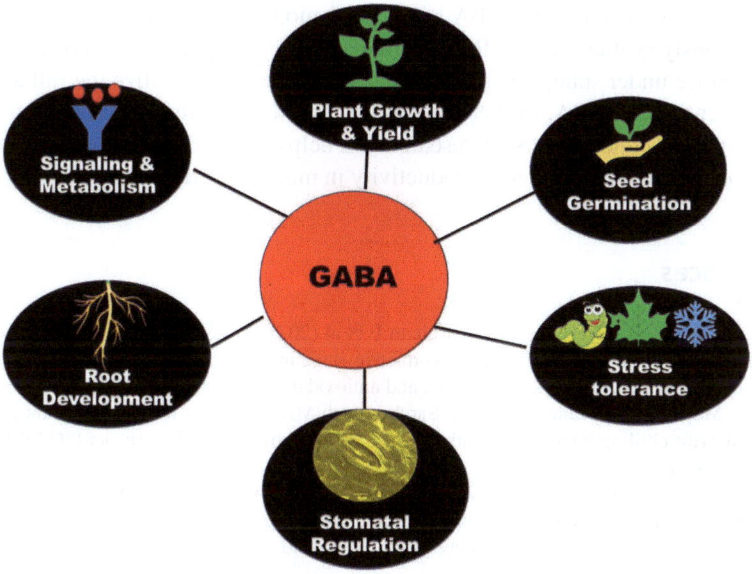

Fig. 1.2 Overview of GABA's multifunctional role as a metabolite and signaling molecule in plants. GABA contributes to various physiological and developmental functions, such as enhancing plant growth and productivity, promoting seed germination, improving tolerance to both biotic and abiotic stresses, supporting root architecture, regulating stomatal activity, and modulating cellular signaling and metabolic pathways

primary and secondary metabolic pathways, underscoring its versatility in plant physiological processes.

Emerging evidence connects GABA accumulation with plant responses to environmental stress, suggesting that GABA is not only a metabolic by-product but also a key player in signaling pathways. It may act directly as a signaling molecule or function to activate secondary messengers that initiate defense responses. These include the induction of ROS, stimulation of antioxidant enzyme activity, interaction with phytohormonal pathways, and the activation of defense mechanisms against fungal pathogens and herbivorous insects.

Under stress conditions, intracellular calcium levels rise, which, in turn, activate the GAD enzyme through a Ca^{2+}/CaM complex. This activation promotes the conversion of glutamate to GABA, leading to changes in transporter activity and the triggering of further physiological responses (Fig. 1.2). However, the underlying molecular mechanisms governing these processes remain to be fully clarified and warrant further investigation.

Beyond its stress-related functions, GABA also influences several aspects of plant development under non-stressful conditions. It plays a role in seed germination, root and shoot elongation, regulation of photosynthetic processes, and nitrogen utilization. Elevated GABA levels have been shown to modulate the expression of genes associated with signaling, metabolism, and growth, ultimately contributing to enhanced plant vigor, improved stress resilience, and increased yield potential.

External application of GABA has been demonstrated to mimic the benefits of endogenously synthesized GABA, offering a promising approach for boosting plant performance under suboptimal conditions. Nevertheless, to realize the full agricultural potential of GABA, further studies are needed across diverse crop species and environmental conditions. Such research will help optimize GABA-based strategies for improving crop quality and productivity in modern agriculture.

References

Abd Elbar Ola H, Amr E, Gniewko N, Reham F et al (2021) Protective effect of γ aminobutyric acid against chilling stress during reproductive stage in tomato plants through modulation of sugar metabolism, chloroplast integrity, and antioxidative defence systems. Front Plant Sci 12

Aghdam MS, Naderi R, Jannatizadeh A, Sarcheshmeh MAA, Babalar M (2016) Enhancement of postharvest chilling tolerance of anthurium cut flowers by γ-aminobutyric acid (GABA) treatments. Sci Hortic 198:52–60

Aghdam MS, Kakavand F, Rabiei V, Zaare-Nahandi F, Razavi F (2019) γ-Aminobutyric acid and nitric oxide treatments preserve sensory and nutritional quality of cornelian cherry fruits during postharvest cold storage by delaying softening and enhancing phenols accumulation. Sci Hortic 246:812–817

Aghdam MS, Flaherty EJ, Shelp BJ (2022) γ-Aminobutyrate improves the postharvest marketability of horticultural commodities: advances and prospects. Front. Plant Sci 13

Akama K, Takaiwa F (2007) C-terminal extension of rice glutamate decarboxylase (OsGAD2) functions as an autoinhibitory domain and overexpression of a truncated mutant results in the accumulation of extremely high levels of GABA in plant cells. J Exp Bot 58(2699):2707

Al-Quraan NA, Al-Share AT (2016) Characterization of the γ-aminobutyric acid shunt pathway and oxidative damage in Arabidopsis thaliana pop2 mutants under various abiotic stresses. Biol Plant 60:132–138

Al-Taher F, Nemzer B (2019) Determination of γ-aminobutyric acid (GABA) content in grains and cruciferous vegetable seeds. J Food Res 8:6

Ansari MI, Chen SCG (2009) Biochemical characterization of γ-aminobutyric acid (GABA): pyruvate transaminase during rice leaf senescence. Inter J Integ Biol 16:27–32

Ansari MI, Lee RH, Chen SCG (2005) A novel senescence-associated gene encoding γ aminobutyric acid (GABA): pyruvate transaminase is upregulated during rice leaf senescence. Physiol Plant 123:1–8

Ansari MI, Hasan S, Jalil SU (2014) Leaf senescence and GABA shunt. Bioinformation 10:730–732

Ansari MI, Jalil SU, Ansari SA et al (2021) GABA shunt: a key player in mitigation of ROS during stress. Plant Growth Regul 94:131–149

Aoyama S, Reyes TH, Guglielminetti L, Lu Y, Morita Y, Sato T, Yamaguchi J (2014) Ubiquitin ligase ATL31 functions in leaf senescence in response to the balance between atmospheric CO_2 and nitrogen availability in Arabidopsis. Plant Cell Physiol 55:293–305

Araki H, Hamada A, Hossain MA, Takahashi T (2012) Waterlogging at jointing and/or after anthesis in wheat induces early leaf senescence and impairs grain filling. Field Crop Res 137:27–36

Asada H, Kawamura Y, Maruyama K, Kume H, Ding RG, Kanabara N et al (1997) Cleft palate and decreased brain γ-aminobutyric acid in mice lacking the 67-kDa isoform of glutamic acid decarboxylase. Proc Natl Acad Sci USA 94:6496–6499

Bach B, Meudec E, Lepoutre JP, Rossignol T, Blondin B, Dequin S, Camarasa C (2009) New insights into γ-aminobutyric acid catabolism: evidence for γ-hydroxybutyric acid and polyhydroxybutyrate synthesis in Saccharomyces cerevisiae. Appl Environ Microbiol 75:4231–4239

Barbosa JM, Singh NK, Cherry JH, Locy RD (2010) Nitrate uptake and utilization is modulated by exogenous gamma-aminobutyric acid in Arabidopsis thaliana seedlings. Plant Physiol Biochem 48(6):443–450

Beuve N, Rispail N, Laine P, Cliquet JB, Ourry A, Le Deunff E (2004) Putative role of γ aminobutyric acid (GABA) as a long-distance signal in up-regulation of nitrate uptake in Brassica napus L. Plant Cell Environ 27:1035–1046

Bhatnagar P, Minocha R, Minocha SC (2002) Genetic manipulation of the metabolism of polyamines in poplar cells. The regulation of putrescine catabolism. Plant Physiol 128:1455–1469

Bouche N, Fromm H (2004) GABA in plants: just a metabolite? Trends Plant Sci 9:110–115

Bouché N, Fait A, Bouchez D, Møller SG, Fromm H (2003) Mitochondrial succinic-semialdehyde dehydrogenase of the γ-aminobutyrate shunt is required to restrict levels of reactive oxygen intermediates in plants. Proc Natl Acad Sci USA 100:6843–6848

Bouchereau A, Aziz A, Larher F, Martin-Tanguy J (1999) Polyamines and environmental challenges: recent development. Plant Sci 140:103–125

Bowery NG, Hill DR, Hudson AL et al (1980) Baclofen decreases neurotransmitter release in the mammalian CNS by an action at a novel GABA receptor. Nature 283:92–94

Bown AW, Shelp BJ (2016) Plant GABA: not just a metabolite. Trends Plant Sci 21:811–813. https://doi.org/10.1016/j.tplants.2016.08.001

Bown AW, Hall DE, MacGregor KB (2002) Insect footsteps on leaves stimulate the accumulation of 4-aminobutyrate and can be visualized through increased chlorophyll fluorescence and superoxide production. Plant Physiol 129:1430–1440

Bown AW, MacGregor KB, Shelp BJ (2006) γ-Aminobutyrate: Defence against invertebrate pests? Trends Plant Sci 11:424–427

Breitkreuz KE, Allan WL, Van Cauwenberghe OR, Jakobs C, Talibi D, Andre B, Shelp BJ (2003) A novel γ-hydroxybutyrate dehydrogenase. J Biol Chem 278:41552–41556

Buet A, Costa ML, Martínez DE, Guiamet JJ (2019) Chloroplast protein degradation in senescing leaves: proteases and lytic compartments. Front Plant Sci 10:747

Cao J, Barbosa JM, Singh NK, Locy RD (2013) GABA shunt mediates thermotolerance in Saccharomyces cerevisiae by reducing reactive oxygen production. Yeast 30:129–144

Carillo P (2018) GABA shunt in durum wheat. Front Plant Sci 9:100

Caroll AD, Fox GG, Laurie S, Phillips R, Ratcliffe RG, Stewart GR (1994) Ammonium assimilation and the role of γ-aminobutyric acid in pH homeostasis in carrot cell suspensions. Plant Physiol 106:513–520

Chen H, Liu T, Xiang L, Hu L, Hu X (2018) GABA enhances muskmelon chloroplast antioxidants to defend against salinity-alkalinity stress. Russ J Plant Physiol 65:674–679

Chou HT, Kwon DH, Hegazy M, Lu CD (2008) Transcriptome analysis of agmatine and putrescine catabolism in Pseudomonas aeruginosa PAO1. J Bacteriol 190:1966–1975

Coleman ST, Fang TK, Rovinsky SA, Turano FJ, Moye-Rowley WS (2001) Expression of a glutamate decarboxylase homologue is required for normal oxidative stress tolerance in Saccharomyces cerevisiae. J Biol Chem 276:244–250

Comparot S, Lingiah G, Martin T (2003) Function and specificity of 14-3-3 proteins in the regulation of carbohydrate and nitrogen metabolism. J Exp Bot 54:595–604

Coruzzi GM, Zhou L (2001) Carbon and nitrogen sensing and signaling in plants: emerging 'matrix effects'. Curr Opin Plant Biol 4:247–253

Crawford LA, Bown AW, Breitkreuz KE, Guinel FC (1994) The synthesis of γ-aminobutyric acid in response to treatment reducing cytosolic pH. Plant Physiol 104:865–871

Dagorn A, Hillion M, Chapalain A, Lesouhaitier O, Duclairoir-Poc C, Vieillard J, Chevalier S, Taupin L, Le Derf F, Feuilloley MGJ (2013) γ-Aminobutyric acid acts as a specific virulence regulator in Pseudomonas aeruginosa. Microbiology 159:339–351

DeFelipe J (1993) Neocortical neuronal diversity: chemical heterogeneity revealed by colocalization studies of classical neurotransmitters, neuropeptides, calcium-binding proteins, and cell surface molecules. Cereb Cortex 3:273–289

Demidchik V, Shabala S (2017) Mechanisms of cytosolic calcium elevation in plants: the role of ion channels, calcium extrusion systems, and NADPH oxidase-mediated 'ROS-Ca^{2+} hub'. Funct Plant Biol 45:9–27

Dhakal R, Bajpai VK, Baek KH (2012) Production of GABA (γ-aminobutyric acid) by microorganisms: a review. Braz J Microbiol 43:1230–1241

Ding J, Yang T, Feng H, Dong M, Slavin M, Xiong S, Zhao S (2016) Enhancing contents of γ-aminobutyric acid (GABA) and other micronutrients in dehulled rice during germination under normoxic and hypoxic conditions. J Agric Food Chem 64:1094–1102
Ding J, Ulanov AV, Dong M, Yang T, Nemzer BV, Xiong S, Zhao S, Feng H (2018) Enhancement of γ-aminobutyric acid (GABA) and other health-related metabolites in germinated red rice (Oryza sativa L.) by ultrasonication. Ultrason Sonochem 40:791–797
Dover S, Halpern YS (1972) Utilization of γ-aminobutyric acid as the sole carbon and nitrogen source by Escherichia coli K-12 mutants. J Bacteriol 109:835–843
Ekema GM, Zheng W, Lu L (2002) Interaction of GABA receptor/channel ρ1 and γ2 subunit. Invest Ophthalmol Vis Sci 43:2326–2333
Escribano MI, Merodio C (2001) Limited ammonium assimilation in cherimoya fruit stored at chilling temperature. Acta Hortic 553:311–312
Fait A, Yellin A, Fromm H (2005) GABA shunt deficiencies and accumulation of reactive oxygen intermediates: insight from Arabidopsis mutants. FEBS Lett 579:415–420
Fait A, Fromm H, Walter D, Galili G, Fernie AR (2008) Highway or byway: the metabolic role of the GABA shunt in plants. Trends Plant Sci 13:14–19
Fait A, Nesi AN, Angelovici R, Lehmann M, Pham PA, Song L, Haslam RP, Napier JA, Galili G, Fernie AR (2011) Targeted enhancement of glutamate-to-γ-aminobutyrate conversion in Arabidopsis seeds affects carbon-nitrogen balance and storage reserves in a development-dependent manner. Plant Physiology 157:1026–1042
Feller U, Fischer A (1994) Nitrogen metabolism in senescing leaves. Crit Rev Plant Sci 13:241–273
Franck C, Lammertyn J, Ho QT, Verboven P, Verlinden B, Nicolaï BM (2007) Browning disorders in pear fruit. Postharvest Biol Technol 43:1–13
Fu D, Sun Y, Yu C, Zheng X, Yu T, Lu H (2017) Comparison of the effects of three types of aminobutyric acids on the control of Penicillium expansum infection in pear fruit. J Sci Food Agric 97:1497–1501
Galvez Rojas RL, Ahn IY, Suárez Mantilla B, Sant'Anna C, Pral EM, Silber AM (2015) The uptake of GABA in Trypanosoma cruzi. J Eukaryot Microbiol 62:629–636
Ge Y, Duan B, Li C, Tang Q, Li X, Wei M et al (2018) γ-Aminobutyric acid delays senescence of blueberry fruit by regulation of reactive oxygen species metabolism and phenylpropanoid pathway. Sci Hortic 240:303–309
Gehring AC, Irving HR, Parish RW (1990) Effects of auxin and abscisic acid on cytosolic calcium and pH in plant cells. Proc Natl Acad Sci USA 87:9645–9649
Ghosh UK, Islam MN, Siddiqui MN, Khan MAR (2021) Understanding the roles of osmolytes for acclimatizing plants to changing environment: a review of potential mechanisms. Plant Signal Behav 16:1913306
Gordeeva A, Zvyagilskaya R, Labas YA (2003) Cross-talk between reactive oxygen species and calcium in living cells. Biochemistry 68:1077–1080
Han S, Nan Y, Qu W, He Y, Ban Q, Lv Y et al (2018) Exogenous γ-aminobutyric acid treatment contributes to regulation of malate metabolism and ethylene synthesis in apple fruit during storage. J Agric Food Chem 66:13473–13482
Han S, Liu H, He Y, Nan Y, Qu W, Rao J (2021) Effects of calcium treatment on malate metabolism and γ-aminobutyric acid (GABA) pathway in postharvest apple fruit. Food Chem 334:127479
Heli Z, Hongyu C, Dapeng B, Yee Shin T, Yejun Z, Xi Z, Yingying W (2022) Recent advances of γ-aminobutyric acid: physiological and immunity function, enrichment, and metabolic pathway. Front Nutr 9:1076223
Himelblau E, Amasino RM (2001) Nutrients mobilized from leaves of Arabidopsis thaliana during leaf senescence. J Plant Physiol 158:1317–1323
Hogema BM, Gupta M, Senephansiri H et al (2001) Pharmacologic rescue of lethal seizures in mice deficient in succinate semialdehyde dehydrogenase. Nat Genet 29:212–216
Hu X, Xu Z, Xu W, Li J, Zhao N, Zhou Y (2015) Application of γ-aminobutyric acid demonstrates a protective role of polyamine and GABA metabolism in muskmelon seedlings under $ca(NO_3)_2$ stress. Plant Physiol Biochem 92:1–10

Jacobs C, Bojasch M, Monch E et al (1981) Urinary excretion of γ-hydroxybutyric acid in a patient with neurological abnormalities: the probability of a new inborn error of metabolism. Clin Chim Acta 111:169–178

Jalil SU, Ansari MI (2020) Physiological role of γ-aminobutyric acid in salt stress tolerance. In: Hasanuzzaman M, Tanveer M (eds) Salt and drought stress tolerance in plants, signaling and communication in plants. Springer, Switzerland, pp 337–350

Jalil SU, Ahmad I, Ansari MI (2016) Phenotypic characterization of GABA-transaminase mutants of Arabidopsis thaliana. Adv Life Sci 5:10005–10008

Jalil SU, Ahmad I, Ansari MI (2017) Functional loss of GABA transaminase (GABA-T) expressed early leaf senescence under various stress conditions in Arabidopsis thaliana. Curr Plant Biol 9–10:11–22

Jalil SU, Khan MAA, Ansari MI (2019a) Tremendous role of GABA shunt metabolic pathway in plant system. J Biol Chem Res 36:188–196

Jalil SU, Khan MQR, Ansari MI (2019b) Role of GABA transaminase in the regulation of development and senescence in Arabidopsis thaliana. Curr Plant Biol 19:100119

Janse van Rensburg HC, Van den Ende W (2020) Priming with γ-aminobutyric acid against Botrytis cinerea reshuffles metabolism and reactive oxygen species: dissecting signalling and metabolism. Antioxidants 9:1174

Ji F, Kanbara N, Obata K (1999) GABA and histogenesis in fetal and neonatal mouse brain lacking both isoforms of glutamic acid decarboxylase. Neurosci Res 33:187–194

Karatzas KAG, Suur L, O'Byrne CP (2012) Characterization of the intracellular glutamate decarboxylase system: analysis of its function, transcription, and role in the acid resistance of various strains of listeria monocytogenes. Appl Environ Microbiol 78:3571

Kaspal M, Kanapaddalagamage MH, Ramesh SA (2021) Emerging roles of γ-aminobutyric acid (GABA) gated channels in plant stress tolerance. Plants (Basel) 10:2178

Katz E, Boo KH, Kim HY, Eigenheer RA, Phinney BS, Shulaev V, Negre-Zakharov F, Sadka R, Blumwald E (2011) Label-free shotgun proteomics and metabolite analysis reveal a significant metabolic shift during citrus fruit development. J Exp Bot 62:5367–5368

Kaur R, Zhawar VK, Kaur G, Asthir B (2020) γ-Aminobutyric acid effect in the alleviation of saline–alkaline stress conditions in rice. Cereal Res Commun 48:1–9

Khan MIR, Jalil SU, Chopra P, Chhillar H, Ferrante A, Khan NA, Ansari MI (2021) Role of GABA in plant growth, development, and senescence. Plant Gene 26:100283

Kinnersley AM, Turano FJ (2000) γ-Aminobutyric acid (GABA) and plant responses to stress. Crit Rev Plant Sci 19:479–509

Krishnan S, Laskowski K, Shukla V, Merewitz EB (2013) Mitigation of drought stress damage by exogenous application of a non-protein amino acid γ-aminobutyric acid on perennial ryegrass. J Am Soc Hortic Sci 138:358–366

Kumar S, Punekar NS (1997) The metabolism of 4-aminobutyrate (GABA) in fungi. Mycol Res 101:403–409

Kumar N, Dubey AK, Upadhyay AK, Gautam A, Ranjan R, Srikishna S, Sahu N, Behera SK, Mallick S (2017) GABA accretion reduces Lsi-1 and Lsi-2 gene expressions and modulates physiological responses in Oryza sativa to provide tolerance towards arsenic. Sci Rep 7:8786. https://doi.org/10.1038/s41598-017-09428-2

Lakhani R, Vogel KR, Till A et al (2014) Defects in GABA metabolism affect selective autophagy pathways and are alleviated by mTOR inhibition. EMBO Mol Med 6:551–566

Lancien M, Roberts MR (2006) Regulation of Arabidopsis thaliana 14-3-3 gene expression by γ-aminobutyric acid. Plant Cell Environ 29:1430–1436

Li JB, Zhang J, Jia HX, Liu B, Sun P, Hu JJ, Wang LJ, Lu MZ (2017) The WUSCHEL-related homeobox 5a (PtoWOX5a) is involved in adventitious root development in poplar. Tree Physiol 38:139–153

Li L, Gu W, Li J, Li C, Xie T, Qu D et al (2018) Exogenously applied spermidine alleviates photosynthetic inhibition under drought stress in maize (Zea mays L.) seedlings associated with changes in endogenous polyamines and phytohormones. Plant Physiol Biochem 129:35–55

Li L, Dou N, Zhang H, Wu C (2021a) The versatile GABA in plants. Plant Signal Behav 16:1862565

Li X, Xu X, Shen F et al (2021b) γ-Aminobutyric acid participates in the adult-phase adventitious rooting recalcitrance. J Plant Growth Regul 40:1981–1991

Ling Y, Chen T, Jing Y, Fan L, Wan Y (2013) γ-Aminobutyric acid (GABA) homeostasis regulates pollen germination and polarized growth in Picea wilsonii. Planta 238:831–843

Liu S, Zhang J, Hu C, Sun X, Xu N (2021) Physiological and transcriptome analysis of γ aminobutyric acid (GABA) in improving Gracilariopsis lemaneiformis stress tolerance at high temperatures. Algal Res 60:102532

Luo HY, Gao HB, Xia QP, Gong BB, Wu XL (2011) Effects of exogenous GABA on reactive oxygen species metabolism and chlorophyll fluorescence parameters in tomato under NaCl stress. Sci Agric Sin 44:753–761

MacRae JI et al (2013) Mitochondrial metabolism of sexual and asexual blood stages of the malaria parasite plasmodium falciparum. BMC Biol 11:67

Maillard A, Diquélou S, Billard V, Laîné P, Garnica M, Prudent M, Garcia-Mina JM, Yvin JC, Ourry A (2015) Leaf mineral nutrient remobilization during leaf senescence and modulation by nutrient deficiency. Front Plant Sci 6:317

Majumdar D, Guha S (1988) Conformation, electrostatic potential and pharmacophoric pattern of GABA (γ-aminobutyric acid) and several GABA inhibitors. J Mol Struct THEOCHEM 180:125–140

Makino Y, Soga N, Oshita S, Kawagoe Y, Tanaka A (2008) Stimulation of γ-aminobutyric acid production in vine-ripe tomato (Lycopersicon esculentum Mill.) fruits under modified atmospheres. J Agric Food Chem 56:7189–7193

Martin DL, Tobin AJ (2000) Mechanisms controlling GABA synthesis and degradation in the brain. In: Martin DL, Olsen RW (eds) GABA in the nervous system: the view at fifty years. Lippincott Williams and Wilkins, Philadelphia, pp 25–41

Martin T, Oswald O, Graham IA (2002) Arabidopsis seedling growth, storage lipid mobilization, and photosynthetic gene expression are regulated by carbon:nitrogen availability. Plant Physiol 128:472–481

Martinez D, Costa M, Guiamet J (2008) Senescence-associated degradation of chloroplast proteins inside and outside the organelle. Plant Biol 10:15–22

McKelvey J, Rai R, Cooper TG (1990) GABA transport in Saccharomyces cerevisiae. Yeast 6:263–270

Medina-Kauwe LK, De Meirleir LT et al (1999) 4-Aminobutyrate aminotransferase (GABA transaminase) deficiency. J Inherit Metab Dis 22:414–427

Merodio C, Muñoz MT, Cura BD, Buitrago D, Escribano M, Isabel IA (1998) Effect of high CO_2 level on the titres of γ-aminobutyric acid, total polyamines, and some pathogenesis related proteins in cherimoya fruit stored at low temperature. J Exp Bot 49:1339–1347. https://doi.org/10.1093/jexbot/49.325.1339

Michaeli S, Fait A, Lagor K, Nunes-Nesi A, Grillich N, Yellin A et al (2011) A mitochondrial GABA permease connects the GABA shunt and the TCA cycle, and is essential for normal carbon metabolism. Plant J. 67:485–498. https://doi.org/10.1111/j.1365-313X.2011.04612.x

Michaeli S, Fromm H (2015) Closing the loop on the GABA shunt in plants: are GABA metabolism and signalling entwined? Front Plant Sci 9:419

Miller AJ, Fan X, Shen Q, Smith SJ (2008) Amino acids and nitrate as signals for the regulation of nitrogen acquisition. J Exp Bot 59:111–119

Moorhead G, Douglas P, Cotelle V, Harthill J, Morrice N, Meek S, Deiting U, Stitt M, Scarabel M, Aitken A, MacKintosh C (1999) Phosphorylation-dependent interactions between enzymes of plant metabolism and 14-3-3 proteins. Plant J 18:1–12

Morse DE, Hooker N, Duncan H, Jensen L (1979) γ-Aminobutyric acid, a neurotransmitter, induces planktonic abalone larvae to settle and begin metamorphosis. Science 204(407):410

Naliwajski MR, Skłodowska M (2018) The relationship between carbon and nitrogen metabolism in cucumber leaves acclimated to salt stress. PeerJ 6:e6043

Nayyar H, Kaur R, Kaur S, Singh R (2014) γ-Aminobutyric acid (GABA) imparts partial protection from heat stress injury to rice seedlings by improving leaf turgor and upregulating osmoprotectants and antioxidants. J Plant Growth Regul 33:408–419

Noe FF, Nickerson WJ (1958) Metabolism of 2-pyrrolidone and γ-aminobutyric acid by Pseudomonas aeruginosa. J Bacteriol 75:674–681

Nunes-Nesi A, Fernie AR, Stitt M (2010) Metabolic and signaling aspects underpinning the regulation of plant carbon-nitrogen interactions. Mol Plant 3:973–996

Oh SJ, Kim HS, Lim ST (2019) Increase of γ-aminobutyric acid contents in rice embryo with protein hydrolysates and pyridoxal-5-phosphate using abiotic stress. J Cereal Sci 89:102803

Ono K, Terashima I, Watanabe A (1996) Interaction between nitrogen deficit of a plant and nitrogen content in the old leaves. Plant Cell Physiol 37:1083–1089

Owens DF, Kriegstein AR (2002) Is there more to GABA than synaptic inhibition? Nat Rev Neurosci. 3(9):715–27. https://doi.org/10.1038/nrn919. PMID: 12209120

Palanivelu R, Brass L, Edlund AF, Preuss D (2003) Pollen tube growth and guidance is regulated by POP2, an Arabidopsis gene that controls GABA levels. Cell 114:47–59

Palma F, Carvajal F, Jiménez-Muñoz R, Pulido A, Jamilena M, Garrido D (2019) Exogenous γ-aminobutyric acid treatment improves the cold tolerance of zucchini fruit during postharvest storage. Plant Physiol Biochem 136:188–195

Park DH, Mirabella R, Bronstein PA et al (2010) Mutations in γ-aminobutyric acid (GABA) transaminase genes in plants or pseudomonas syringae reduce bacterial virulence. Plant J 64:318–330

Pedreschi R, Hertog M, Robben J, Lilley KS, Karp NA, Baggerman G, Vanderleyden J, Nicolaï B (2009) Gel-based proteomics approach to the study of metabolic changes in pear tissue during storage. J Agric Food Chem 57:6997–7004

Podlešáková K, Ugena L, Spíchal L, Doležal K, De Diego N (2019) Phytohormones and polyamines regulate plant stress responses by altering the GABA pathway. New Biotechnol 48:53–65

Qi H, Kang D, Zeng W, Jawad Hassan M, Peng Y, Zhang X, Zhang Y, Feng G, Li Z (2021) Alterations of endogenous hormones, antioxidant metabolism, and aquaporin gene expression in relation to γ-aminobutyric acid-regulated thermotolerance in white clover. Antioxidants 10:1099

Ramesh SA, Stephen DT, Xu B, Bose J, Kaur S, Conn V, Domingos P, Ullah S, Wege S, Shabala S, Feijó JA, Ryan PR, Gilliham M (2015) GABA signaling modulates plant growth by directly regulating the activity of plant-specific anion transporters. Nat Commun 6:7879

Ramos F, Guezzar M, Grenson M, Wiame JM (1985) Mutations affecting the enzymes involved in the utilization of 4-aminobutyric acid as a nitrogen source by the yeast Saccharomyces cerevisiae. Eur J Biochem 149:401–404

Ramos-Ruiz R, Martinez F, Knauf-Beiter G, Tejada Moral M (2019) The effects of GABA in plants. Cogent Food Agric 5:1

Ramputh AI, Bown AW (1996) Rapid γ-aminobutyric acid synthesis and the inhibition of the growth and development of oblique banded leaf roller larvae. Plant Physiol 111(1349):1352

Reed LJ (1950) The occurrence of γ-aminobutyric acid in yeast extract; its isolation and identification. J Biol Chem 183:451–458

Ren T, Zheng P, Zhang K, Liao J, Xiong F, Shen Q, Ma Y, Fang W, Zhu X (2021) Effects of GABA on polyphenol accumulation and antioxidant activities in tea plants (Camellia sinensis L.) under heat-stress conditions. Plant Physiol Biochem 159:363–371

Renault H, Roussel V, El Amrani A, Arzel M, Renault D, Bouchereau A et al (2010) The Arabidopsis pop2-1 mutant reveals the involvement of GABA transaminase in salt stress tolerance. BMC Plant Biol 10:20

Renault H, El Amrani A, Palanivelu R, Updegraff EP, Yu A, Renou S (2011) GABA accumulation causes cell elongation defects and a decrease in expression of genes encoding secreted and cell wall-related proteins in Arabidopsis thaliana. Plant Cell Physiol 52:894–908

Rezaei-Chiyaneh E, Seyyedi SM, Ebrahimian E, Moghaddam SS, Damalas CA (2018) Exogenous application of γ-aminobutyric acid (GABA) alleviates the effect of water deficit stress in black cumin (Nigella sativa L.). Ind Crop Prod 112:741–748

Ribak CE, Yan XX (2000) GABA neurons in the neocortex. In: Martin DL, Olsen RW (eds) GABA in the nervous system: the view at fifty years. Lippincott Williams and Wilkins, Philadelphia, pp 357–368

Rico A, Preston GM (2008) Pseudomonas syringae pv. Tomato DC3000 uses constitutive and apoplast-induced nutrient assimilation pathways to catabolize nutrients that are abundant in the tomato apoplast. Mol Plant-Microbe Interact 21:269–282
Roberts E (1984) GABA-Related phenomena, models of nervous system function, and seizures. Ann. Neurol. 16, S77–S89. https://doi.org/10.1002/ana.410160713
Roberts E (1986) Failure of GABAergic inhibition: a key to local and global seizures. Adv Neurol 44:319–341
Roberts JKM, Hooks MA, Miaullis AP, Edwards S, Webster C (1992) Contribution of malate and amino acid metabolism to cytoplasmic pH regulation in hypoxic maize root tips studied using nuclear magnetic resonance spectroscopy. Plant Physiol 98:480–487
Roettger VR, Amara SG (1999) GABA and glutamate transporters: therapeutic and etiological implications for epilepsy. Adv Neurol 79:551–560
Sakamoto W, Takami T (2014) Nucleases in higher plants and their possible involvement in DNA degradation during leaf senescence. J Exp Bot 65:3835–3843
Sapse AM (2000) Molecular orbital calculations for amino acids and peptides. Birkhäuser, ISBN 978-0-8176-3893-1
Sato T, Maekawa S, Yasuda S, Domeki Y, Sueyoshi K, Fujiwara M, Fukao Y, Goto DB, Yamaguchi J (2011) Identification of 14-3-3 proteins as a target of ATL31 ubiquitin ligase, a regulator of the C/N response in Arabidopsis. Plant J 68:137–146
Satyanarayan V, Nair PM (1990) Metabolism, enzymology, and possible roles of 4 aminobutyrate in higher plants. Phytochemistry 29:367–375
Scholz SS, Reichelt M, Mekonnen DW, Ludewig F, Mithöfer A (2015) Insect herbivory elicited GABA accumulation in plants is wound-induced, direct, systemic, and jasmonate dependent Defence response. Front Plant Sci 6:1128. https://doi.org/10.3389/fpls.2015.01128
Schwacke R, Gralath S, Breitkreuz KE, Stransky E, Stransky H, Frommer WB, Rentsch D (1999) LeProT1, a transporter for proline, glycine betaine, and γ-aminobutyric acid in tomato pollen. Plant Cell 11:377–392
Seifi HS, Curvers K, De Vleesschauwer D, Deleare I, Aziz A, Hofte M (2013) Concurrent overactivation of the cytosolic glutamine synthetase and the GABA shunt in the ABA deficient sitiens mutant tomato leads to resistance against Botrytis cinerea. New Phytol 199:490–550
Shang H, Cao S, Yang Z, Cai Y, Zheng Y (2011) Effect of exogenous γ-aminobutyric acid treatment on proline accumulation and chilling injury in peach fruit after long-term cold storage. J Agric Food Chem 59:1264–1268
Shao Q, Wang H, Guo H, Zhou A, Huang Y, Sun Y et al (2014) Effects of shade treatments on photosynthetic characteristics, chloroplast ultrastructure, and physiology of Anoectochilus roxburghii. PLoS One 9:e85996
Shekari A, Naghshiband Hassani R, Soleimani Aghdam M (2021) Exogenous application of GABA retards cap browning in Agaricus bisporus and its possible mechanism. Postharvest Biol Technol 174:111434
Shelp BJ, Bown AW, Mclean MD (1999) Metabolism and function of γ-aminobutyric acid. Trends Plant Sci 4:446–452
Shelp BJ, Bown AW, Faure D (2006) Extracellular -aminobutyrate mediates communication between plantsand other organisms. Plant Physiol 142(4):1350–1352. https://doi.org/10.1104/pp.106.088955
Shelp BJ, Bozzo GG, Trobacher CP, Zarei A, Deyman KL, Brikis CJ (2012) Hypothesis/review: contribution of putrescine to 4-aminobutyrate (GABA) production in response to abiotic stress. Plant Sci 193–194:130–135
Shelp BJ, Aghdam MS, Flaherty EJ (2021) γ-Aminobutyrate (GABA) regulated plant Defence: mechanisms and opportunities. Plants 10:1939
Sheng L, Shen D, Luo Y, Sun X, Wang J, Luo T et al (2017) Exogenous γ-aminobutyric acid treatment affects citrate and amino acid accumulation to improve fruit quality and storage performance of postharvest citrus fruit. Food Chem 216:138–145

Shi SQ, Shi Z, Jiang ZP, Qi LW, Sun XM, Li CX, Liu JF, Xiao WF, Zhang SG (2010) Effects of exogenous GABA on gene expression of Caragana intermedia roots under NaCl stress: regulatory roles for H_2O_2 and ethylene production. Plant Cell Environ 33:149–162

Signorelli S, Dans PD, Coitiño EL, Borsani O, Monza J (2015) Connecting proline and γ aminobutyric acid in stressed plants through non-enzymatic reactions. PLoS One 10:e0115349

Singh P, Choudhary KK, Chaudhary N, Gupta S, Sahu M, Tejaswini B, Sarkar S (2022) Salt stress resilience in plants mediated through osmolyte accumulation and its crosstalk mechanism with phytohormones. Front Plant Sci 13:1006617. https://doi.org/10.3389/fpls.2022.1006617

Siragusa S, de Angelis M, di Cagno R, Rizzello CG, Coda R, Gobbetti M (2007) Synthesis of γ-aminobutyric acid by lactic acid bacteria isolated from a variety of Italian cheeses. Appl Environ Microbiol 73:7283–7290

Sita K, Kumar V (2020) Role of γ-aminobutyric acid (GABA) against abiotic stress tolerance in legumes: a review. Plant Physiol Rep 25:1–10

Steward FC, Thompson JF, Dent CE (1949) γ-Aminobutyric acid: a constituent of potato tubers? Science 110:439–440

Sulpice R, Nikoloski Z, Tschoep H, Antonio C, Kleessen S, Larhlimi A, Selbig J, Ishihara H, Gibon Y, Fernie AR et al (2013) Impact of the carbon and nitrogen supply on relationships and connectivity between metabolism and biomass in a broad panel of Arabidopsis accessions. Plant Physiol 162:347–363

Sun X, Zhu A, Liu S, Sheng L, Ma Q, Zhang L, Nishawy EM, Zeng Y, Xu J, Ma Z, Cheng Y, Deng X (2013) Integration of metabolomics and subcellular organelle expression microarray to increase understanding of organic acid changes in post-harvest citrus fruit. J Integr Plant Biol 55:1038–1053

Tcherkez G, Mahé A, Gauthier P, Mauve C, Gout E, Bligny R (2009) In folio respiratory fluxomics revealed by ^{13}C isotopic labeling and H/D isotope effects highlight the noncyclic nature of the tricarboxylic acid "cycle" in illuminated leaves. Plant Physiol 151:620–630

Vijayakumari K, Puthur JT (2015) γ-Aminobutyric acid (GABA) priming enhances osmotic stress tolerance in Piper nigrum Linn. Plants subjected to PEG-induced stress. Plant Growth Regul 78:57–67

Wallace W, Secor J, Schrader LE (1984) Rapid accumulation of γ-aminobutyric acid and alanine in soybean leaves in response to an abrupt transfer to lower temperature, darkness, or mechanical manipulation. Plant Physiol 75:170–175

Wang X, Cao J, Qiao J, Pan J, Zhang S, Li Q, Wang Q, Gong B, Shi J (2022) GABA keeps nitric oxide in balance by regulating GSNOR to enhance disease resistance of harvested tomato against Botrytis cinerea. Food Chem 392:133299

Wani SH, Singh NB, Haribhushan A, Mir JI (2013) Compatible solute engineering in plants for abiotic stress tolerance: role of glycine betaine. Curr Genomics 14:157–165

Weston E, Thorogood K, Vinti G, López-Juez E (2000) Light quantity controls leaf-cell and chloroplast development in Arabidopsis thaliana wild type and blue-light-perception mutants. Planta 211:807–815

Wu X, Jia Q, Ji S, Gong B, Li J, Lü G, Gao H (2020) Gamma-aminobutyric acid (GABA) alleviates salt damage in tomato by modulating Na+ uptake, the GAD gene, amino acid synthesis and reactive oxygen species metabolism. BMC Plant Biol 20(1):1–21

Wu Q, Su N, Huang X, Cui J, Shabala L, Zhou M, Yu M, Shabala S (2021) Hypoxia-induced increase in GABA content is essential for restoration of membrane potential and preventing ROS-induced disturbance to ion homeostasis. Plant Commun 2:100188

Xiang L, Hu L, Xu W, Zhen A, Zhang L, Hu X (2016) Exogenous γ-aminobutyric acid improves the structure and function of photosystem II in muskmelon seedlings exposed to salinity-alkalinity stress. PLoS One 11:e0164847

Xie T, Ji J, Chen W, Yue J, Du C, Sun J, Shi S (2020) GABA negatively regulates adventitious root development. J Exp Bot 71:1459–1474

Xu B, Long Y, Feng X, Zhu X, Gilliham M (2021) GABA signalling modulates stomatal opening to enhance plant water use efficiency and drought resilience. Nat Commun 12(1):13

Yang J, Sun C, Zhang Y, Fu D, Zheng X, Yu T (2017) Induced resistance in tomato fruit by γ aminobutyric acid for the control of Alternaria rot caused by Alternaria alternata. Food Chem 221:1014–1020

Ying L, Xiaojun Y, Kaikai M (2022) Physiological effects of γ-aminobutyric acid application on cold tolerance in Medicago ruthenica. Front. Plant Sci 13:xxxxx

Yu GH, Zou J, Feng J, Peng XB, Wu JY, Wu YL, Palanivelu R, Sun MX (2014) Exogenous γ-aminobutyric acid (GABA) affects pollen tube growth via modulating putative Ca2+-permeable membrane channels and is coupled to negative regulation on glutamate decarboxylase. J Exp Bot 65(12):3235–48. https://doi.org/10.1093/jxb/eru171

Yu C, Zeng L, Sheng K, Chen F, Yu T (2014a) γ-Aminobutyric acid induces resistance against Penicillium expansum by priming of Defence responses in pear fruit. Food Chem 159(29):37

Yu GH, Zou J, Feng J, Peng XB, Wu JY, Wu YL, Palanivelu R, Sun MX (2014b) Exogenous γ-aminobutyric acid (GABA) affects pollen tube growth via modulating putative Ca^{2+} permeable membrane channels and is coupled to negative regulation on glutamate decarboxylase. J Exp Bot 65:3235–3248

Yuan F, Yang H, Xue Y, Kong D, Ye R, Li C, Zhang J, Theprungsirikul L, Shrift T, Krichilsky B (2014) OSCA1 mediates osmotic-stress-evoked Ca^2+ increases vital for osmosensing in Arabidopsis. Nature 514:367–371

Yue JY, Du CJ, Ji J, Xie TT, Chen W, Chang EM, Chen LZ, Jiang ZP, Shi SQ (2018) Inhibition of α-ketoglutarate dehydrogenase activity affects adventitious root growth in poplar via changes in the GABA shunt. Planta 248:963–979

Zarei A, Chiu GZ, Yu G, Trobacher CP, Shelp BJ (2017) Salinity-regulated expression of genes involved in GABA metabolism and signaling. Botany 95:621–627

Zhen A, Zhang Z, Jin X, Liu T, Ren W, Hu X (2018) Exogenous GABA application improves NO_3^--N absorption and assimilation in $ca(NO_3)_2$-treated muskmelon seedlings. Sci Hortic 227:117–123

Zhou M, Hassan MJ, Peng Y, Liu L, Liu W, Zhang Y, Li Z (2021) γ-Aminobutyric acid (GABA) priming improves seed germination and seedling stress tolerance associated with enhanced antioxidant metabolism, DREB expression, and dehydrin accumulation in white clover under water stress. Front Plant Sci 12:776939

Zushi K, Matsuzoe N (2007) Salt stress-enhanced γ-aminobutyric acid (GABA) in tomato fruit. Acta Hortic 761:431–435

GABA Synthesis, Signaling, and Transportation

2

Abstract

GABA synthesis, signaling, and transport in plants elucidate its diverse roles in plant physiology, including stress response, growth, and modulation of physiological processes. Unlike animals, plants have distinct pathways for GABA biosynthesis. The primary enzymatic route engages glutamate decarboxylase (GAD) and the GABA shunt pathway for GABA biosynthesis. GAD activity, regulated by calcium (Ca^{2+}) and calmodulin (CaM), performs decarboxylation of glutamate to produce GABA. The nonenzymatic alternative pathway, such as polyamine (PA) breakdown for GABA synthesis, also exists, particularly under oxidative stress. The GABA shunt pathway connects GABA metabolism to the Tricarboxylic Acid cycle (TCA) through GABA, GABA transaminase (GABA-T), and succinic semialdehyde dehydrogenase (SSADH). The GABA signaling pathways modulate various physiological processes and stress responses and operate through GABA receptors, ion currents, etc. GABA functions as a signaling molecule, modulating ion transport, stomatal dynamics, and root growth. The aluminum-activated malate transporter (ALMT) protein family serves as GABA receptors, linking GABA signaling to ion homeostasis and stress adaptation. GABA cross talk with hormones such as abscisic acid (ABA), ethylene and auxins enhances plant resilience. GABA transporters (GATs) regulate its intracellular balance and redistribution, ensuring efficient signaling under stress. By integrating biosynthesis, signaling, and transport, GABA plays a central role in plant adaptation and development, offering potential strategies for improving stress tolerance and productivity.

Keywords

Plant defense · Signaling pathways · Stress response · Transporter

S. U. Jalil et al., *GABA Signaling System and Abiotic Stress Tolerance in Plants*,
https://doi.org/10.1007/978-981-95-0437-4_2

Highlights

- Enzymatic pathways for GABA biosynthesis involving glutamate decarboxylase (GAD) and the GABA shunt, along with nonenzymatic synthesis through polyamine (PA) degradation.
- Regulation of ion transport, stomatal dynamics, and root growth through GABA signaling, contributing to stress adaptation.
- Identification of aluminum-activated malate transporter (ALMT) proteins as GABA receptors and their role in maintaining ion homeostasis and plant responses to environmental stimuli.
- GABA can modulate hormone biosynthesis, transport, or signal transduction for modulating developmental processes and enhancing plant stress responses.
- Role of GABA transporters (GATs) in intracellular balance, redistribution, and systemic signaling under varying conditions.
- Potential applications of GABA-related pathways in enhancing stress tolerance and productivity through metabolic and genetic approaches.

2.1 Introduction

GABA plays an essential role in plants by supporting their growth, development, and response to environmental stresses, such as drought, salinity, and pathogen attacks (Michaeli and Fromm 2015). Over the years, studies have greatly expanded our understanding of how GABA is synthesized, how it signals within the plant system, and how it is transported across different tissues (Li et al. 2020).

In contrast to animals, where GABA is primarily produced in the nervous system through enzymatic processes, plants possess both enzymatic and nonenzymatic mechanisms to produce this molecule. One of the main enzymes involved is glutamate decarboxylase (GAD), which converts glutamate into GABA (Baum et al. 1993; Michaeli and Fromm 2015). Once formed, GABA can enter the GABA shunt pathway, where it is metabolized by GABA transaminase (GABA-T) and succinic semialdehyde dehydrogenase (SSADH) to generate succinate, an important component of the Tricarboxylic Acid cycle (TCA) (Fait et al. 2008; Ansari et al. 2014; Michaeli and Fromm 2015).

Beyond this classical route, GABA can also be produced through the degradation of polyamines (PAs) such as putrescine and spermidine (Bhatnagar et al. 2002; Shelp et al. 2012) or via a nonenzymatic process involving proline under oxidative stress conditions (Signorelli et al. 2015). GABA metabolism is not limited to its synthesis; it is also broken down in catabolic pathways (Bach et al. 2009) and contributes to the formation of other bioactive compounds. Additionally, it acts as a key signaling molecule that helps plants cope with environmental stressors (Ansari et al. 2021; Khan et al. 2021).

In plants, GABA acts as a messenger that regulates a wide range of physiological activities (Kinnersley and Turano 2000; Bouche and Fromm 2004; Ansari and Chen 2009; Clark et al. 2009). While in animals it serves as a neurotransmitter, in plants, it functions through specific receptors and transport systems that influence ion movement, electrical signals across membranes, and gene activity (Bouché et al. 2003; Ramesh et al. 2015). Under stress, GABA levels rise, helping trigger protective mechanisms like stomatal regulation, ion homeostasis, enhanced antioxidant activity, and the activation of stress-related genes (Bouche et al. 2003; Fait et al. 2005). One key aspect of its signaling involves the movement of calcium (Ca^{2+}) ions, which act as secondary messengers in stress response pathways (Baum et al. 1993; Yu et al. 2014).

The distribution of GABA within and between plant cells is managed by various transporter proteins. These include aluminum-activated malate transporters (ALMTs), GABA transporters (GATs), bidirectional amino acid transporters (BATs), and cationic amino acid transporters (CATs), which move GABA into cellular compartments such as the cytoplasm, vacuole, and mitochondria (Ramesh et al. 2015; Meyer et al. 2006; Dündar and Bush 2009; Snowden et al. 2015; Yuan et al. 2023). These transporters are finely regulated according to environmental conditions and developmental stages, allowing GABA levels to rise when needed, such as during stress.

Studying GABA biosynthesis, signaling, and transport has revealed its vital role in plant physiology. Manipulating these pathways may offer innovative strategies to improve crop stress tolerance and yield. However, many aspects of how GABA integrates with broader plant signaling networks remain unclear and require further investigation.

2.2 Pathways and Regulation of GABA Biosynthesis and Metabolism

GABA production occurs via multiple interconnected pathways, each involving distinct enzymatic reactions and precursor molecules. The principal pathway of GABA synthesis is the decarboxylation of glutamate, catalyzed by GAD. However, alternative pathways also contribute to GABA accumulation, including the polyamine degradation pathway, where putrescine serves as a precursor, and the GABA shunt, which links GABA metabolism to the TCA cycle. These pathways are tightly regulated by several biochemical and environmental factors, ensuring a dynamic balance between GABA synthesis and degradation.

2.2.1 GABA Shunt Pathway

The GABA shunt represents a key metabolic route in plants, comprising a sequence of enzymatic reactions that link carbon and nitrogen (C and N) metabolism. This pathway begins with the decarboxylation of glutamate, catalyzed by GAD (Fig. 2.1), resulting in the formation of GABA in the cytosol (Baum et al. 1993; Michaeli and Fromm 2015). This reaction is irreversible and involves the removal of a carboxyl group, releasing CO_2.

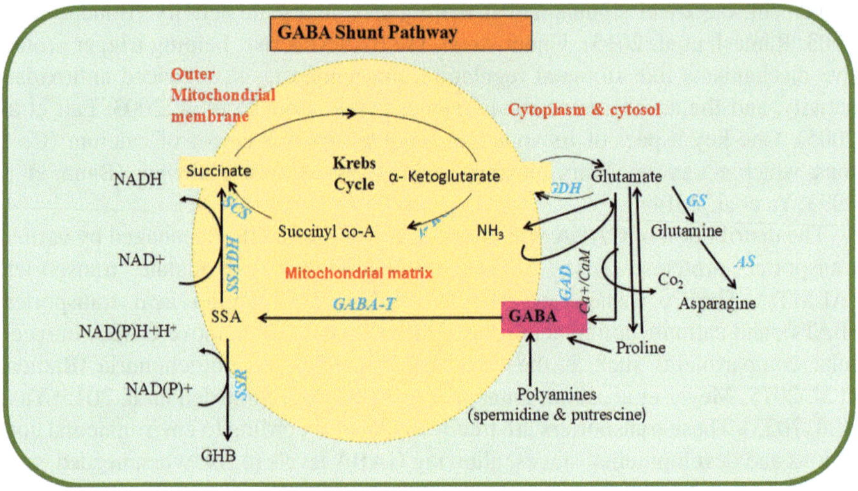

Fig. 2.1 Simplified representation of the GABA shunt pathway in plants, highlighting its role in bridging carbon and nitrogen metabolism. The pathway functions across cellular compartments, including the cytosol and mitochondrial matrix, and involves the biosynthesis and further metabolism of GABA

Following its synthesis, GABA is transported into the mitochondria, where it undergoes further metabolic conversion. Inside the mitochondrial matrix, GABA is transaminated by GABA-T, producing succinic semialdehyde (SSA). This step may involve the transfer of the amino group to acceptors such as pyruvate, α-ketoglutarate, or glyoxylate (Clark et al. 2009; Shelp et al. 2012; Michaeli and Fromm 2015; Jalil et al. 2016, 2019a, b). SSA is then oxidized by SSADH to yield succinate, which feeds into the TCA cycle and contributes to cellular respiration (Barbosa et al. 2010).

Studies on SSADH have also suggested that its regulation may be influenced by calmodulin (CaM)-mediated signaling, indicating a potential link between GABA metabolism and calcium signaling pathways (Busch and Fromm 1999). In addition, an alternative metabolic route for SSA conversion has been identified in various organisms, including animals, plants, and *Escherichia coli*, wherein SSA is reduced to gamma-hydroxybutyrate (GHB) by 4-Hydroxybutyrate Dehydrogenase (GHBDH) (Andriamampandry et al. 1998; Breitkreuz et al. 2003; Saito et al. 2009). Evidence suggests that this alternate pathway becomes active under stress conditions, such as low oxygen, intense light, or water and heat stress, pointing to a possible role for GHBDH in plant abiotic stress responses. The activity of this alternative pathway appears to be modulated by the cell's redox status and Nicotinamide Adenine Dinucleotide Phosphate (NADPH)-dependent glyoxylate reductases (Breitkreuz et al. 2003; Allan et al. 2008).

The genetic characterization of GAD in plants has highlighted the presence of a calcium/calmodulin-binding domain (CaMBD), initially identified in *Petunia* (Baum et al. 1993). This domain suggests that GAD activity may be regulated by intracellular calcium signals. Subsequent research has identified GAD homologs across a range of plant species (Ling et al. 1994; Snedden et al. 1995). Functional analyses in transgenic tobacco expressing GAD lacking the CaMBD have demonstrated the critical role of this domain. The altered GAD led to disrupted glutamate metabolism and GABA accumulation, which caused abnormal plant development. These phenotypes included reduced stem elongation due to limited cell expansion in the cortex, accumulation of GABA to high levels, and significantly reduced glutamate pools (Baum et al. 1993).

As for GAD activity regulation, the first identified plant GAD isoform (OsGAD2) from rice roots lacks a functional CaM-binding domain at its C-terminal region. This suggests that OsGAD2 operates independently of CaM regulation. In contrast, the other rice GAD isoform (OsGAD1) specifically functions in a Ca^{2+}/CaM-dependent manner (Akama and Takaiwa 2007). This distinction between the two isoforms indicates that rice possesses both CaM-independent and CaM-dependent mechanisms for GABA biosynthesis, allowing for a more flexible and responsive regulation of GABA levels under varying physiological and environmental conditions.

2.2.2 Polyamine Degradation Pathway

GABA biosynthesis can potentially occur through the breakdown of PAs, particularly putrescine (Put) and spermidine (Spd) (Bhatnagar et al. 2002; Shelp et al. 2012). The action of ornithine decarboxylase, eliciting the decarboxylation of ornithine, can directly produce Put. Alternatively, the arginine decarboxylase may indirectly produce the polyamine Put from arginine, an essential amino acid. Spd, on the other hand, is synthesized by the reaction of Put with decarboxylated S-adenosylmethionine, catalyzed by the enzyme Spd synthase. The catabolism of PAs is primarily catalyzed by two types of amine oxidases: copper-dependent diamine oxidase (DAO), responsible for the degradation of Put and Spd into Δ^1-pyrroline (PYr), and Flavin Adenine Dinucleotide (FAD)-dependent polyamine oxidase (PAO), which also participates in the catabolism of PAs (Kusano et al. 2015). Δ^1-pyrroline is then utilized in the synthesis of GABA through the action of pyrroline dehydrogenase. Thus, the levels of GABA are influenced by changes in polyamine metabolism (Fig. 2.2).

PAs work in conjunction with GABA to influence the Reactive Oxygen Species (ROS) balance in plants. The degradation of these PAs, primarily through the action of DAO, contributes to the generation of GABA (Shelp et al. 2012). Under stress conditions, the activities of both GAD and DAO are often upregulated, promoting GABA synthesis (Bouché et al. 2003; Bouché and Fromm 2004). This process also leads to the production of hydrogen peroxide (H_2O_2), which acts as a signaling molecule to initiate antioxidant defense mechanisms (Gupta et al. 2016).

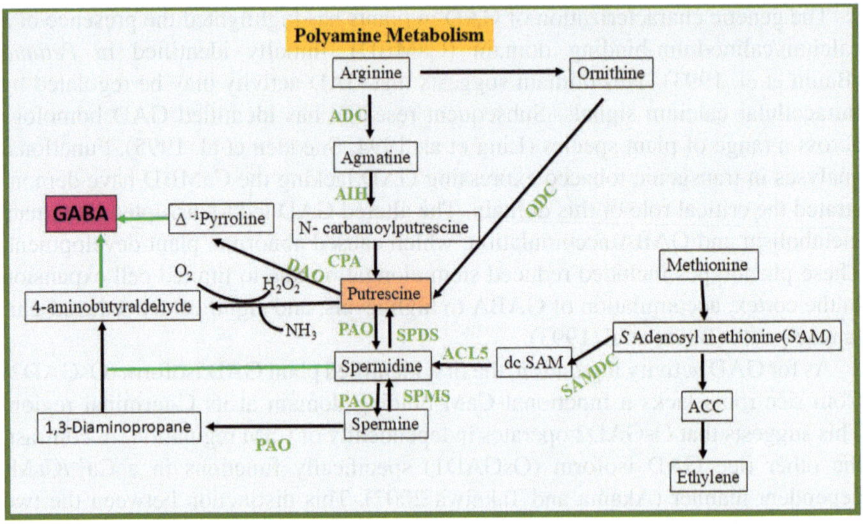

Fig. 2.2 The figure illustrates the interconnection between polyamine metabolism and GABA synthesis in plants. Arginine is metabolized into putrescine through intermediates such as agmatine and N-carbamoyl putrescine. Putrescine acts as a precursor for the synthesis of spermidine and spermine, which are further degraded by polyamine oxidases (PAOs) into 1,3-diaminopropane. Putrescine also undergoes oxidative deamination via diamine oxidase (DAO), producing hydrogen peroxide (H_2O_2), ammonia (NH_3), and 4-aminobutyraldehyde, which is subsequently converted into GABA via Δ^1-pyrroline. Additionally, S-adenosyl methionine (SAM) contributes to polyamine biosynthesis and ethylene production, highlighting the complex metabolic interplay between polyamines and GABA in stress responses and nitrogen metabolism

According to Minocha et al. (2014), PAs are critical in strengthening antioxidant defenses during stress. Since PA breakdown contributes directly to GABA formation, a close relationship exists between the internal levels of PAs and GABA in plant cells (Wang et al. 2014a). The concentrations of these molecules are highly responsive to environmental cues, often increasing under hypoxic conditions and decreasing once the stress subsides, indicating a coordinated response mechanism (Song et al. 2012).

The major polyamines in plants—Put, Spd, and spermine—act as secondary messengers, orchestrating responses to diverse stress factors. The application of external GABA has been shown to protect muskmelon seedlings from calcium nitrate–induced damage by influencing PA biosynthesis and degradation (Hu et al. 2015). During PA catabolism, DAO and PAO contribute to GABA production. This GABA is further processed into succinate via transamination and oxidation, which then enters the TCA cycle (Duan et al. 2008). However, in germinating fava beans under hypoxic conditions, only about one-third of GABA originates from PA degradation (Yang et al. 2013).

Interestingly, exogenous GABA can inhibit the expression and activity of enzymes like DAO and PAO, particularly those bound to the cell wall. This suppression has a stronger effect on PA breakdown than on its synthesis in melon plants

(Wang et al. 2014b). This suggests that elevated GABA levels from PA oxidation might initiate a negative feedback loop, limiting further GABA synthesis in these plants (Wang et al. 2014b). Such interactions highlight the dynamic reprogramming of PA and GABA metabolism under stress, especially cold conditions (Xu et al. 2022). Studies also suggest that both PAs and GABA are involved in mitigating cold stress in various plant species (Baier et al. 2019). GABA may help stabilize PA levels, thereby reducing damage from hypoxia (Wang et al. 2014b).

When applied externally, GABA raises its own internal concentration and enhances drought resistance in white clover by regulating the GABA shunt, polyamine, and proline metabolism. This leads to increased relative water content, reduced electrolyte leakage, and less lipid peroxidation and leaf wilting. Additionally, GABA treatment enhances the activity of enzymes such as GABA-T and α-ketoglutarate dehydrogenase, while suppressing GAD, which results in increased levels of glutamate and GABA (Yong et al. 2017). GABA also promotes the synthesis of Put and Spd, suppresses their degradation, and boosts drought-induced proline synthesis by activating enzymes like P5C synthetase and proline dehydrogenase (ProDH). Meanwhile, it downregulates ornithine-δ-aminotransferase, leading to increased proline accumulation under drought (Yong et al. 2017).

PAs also function as signaling molecules that influence gene expression, such as catalase in the green alga *Ulva fasciata* (Sung et al. 2011). In higher plants, PAs regulate ROS accumulation during pathogen interactions (Lou et al. 2016; Seifi and Shelp 2019). Spermine breakdown in crops like tobacco and cotton can trigger hypersensitive responses. Furthermore, external spermine treatment induces H_2O_2 generation, which plays a role in signaling during viral infection, as shown in *Arabidopsis* infected with cucumber mosaic virus (Mo et al. 2015; Yoda et al. 2003). Conversely, when putrescine catabolism is blocked through acetylation, H_2O_2 production decreases, weakening the defense response in *Arabidopsis thaliana* against *Pseudomonas syringae* (Lou et al. 2016).

GABA is also involved in fine-tuning antioxidant systems during stress (Seifi et al. 2013; Song et al. 2010; Wang et al. 2014b). The combined breakdown of PAs not only produces GABA but also generates H_2O_2, both of which contribute to ROS regulation (Gupta et al. 2016). Given the strong biochemical ties between GABA and PAs in stress conditions, targeted adjustment of metabolic fluxes between these two pathways may offer effective ways to modulate ROS signaling under adverse environments (Hatmi et al. 2015; Podlešáková et al. 2019; Wang et al. 2014b).

2.2.3 GABA Biosynthesis from Proline

Apart from the known enzymatic pathways, GABA can also be synthesized nonenzymatically from proline under oxidative stress conditions (Fig. 2.3) (Signorelli et al. 2015; Michaeli and Fromm 2015). The buildup of proline in plant cells is commonly observed during exposure to abiotic stress (Zhao et al. 2009). Environmental challenges, such as drought, cold, and high salinity, often trigger a significant increase in intracellular proline levels (Fancy et al. 2017; Zarattini and Forlani

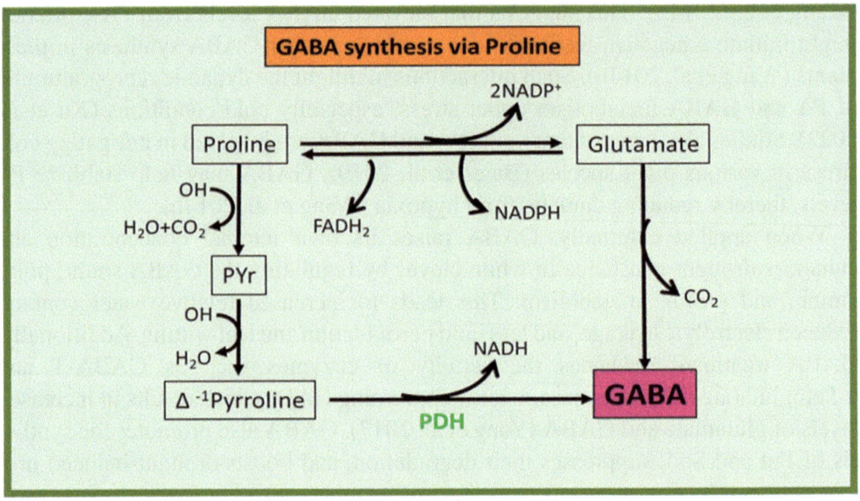

Fig. 2.3 The metabolic pathway of GABA synthesis via proline. Proline undergoes oxidation to pyrroline (PYr) and then to Δ^1-pyrroline-5-carboxylate (P5C) through nonenzymatic reactions involving proline dehydrogenase (ProDH), with the generation of Flavin Adenine Dinucleotide ($FADH_2$). P5C is further converted into glutamate, facilitated by pyrroline-5-carboxylate dehydrogenase (P5CDH), using $NADP^+$, which gets reduced to NADPH. Glutamate is then decarboxylated by glutamate decarboxylase (GAD), releasing CO_2 and forming GABA. This pathway links proline metabolism to the GABA shunt, emphasizing its significance in plant responses to stress, redox homeostasis, and nitrogen assimilation

2017). These conditions also elevate ROS, which, along with nitric oxide (NO), are tightly linked to proline metabolism (Ruan et al. 2004; He and He 2018).

Proline biosynthesis in plants proceeds via two main metabolic routes: the glutamate pathway and the ornithine pathway. Under osmotic stress or reduced NO availability, the glutamate pathway becomes the predominant route for proline production (Delauney and Verma 1993). In this pathway, glutamate is reduced in a two-step process to form proline, with pyrroline-5-carboxylate (P5C) acting as a key intermediate. The enzyme P5C synthetase catalyzes the first committed step in this sequence (Peng et al. 1996). Importantly, glutamate is a precursor not only for proline but also for GABA, creating a biochemical link between the two compounds (Biancucci et al. 2015). Once glutamate is formed, it can be converted to GABA through the action of GAD.

In the ornithine pathway, the amino acid ornithine undergoes transamination catalyzed by δ-ornithine aminotransferase (δ-OAT), resulting in the formation of either glutamate-γ-semialdehyde (GSA) or α-keto-δ-aminovalerate. GSA spontaneously cyclizes to form P5C, which is then reduced to proline by the enzyme P5C reductase. Alternatively, α-keto-δ-aminovalerate forms pyrroline-2-carboxylate, which is reduced to proline via pyrroline-2-carboxylate reductase (Trovato et al. 2008).

Under oxidative stress, hydroxyl radicals (•OH) can abstract hydrogen atoms from proline, initiating its decarboxylation and the generation of pyrroline. Further oxidation of this intermediate results in Δ^1-pyrroline, which serves as a substrate for pyrroline dehydrogenase, ultimately producing GABA (Signorelli et al. 2015). This route highlights a potential nonenzymatic mechanism linking stress-induced proline accumulation to GABA biosynthesis, reinforcing the metabolic connection between these two stress-responsive molecules.

Studies have demonstrated that NO, GABA, and proline can enhance the cold tolerance in tea plants, and there may exist signaling pathways and biosynthetic links among them that enhance their functions (Wang et al. 2020).

Overall, the biosynthesis of GABA in plants occurs through multiple pathways, including the decarboxylation of L-glutamate, oxidative deamination of Put, and oxidation of spermidine, and potentially through proline during oxidative stress. The regulation and contribution of each pathway to GABA accumulation in different plant tissues and under various conditions help us understand GABA metabolism and its functional significance in plant biology. Additional research is required to elucidate the enzymes, regulatory mechanisms, and environmental factors governing these pathways and their interactions.

2.3 Plant GABA Signaling Network

Plants are equipped with highly sophisticated signaling systems that enable them to detect and respond to a wide range of internal and external stimuli (Lamers et al. 2020). While GABA is well-known in animals for its role in neurotransmission through various types of GABA receptors (Medina and Chudotvorova 2006; Buddhala et al. 2009; Lorenz-Guertin and Jacob 2018), research has also uncovered its significance in microbial signaling. Chevrot et al. (2006) demonstrated GABA's involvement in bacterial quorum sensing. These discoveries have prompted researchers to explore GABA's function as a signaling molecule in the plant kingdom as well.

In plants, GABA serves as a dynamic signaling agent, participating in a wide array of physiological and developmental processes. It engages with specific receptors and activates downstream signaling pathways, thereby influencing growth, adaptation to stress, and interaction with the environment (Michaeli and Fromm 2015; Bown and Shelp 2016; Ramesh et al. 2017). This signaling occurs through both intracellular and extracellular pathways, involving receptors, ion channels, and second messengers (Bouché et al. 2003; Ramesh et al. 2015).

GABA binds to particular receptor proteins in plants, including those analogous to animal glutamate receptors. Notably, one plant-specific family of anion channels, ALMTs, has emerged as a unique component of GABA signaling (Ramesh et al. 2015). ALMTs are membrane-localized transporters that display tissue-specific expression patterns and have been implicated in mediating GABA responses (Kollmeier et al. 2001; Pineros and Kochian 2001).

Genetic studies have identified members of this family, such as TaALMT1 in wheat (Sasaki et al. 2004) and AtALMT1 in *Arabidopsis* (Hoekenga et al. 2006). Ramesh et al. (2015) further provided evidence supporting the involvement of these ALMT proteins in GABA-mediated signaling, particularly in wheat roots, highlighting their role in regulating anion fluxes and responding to environmental signals.

2.3.1 GABA Signaling and Modulation of Physiological Processes

GABA plays a central role in regulating numerous physiological processes related to plant growth and development. Its signaling functions extend to key cellular activities, including division, elongation, and differentiation, which collectively determine plant structure and reproductive success. One key demonstration of GABA's signaling role is observed in pollen tube guidance, where a concentration gradient of GABA from the stigma to the ovule assists in directing the tube toward the ovule, facilitating fertilization (Palanivelu et al. 2003). In *Arabidopsis*, disruption of this gradient in a GABA-T mutant (pop2) results in the failure of pollen tubes to reach the ovule, leading to sterility (Palanivelu et al. 2003).

GABA helps coordinate the interaction between the pistil and the growing pollen tube (Palanivelu et al. 2003; Yue et al. 2014). Evidence suggests that low levels of GABA promote pollen tube elongation, whereas elevated levels inhibit it, revealing a biphasic regulatory effect (Yue et al. 2014). In tobacco, a decreasing GABA gradient from the stigma to the ovary is proposed to guide the pollen tubes during fertilization. Excessive GABA levels, however, can negatively impact pollen tube direction and fertilization success (Palanivelu et al. 2003; Renault et al. 2011). This concentration-dependent effect has been confirmed in several species (Ramesh et al. 2015; Ling et al. 2013). For instance, in *Picea wilsonii*, pollen treated with an optimal GABA concentration shows enhanced tube elongation, while deviations above or below this threshold suppress growth (Ling et al. 2013). In both *Arabidopsis* and grapevine, muscimol (a GABA receptor agonist) suppresses pollen tube development, while bicuculline (a GABA antagonist) mitigates this inhibition (Ramesh et al. 2015).

Although the precise mechanisms by which GABA modulates pollen tube growth remain unclear, it is believed that ion transport across the plasma membrane plays a role—possibly involving ALMTs or other channels (Ramesh et al. 2015; Gutermuth et al. 2013). High GABA concentrations may impair Ca^{2+} influx by affecting Ca^{2+} channels, as well as GAD function, which is subject to feedback regulation by the CaM system (Yu et al. 2014). Pollen tube elongation depends on ion fluxes, such as calcium and chloride oscillations, in the tip. GABA increases cytosolic Ca^{2+} in tobacco pollen protoplasts, which activates GAD and may influence GABA biosynthesis. This feedback regulation affects ion movement, thereby fine-tuning pollen tube growth (Yu et al. 2014). However, further molecular-level studies are needed to fully elucidate this signaling pathway.

GABA signaling is also involved in fruit ripening by modulating ethylene production. In tomatoes, GABA accumulates during the mature green stage but declines as the fruit ripens and seeds develop (Takayama and Ezura 2015). Reducing the expression of SlGAD genes lowers GABA content without significantly affecting plant growth (Takayama et al. 2015). On the other hand, overexpression of SlGAD2 increases GABA levels but negatively affects growth, flowering, and fruit production (Nonaka et al. 2017). Interestingly, the targeted overexpression of SlGAD3 in tomato fruit using the E8 promoter boosts GABA content without affecting morphology. However, modifying the SlGAD3 gene to delay ethylene production leads to incomplete fruit ripening and reduced ethylene sensitivity (Nonaka et al. 2017).

GABA also supports seed germination and the development of both primary and adventitious roots. Applying GABA externally improves germination in many species (Li et al. 2016; Zhou et al. 2021). Internally, GABA levels increase during germination in species such as soybean (Matsuyama et al. 2009), oats (Xu et al. 2010), barnyard millet (Sharma et al. 2016), adlay (Xu et al. 2017), rice (Zhao et al. 2017), wheat (Kim et al. 2018; AL-Quraan et al. 2019), barley (Sheng et al. 2018; AL-Quraan et al. 2019), durum wheat (Samarah et al. 2023), Chinese wild rice (Chu et al. 2019), and white clover (Cheng et al. 2018; Zhou et al. 2021). GABA stimulates the expression of genes involved in the degradation of seed storage compounds such as α-amylase and helps overcome dormancy while enhancing seedling vigor (Sheng et al. 2018). In salinity-stressed mung bean seeds, GABA may regulate respiration during germination (Ji et al. 2020).

Regarding root development, GABA affects both primary and adventitious root growth. Elevated levels of GABA can inhibit primary root and hypocotyl elongation (Renault et al. 2011). While internal GABA promotes germination in soybean and mung bean, external GABA suppresses both germination and primary root elongation in Chinese chestnut, possibly by disrupting the C and N balance to sustain dormancy (Du et al. 2020). Likewise, increased GABA levels through the inhibition of α-ketoglutarate dehydrogenase, overexpression of PagGAD2, or external application hinder adventitious root formation in poplar (Yue et al. 2018; Xie et al. 2020). These effects are associated with altered GABA shunt activity, shifts in genes linked to cell-wall carbon metabolism, and changes in hormone-signaling pathways (Yue et al. 2018; Xie et al. 2020).

GABA signaling is also involved in nitrate uptake in *Arabidopsis* (Barbosa et al. 2000, 2010) and *Brassica napus* (Beuve et al. 2004), nodule formation in *Medicago* (Sulieman and Schulze 2010; Sulieman 2011), and leaf senescence in *Arabidopsis* (Jalil et al. 2017). Endogenous GABA concentrations show oscillations linked to light cycles in *Arabidopsis*, suggesting a connection between GABA signaling and circadian rhythm (Dubois et al. 2017; Pelvan et al. 2021). Furthermore, GABA can be transported over long distances through the xylem and phloem (Shelp et al. 2012), highlighting its systemic signaling potential. Modifying GABA signaling through genetic or chemical means holds promise for enhancing plant resilience and productivity in agriculture.

2.3.2 GABA Signaling Cross Talk and Abiotic Stress

Environmental stresses are known to enhance GABA biosynthesis and its release within plant systems. In such scenarios, GABA functions as a crucial signaling molecule that modulates numerous physiological and biochemical responses, helping the plant mitigate the damaging effects of stress. The disruption of the GABA shunt, such as through the functional knockout of the SSADH gene, leads to an overaccumulation of ROS, tissue necrosis, reduced growth, and increased susceptibility to stress (Bouché et al. 2003; Fait et al. 2005). Despite these findings, the precise role of GABA in plant systems is still under investigation. It is hypothesized that GABA either functions as an intermediate metabolite feeding into the TCA cycle or as a messenger molecule through interactions with specific receptors. Once produced, GABA can be directed to different intracellular compartments or moved systemically throughout the plant.

When exposed to abiotic stress, plants typically show a rise in GABA levels, which correlates with enhanced stress tolerance. GABA contributes to stress mitigation through multiple mechanisms. One of its primary roles is as an osmoprotectant, aiding in cellular hydration and membrane integrity to minimize stress-related damage. Moreover, GABA has the ability to scavenge ROS, thus helping reduce oxidative damage.

In terms of signaling, GABA interacts with specific channels such as ALMTs, which are unique to plants and serve as multigenic anion transporters. These transporters, found in the plasma membrane, display tissue-specific expression patterns (Ramesh et al. 2015; Kollmeier et al. 2001; Pineros and Kochian 2001). Ramesh et al. (2015) demonstrated that ALMTs play a role in mediating GABA signaling in wheat roots. Interestingly, GABA serves as a negative regulator of ALMT activity, influencing various growth-related processes, including the development of roots and pollen tubes.

A noteworthy relationship has been observed between GABA accumulation and stress types such as acidosis and aluminum (Al^{3+}) toxicity. Acidosis, a form of abiotic stress, triggers GABA accumulation (Crawford et al. 1994) and often occurs alongside Al^{3+} stress (Sasaki et al. 2004). ALMTs have the ability to influence membrane potential through their electrogenic activity. GABA modulates the movement of anions via these channels, leading to changes in the membrane's electrical properties. Since plants typically maintain a net positive anion balance, GABA-induced alterations in anion flux can result in depolarization of the cell membrane, ultimately affecting downstream signaling (Ramesh et al. 2015).

In addition to ALMTs, GABA-binding domains are conserved in Guard Cell Outward-Rectifying K^+ channel (GORK) channels, particularly in root epidermal cells (Adem et al. 2020). Studies on the *atgork1* mutant, which lacks this domain, reveal enhanced tolerance to waterlogging, improved potassium (K^+) retention, and increased Ca^{2+} signaling under hypoxic conditions. Notably, this mutant does not exhibit any significant changes in K^+ efflux when exposed to GABA (Adem et al. 2020; Wang et al. 2017). The accumulation of GABA in response to stress may lead to reduced efflux of both malate and K^+ from root cells (Gilliham and Tyerman

2016; Ramesh et al. 2018; Long et al. 2020; Bown and Shelp 2020). While the GABA-binding domain appears to be conserved across most ALMTs (Ramesh et al. 2017), further studies are necessary to determine whether these transporters also facilitate GABA movement directly, in addition to anions.

GABA signaling has a significant impact on ion transport processes, particularly calcium (Ca^{2+}) and potassium (K^+) channels, which play crucial roles in plant stress responses (Fig. 2.4). By modulating the activity of these ion channels, GABA can bring about changes in cellular ion fluxes and maintain ion homeostasis. This modulation is particularly important in regulating stomatal closure, leading to enhanced water use efficiency and improved drought tolerance. Notably, the functioning of

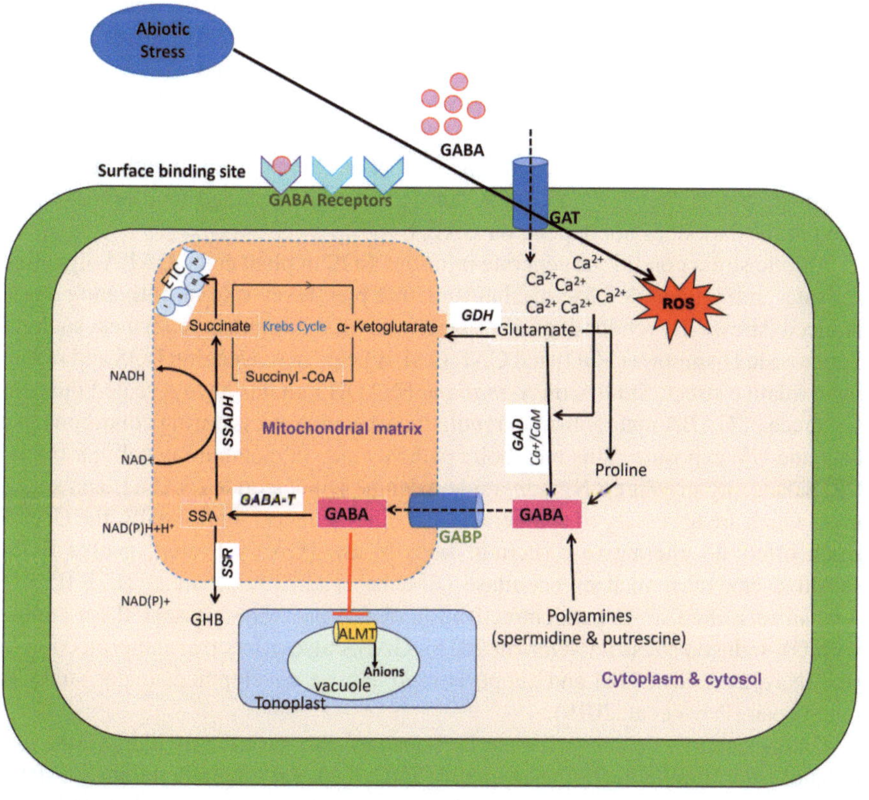

Fig. 2.4 Schematic representation of GABA-mediated mechanisms in plant adaptation to abiotic stress. Environmental stressors lead to elevated levels of ROS and Ca^{2+} influx, which activate GAD to convert glutamate into GABA. GABA is then transported through specific transporters (GAT and GA Binding Protein (GABP)) and sequestered into the vacuole via ALMT, contributing to ion balance and stomatal control. Through the GABA shunt, GABA is metabolized by GABA-T into succinate, feeding into the Krebs cycle to support energy production. GABA also participates in cross talk with polyamine and proline metabolism, as well as ROS scavenging pathways, reinforcing plant resilience. Additionally, GABA signaling is perceived by receptors that initiate downstream defense responses. This coordinated network supports redox balance, energy metabolism, and cellular stability during abiotic stress

stomata is facilitated by ALMTs and GORK (Saito and Uozumi 2019). During stomatal opening, the AtALMT6 tonoplast protein mediates the influx of malate, and the AtALMT9 tonoplast protein, activated by malate, regulates chloride ions into the guard cell vacuole. During stomatal closure, two proteins, AtALMT12 and AtGORK, help move malate and potassium ions out of the cells. When plants face drought, higher levels of ABA increase the activity of GORK, which leads to stomatal closure and improved drought resistance. AtALMT6 and AtALMT4, on the other hand, help release malate from the vacuole during stomatal opening. Additionally, calcium channels raise calcium levels in the cytosol under stress conditions (Saito and Uozumi 2019; Demidchik and Shabala 2017). Research on *Arabidopsis* mutants with altered GABA levels has helped us understand the role of GABA in drought resistance. For example, the atgad1/2 double mutant, which struggles with stomatal closure, is more sensitive to drought (Mekonnen et al. 2016). Mutants like atgad2 and atalmt9 have lower GABA levels and slower transitions from light to dark, showing that GABA buildup during drought helps prevent stomatal opening in response to light. Interestingly, GABA does not affect stomatal opening under constant light conditions (Xu et al. 2021). Even though the almt12 mutant responds normally to GABA during stomatal opening, the closing of stomata in the dark does not depend on GABA.

Abiotic stress conditions generate injurious ROS in plant cells. GABA signaling activates antioxidant defense mechanisms that play a key role in mitigating ROS-induced stress. By endorsing the expression of antioxidant enzymes such as Superoxide Dismutase (SOD) and CAT, GABA helps in scavenging ROS and reducing oxidative stress. Studies on *A. thaliana* SSADH mutants have revealed the consequence of GABA metabolism in regulating oxidative stress during conditions like heat and UV exposure. This metabolic pathway aids in maintaining cellular oxidative balance by providing Nicotinamide Adenine Dinucleotide (NADH) and succinate, which are necessary to counteract ROS intermediates. Additionally, the GABA shunt offers an alternative to certain steps in the TCA cycle that involve ROS-sensitive enzymes, such as aconitase (Bouché et al. 2003; Fait et al. 2005). In *Arabidopsis*, the Ca^{2+}-dependent activation of this pathway is essential for coping with UV-induced stress (Al-Quraan 2015). GABA also offers thermal protection in rice (Nayyar et al. 2014) and supports reproductive development under stress in mung bean (Priya et al. 2019).

GABA enhances plant resilience by boosting osmoprotectants like proline and trehalose and stimulating both enzymatic (SOD, CAT, Ascorbate Peroxidase (APX), and Glutathione Reductase (GR)) and nonenzymatic (ascorbate and glutathione) antioxidants. In mung beans, GABA improves photosynthetic efficiency and carbon assimilation, leading to decreased lipid peroxidation and reduced H_2O_2 levels.

GABA also integrates into broader plant signaling by interacting with nitric oxide (NO), a critical molecule in stress signaling. GABA can influence NO levels and its signaling cascade, which impacts gene expression, stomatal regulation, root development, and programmed cell death during stress (Wendehenne et al. 2001; Palavan-Unsal and Arisan 2009). NO is synthesized through three main pathways in plants: NO-associated (NOA) protein, nitrate reductase (NR), and nitrite reductase

(NiR) (Pedroso et al. 2000; Yamasaki et al. 1999; Yamasaki and Sakihama 2000; Stöhr et al. 2001). These pathways utilize L-arginine and NADPH for NO production (Foresi et al. 2010), and in *Arabidopsis*, a reduction in arginase activity increases NO synthesis from arginine (Flores et al. 2008). The association between NO and proline has also been noted, particularly under stress (Ruan et al. 2004; He and He 2018). Proline biosynthesis occurs via glutamate or ornithine pathways (Delauney and Verma 1993), with the glutamate route predominating under osmotic or low-NO conditions. Glutamate serves as a common precursor for both proline and GABA (Biancucci et al. 2015), linking the metabolic networks of proline, arginine, NO, and GABA. NO, in turn, regulates plant development and stress adaptation (Astier et al. 2018).

Several studies have documented NO's role in triggering antioxidant responses. For example, spermidine-induced NO production through the NOA and NR pathways supports antioxidant defenses in white clover during water stress (Peng et al. 2016). In creeping bentgrass, GABA-induced NO synthesis through NR and NOA pathways boosts the antioxidant activity and contributes to metabolic balance under drought (Tang et al. 2020). Similarly, NO has been found to participate in GABA-mediated mitigation of arsenic (As^{5+}) toxicity in tomato and eggplant seedlings (Suhel et al. 2023). These signaling connections also improve cold tolerance in tea plants, where NO modulates proline and GABA metabolism (Wang et al. 2020).

Beyond NO, GABA also intersects with hormone-signaling networks. It can influence both the levels and activity of stress-related hormones such as abscisic acid (ABA), ethylene, and jasmonic acid (JA). These interactions enhance hormone sensitivity or promote synthesis during stress responses. GABA has been identified as a stress-responsive metabolite that is tightly connected with hormonal regulation (Kinnersley and Turano 2000; Busov et al. 2006; Urano et al. 2009; Merewitz et al. 2012; Jespersen et al. 2015) and oxidative signaling (Xing et al. 2007; Al-Quraan et al. 2011; Sweetlove et al. 2002; Luo et al. 2015; Janzen et al. 2001).

For example, GABA stimulates ethylene biosynthesis in sunflower (*Helianthus annuus*) and *Stellaria longipes* (Kathiresan et al. 1997). Conversely, GABA application reduces ethylene production in *Caragana intermedia* roots under salt stress (Shi et al. 2010). Ethylene also regulates GABA metabolism: ethylene inhibitors such as AVG reduce GABA levels in creeping bentgrass under heat stress (Jespersen et al. 2015). In wheat, AVG and AIB enhance Al^{3+}-activated malate release, whereas ethylene donors inhibit this response in tobacco expressing TaALMT1 (Tian et al. 2014). Under Al^{3+} stress, reduced GABA correlates with increased malate efflux, suggesting that ethylene inhibitors might modulate GABA concentrations or ALMT function to support this process (Ramesh et al. 2015; Tian et al. 2014). These findings imply a complex GABA–ethylene cross talk in controlling ALMT1 activity and malate transport.

Other hormones also influence GABA metabolism. In *Arabidopsis*, the disruption of the NCED3 gene—crucial for ABA biosynthesis—results in elevated GABA levels under dehydration stress (Kochian 1995; Schwartz et al. 1997; Luchi et al. 2001). In poplar, the overexpression of DELLA proteins GAI and RGL1 dramatically increases gibberellin levels and leads to a threefold rise in GABA (Busov et al.

2006). Chilling induces GABA accumulation in *Eriobotrya japonica*, and this effect is intensified with methyl jasmonate (MeJA) treatment (Cao et al. 2013). Although the role of GABA in herbivory defense remains unclear, both GABA and MeJA appear to enhance protection against insect attack (Scholz et al. 2015). In *Arabidopsis*, the gad1/gad2/pop2 triple mutant exhibits elevated GABA and increased resistance to *Spodoptera littoralis*, despite no change in JA or JA-Ile levels (Scholz et al. 2015; Fonseca et al. 2009). This suggests that GABA may interact with ABA, Gibberellic Acid (GA), and JA, possibly in a stimulus-specific manner.

GABA signaling plays a dynamic and integrative role in plant adaptation to abiotic stress. Elucidating its molecular mechanisms and signaling cross-links with hormones and NO will provide deeper insights into how plants maintain resilience. Advancing this knowledge could contribute to developing GABA-based strategies for improving stress tolerance in crops and ensuring stable agricultural productivity in the face of environmental challenges.

2.3.3 GABA-Mediated Signaling and Microbe Interactions

GABA signaling provides plants with an adaptive advantage by facilitating interactions not only with their environment but also with a variety of microorganisms and insect pests (Shelp et al. 2006). GABA has emerged as a molecular signal involved in cross-kingdom communication, mediating exchanges between plants and animals, fungi, bacteria, and even neighboring plants (Shelp et al. 2006; MacGregor et al. 2003; Lang et al. 2016). One of the earliest documented examples of this phenomenon comes from marine ecosystems. In crustose red algae like *Lithothamnium* species, surface-associated GABA-like molecules were found to aid in the settlement and transformation of *Haliotis rufescens* larvae. These GABA-like compounds act as developmental cues, triggering metamorphosis (Morse et al. 1979; Morse and Morse 1984). Further research identified GABA receptors in *H. rufescens*, confirming their role in metamorphic signaling (Trapido-Rosenthal and Morse 1986).

In terrestrial systems, GABA has been shown to impact plant–insect interactions. Elevated GABA levels in plants can negatively affect insect larvae by reducing their growth and survival (Ramputh and Bown 1996). Insects such as the tobacco budworm and the oblique-banded leafroller cause damage to tobacco and soybean leaves, which stimulates an increase in GABA content in plant tissues (Bown et al. 2002). This GABA surge appears to be triggered by the mechanical injury caused by the larval foot structures (Hall et al. 2002). Studies using transgenic tobacco plants engineered to overproduce GABA by overexpressing specific GAD genes have shown that GABA-enriched foliage can suppress larval feeding behaviors (MacGregor et al. 2003). Moreover, GABA accumulation also hinders the normal development of root-knot nematodes (McLean et al. 2003). These inhibitory effects might be due to GABA's function as a neurotransmitter in animals, where it activates chloride channels at neuromuscular junctions in invertebrates, thus interfering with motor control and development (Bown et al. 2006). These findings support the idea that GABA contributes to plant defense by disrupting the physiology of herbivorous pests.

Beyond insect resistance, GABA is also implicated in plant–fungal interactions. In the case of the tomato–the *Cladosporium fulvum* pathosystem—the fungus remains confined to the apoplast—the extracellular space—and depends on nutrients present in this compartment (Oliver and Solomon 2004). During infection, there is a notable increase in GABA concentration in the apoplast. This accumulation is associated with elevated activity of glutamate decarboxylase (GAD), likely induced by a drop in apoplastic pH and a rise in cytosolic Ca^{2+} levels during fungal attack (Solomon and Oliver 2001). Interestingly, the fungus itself expresses GABA transaminase (GABA-T) and succinic semialdehyde dehydrogenase (SSADH), which implies that it utilizes GABA as a nutrient source (Solomon and Oliver 2002). In this interaction, the pathogen manipulates host metabolism to enhance GABA production, which not only serves as a nutrient but may also signal the activation of fungal metabolic enzymes necessary for colonization.

GABA is equally significant in plant–bacterial communication. For example, in the plant pathogen *Agrobacterium tumefaciens*, the exposure to GABA or extracts from wounded tomato stems triggers the expression of the AttM lactonase enzyme, which degrades N-(3-oxooctanoyl) homoserine lactone—an important quorum-sensing molecule (Chevrot et al. 2006). This external GABA enters the bacterial cells through an ABC transporter. Wounding-induced accumulation of GABA in tomato stems leads to reduced virulence in *A. tumefaciens*. GABA also impacts the horizontal gene transfer of the Ti plasmid in *A. tumefaciens* by inducing BlcC lactonase expression, thereby interfering with quorum sensing and virulence regulation (Lang et al. 2016). These observations suggest that plants may use GABA as a molecular tool to control bacterial behavior, including pathogenicity and gene transfer. Overall, GABA signaling in plants enables interactions with the environment, influencing symbiotic associations, defense responses, and communication with various organisms. Thus, a comprehensive approach aimed at understanding the mechanisms and pathways would resolve the complexity of GABA signaling networks and provide potential avenues for crop improvement and sustainable agriculture.

2.4 GABA Gateways in Plant Cells

The transport of GABA within plant systems is governed by a coordinated network of specialized transporters that regulate its movement, concentration, and localization. These transport mechanisms are vital for modulating plant development, physiological responses, and adaptation to stress. Depending on their location within the cell and their expression profiles, different transporters have distinct roles. Some are responsible for importing GABA from the external environment, such as from surrounding tissues or the rhizosphere (Yuan et al. 2023), while others facilitate its export or redistribution throughout the plant. These systems help maintain internal

GABA equilibrium in response to developmental signals and environmental changes, including stress, hormonal cues, and circadian rhythms.

The existence of GABA transporters was first confirmed in *Arabidopsis thaliana*, where it was observed that the plant could thrive using GABA as its only nitrogen source (Breitkreuz et al. 1999). GABA is transported across cellular membranes and into various organelles through distinct transporter proteins. Among these, the key transporters include ALMTs, GATs, BATs, and CATs, each operating at specific cellular sites (Ramesh et al. 2015; Meyer et al. 2006; Dündar and Bush 2009; Snowden et al. 2015).

ALMTs represent a group of membrane-embedded anion transporters that can be activated by aluminum ions (Ramesh et al. 2018). Genomic analyses have identified 12 ALMT-related genes in plants (Ma et al. 2016). Although these proteins mainly transport malate and anions, some, like TaALMT1, also facilitate GABA movement across the plasma membrane in species such as *Arabidopsis*, wheat, rice, and rape (Ramesh et al. 2018).

GABA transporters (GATs) are another class of transmembrane proteins belonging to the AAAP (amino acid/auxin permease) gene family (Meyer et al. 2006; Batushansky et al. 2015; Ma et al. 2016). Four GAT genes, GAT1 through GAT4, have been identified in plants. GAT1, in particular, localizes to the plasma membrane and facilitates the uptake of GABA from the apoplast into the cytoplasm. Unlike ALMT1, the GAT1-mediated transport is not inhibited by aluminum ions (Long et al. 2020). Homologs of GAT1 have been characterized in several species, including *Arabidopsis*, rice, and potato, with most studies focusing on AtGAT1 in *Arabidopsis*. Although GAT1 plays a major role in GABA uptake, its absence can be partially compensated for by other transporters.

In yeast, additional low-affinity GABA transporters such as amino acid permease (AAP3) and proline transporter 2 (ProT2) have been discovered, and their homologs have also been found in plants like *Arabidopsis*, potato, and rice (Young et al. 1999; Wipf et al. 2002; Tian et al. 2020).

Bidirectional amino acid transporters (BATs) are localized to mitochondrial membranes and are capable of transporting amino acids in both directions (Wolf-Nicolas et al. 1998; Dündar and Bush 2009). There are seven BAT genes reported in various plant species, including *Arabidopsis*, rice, and potato (Ma et al. 2016; Tian et al. 2020), and BAT1, in particular, is involved in the transport of amino acids (Chen et al. 2001).

CATs are mostly localized to the tonoplast (vacuolar membrane) and are part of the amino acid–polyamine–choline (APC) transporter family (Snowden et al. 2015; Carter et al. 2004; Schmidt et al. 2007). Among the nine known CAT genes in plants, CAT9 plays a major role in bidirectional GABA transport between the cytoplasm and the vacuole. CAT9 homologs have been identified in *Arabidopsis*, tomato, rice, and potato, and the involvement of SlCAT9 in GABA movement in tomato has been experimentally verified (Snowden et al. 2015; Whiteman et al. 2008).

The activity of these GABA transporters is regulated by developmental stages and environmental stimuli (Ramesh et al. 2017). Depending on the tissue type and developmental phase, different transporters are expressed to ensure that GABA is delivered where it is most needed. Environmental conditions, such as light intensity, temperature changes, and stress, along with hormone signals, modulate the expression and function of these transporters. Hormones, including ABA, ethylene, and auxin, have been shown to influence GABA transport dynamics in response to external and internal signals (Podlešáková et al. 2019). GABA itself also acts as a signaling molecule and participates in feedback regulation of its own transport by affecting hormone-signaling pathways.

Additionally, GABA transporters are subject to regulatory control at both transcriptional and posttranscriptional levels. These regulatory mechanisms help adjust transporter activity and localization based on the plant's immediate needs (Pratelli and Pilot 2014). Overall, the movement of GABA in plants is a tightly regulated and highly dynamic process. Understanding how GABA transporters operate and how they are integrated into broader plant physiological networks remains a key area of research. Ongoing studies continue to uncover the roles of these transport proteins and their importance in linking GABA signaling with plant growth, metabolism, and environmental adaptation.

2.5 Conclusion

Recent insights into GABA biosynthesis, transport, and signaling highlight its essential involvement in controlling various physiological functions in plants. By counteracting the adverse effects of various stressors, GABA enhances plant resilience, promotes growth, and contributes to economic returns. However, several gaps remain in elucidating the precise molecular mechanisms underlying GABA action. Future research should focus on unraveling GABA's interactions with plant hormones, its role in gene expression regulation, and its impact on metabolic pathways that govern stress tolerance and crop productivity. A comprehensive approach integrating physiological, biochemical, and molecular studies will be essential to fully exploit GABA's potential in plant adaptation strategies. Advancements in genetic engineering, including genome editing and transgenic approaches, hold promise for manipulating GABA metabolism to develop stress-resilient crops. Similarly, breeding programs utilizing marker-assisted selection can enhance GABA-related traits in commercially important crops. Combining the above-mentioned approaches offers promising potential for breeding new cultivars with enhanced GABA metabolism, leading to improved stress resilience, higher productivity, and more sustainable agricultural practices.

References

Adem GD, Chen G, Shabala L, Chen Z-H, Shabala S (2020) GORK channel: a master switch of plant metabolism? Trends Plant Sci 25:434–445

Akama K, Takaiwa F (2007) C-terminal extension of rice glutamate decarboxylase (OsGAD2) functions as an autoinhibitory domain and overexpression of a truncated mutant results in the accumulation of extremely high levels of GABA in plant cells. J Exp Bot 58:2699–2707

Allan WL, Simpson JP, Clark SM, Shelp BJ (2008) Gamma-hydroxybutyrate accumulation in Arabidopsis and tobacco plants is a general response to abiotic stress: putative regulation by redox balance and glyoxylate reductase isoforms. J Exp Bot 59:2555–2564

AL-Quraan NA, AL-Ajlouni ZI, Obedat DI (2019) The GABA shunt pathway in germinating seeds of wheat (*Triticum aestivum* L.) and barley (Hordeum vulgare L.) under salt stress. Seed Sci Res 29:1–11. https://doi.org/10.1017/S0960258519000230

Al-Quraan NA, Locy RD, Singh NK (2011) Implications of paraquat and hydrogen peroxide-induced oxidative stress treatments on the GABA shunt pathway in Arabidopsis thaliana calmodulin mutants. Plant Biotechnol Rep 5(3):225–234

Al-Quraan NA (2015) GABA shunt deficiencies and accumulation of reactive oxygen species under UV treatments: insight from Arabidopsis thaliana calmodulin mutants. Acta Physiol Plant 37(4)

Andriamampandry C, Siffert JC, Schmitt M et al (1998) Cloning of a rat brain succinic semialdehyde reductase involved in the synthesis of the neuromodulator gamma-hydroxybutyrate. Biochem J 334(Pt 1):43–50. https://doi.org/10.1042/bj3340043. PMID: 9693100; PMCID: PMC1219659

Ansari MI, Hasan S, Jalil SU (2014) Leaf senescence and GABA shunt. Bioinformation 10:730–732

Ansari MI, Chen SCG (2009) Biochemical characterization of γ-aminobutyric acid (GABA): pyruvate transaminase during rice leaf senescence. Int J Integr Biol 16:27–32

Ansari MI, Jalil SU, Ansari SA et al (2021) GABA shunt: a key-player in mitigation of ROS during stress. Plant Growth Regul 94:131–149

Astier J, Gross I, Durner J (2018) Nitric oxide production in plants: an update. J Exp Bot 69:3401–3411. https://doi.org/10.1093/jxb/erx420

Bach B, Meudec E, Lepoutre JP, Rossignol T, Blondin B, Dequin S, Camarasa C (2009) New insights into γ-aminobutyric acid catabolism: evidence for γ-hydroxybutyric acid and polyhydroxybutyrate synthesis in *Saccharomyces cerevisiae*. Appl Environ Microbiol 75(13):4231–4239

Baier M, Bittner A, Prescher A, van Buer J (2019) Preparing plants for improved cold tolerance by priming. Plant Cell Environ 42(3):782–800. https://doi.org/10.1111/pce.13394

Barbosa JM, Singh NK, Cherry JH, Locy RD (2010) Nitrate uptake and utilization is modulated by exogenous γ-aminobutyric acid in *Arabidopsis thaliana* seedlings. Plant Physiol Biochem 48(6):443–450

Barbosa JM, Singh NK, Cherry JH, Locy RD (2000) GABA increases the rate of nitrate uptake and utilization in Arabidopsis roots. Plant Biol 2000:133–166

Batushansky A, Kirma M, Grillich N, Pham PA, Rentsch D, Galili G, Fernie AR, Fait A (2015) The transporter GAT1 plays an important role in GABA-mediated carbon-nitrogen interactions in *Arabidopsis*. Front Plant Sci 6:785. https://doi.org/10.3389/fpls.2015.00785

Baum G, Chen Y, Arazi T, Takatsuji H, Fromm H (1993) A plant glutamate decarboxylase containing a calmodulin binding domain. Cloning, sequence, and functional analysis. J Biol Chem 268:19610–19617

Beuve N, Rispail N, Laine P, Cliquet JB, Ourry A, Le Deunff E (2004) Putative role of γ-aminobutyric acid (GABA) as a long-distance signal in up-regulation of nitrate uptake in *Brassica napus* L. Plant Cell Environ 27:1035–1046

Bhatnagar P, Minocha R, Minocha SC (2002) Genetic manipulation of the metabolism of polyamines in poplar cells: the regulation of putrescine catabolism. Plant Physiol 128:1455–1469

Biancucci M et al (2015) Role of proline and GABA in sexual reproduction of angiosperms. Front Plant Sci 6:680

Bouché N, Lacombe B, Fromm H (2003) GABA signaling: a conserved and ubiquitous mechanism. Trends Cell Biol. Dec;13(12):607–610. https://doi.org/10.1016/j.tcb.2003.10.001. Erratum in: Trends Cell Biol. 2004 Feb;14(2): 63. PMID: 14624837

Bouché N, Fromm H (2004) GABA in plants: just a metabolite? Trends Plant Sci. Mar;9(3):110–115. https://doi.org/10.1016/j.tplants.2004.01.006

Bown AW, Hall DE, MacGregor KB (2002) Insect footsteps on leaves stimulate the accumulation of 4-aminobutyrate and can be visualized through increased chlorophyll fluorescence and superoxide production. Plant Physiol 129:1430–1440

Bown AW, MacGregor KB, Shelp BJ (2006) γ-Aminobutyrate: defense against invertebrate pests? Trends Plant Sci 11:424–427

Bown AW, Shelp BJ (2016) Plant GABA: not just a metabolite. Trends Plant Sci 21(10):811–813. https://doi.org/10.1016/j.tplants.2016.08.001

Bown AW, Shelp BJ (2020) Does the GABA shunt regulate cytosolic GABA? Trends Plant Sci 25:422–424

Breitkreuz KE, Shelp BJ, Fischer WN, Schwacke R, Rentsch D (1999) Identification and characterization of GABA, proline and quaternary ammonium compound transporters from Arabidopsis thaliana. FEBS Lett. May 7;450(3):280–4. https://doi.org/10.1016/s0014-5793(99)00516-5

Breitkreuz KE, Allan WL, Van Cauwenberghe OR, Jakobs C, Talibi D, Andre B, Shelp BJ (2003) A novel γ-hydroxybutyrate dehydrogenase. J Biol Chem 278:41552–41556

Buddhala C, Hsu CC, Wu JY (2009) A novel mechanism for GABA synthesis and packaging into synaptic vesicles. Neurochem Int 55(1-3):9–12. https://doi.org/10.1016/j.neuint.2009.01.020. Epub 2009 Feb 6

Busch KB and Fromm H (1999) Plant succinic semialdehyde dehydrogenase. Cloning, purification, localization in mitochondria, and regulation by adenine nucleotides. Plant Physiol. 121: 589–597.

Busov V, Meilan R, Pearce DW, Rood SB, Ma CP, Tschaplinski TJ, Strauss SH (2006) Transgenic modification of gai or rgl1 causes dwarfing and alters gibberellins, root growth, and metabolite profiles in Populus. Planta 224(2):288–299. https://doi.org/10.1007/s00425-005-0213-9

Cao J, Barbosa JM, Singh NK, Locy RD (2013) GABA shunt mediates thermotolerance in *Saccharomyces cerevisiae* by reducing reactive oxygen production. Yeast 30(4):129–144

Carter C, Pan S, Zouhar J, Avila EL, Girke T, Raikhel NV (2004) The vegetative vacuole proteome of *Arabidopsis thaliana* reveals predicted and unexpected proteins. Plant Cell 16:3285–3303. https://doi.org/10.1105/tpc.104.027078

Chen L, Ortiz-Lopez A, Jung A, Bush DR (2001) ANT1, an aromatic and neutral amino acid transporter in *Arabidopsis*. Plant Physiol 125:1813–1820. https://doi.org/10.1104/pp.125.4.1813

Cheng B, Li Z, Liang L, Cao Y, Zeng W, Zhang X et al (2018) The γ-aminobutyric acid (GABA) alleviates salt stress damage during seeds germination of white clover associated with Na+/K+ transportation, dehydrins accumulation, and stress-related genes expression in white clover. Int J Mol Sci 19:2520. https://doi.org/10.3390/ijms19092520

Chevrot R, Rosen R, Haudecoeur E, Cirou A, Shelp BJ, Ron E, Faure D (2006) GABA controls the level of quorum-sensing signal in Agrobacterium tumefaciens. Proc Natl Acad Sci U S A 103(19):7460–7464. https://doi.org/10.1073/pnas.0600313103. Epub 2006 Apr 27. PMID: 16645034; PMCID: PMC1464361

Chu C, Yan N, Du Y, Liu X, Chu M, Shi J, Zhang H, Liu Y, Zhang Z (2019) iTRAQ-based proteomic analysis reveals the accumulation of bioactive compounds in Chinese wild rice (*Zizania latifolia*) during germination. Food Chem 289:635–644. https://doi.org/10.1016/j.foodchem.2019.03.092

Clark SM, Di Leo R, Dhanoa PK, Van Cauwenberghe OR, Mullen RT, Shelp J (2009) Biochemical characterization, mitochondrial localization, expression, and potential functions for an Arabidopsis γ-aminobutyrate transaminase that utilizes both pyruvate and glyoxylate. J Exp Bot 60:1743–1757

Crawford LA, Bown AW, Breitkreuz KE, Guinel FC (1994) The synthesis of γ-aminobutyric acid in response to treatment reducing cytosolic pH. Plant Physiol 104:865–871

Delauney AJ, Verma DPS (1993) Proline biosynthesis and osmoregulation in plants. Plant J 4:215–223

Demidchik V, Shabala S (2017) Mechanisms of cytosolic calcium elevation in plants: the role of ion channels, calcium extrusion systems and NADPH oxidase-mediated 'ROS-Ca^{2+} hub'. Funct Plant Biol 45:9–27

Du C, Chen W, Wu Y, Wang G, Zhao J, Sun J, Ji J, Yan D, Jiang Z, Shi S (2020) Effects of GABA and Vigabatrin on the Germination of Chinese Chestnut Recalcitrant Seeds and Its Implications for Seed Dormancy and Storage. Plants 9:449. https://doi.org/10.3390/plants9040449

Duan JJ, Li J, Guo SR, Kang YY (2008) Exogenous spermidine affects polyamine metabolism in salinity-stressed Cucumis sativus roots and enhances short-term salinity tolerance. J Plant Physiol 165:1620–1635. https://doi.org/10.1016/j.jplph.2007.11.006

Dubois M, Claeys H, Van Den Broeck L, Inzé D (2017) Time of day determines Arabidopsis transcriptome and growth dynamics under mild drought. Plant Cell Environ 40(2):180–189. https://doi.org/10.1111/pce.12809

Dündar E, Bush DR (2009) BAT1, a bidirectional amino acid transporter in *Arabidopsis*. Planta 229:1047–1056. https://doi.org/10.1007/s00425-009-0892-8

Fait A, Fromm H, Walter D, Galili G, Fernie AR (2008) Highway or byway: the metabolic role of the GABA shunt in plants. Trends Plant Sci 13:14–19

Fait A, Yellin A, Fromm H (2005) GABA shunt deficiencies and accumulation of reactive oxygen intermediates: insight from *Arabidopsis* mutants. FEBS Lett 579:415–420

Fancy NN, Bahlmann AK, Loake GJ (2017) Nitric oxide function in plant abiotic stress. Plant Cell Environ 40:462–472

Flores T et al (2008) Arginase-negative mutants of *Arabidopsis* exhibit increased nitric oxide signaling in root development. Plant Physiol 147:1936–1946

Fonseca S, Chini A, Hamberg M, Adie B, Porzel A, Kramell R, Miersch O, Wasternack C, Solano R (2009) (?)-7-iso-Jasmonoyl-L-isoleucine is the endogenous bioactive jasmonate. Nat Chem Biol 5(5):344–350. https://doi.org/10.1038/nchembio.161

Foresi N et al (2010) Characterization of a nitric oxide synthase from the plant kingdom: NO generation from the green alga ostreococcus tauri is light irradiance and growth phase dependent. Plant Cell 22:3816–3830

Gilliham M, Tyerman SD (2016) Linking metabolism to membrane signaling: the GABA–malate connection. Trends Plant Sci 21:296–301

Gupta K, Sengupta A, Chakraborty M, Gupta B (2016) Hydrogen peroxide and polyamines act as double edged swords in plant abiotic stress responses. Front Plant Sci 7(9):1343

Gutermuth T, Lassig R, Portes MT, Maierhofer T, Romeis T, Borst JW, Hedrich R, Feijó JA, Konrad KR (2013) Pollen tube growth regulation by free anions depends on the interaction between the anion channel SLAH3 and calciumdependent protein kinases CPK2 and CPK20. Plant Cell 25(11):4525–4543. https://doi.org/10.1105/tpc.113.118463

Hall DE, MacGregor KB, Nijsse J, Bown AW (2002) Footsteps from insect larvae damage leaf surfaces and initiate rapid responses. Eur J Plant Pathol 110:441–447

Hatmi S, Gruau C, Trotel-Aziz P, Villaume S, Rabenoelina F, Baillieul F, Aziz A (2015) Drought stress tolerance in grapevine involves activation of polyamine oxidation contributing to improved immune response and low susceptibility to Botrytis cinerea. J Exp Bot 66(3):775–787

He HY, He LF (2018) Regulation of gaseous signaling molecules on proline metabolism in plants. Plant Cell Rep 37:387–392

Hoekenga OA, Maron LG, Piñeros MA, Cançado GMA, Kobayashi JOY, Ryan PR, Dong B, Delhaize E, Sasaki T, Matsumoto H, Yamamoto Y, Koyama H, Kochian LV (2006) AtALMT1, which encodes a malate transporter, is identified as one of several genes critical for aluminium tolerance in Arabidopsis. Proc Natl Acad Sci USA 103:9738–9743

Hu X, Xu Z, Xu W, Li J, Zhao N, Zhou Y (2015) Application of γ-aminobutyric acid demonstrates a protective role of polyamine and GABA metabolism in muskmelon seedlings under $Ca(NO_3)_2$ stress. Plant Physiol Biochem 92:1–10

Jalil SU, Ahmad I, Ansari MI (2017) Functional loss of GABA transaminase (GABA-T) expressed early leaf senescence under various stress conditions in *Arabidopsis thaliana*. Curr Plant Biol 9–10:11–22

Jalil SU, Ahmad I, Ansari MI (2016) Phenotypic characterization of GABA-transaminase mutants of *Arabidopsis thaliana*. Adv Life Sci 5:10005–10008

Jalil SU, Khan MAA, Ansari MI (2019a) Tremendous role of GABA shunt metabolic pathway in plant system. J Biol Chem Res 36:188–196

Jalil SU, Khan MQR, Ansari MI (2019b) Role of GABA transaminase in the regulation of development and senescence in *Arabidopsis thaliana*. Curr Plant Biol 19:100119. https://doi.org/10.1016/j.cpb.2019.100119

Janzen DJ, Allen LJ, MacGregor KB, Bown AW (2001) Cytosolic acidification and gamma-aminobutyric acid synthesis during the oxidative burst in isolated Asparagus sprengeri mesophyll cells. Can J Bot 79(4):438–443

Jespersen D, Yu J, Huang B (2015) Metabolite responses to exogenous application of nitrogen, cytokinin, and ethylene inhibitors in relation to heat-induced senescence in creeping bentgrass. PLoS One 10(3):e0123744. https://doi.org/10.1371/journal.pone.0123744

Ji J, Shi S, Chen W, Xie T, Du C, Sun J, Shi Z, Gao R, Jiang Z, Xiao W (2020) Effects of exogenous γ-aminobutyric acid on the regulation of respiration and protein expression in germinating seeds of mungbean (Vigna radiata) under salt conditions. Electron J Biotechnol 47:1–9. https://doi.org/10.1016/j.ejbt.2020.05.005

Kathiresan A, Tung P, Chinnappa CC, Reid DM (1997) Gammaaminobutyric acid stimulates ethylene biosynthesis in sunflower. Plant Physiol 115(1):129–135. https://doi.org/10.1104/pp.115.1.129

Khan MIR, Jalil SU, Chopra P, Chhillar H, Ferrante A, Khan NA, Ansari MI (2021) Role of GABA in plant growth, development, and senescence. Plant Gene 26:100283

Kim MJ, Kwak HS, Kim SS (2018) Effects of germination on protein, γ-aminobutyric acid, phenolic acids, and antioxidant capacity in wheat. Molecules 23(9). https://doi.org/10.3390/molecules23092244

Kinnersley AM, Turano FJ (2000) γ-Aminobutyric acid (GABA) and plant responses to stress. Crit Rev Plant Sci 19:479–509

Kochian LV (1995) Cellular mechanisms of aluminum toxicity and resistance in plants. Annu Rev Plant Biol 46(1):237–260

Kollmeier M, Dietrich P, Bauer CS, Horst WJ, Hedrich R (2001) Aluminum activates a citrate-permeable anion channel in the aluminum-sensitive zone of the maize root apex. A comparison between an aluminum-sensitive and an aluminum- resistant cultivar. Plant Physiol 126:397–410

Kusano T, Kim DW, Liu T, Berberich T (2015) Polyamine catabolism in plants. In: Kusano T, Suzuki H (eds) Polyamines. Springer, Tokyo. https://doi.org/10.1007/978-4-431-55212-3_6

Lang J, Gonzalez-Mula A, Taconnat L, Clement G, Faure D (2016) The plant GABA signaling downregulates horizontal transfer of the agrobacterium tumefaciens virulence plasmid. New Phytol 210(3):974–983. https://doi.org/10.1111/nph.13813

Lamers J, van der Meer T, Testerink C (2020) How plants sense and respond to stressful environments. Plant Physiol 182(4):1624–1635. https://doi.org/10.1104/pp.19.01464. Epub 2020 Mar 4. PMID: 32132112; PMCID: PMC7140927

Ling V, Snedden WA, Shelp BJ, Assmann SM (1994) Analysis of a soluble calmodulin binding protein from fava bean roots: identification of glutamate decarboxylase as a calmodulin-activated enzyme. Plant Cell 6:1135–1143

Li W, Liu J, Ashraf U, Li G, Li Y, Lu W, Gao L, Han F, Hu J (2016) Exogenous γ-aminobutyric acid (GABA) application improved early growth, net photosynthesis, and associated physio-biochemical events in maize. Front Plant Sci 7:919. https://doi.org/10.3389/fpls.2016.00919

Li Y, Liu B, Peng Y, Liu C, Zhang X, Zhang Z, Liang W, Ma F, Li C (2020) Exogenous GABA alleviates alkaline stress in Malus hupehensis by regulating the accumulation of organic acids. Sci Hortic 261:108982

Ling Y, Chen T, Jing Y, Fan L, Wan Y (2013) γ-Aminobutyric acid (GABA) homeostasis regulates pollen germination and polarized growth in *Picea wilsonii*. Planta 238(5):831–843

Long Y, Tyerman SD, Gilliham M (2020) Cytosolic GABA inhibits anion transport by wheat ALMT1. New Phytol 225:671–678

Lorenz-Guertin JM, Jacob TC (2018) GABA type a receptor trafficking and the architecture of synaptic inhibition. Dev Neurobiol 78(3):238–270. https://doi.org/10.1002/dneu.22536. Epub 2017 Sep 19. PMID: 28901728; PMCID: PMC6589839

Lou YR, Bor M, Yan J, Preuss AS, Jander G (2016) Arabidopsis NATA1 acetylates putrescine and decreases defense-related hydrogen peroxide accumulation. Plant Physiol 171(2):1443–1455

Luchi S, Kobayashi M, Taji T, Naramoto M, Seki M, Kato T, Tabata S, Kakubari Y, Yamaguchi-Shinozaki K, Shinozaki K (2001) Regulation of drought tolerance by gene manipulation of 9-cis-epoxycarotenoid dioxygenase, a key enzyme in abscisci acid biosynthesis in Arabidopsis. Plant J 27:325–333

Luo F, Wang Q, Yin C, Ge Y, Hu F, Huang B, Zhou H, Bao G, Wang B, Lu R, Li Z (2015) Differential metabolic responses of Beauveria bassiana cultured in pupae extracts, root exudates and its interactions with insect and plant. J Invertebr Pathol 130:154–164. https://doi.org/10.1016/j.jip.2015.01.003

Ma H, Cao X, Shi S, Li S, Gao J, Ma Y, Zhao Q, Chen Q (2016) Genome-wide survey and expression analysis of the amino acid transporter superfamily in potato (*Solanum tuberosum* L.). Plant Physiol Biochem 107:164–177. https://doi.org/10.1016/j.plaphy.2016.06.007

Matsuyama A, Yoshimura K, Shimizu C, Murano Y, Takeuchi H, Ishimoto M (2009) Characterization of glutamate decarboxylase mediating γ-amino butyric acid increase in the early germination stage of soybean (*Glycine max* [L.] Merr). J Biosci Bioeng 107:538–543. https://doi.org/10.1016/j.jbiosc.2009.01.012

MacGregor KB, Shelp BJ, Peiris SE, Bown AW (2003) Overexpression of glutamate decarboxylase in transgenic tobacco deters feeding by phytophagous insect larvae. J Chem Ecol 29:2177–2182

McLean MD, Yevtushenko DP, Deschene D, Van Cauwenberghe OR, Makhmoudova A, Potter JW, Bown AW, Shelp BJ (2003) Overexpression of glutamate decarboxylase in transgenic tobacco plants confers resistance to the northern root-knot nematode. Mol Breed 11:277–285

Medina I, Chudotvorova I (2006) GABA neurotransmission and neural cation-chloride co-transporters: actions beyond ion transport. Crit Rev Neurobiol 18(1–2):105–112

Mekonnen DW, Flügge U-I, Ludewig F (2016) Gamma-aminobutyric acid depletion affects stomata closure and drought tolerance of *Arabidopsis thaliana*. Plant Sci 245:25–34

Merewitz EB, Du H, Yu W, Liu Y, Gianfagna T, Huang B (2012) Elevated cytokinin content in ipt transgenic creeping bentgrass promotes drought tolerance through regulating metabolite accumulation. J Exp Bot 63(3):1315–1328. https://doi.org/10.1093/jxb/err372

Meyer A, Eskandari S, Grallath S, Rentsch D (2006) AtGAT1, a high affinity transporter for gamma-aminobutyric acid in *Arabidopsis thaliana*. J Biol Chem 281:7197–7204. https://doi.org/10.1074/jbc.M510766200

Michaeli S, Fromm H (2015) Closing the loop on the GABA shunt in plants: are GABA metabolism and signalling entwined? Front Plant Sci 9:419

Minocha R, Majumdar R, Minocha SC (2014) Polyamines and abiotic stress in plants: a complex relationship. Front Plant Sci 5:175. https://doi.org/10.3389/fpls.2014.00175

Mo H, Wang X, Zhang Y, Zhang G, Zhang J, Ma Z (2015) Cotton polyamine oxidase is required for spermine and camalexin signalling in the defence response to Verticillium dahliae. Plant J 83:962–975

Morse DE, Hooker N, Duncan H, Jensen L (1979) γ-Aminobutyric acid, a neurotransmitter, induces planktonic abalone larvae to settle and begin metamorphosis. Science 204:407–410

Morse ANC, Morse DE (1984) Recruitment and metamorphosis of *Haliotis* larvae induced by molecules uniquely available at the surface of crustose red algae. J Exp Mar Biol Ecol 75:191–215

Nayyar H, Kaur R, Kaur S, Singh R (2014) γ-Aminobutyric acid (GABA) imparts partial protection from heat stress injury to rice seedlings by improving leaf turgor and upregulating osmoprotectants and antioxidants. J Plant Growth Regul 33:408–419

Nonaka S, Arai C, Takayama M, Matsukura C, Ezura H (2017) Efficient increase of γ-aminobutyric acid (GABA) content in tomato fruits by targeted mutagenesis. Sci Rep 7(1):7057. https://doi.org/10.1038/s41598-017-06400-y

Oliver RP, Solomon PS (2004) Does the oxidative stress used by plants for defence provide a source of nutrients for pathogenic fungi? Trends Plant Sci 9:472–473

Palavan-Unsal N, Arisan D (2009) Nitric oxide signalling in plants. Bot Rev 75:203–229

Palanivelu R, Brass L, Edlund AF, Preuss D (2003) Pollen tube growth and guidance is regulated by POP2, an *Arabidopsis* gene that controls GABA levels. Cell 114(1):47–59

Pedroso MC, Magalhaes JR, Durzan D (2000) Nitric oxide induces cell death in *Taxus* cells. Plant Sci 157:178–180. https://doi.org/10.1016/S0168-9452(00)00278-8

Pelvan A, Bor M, Yolcu S, Özdemir F, Türkan I (2021) Day and night fluctuations in GABA biosynthesis contribute to drought responses in *Nicotiana tabacum* L. Plant Signal Behav 16(5):1899672. https://doi.org/10.1080/15592324.2021.1899672. Epub 2021 Mar 11. PMID: 33704006; PMCID: PMC8078508

Peng Z, Lu Q, Verma DPS (1996) Reciprocal regulation of Delta(1)-pyrroline-5-carboxylate synthetase and proline dehydrogenase genes controls proline levels during and after osmotic stress in plants. Mol Gen Genet 253:334–341

Peng D, Wang X, Li Z et al (2016) NO is involved in spermidine-induced drought tolerance in white clover via activation of antioxidant enzymes and genes. Protoplasma 253:1243–1254. https://doi.org/10.1007/s00709-015-0880-8

Pineros MA, Kochian LV (2001) A patch-clamp study on the physiology of aluminum toxicity and aluminum tolerance in maize. Identification and characterization of Al + induced anion channels. Plant Physiol 125:292–305

Podlešáková K, Ugena L, Spíchal L, Doležal K, De Diego N (2019) Phytohormones and polyamines regulate plant stress responses by altering GABA pathway. New Biotechnol 48:53–65

Pratelli R, Pilot G (2014) Regulation of amino acid metabolic enzymes and transporters in plants. J Exp Bot 65(19):5535–5556. https://doi.org/10.1093/jxb/eru320

Priya M, Sharma L, Kaur R, Bindumadhava H, Nair RM, Siddique KHM, Nayyar H (2019) GABA (γ-aminobutyric acid), as a thermo-protectant, to improve the reproductive function of heat-stressed mungbean plants. Sci Rep 9:7788

Ramesh SA, Kamran M, Sullivan W, Chirkova L, Okamoto M, Degryse F, McLaughlin M, Gilliham M, Tyerman SD (2018) Luminum activated malate transporters can facilitate GABA transport. Plant Cell 30:1147–1164

Ramesh SA, Stephen DT, Xu B et al (2015) GABA signalling modulates plant growth by directly regulating the activity of plant-specific anion transporters. Nat Commun 6:7879

Ramesh SA, Tyerman SD, Gilliham M, Xu B (2017) γ-Aminobutyric acid (GABA) signalling in plants. Cell Mol Life Sci 74:1577–1603

Ramputh AI, Bown AW (1996) Rapid γ-aminobutyric acid synthesis and the inhibition of the growth and development of oblique-banded leaf roller larvae. Plant Physiol 111:1349–1352

Renault H, Amrani AE, Palanivelu R, Updegraff EP, Yu A, Renou J, Preuss D, Bouchereau A, Deleu C (2011) GABA accumulation causes cell elongation defects and a decrease in expression of genes encoding secreted and cell wall-related proteins in *Arabidopsis thaliana*. Plant Cell Physiol 52:894–908

Ruan HH, Shen WB, Xu LL (2004) Nitric oxide involved in the abscisic acid induced proline accumulation in wheat seedling leaves under salt stress. Acta Bot Sin 46:1307–1315

Samarah NH, AL-Quraan Nisreen A, Al-Wraikat Batool S (2023) Ultrasonic treatment to enhance seed germination and vigour of wheat (*Triticum durum*) in association with γ-aminobutyric acid (GABA) shunt pathway. Funct Plant Biol 50:277–293

Sasaki T, Yamamoto Y, Ezaki B, Katsuhara M, Ahn SJ, Ryan PR, Delhaize E, Matsumoto H (2004) A wheat gene encoding an aluminum-activated malate transporter. Plant J 37:645–653

Saito N, Robert M, Kochi H, Matsuo G, Kakazu Y, Soga T, Tomita M (2009) Metabolite profiling reveals YihU as a novel hydroxybutyrate dehydrogenase for alternative succinic semialdehyde metabolism in Escherichia coli. J Biol Chem 284:16442–16451

Saito S, Uozumi N (2019) Guard cell membrane anion transport systems and their regulatory components: an elaborate mechanism controlling stress-induced stomatal closure. Plants 8:9

Schmidt UG, Endler A, Schelbert S, Brunner A, Schnell M, Neuhaus HE, Marty-Mazars D, Marty F, Baginsky S, Martinoia E (2007) Novel tonoplast transporters identified using a proteomic approach with vacuoles isolated from cauliflower buds. Plant Physiol 145:216–229. https://doi.org/10.1104/pp.107.096917

Scholz SS, Reichelt M, Mekonnen DW, Ludewig F, Mithöfer A (2015) Insect herbivory-elicited GABA accumulation in plants in wound-induced, direct, systemic and jasmonate-dependent defense response. Front Plant Sci 6:1128. https://doi.org/10.3389/fpls.2015.01128

Schwartz SH, LeonKloosterziel KM, Koornneef M, Zeevaart JAD (1997) Biochemical characterization of the aba2 and aba3 mutants in Arabidopsis thaliana. Plant Physiol 114(1):161–166. https://doi.org/10.1104/pp.114.1.161

Seifi HS, Shelp BJ (2019) Spermine differentially refines plant defense responses against biotic and abiotic stresses. Front Plant Sci 10(2):117. https://doi.org/10.3389/fpls.2019.00117

Seifi HS, Curvers K, De Vleesschauwer D, Delaere I, Aziz A, Höfte M (2013) Concurrent overactivation of the cytosolic glutamine synthetase and the GABA shunt in the ABA-deficient sitiens mutant of tomato leads to resistance against Botrytis cinerea. New Phytologist, 199(2):490–504

Sharma S, Saxena DC, Riar CS (2016) Analysing the effect of germination on phenolics, dietary fibres, minerals and γ-aminobutyric acid contents of barnyard millet (*Echinochloa frumentaceae*). Food Biosci 13:60–68. https://doi.org/10.1016/j.fbio.2015.12.007

Shelp BJ, Bozzo GG, Trobacher CP, Zarei A, Deyman KL, Brikis CJ (2012) Hypothesis/review: contribution of putrescine to 4-aminobutyrate (GABA) production in response to abiotic stress. Plant Sci 193–194:130–135

Shelp BJ, Bown AW, Faure D (2006) Extracellular γ-aminobutyrate mediates communication between plants and other organisms. Plant Physiol 142(4):1350–1352. https://doi.org/10.1104/pp.106.088955

Sheng Y, Xiao H, Guo C, Wu H, Wang X (2018) Effects of exogenous gamma-aminobutyric acid on α-amylase activity in the aleurone of barley seeds. Plant Physiol Biochem 127:39–46

Shi SQ, Shi Z, Jiang ZP, Qi LW, Sun XM, Li CX, Liu JF, Xiao WF, Zhang SG (2010) Effects of exogenous GABA on gene expression of *Caragana intermedia* roots under NaCl stress: regulatory roles for H_2O_2 and ethylene production. Plant Cell Environ 33:149–162

Signorelli S, Dans PD, Coitiño EL, Borsani O, Monza J (2015) Connecting proline and γ-aminobutyric acid in stressed plants through non-enzymatic reactions. PLoS One 10:e0115349

Snedden WW, Arazi T, Fromm H, Shelp BJ (1995) Calcium/calmodulin activation of soybean glutamate decarboxylase. Plant Physiol 108:543–549

Snowden CJ, Thomas B, Baxter CJ, Smith JA, Sweetlove LJ (2015) A tonoplast Glu/Asp/GABA exchanger that affects tomato fruit amino acid composition. Plant J 81:651–660. https://doi.org/10.1111/tpj.12766

Solomon PS, Oliver RP (2001) The nitrogen content of the tomato leaf apoplast increases during infection by Cladosporium fulvum. Planta 213:241–249

Solomon PS, Oliver RP (2002) Evidence that γ-aminobutyric acid is a major nitrogen source during Cladosporium fulvum infection of tomato. Planta 214:414–420

Song H, Xu X, Wang H, Tao Y (2010) Exogenous γ-aminobutyric acid alleviates oxidative damage caused by aluminium and proton stresses on barley seedlings. Journal of the Sciences of Food and Agriculture, 90(9):1410–1416

Song S, Li J, Gao H, Li QY, Yang LW, Gong RJ (2012) Effect of exogenous γ-aminobutyric acid on inorganic nitrogen metabolism and mineral elements contents of melon seedling under hypoxia stress. Acta Hort Sin 39:695–704

Stöhr C, Strube F, Marx G, Ullrich WR, Rockel P (2001) A plasma membrane-bound enzyme of tobacco roots catalyses the formation of nitric oxide from nitrite. Planta 212:835–841. https://doi.org/10.1007/s004250000447

Suhel M, Husain T, Prasad SM et al (2023) GABA requires nitric oxide for alleviating arsenate stress in tomato and Brinjal seedlings. J Plant Growth Regul 42:670–683. https://doi.org/10.1007/s00344-022-10576-7

Sulieman S, Schulze J (2010) Phloem-derived gammaaminobutyric acid (GABA) is involved in upregulating nodule N2 fixation efficiency in the model legume Medicago truncatula. Plant Cell Environ 33(12):2162–2172. https://doi.org/10.1111/j.1365-3040.2010.02214.x

Sulieman S (2011) Does GABA increase the efficiency of symbiotic N2 fixation in legumes? Plant Signal Behav 6(1):32–36

Sung MS, Chow TJ, Lee TM (2011) Polyamine acclimation alleviates hyper salinity-induced oxidative stress in a marine green microalga, Ulva fasciata, by modulation of antioxidative enzyme gene expression. J Phycol 47(3):538–547

Sweetlove LJ, Heazlewood JL, Herald V, Holtzapffel R, Day DA, Leaver CJ, Millar AH (2002) The impact of oxidative stress on Arabidopsis mitochondria. Plant J 32(6):891–904. https://doi.org/10.1046/j.1365-313x.2002.01474.x

Takayama M, Ezura H (2015) How and why does tomato accumulate a large amount of GABA in the fruit? Front Plant Sci 6:612

Takayama M, Koike S, Kusano M, Matsukura C, Saito K, Ariizumi T, Ezura H (2015) Tomato glutamate decarboxylase genes *SlGAD2* and *SlGAD3* play key roles in regulating γ-aminobutyric acid levels in tomato (*Solanum lycopersicum*). Plant Cell Physiol 56(8):1533–1545

Tang M, Li Z, Luo L, Cheng B, Zhang Y, Zeng W, Peng Y (2020) Nitric oxide signal, nitrogen metabolism, and water balance affected by γ-aminobutyric acid (GABA) in relation to enhanced tolerance to water stress in creeping bentgrass. Int J Mol Sci 21(20):7460. https://doi.org/10.3390/ijms21207460. PMID: 33050389; PMCID: PMC7589152

Tian R, Yang Y, Chen M (2020) Genome-wide survey of the amino acid transporter gene family in wheat (*Triticum aestivum* L.): identification, expression analysis and response to abiotic stress. Int J Biol Macromol 162:1372–1387. https://doi.org/10.1016/j.ijbiomac.2020.07.302

Tian Q, Zhang X, Ramesh S, Gilliham M, Tyerman S, Zhang W (2014) Ethylene negatively regulates aluminium-induced malate efflux from wheat roots and tobacco cells transformed with TaALMT1. J Exp Bot 65(9):2415–2426. https://doi.org/10.1093/jxb/eru123

Trapido-Rosenthal HG, Morse DE (1986) Availability of chemosensory receptors is down-regulated by habituation of larvae to a morphogenetic signal. Proc Natl Acad Sci USA 83:7658–7662

Trovato M, Mattioli R, Costantino P (2008) Multiple roles of proline in plant stress tolerance and development. Rend. Lincei: 325–346. https://doi.org/10.1007/s12210-008-0022-8

Urano K, Maruyama K, Ogata Y, Morishita Y, Takeda M, Sakurai N, Suzuki H, Saito K, Shibata D, Kobayashi M, Yamaguchi-Shinozaki K, Shinozaki K (2009) Characterization of the ABA-regulated global responses to dehydration in Arabidopsis by metabolomics. Plant J 57(6):1065–1078. https://doi.org/10.1111/j.1365-313X.2008.03748.x

Wang Y, Luo Z, Huang X, Yang K, Gao S, Du R (2014a) Effect of exogenous γ-aminobutyric acid (GABA) treatment on chilling injury and antioxidant capacity in banana peel. Sci Hortic 168(3):132–137

Wang CY, Li JR, Xia QP, Wu XL, Gao HB (2014b) Influence of exogenous γ-aminobutyric acid (GABA) on GABA metabolism and amino acid contents in roots of melon seedling under hypoxia stress. J Appl Ecol 25:2011–2018

Wang F, Chen Z-H, Liu X, Colmer TD, Shabala L, Salih A, Zhou M, Shabala S (2017) Revealing the roles of GORK channels and NADPH oxidase in acclimation to hypoxia in Arabidopsis. J Exp Bot 68:3191–3204

Wang Y, Xiong F, Nong S et al (2020) Effects of nitric oxide on the GABA, polyamines, and proline in tea (*Camellia sinensis*) roots under cold stress. Sci Rep 10:12240. https://doi.org/10.1038/s41598-020-69253-y

Wendehenne D, Pugin A, Klessig DF, Durner J (2001) Nitric oxide: comparative synthesis and signaling in animal and plant cells. Trends Plant Sci 6:177–183

Whiteman SA, Serazetdinova L, Jones AM, Sanders D, Rathjen J, Peck SC, Maathuis FJ (2008) Identification of novel proteins and phosphorylation sites in a tonoplast enriched membrane fraction of *Arabidopsis thaliana*. Wiley-VCH Verlag GMBH 8:3536–3547. https://doi.org/10.1002/pmic.200701104

Wipf D, Ludewig U, Tegeder M, Rentsch D, Koch W, Frommer WB (2002) Conservation of amino acid transporters in fungi, plants and animals. Trends Biochem Sci 27:139–147. https://doi.org/10.1016/S0968-0004(01)02054-0

Wolf-Nicolas F, Bruno A, Doris R, Sylvia K, Mechthild T, Kevin B, Wolf BF (1998) Amino acid transport in plants. Trends Plant Sci 3:188–195

Xie T, Ji J, Chen W, Yue J, Du C, Sun J, Shi S (2020) GABA negatively regulates adventitious root development. J Exp Bot 71:1459–1474

Xing SG, Jun YB, Hau ZW, Liang LY (2007) Higher accumulation of c-aminobutyric acid induced by salt stress through stimulating the activity of diamine oxidases in Glycine max (L.) Merr. roots. Plant Physiol Biochem 45(8):560–566

Xu B, Long Y, Feng X, Zhu X, Gilliham M (2021) GABA signalling modulates stomatal opening to enhance plant water use efficiency and drought resilience. Nat Commun 12:1952

Xu L, Chen L, Ali B, Yang N, Chen Y, Wu F, Jin Z, Xu X (2017) Impact of germination on nutritional and physicochemical properties of adlay seed (*Coixlachryma-jobi* L.). Food Chem 229:312–318

Xu JG, Hu QP, Duan JL, Tian CR (2010) Dynamic changes in gamma-aminobutyric acid and glutamate decarboxylase activity in oats (*Avena nuda* L.) during steeping and germination. J Agric Food Chem 58(17):9759–9763. https://doi.org/10.1021/jf101268a

Xu M, Yang Q, Bai G, Li P, Yan J (2022) Polyamine pathways interconnect with GABA metabolic processes to mediate the low-temperature response in plants. Front Plant Sci 13:1035414. https://doi.org/10.3389/fpls.2022.1035414. PMID: 36340403; PMCID: PMC9630558

Yamasaki H, Sakihama Y (2000) Simultaneous production of nitric oxide and peroxynitrite by plant nitrate reductase: in vitro evidence for the NR-dependent formation of active nitrogen species. FEBS Lett 468:89–92. https://doi.org/10.1016/S0014-5793(00)01203-5

Yamasaki H, Sakihama Y, Takahashi S (1999) An alternative pathway for nitric oxide production in plants: new features of an old enzyme. Trends Plant Sci 4:128–129. https://doi.org/10.1016/S1360-1385(99)01393-X

Yang R, Guo Q, Gu Z (2013) GABA shunt and polyamine degradation pathway on g-aminobutyric acid accumulation in germinating fava bean (Vicia faba L.) under hypoxia. Food Chem 136:152159. https://doi.org/10.1016/j.foodchem.2012.08.008

Yoda H, Yamaguchi Y, Sano H (2003) Induction of hypersensitive cell death by hydrogen peroxide produced through polyamine degradation. Plant Physiol 132(4):1973–1981

Yong B, Xie H, Li Z, Li YP, Zhang Y, Nie G, Zhang XQ, Ma X, Huang LK, Yan YH (2017) Exogenous application of GABA improves PEG-induced drought tolerance positively associated with GABA-shunt, polyamines, and proline metabolism in white clover. Front Physiol 8:1107. https://doi.org/10.3389/fphys.2017.01107

Young GB, Jack DL, Smith DW, Saier MH (1999) The amino acidauxinproton symport permease family. Biochim Biophys Acta 1415:306–322. https://doi.org/10.1016/S0005-2736(98)00196-5

Yuan D, Wu X, Gong B, Huo R, Zhao L, Li J, Lü G, Gao H (2023) GABA metabolism, transport and their roles and mechanisms in the regulation of abiotic stress (hypoxia, salt, drought) resistance in plants. Metabolites 13(3). https://doi.org/10.3390/metabo13030347

Yue X, Gao XQ, Wang F, Dong Y, Li X, Zhang XS (2014) Transcriptional evidence for inferred pattern of pollen tube stigma metabolic coupling during pollination. PLoS One 9(9):e107046. https://doi.org/10.1371/journal.pone.0107046

Yue J, Du C, Ji J (2018) Inhibition of α-ketoglutarate dehydrogenase activity affects adventitious root growth in poplar via changes in GABA shunt. Planta 248:963–979. https://doi.org/10.1007/s00425-018-2929-3

Yu GH, Zou J, Feng J, Peng XB, Wu JY, Wu YL, Palanivelu R, Sun MX (2014) Exogenous γ-aminobutyric acid (GABA) affects pollen tube growth via modulating putative Ca^{2+}-permeable membrane channels and is coupled to negative regulation on glutamate decarboxylase. J Exp Bot 65(12):3235–3248

Zarattini M, Forlani G (2017) Toward unveiling the mechanisms for transcriptional regulation of proline biosynthesis in the plant cell response to biotic and abiotic stress conditions. Front Plant Sci 8:927

Zhao GC, Xie MX, Wang YC, Li JY (2017) Molecular mechanisms underlying γ-aminobutyric acid (GABA) accumulation in giant embryo rice seeds. J Agric Food Chem 65:4883–4889. https://doi.org/10.1021/acs.jafc.7b00013

Zhao MG, Chen L, Zhang LL, Zhang WH (2009) Nitric reductase-dependent nitric oxide production is involved in cold acclimation and freezing tolerance in Arabidopsis. Plant Physiol 151:755–767

Zhou M, Hassan MJ, Peng Y, Liu L, Liu W, Zhang Y, Li Z (2021) γ-Aminobutyric acid (GABA) priming improves seed germination and seedling stress tolerance associated with enhanced antioxidant metabolism, DREB expression, and dehydrin accumulation in white clover under water stress. Front Plant Sci 12:776939

ROS Generation and Consequences in Relation to Stress Conditions in Plant Cells

3

Abstract

Reactive oxygen species (ROS) play a crucial role in plant cell signaling and stress responses. Under normal conditions, ROS are produced at low levels and serve as signaling molecules involved in various physiological processes. However, during stress conditions, the balance between ROS production and detoxification is disrupted, leading to an excessive accumulation of ROS in plant cells. This chapter provides an overview of the generation and consequences of ROS in plant cells during stress conditions. We discuss the sources of ROS production, including both enzymatic and nonenzymatic processes, and the factors that contribute to their accumulation. Furthermore, we explore the impact of ROS on plant cells, including oxidative damage to cellular components, activation of defense mechanisms, and modulation of gene expression. Understanding the generation and consequences of ROS during stress conditions is crucial for developing strategies to enhance plant stress tolerance and improve crop productivity.

Keywords

Antioxidant defense · ROS · Oxidative stress · Plant stress response · Signaling pathways

Highlights

- Reactive oxygen species (ROS) act as signaling molecules under normal conditions but accumulate excessively during stress, disrupting cellular balance.
- Stress conditions enhance ROS production through enzymatic and nonenzymatic processes, leading to oxidative stress.

S. U. Jalil et al., *GABA Signaling System and Abiotic Stress Tolerance in Plants*,
https://doi.org/10.1007/978-981-95-0437-4_3

- Excess ROS cause oxidative damage to cellular components, including lipids, proteins, and DNA.
- ROS trigger defense responses and modulate gene expression to help plants cope with stress.
- Understanding ROS dynamics during stress is essential for developing strategies to enhance plant stress tolerance and crop productivity.

3.1 Introduction

Plants very often encounter a range of environmental stressors, including intense light, temperature fluctuations, drought, salinity, and pathogen invasion (Hasanuzzaman et al. 2020; He et al. 2018; Sachdev et al. 2021). Exposure to adverse conditions results in the excessive production of reactive oxygen species (ROS) within plant cells. ROS, including $O^{2\bullet-}$, H_2O_2, •OH, $_1O^2$, ROO•, and RO•, function not only as signaling molecules but also as key mediators in stress response pathways. However, under severe stress, the balance between •S generation and scavenging mechanisms becomes disrupted, leading to the accumulation of ROS and the onset of oxidative stress (Maurya 2020). The ROS production intervenes in multiple physiological pathways, such as photosynthesis, respiration, and enzymatic reactions, with disruptions in electron transport chains, antioxidant defense imbalances, and the activation of Nicotinamide Adenine Dinucleotide Phosphate (NADPH) oxidases (Sachdev et al. 2021).

Plants fight a multifront war against the ROS produced through various pathways during stress, including photosynthesis, respiration, and specific enzymatic reactions. Like a well-organized defense mechanism, the plants rely on different systems to manage the ROS onslaught. However, disruptions in these systems—whether through electron transport chain malfunctions or imbalances in antioxidant defenses—can lead to a dangerous buildup of ROS, further aggravating stress, as Sachdev et al. (2021) point out.

The aftermath of this molecular battle can be devastating. ROS, if left unchecked, can wreak havoc on cellular components. Lipids, proteins, and nucleic acids are especially vulnerable, leading to compromised cell membranes, malfunctioning enzymes, and unstable DNA. The result is a cascade of damages that impairs not only cellular function but also the overall growth and development of the plant, which Juan et al. (2021) have emphasized. The damage remains persistent—ROS can also trigger programmed cell death (PCD) pathways, adding to the plant's woes by initiating self-destruct sequences in affected cells, a phenomenon noted by Bhuyan et al. (2020) and Shah et al. (2019).

But the plant is not defenseless and engages a robust antioxidant defense system—its own elite force trained to neutralize ROS and prevent oxidative damage. This system includes enzymatic antioxidants like superoxide dismutase (SOD), Catalase (CAT), and Peroxidases (PODs), which are the frontline warriors that scavenge ROS and keep their levels in check. Nonenzymatic antioxidants such as ascorbic acid, glutathione, and tocopherols act as reinforcements, ensuring that any ROS that slip through the first line of defense are promptly dealt with (Hasanuzzaman et al. 2020), Wongshaya et al. 2020; Kumar et al. 2017).

Interestingly, the story of ROS in plants is not just about destruction. These molecules also play a role in signaling, helping plants adapt to their environment (Jalil and Ansari, 2020). Controlled ROS production is involved in redox signaling, which regulates the plant's growth and development even during stress. ROS facilitate cell-to-cell communication through NADPH oxidase, also known as Respiratory Burst Oxidase Homolog (RBOH), and activate defense-responsive genes by modifying protein structures (Shah et al. 2019). Understanding how plants regulate and signal with ROS during stress is like decoding the combat strategies of the adversaries. By genetically manipulating antioxidant enzymes or modulating ROS-signaling pathways, scientists can potentially enhance a plant's resilience to environmental stresses. Moreover, applying exogenous antioxidant compounds or employing priming techniques can effectively reduce ROS-induced damage, offering new ways to bolster plant defenses.

3.2 ROS Generation During Stress Conditions

ROS generation is a hallmark of stress conditions that plants experience and respond to. Various endogenous physio-biochemical disturbances arising from stress conditions contribute to ROS production.

3.2.1 Enzymes of ROS Production

ROS in plants are generated through various enzymatic reactions occurring in different cellular compartments. These enzymatic sources of ROS play essential roles in cellular processes and signaling.

3.2.1.1 RBOHs

RBOHs, also known as NADPH oxidases, are a class of enzymes that play a crucial role in generating ROS in plants (Gui et al. 2022). They serve as key enzymes in the ROS production pathway and are essential components of the ROS-signaling network (Marino et al. 2012; Wang et al. 2020). RBOHs are primarily localized in the plasma membrane and are responsible for the production of $O_2^{\bullet-}$ by transferring electrons from NADPH to molecular oxygen (Benkő et al. 2022). The generated $O_2^{\bullet-}$, on conversion, leads to other ROS, such as H_2O_2, either through the action of superoxide dismutases or spontaneous dismutation (Suzuki et al. 2011; Kadota et al. 2015). Various factors, including Ca^{2+}, phosphorylation, and the presence of regulatory proteins, regulate the activity of RBOHs. Ca^{2+} activates RBOHs by binding to specific EF-hand (a calcium-binding domain) motifs present in their structure. Phosphorylation events mediated by protein kinases also modulate the activity of RBOHs, with phosphorylation often enhancing their ROS-generating capacity (Sachdev et al. 2021). Both abiotic and biotic stress responses engage RBOHs, which also play various roles in plant development (Chapman et al. 2019; Kadota et al. 2015; Torres and Dangl 2005). Of the ten isoforms of RBOHs, namely A–J, in *Arabidopsis thaliana*, RBOHD and RBOHF perform crucial roles in both biotic and abiotic stress responses, while others, such as RBOHC, RBOHH, and RBOHJ, are closely related to plant developmental processes (Farvardin et al. 2020). Further, RBOH in *Arabidopsis thaliana* participates in H_2O_2 production in the apoplastic space of the plasma membrane during hypoxia conditions (Wang et al. 2017).

3.2.1.2 XDH/XOD and AO

Plants have a fascinating set of enzymes that help them manage important life processes and also respond to stress. Two such enzymes are xanthine oxidase (XOD) and Aldehyde Oxidase (AO), both of which contain special cofactors like molybdenum and flavin to carry out their functions (Janků et al. 2019). XOD plays a key role in the breakdown of purines, which are important building blocks of DNA and molecules like Adenosine Triphosphate (ATP), while AO is mainly involved in the production of plant hormones and the detoxification of harmful chemicals known as xenobiotics.

Interestingly, during their normal activities, both XOD and AO unintentionally produce reactive oxygen species (ROS), which are molecules like superoxide (O_2•–) and hydrogen peroxide (H_2O_2). XOD catalyzes the oxidation of compounds like xanthine and hypoxanthine, producing superoxide as a by-product. Similarly, AO catalyzes the oxidation of aldehydes into carboxylic acids, using molecular oxygen as a co-substrate, and in doing so, it releases both superoxide and hydrogen peroxide into the surrounding tissues (Janků et al. 2019). This generation of ROS by AO can occur in different parts of the plant, depending on where the enzyme is active.

Under normal conditions, the ROS produced by XOD and AO are kept under control by the plant's antioxidant systems. However, under stressful conditions, like drought, salinity, or pathogen attack, the activities of XOD and AO can increase, leading to a higher buildup of ROS. This elevated ROS production can be beneficial because ROS act as signals that trigger plant defense mechanisms. But if too much ROS accumulate and the antioxidants cannot keep up, it can lead to oxidative stress, damaging plant cells and disturbing normal functions (Janků et al. 2019).

Experimental studies have provided fascinating insights into these processes. For example, in in vitro experiments, it has been shown that XOD from tomato and *Arabidopsis* mainly produces superoxide and not hydrogen peroxide, unlike the animal version of XOD, which generates both types of ROS (Yesbergenova et al. 2005). Further evidence comes from genetic studies: when the Atxdh1 gene (which codes for XOD) was disrupted in *Arabidopsis* plants using T-DNA insertion or RNA interference, these plants lost their ability to generate superoxide, clearly demonstrating that plant XOD naturally produces superoxide. On the other hand, AO in plants predominantly produces hydrogen peroxide, and surprisingly, this activity is not affected by inhibitors like diphenyl iodonium, which usually block flavin-containing enzymes (Yesbergenova et al. 2005).

Stress conditions such as drought or the external application of abscisic acid (ABA) further boost the production of ROS and the expression of AO and XOD enzymes in both leaves and roots. This suggests that during environmental stress, AO and XOD become significant contributors to the buildup of ROS in plant tissues (Yesbergenova et al. 2005).

Recent studies have also uncovered more specific and exciting roles of different versions (isoforms) of AO and XOD. In tobacco plants infected with tomato bushy stunt virus, certain AO isoforms, namely AO2 and AO3, have been found to actively produce superoxide, helping the plant mount a stronger defense against the virus (Yergaliyev et al. 2016). Another special form of AO, called AsAO4, plays a different but equally important role: it helps detoxify harmful aldehydes and delays the aging of siliques under stress conditions by generating both superoxide and hydrogen peroxide, depending on the presence of aldehydes (Srivastava et al. 2017).

Moreover, the enzyme XDH1 in *Arabidopsis* shows an even more complex behavior during plant immune responses, especially against powdery mildew infection. In the outer layer of leaves (epidermal cells), XDH1 works together with other enzymes like RBOHD and RBOHF (both NADPH oxidases) to generate superoxide. This ROS production helps restrict the spread of fungal structures known as haustoria inside infected cells by causing local accumulation of hydrogen peroxide (Ma et al. 2016). However, in other parts of the plant, XDH1 acts differently; it produces uric acid instead, which helps remove excess hydrogen peroxide from stressed chloroplasts, thus protecting the plant cells from oxidative damage caused by biotic stress (Ma et al. 2016).

3.2.1.3 PAOs

Polyamine oxidases (PAOs) are enzymes that facilitate the oxidation of polyamines, including putrescine, spermidine, and spermine. The enzymatic process leads to the production of H_2O_2 and aldehydes. PAOs play essential roles in various physiological processes, such as cell growth, development, and stress responses. Plant PAOs can be divided into two classes based on their functions in polyamine catabolism. The first class of PAOs is responsible for the terminal catabolism of polyamines. They catalyze the oxidation of spermidine or spermine, resulting in the production of H_2O_2. This class of PAOs primarily operates extracellularly, generating H_2O_2 as a by-product (Benkő et al. 2022). The second class of PAOs catalyzes back-conversion reactions of polyamines, converting spermine to spermidine and spermidine to putrescine. These PAOs also generate H_2O_2 as a by-product of their catalytic activity. The back-conversion pathway mainly occurs intracellularly, within the cytoplasm and peroxisomes (Wang et al. 2019a, b). PAOs are distributed in various cellular compartments, including the cytosol, peroxisomes, and apoplast—the space within the cell wall but outside the cell membrane (Moschou et al. 2012; Tavladoraki et al. 2016). Recent evidence suggests that PAOs and the catabolic products of polyamines play crucial signaling roles in various cellular and developmental processes (Chen et al. 2018; Yu et al. 2019a, b; Wang et al. 2019a, b). These roles are likely mediated through the regulation of polyamine homeostasis and the generation of H_2O_2.

H_2O_2 produced by polyamine (PA)-decomposing enzymes can serve as important signaling molecules, enhancing antioxidative defense reactions. However, when H_2O_2 levels surpass a certain threshold, they can exacerbate cellular damage and potentially induce PCD (Wang et al. 2019a, b; Tavladoraki et al. 2012). For instance, studies have shown that tobacco cells secrete spermidine into the apoplast, where PAO oxidizes it, resulting in the generation of H_2O_2 that promotes PCD (Moschou et al. 2008a, b). In citrus plants (*Citrus x sinensis*), the apoplastic PAO (CsPAO4) reportedly produces H_2O_2 and perpetrates oxidative damage under salt stress (Wang and Liu, 2016; Wang et al. 2017). In tomato plants, PAO enzymes involved in PA catabolism exhibit stronger responses to sublethal doses of the stress hormone, salicylic acid, indicating their role in salinity tolerance signaling (Takács et al. 2016). The downregulation of PAO-coding genes in tobacco promotes thermotolerance, likely due to reduced heat-induced H_2O_2 production (Mellidou et al. 2017). Additionally, PAO activity contributes to aluminum- or selenium-induced oxidative stress, highlighting its prooxidant role during severe stress conditions (Yu et al. 2018; Wang et al. 2019a, b).

PAs and the enzymes involved in their metabolism become highly active when plants are attacked by various types of pathogens. These include both biotrophic pathogens, which feed on living plant tissues, and necrotrophic pathogens, which kill plant cells and then consume them (Walters 2003; Marina et al. 2008; Pál and Janda 2017). For instance, in tobacco plants, an infection by the biotrophic pathogen *Pseudomonas syringae* leads to an increase in the activity of polyamine oxidase (PAO), an important enzyme that helps break down polyamines (Moschou et al. 2009). Along with this rise in PAO activity, the plant also secretes more spermine, a type of polyamine. Together, these changes cause a significant buildup of H_2O_2 in the apoplast, the space outside the plant cell membranes.

Interestingly, when PAO activity is blocked using specific inhibitors, the spermine-triggered disease resistance in the apoplast becomes weaker (Marina et al. 2008; Moschou et al. 2009). This shows that the H_2O_2 produced by PAO during polyamine breakdown plays a crucial role in strengthening plant defenses during biotrophic infections. In fact, the generation of H_2O_2 through PAO activity is believed to contribute to the hypersensitive response (HR), a defense mechanism where infected plant cells quickly die to prevent the spread of pathogens. This type of response has been observed in tobacco plants infected with tobacco mosaic virus or *Pseudomonas cichorii* as well as in barley plants infected with powdery mildew (Cowley and Walters 2002; Yoda et al. 2003).

A similar defense-related process is triggered by a protein called cryptogein, secreted by the oomycete *Phytophthora cryptogea*, which induces HR in tobacco. When scientists blocked the gene responsible for making the apoplastic PAO enzyme in tobacco, they observed that polyamine degradation was prevented, cryptogenic-induced H_2O_2 production stopped, and the typical HR-associated cell death did not occur (Yoda et al. 2006). Furthermore, when tobacco plants were genetically engineered to overexpress the PAO enzyme from corn, they maintained high levels of H_2O_2. In some cases, this overproduction of H_2O_2 led to programmed cell death (PCD), suggesting that PAO-generated H_2O_2 not only acts as a cytotoxic molecule but also has an important signaling function during the HR (Moschou et al. 2008a).

This signaling role of H_2O_2 produced from spermine breakdown is also thought to be important in the transcriptional responses of *Arabidopsis* plants to infections by cucumber mosaic virus, another pathogen that induces HR (Mitsuya et al. 2009).

3.2.1.4 PODs and LOXs

PODs primarily reside in the plant cell wall and serve as a significant source of apoplastic ROS (Lazzarotto et al. 2015). These PODs exhibit a wide substrate specificity and can function either by consuming ROS or by generating ROS through peroxidation or hydroxylation cycles, respectively (Berglund et al. 2002; Lüthje and Martinez-Cortes 2018; O'Brien et al. 2012; Francoz et al. 2015). In the ROS-consuming peroxidase mode, an oxygen atom from H_2O_2 is transferred to the heme (ferric) group of the enzymes, forming the first intermediate. Subsequently, an electron from a reducing substrate generates the second intermediate, and the cycle concludes with the regeneration of the ferric enzyme (Berglund et al. 2002). The reduction of H_2O_2 by PODs occurs concurrently with the oxidation of electron donors such as phenolic compounds (Veitch 2004). On the other hand, PODs produce ROS through the oxidase cycle, leading to the formation of the third intermediate. Plant class III PODs are responsible for generating apoplastic ROS and are involved in stress signaling triggered by pathogen attacks, wounds, and various abiotic stress factors (Shigeto and Tsutsumi 2016). The apoplastic oxidative burst results from the ROS-producing activities of multiple enzyme groups, including cell wall PODs, plasma membrane NADPH oxidases, amine oxidases, lipoxygenases, oxalate oxidases, and quinone reductases (Survila et al. 2016; Kärkönen and Kuchitsu 2015).

Researchers have proposed PODs as alternative ROS producers, distinct from NADPH oxidases (Apel and Hirt 2004; Bindschedler et al. 2006). For example, in suspension cultured cells of French beans (*Phaseolus vulgaris*) treated with a cell wall elicitor from *Colletotrichum lindemuthianum*, the fungus causing anthracnose disease, PODs rather than NADPH oxidases are primarily responsible for ROS production (Bolwell et al. 1998, 1999). Additionally, under salinity stress, increased activity of cell wall peroxidase and NADH peroxidase in rice seedlings has been linked to reduced root growth and the accumulation of H_2O_2 in the cell wall (Lin and Kao 2001). Various abiotic stresses, such as exposure to ground-level O_3, induce an oxidative burst in plant cells, leading to the production and accumulation of H_2O_2 and $O_2^{\bullet-}$ in the apoplast, resulting in necrosis and cell death (Langebartels et al. 2002). Cell wall–associated PODs reportedly produce H_2O_2 in *Arabidopsis* under potassium (K^+) deficiency conditions (Martinez et al. 1998).

In a study by Kim et al. (2010), a peroxidase called RCI3, belonging to class III PODs, contributes to ROS production as the root response of *Arabidopsis* to potassium deficiency. RCI3 exhibits peroxidase activity specifically in root cells (Llorente et al. 2002). As a matter of fact, class III PODs target the outside of plant cells or the vacuole through the endoplasmic reticulum (Welinder et al. 2002). Cell wall–localized PODs have been identified as playing a role in ROS production (Bindschedler et al. 2006; Choi et al. 2007; Pnueli et al. 2003). Transgenic *Arabidopsis* plants expressing an antisense French bean peroxidase type 1 (FBP1) cDNA exhibit impaired oxidative burst and increased susceptibility to fungal and bacterial pathogens compared to wild-type plants (Bindschedler et al. 2006). Moreover, the overexpression of $CaPO_2$, a novel extracellular peroxidase gene from pepper (*Capsicum annuum*) in *Arabidopsis* resulted in elevated H_2O_2 concentrations in response to *Pseudomonas syringae* pv. tomato infection (Choi et al. 2007).

Lipoxygenase (LOXs) are a family of enzymes found in plants that serve crucial roles in numerous physiological and biochemical processes. Among their many functions, LOXs are particularly renowned for their involvement in the generation of ROS within plants. By catalyzing the oxygenation of polyunsaturated fatty acids (PUFAs), such as linoleic acid and α-linolenic acid, LOXs give rise to hydro-peroxy fatty acids, which are commonly referred to as lipid peroxides.

The process of lipoxygenase-mediated oxygenation entails the incorporation of molecular oxygen into the fatty acid substrate, leading to the formation of a fatty acid hydroperoxide. Subsequently, these lipid peroxides can engage in nonenzymatic reactions that culminate in the production of ROS, including $^1O_2$ and lipid-derived radicals. It is important to note that under stress conditions, lipoxygenase-induced hydro-peroxidation of polyunsaturated fatty acids derived from cell membranes represents an additional potential source of ROS (Karuppanapandian et al. 2011). The liberated polyunsaturated fatty acids can serve as substrates for lipoxygenase, thereby triggering the synthesis of hydro-peroxy polyunsaturated fatty acids, also known as lipid hydroperoxides (Fig. 3.1) (Prasad et al. 2017).

Under reducing conditions, transition metals reduce polyunsaturated fatty acids to produce lipid alkoxyl radicals, which may subsequently abstract hydrogen from other polyunsaturated fatty acids, leading to the generation of additional lipid radicals and hydroxyl polyunsaturated fatty acids (Miyamoto et al. 2014). Conversely, under oxidizing conditions, polyunsaturated fatty acids get oxidized to form lipid peroxyl radicals in the presence of oxidized transition metals. The cyclization of lipid peroxyl radicals is known to give rise to a cyclic endoperoxide, also referred to as dioxetane, while the recombination of two lipid peroxyl radicals yield a linear tetroxide (Pospíšil et al. 2014; Miyamoto et al. 2014). These high-energy intermediates subsequently decompose into triplet carbonyls ($3\ L = O^*$), which can transfer triplet energy to pigments, resulting in the formation of excited pigments, or to molecular oxygen, leading to the production of $_1O^2$. Additionally, the tetroxide (Fig. 3.1) can directly decompose into $_1O^2$ through the Russell mechanism during enzymatic lipid peroxidation (Cilento and Adam 1995; Miyamoto et al. 2014; Fedorova and Trofimov 2007).

3.2.1.5 Disparate Enzymes and Metal Ions

Under stressful conditions, plants produce ROS within peroxisomes, which are small organelles that help manage cellular metabolism. Various enzymes in peroxisomes are responsible for generating these ROS. For example, the photorespiration occurring in the chloroplast produces a molecule called glycolate. When glycolate reaches the peroxisomes, the enzyme glycolate oxidase converts it into glyoxylate and H_2O_2, a type of ROS. Similarly, during the breakdown of fatty acids through

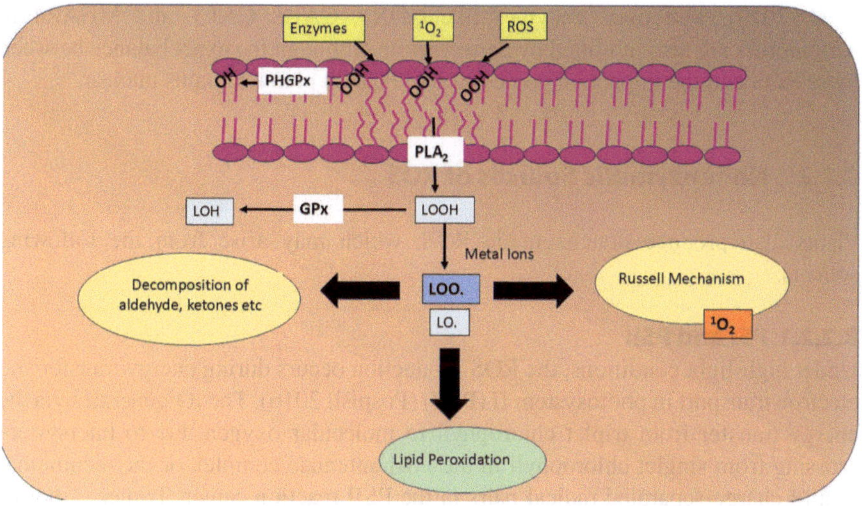

Fig. 3.1 Lipid peroxidation in membranes triggered by ROS and singlet oxygen ($^1O_2$), leading to the formation of lipid hydroperoxides (LOOHs). Under reducing conditions, LOOH is converted into lipid alkoxyl radicals (LO•), while under oxidizing conditions, it forms lipid peroxyl radicals (LOO•). These radicals promote membrane damage and lipid peroxidation. Enzymes like GPx and PHGPx help detoxify LOOH to LOH. LOO• can form high-energy intermediates that decompose into singlet oxygen ($^1O_2$) via the Russell mechanism, intensifying oxidative stress

β-oxidation, the enzyme Acyl-CoA Oxidase (ACX) generates H_2O_2, alongside acetyl-Coenzyme A (CoA), a crucial energy molecule (del Río et al. 1998). Another enzyme, XOD, converts xanthine and hypoxanthine into uric acid and ROS, namely $O2^{•-}$. Further, urate oxidase breaks down uric acid, producing more H_2O_2 (Corpas et al. 2019). Additional compounds like sarcosine and sulfite are also oxidized by the enzymes sarcosine oxidase and sulfite oxidase in the peroxisomes, generating H_2O_2 (Nyathi and Baker 2006). Additionally, other enzymes, like flavin oxidases, oxalate oxidase, and SOD, further contribute to ROS production across different cellular compartments, generating various ROS, such as H_2O_2, singlet oxygen, and lipid radicals. These reactive molecules are crucial for plant stress responses, serving as signals and playing key roles in defense mechanisms.

Another key enzyme, SOD, plays a significant role in converting $O^{2•-}$ into H_2O_2. This process involves a metal ion that acts as an intermediate carrier of electrons (Prakash Sanyal et al. 2018). Specifically, the copper/zinc form of SOD (AtCSD3) in the model plant *Arabidopsis* is involved in this conversion within peroxisomes (Huang et al. 2012). However, the activity of AtCSD3 can be significantly reduced by a compound called peroxynitrite, which is formed by a rapid reaction between superoxide and nitric oxide (Holzmeister et al. 2015; Corpas et al. 2019). This reduction in activity can be even more pronounced under stress conditions (Corpas et al. 2019). Moreover, enzymes like GOX1, CAT2, CAT3, and MDAR4 in *Arabidopsis* are also inhibited by peroxynitrite, although the exact balance between these ROS and the impact on processes like PCD in plants remains unclear.

3.2.2 Nonenzymatic Sources of ROS

Plants also produce nonenzymatic ROS, which may arise from the following sources:

3.2.2.1 PSI and PSII

Under high-light conditions, the ROS production occurs during energy transfer and electron transport in photosystem II (PSII) (Pospíšil 2016). The $_1O^2$generates via the energy transfer from triplet chlorophyll to molecular oxygen due to intersystem crossing from singlet chlorophyll in the PSII antennae complex or the recombination of charge-separated radical pairs in the PSII reaction center. Triplet carbonyl, formed through lipid peroxidation, also contributes to $_1O^2$ production by transferring energy to molecular oxygen. On the electron acceptor side of PSII, electron leakage results in the formation of $O^{2\bullet-}$, which converts to H_2O_2 and further reduces to hydroxyl radical by nonheme iron. Meanwhile, incomplete water oxidation on the electron donor side produces H_2O_2, reduced by manganese to hydroxyl radical (Pospíšil 2016; Krieger-Liszkay and Shimakawa 2022). Under elevated temperatures, lipoxygenase elicits the dark production of $_1O^2$ during lipid peroxidation, and incomplete water oxidation again leads to the formation of H_2O_2, reduced by manganese to hydroxyl radical. This understanding of the molecular basis for ROS production by PSII provides valuable insights into the mechanisms enabling plant survival under adverse environmental conditions.

Chloroplasts, responsible for converting light energy into chemical bond energy, utilize the light absorption by chlorophyll molecules in photosystems I and II (PSI and PSII). This triggers a series of redox reactions along the thylakoid membrane, leading to water oxidation, reduction of Nicotinamide Adenine Dinucleotide Phosphate $(NADP)^+$ to NADPH, and the establishment of a proton gradient (Fig. 3.2) (Khorobrykh et al. 2020; Krieger-Liszkay and Shimakawa 2022). Within chloroplasts, PSII performs the photolysis of water and generates O_2, whose concentration inside intact chloroplasts depends on external O_2 concentrations. The $_1O^2$ formation mainly occurs through the interaction of O_2 with the triplet excited state of chlorophyll. The reduction of O_2 by electron carriers in the photosynthetic electron transfer chain produces $O_2^{\bullet-}$ and H_2O_2 (Khorobrykh et al. 2020). PSII, as a water–plastoquinone oxidoreductase embedded in the thylakoid membrane, catalyzes light-driven H_2O oxidation to O_2 and plastoquinone reduction (Nelson and Junge 2015; Suga et al. 2015; Najafpour et al. 2016). The primary charge separation in PSII involves the formation of radical pairs and subsequent stabilization, leading to the liberation of O_2 into the lumen (Vogt et al. 2015). The overall process highlights the essential role of light in ROS production within chloroplasts, where ROS

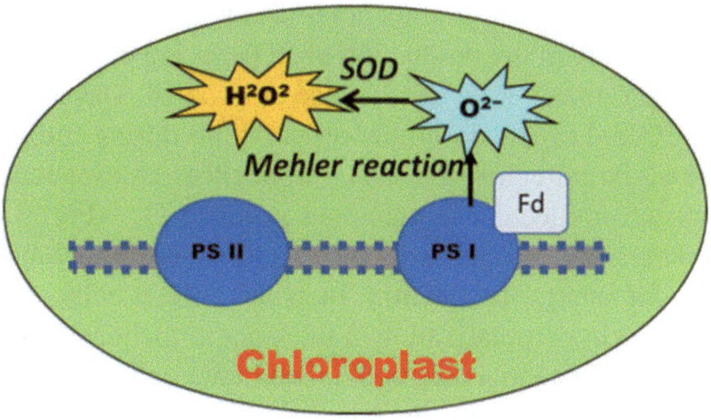

Fig. 3.2 Generation of ROS in the chloroplast, particularly through the Mehler reaction. In photosystem I (PSI), ferredoxin (Fd) donates electrons to O_2, leading to the formation of superoxide radicals (O_2^-). The SOD enzyme converts O_2^- into H_2O_2, preventing oxidative damage. These ROS play a role in redox signaling but can also contribute to oxidative stress if not properly regulated. Photosystem II (PSII) is also involved in the electron transport process, linking it to ROS dynamics in the chloroplast

can be interconverted through interactions with redox-active compounds (Pospíšil 2016).

3.2.2.2 ETCs

The Electron Transport Chain (ETC) is an important part of cellular respiration in eukaryotic cells, including plants. It generates cellular energy in the form of ATP but can also produce ROS as by-products. The ETC consists of protein complexes in the inner mitochondrial membrane that transfer electrons sequentially. However, electron leakage can occur, leading to the production of ROS, particularly the superoxide radical. Factors such as high electron flow, redox imbalance, membrane potential disturbances, and deficient antioxidant defenses contribute to increased ROS production. Cells have antioxidant defense mechanisms to minimize ROS damage, including enzymes like SOD and nonenzymatic antioxidants. While ROS can damage cells, they also function as signaling molecules in physiological processes. Understanding ROS regulation and the interplay with antioxidants is crucial for cellular redox balance and plant responses to the environment.

Under abiotic stress conditions, NADH and/or succinate generated during photorespiration and the Krebs cycle can enter the ETC at Complex I and Complex II, respectively. At Complex I, NADH is oxidized to NAD^+ and H^+ ions, releasing an electron. Similarly, at Complex II, succinate is converted to fumarate, releasing an electron. These electrons are transferred to Ubiquinone (UQ), generating superoxide and H_2O_2 as by-products. Complex III and Complex IV further facilitate electron transfers, with oxygen being oxidized to water without ROS production at Complex IV. The movement of electrons through the ETC also leads to the pumping of protons across the inner mitochondrial membrane, creating an electrochemical gradient utilized by Complex V (ATP synthase) to produce ATP.

Disruptions in the ETC within mitochondria can lead to an excessive reduction of the UQ pool, resulting in the production of ROS during drought or salinity stress (Miller et al. 2010; Yamane et al. 2012; Bose et al. 2014; Das and Roychoudhury 2014). Hu et al. (2008) have made similar observations during chilling stress. When subjected to stress, the mitochondrial ETC experiences an over-reduction, causing electrons to leak into molecular oxygen and generate $O_2^{\bullet -}$ (Fig. 3.3). Heat stress places additional pressure on plant cells by affecting their mitochondria, the organelles responsible for energy production. In winter wheat cells, exposure to high temperatures causes a hyperpolarization of the inner mitochondrial membrane. This means that the electric potential across the membrane becomes even stronger than normal. Such a change disrupts the delicate balance of electron flow in the respiratory electron transport chain, leading to its over-reduction. When the electron carriers are excessively reduced, they can no longer transfer electrons efficiently, and instead, electrons leak out and interact with oxygen molecules, resulting in the production of ROS. This excessive ROS generation under heat stress can damage cellular structures, impair normal metabolism, and trigger further stress responses in the plant (Fedyaeva et al. 2014). The succinate dehydrogenase complex (Complex II) indirectly contributes to the generation of ROS in mitochondria by redirecting the electron flow toward Complex I when there is a scarcity of NAD^+-linked substrates (Turrens 2003). Under excessive light, the glycine dehydrogenase complex (GDC) may serve as the primary substrate, producing NADP and donating an electron to Complex I, thus initiating the ETC and potentially resulting in the production of $O_2^{\bullet -}$ (Schertl and Braun 2014). Gao et al. (2008) have reported various effects from ROS generation, including the activation of caspase-like protease, DNA

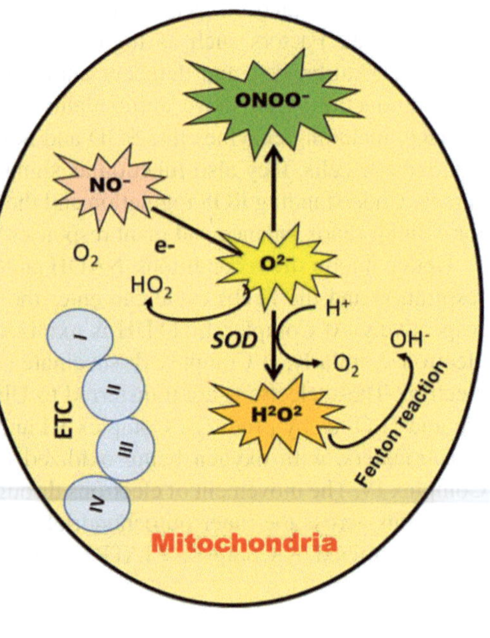

Fig. 3.3 Generation of ROS and reactive nitrogen species (RNS) in the mitochondria. The ETC generates superoxide radicals (O_2^-) through electron leakage. SOD converts O_2^- into H_2O_2, which can further participate in the Fenton reaction, producing highly reactive hydroxyl radicals (OH^-). Additionally, nitric oxide (NO•) reacts with O_2^- to form peroxynitrite ($ONOO^-$), a potent oxidant. These reactive species play crucial roles in cellular signaling and oxidative stress responses

laddering, nucleus fragmentation, and PCD in *Arabidopsis* under excessive UV radiation exposure. Thus, understanding ROS regulation and the conventional and alternative pathways is crucial for plant responses to abiotic stress and maintaining cellular redox balance.

3.2.2.3 Fatty Acid Oxidation and Photorespiration

Peroxisomes are organelles crucial for metabolic processes such as fatty acid oxidation and photorespiration in plants. These activities through ACX and enzyme activities like flavin oxidase, urate oxidase, and XOD in peroxisomes contribute to ROS generation (del Río et al. 1998; Corpas et al. 2019).

β-oxidation is a key process responsible for fatty acid breakdown, generating acetyl-CoA that enters mitochondria for aerobic ATP production (Huang et al. 2022). The ACX, a flavoprotein oxidase, initiates peroxisomal β-oxidation by transferring two electrons from acetyl-CoA to Flavin Adenine Dinucleotide (FAD) and then to O_2, forming H_2O_2 (Baker et al. 2006). Increased fatty acid flux into peroxisomes for β-oxidation leads to ROS accumulation, and vice versa (Fig. 3.4). (Yu et al. 2019a, b).

Photorespiration generates ROS by being initiated in chloroplasts under conditions of low CO_2 and high O_2. The oxygenation of ribulose-1,5-bisphosphate in photorespiration leads to glycolate production. This glycolate is transported to peroxisomes, where glycolate oxidase oxidizes it, producing H_2O_2. Thus, photorespiration directly contributes to ROS in plant cells. Under abiotic stresses, like flooding, drought, salinity, and high irradiance, photorespiration is autonomously triggered in chloroplasts and generates copious glycolate (Voss et al. 2013; Suo et al. 2017) that gets transported to peroxisomes for H_2O_2 production (Cruz 2008; Das and Roychoudhury 2014). Salinity stress enhances photorespiration and H_2O_2 levels (Yamane et al. 2012; Mittova et al. 2003). Drought conditions contribute to over

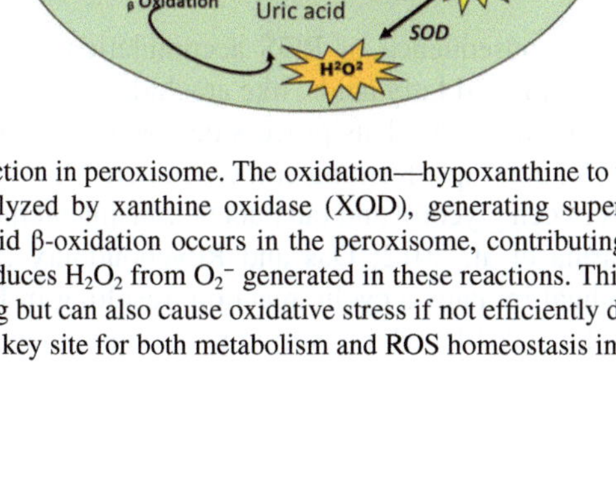

Fig. 3.4 ROS production in peroxisome. The oxidation—hypoxanthine to xanthine and xanthine to uric acid—is catalyzed by xanthine oxidase (XOD), generating superoxide radicals (O_2^-). Additionally, fatty acid β-oxidation occurs in the peroxisome, contributing to metabolic energy production. SOD produces H_2O_2 from O_2^- generated in these reactions. This H_2O_2 plays a crucial role in redox signaling but can also cause oxidative stress if not efficiently detoxified. The peroxisome thus serves as a key site for both metabolism and ROS homeostasis in plant cells

70% of H_2O_2 generation through photorespiration (Voss et al. 2013). Peroxisomes also produce H_2O_2 during prolonged darkness and exposure to cadmium (Cd) and 2,4-dichlorophenoxyacetic acid (2,4-D) (Ortega-Galisteo et al. 2012). Abiotic stresses induce peroxisome proliferation, and $O_2^{\bullet-}$ is produced via membrane-localized NADH-dependent electron transport chains and XOD-mediated oxidation of xanthine (Maurya 2020; Das and Roychoudhury 2014; del Río and Lopez-Huertas 2016). Cd stress increases H_2O_2 production through glycolate oxidase and reduces CAT activity (Ortega-Galisteo et al. 2012). Additionally, Cd stress induces peroxisome proliferation, accelerates $O_2^{\bullet-}$ and H_2O_2 production, and alters endogenous proteins in plants (Romero-Puertas et al. 1999; Romero-Puertas et al. 2002).

3.3 Consequences of ROS Accumulation

ROS act as double-edged swords, functioning as signaling molecules in low concentrations and becoming toxic in excess (Das and Roychoudhury 2014). When ROS levels exceed the antioxidant defense capacity of plants, oxidative stress occurs, leading to the damage of cellular components, including lipids, proteins, and nucleic acids (Sharma et al. 2012).

3.3.1 Oxidative Damage to Cellular Components

Elevated ROS levels in adverse conditions cause oxidative damage to essential cellular components, including lipids, proteins, and nucleic acids. This leads to lipid peroxidation, disrupting cell membranes; oxidative protein modifications, affecting enzyme function; chlorophyll degradation; and the activation of apoptosis and PCD pathways in severe cases (Shah et al. 2019; Roychowdhury et al. 2019). Nucleic acid damage may result in mutations and genomic instability. The extent of damage depends on factors like biomolecule concentration, location in relation to ROS generation, reaction rate, secondary incidents, and the effectiveness of ROS-scavenging or repair systems (Davies 2005). These effects collectively contribute to cellular dysfunction and senescence and impact overall plant health and productivity.

3.3.2 Lipid Peroxidation

Lipid peroxidation, a consequence of ROS accumulation, targets cellular membranes rich in polyunsaturated fatty acids, like arachidonic acid, linolenic acid, and linoleic acid (Anjum et al. 2015). This process begins with the •OH radical attacking fatty acid methylene groups, forming carbon-centered lipid radicals and breaking ester linkages between glycerol and fatty acids, causing membrane phospholipid disintegration (Sharma et al. 2012; Das and Roychoudhury 2014; Anjum et al. 2015). The •OH radical initiates a cyclic reaction, resulting in PUFA peroxidation (Das and Roychoudhury 2014), which is lipid peroxidation involving three stages:

initiation, propagation, and termination (Anjum et al. 2015). Initiation involves ROS production from O_2 reduction, triggering a cascade of reactions that form lipid radicals and eventually lipid dimmers (Das and Roychoudhury 2014). Lipid peroxidation compromises membrane stability, affecting permeability, electrolyte leakage, and enzyme and receptor deactivation and promoting nucleic acid and protein oxidation. This process primarily causes altered fluidity and disrupted transport processes (Sachdev et al. 2021; Smirnoff and Wheeler, 2000).

Lipid peroxidation impacts various cellular components and functions. Membrane damage can lead to leakage of ions and molecules, disrupting cellular homeostasis. Lipid peroxidation–derived aldehydes, such as Malondialdehyde (MDA), contribute to protein carbonylation, DNA damage, and cross-linking of macromolecules, all of which collectively contribute to cellular dysfunction and senescence. Abiotic stresses enhance ROS production, exacerbating lipid peroxidation, leading to cellular and organelle damage. For instance, excess aluminum induces lipid peroxidation and DNA damage in rice roots (Meriga et al. 2004), while arsenic stress prompts H_2O_2 accumulation and lipid peroxidation in common bean seedlings (Talukdar 2013). Markers like MDA indicate oxidative damage severity, observed in tomato plants exposed to salinity (Martinez et al. 2016), heat, and combined stresses, as well as in *Solanum tuberosum* under elevated O_3 (Kumari et al. 2015). Elevated temperature in *Phalaenopsis* induces membrane lipid peroxidation, causing loss of integrity and K^+ leakage (Ali et al. 2005). Understanding such mechanisms can enhance plant cell tolerance against oxidative damage. Plants utilize specific mechanisms to counteract stress-induced oxidative damage, including the activation of stress-responsive genes, accumulation of compatible solutes, and upregulation of antioxidant enzyme activities.

3.3.3 Protein Oxidation

Proteins are vital in helping plants cope with stressful environmental conditions by adjusting their physiological processes (Kosova et al. 2018). However, when proteins change shape or clump together, it can affect their ability to function properly, such as in enzyme activity or binding (Oracz et al. 2007). Proteins are particularly vulnerable to damage from ROS, which are highly reactive molecules. Due to their abundance in cells and their tendency to react quickly, proteins are often the first to be affected by oxidation (de Dios Alché 2019). ROS can directly or indirectly cause protein damage; some cause minor, selective harm, while others, like •OH, cause prevalent and nonspecific damage (de Dios Alché 2019).

The protein backbone can be attacked by non-radical oxidants, leading to limited damage because the oxidation occurs slowly. In contrast, radicals cause extensive damage by quickly reacting with the protein, particularly by removing hydrogen atoms from certain sites, creating more stable but damaging radicals (de Dios Alché 2019). ROS can directly oxidize proteins by targeting amino acids like cysteine and methionine, especially those containing sulfur or thiol groups. This can lead to various changes in the protein, such as carbonylation, nitrosylation, disulfide bond formation, and glutathionylation, all of which can alter the protein function (Das and Roychoudhury 2014).

Indirect protein oxidation can also occur through by-products of lipid peroxidation (Das and Roychoudhury 2014). Oxidized proteins become more prone to being broken down, often tagged for degradation through a process called ubiquitination, which leads to their destruction by the proteasome, a protein complex responsible for degrading unneeded or damaged proteins (Moller et al. 2007). Protein oxidation can be either irreversible or reversible; for example, $O^{2\bullet-}$ can cause irreversible damage to enzymes with iron–sulfur centers, leading to functional loss (Das and Roychoudhury 2014).

One of the most severe and irreversible forms of protein damage is carbonylation, where proteins undergo oxidative cleavage, making them a reliable marker for assessing oxidative damage under stress (Moller et al. 2007). For instance, heat shock proteins lose their ability to maintain protein function when oxidized, especially in dehydrated seeds (Sweetlove and Møller 2009). Studies have shown a significant increase in carbonylated proteins in various plant organelles, including chloroplasts, mitochondria, the nucleus, cytosol, and peroxisomes, under stress conditions, like drought, salinity, and exposure to high CO_2 levels (Bartoli et al. 2004; Qiu et al. 2008; Ferreira-Silva et al. 2012; Karuppanapandian and Kim 2013).

3.3.4 Nucleic Acid Damage

Oxidative stress is a significant threat to plants, with one of the most critical consequences being the damage it causes to nucleic acids—DNA and RNA. These essential molecules, which store and transmit genetic information, are particularly vulnerable to oxidative modifications due to the highly reactive nature of ROS, such as $O2^{\bullet-}$ and OH• (Hollosy 2002; Sharma et al. 2012).

Plant DNA is especially prone to ROS attack. When ROS interact with DNA, they can cause various types of damage, including changes to the DNA bases (like the formation of 8-oxoguanine) and breaks in the DNA strands (Das and Roychoudhury 2014). These alterations can interfere with DNA replication, potentially leading to mutations when cells divide. If these DNA lesions go unrepaired, they can destabilize the plant genome, increasing the risk of further genetic abnormalities (Das and Roychoudhury 2014). For example, in sensitive wheat genotypes, drought conditions result in ROS-induced DNA fragmentation, contributing to PCD in leaves (Hameed et al. 2013). Among all ROS, the •OH is particularly damaging to DNA because of its ability to react with both purine and pyrimidine bases as well as the sugar backbone of the DNA (Soares et al. 2019). This radical can cause breaks in the DNA strand and initiate the formation of harmful products like 8-hydroquinine and thymine glycol (Das and Roychoudhury 2014). Additionally, the •OH radical can cause cross-linking between DNA and proteins, creating complex lesions such that the plant repair system experiences extreme difficulty in fixing the same. If these cross-links are not repaired before DNA replication or transcription, they can have lethal effects on the plant cell (Das and Roychoudhury

2014). Lipid radicals can also indirectly oxidize DNA. One such product, MDA, reacts with guanine in DNA to form harmful adducts that further compromise DNA integrity (Roldan-Arjona and Ariza 2009; Fink et al. 1997).

RNA, while less stable than DNA, is also susceptible to oxidative damage by ROS (Mittler 2002). This damage can disrupt RNA transcription and translation, leading to errors in gene expression and protein synthesis. For instance, oxidative stress in plants induces RNA oxidation (Mittler 2002). Such damage has been observed in soybean seedlings exposed to cadmium stress (Chmielowska-Bak et al. 2018) and in the degradation of messenger Ribonucleic Acid (mRNA) during drought-induced desiccation in *Lindernia subracemosa* (Dinakar and Bartels 2012). The impairment of RNA functions due to ROS can disturb critical cellular processes, reducing metabolic efficiency, causing genetic instability, and disrupting the overall balance within plant cells (Soares et al. 2019).

The repercussions of nucleic acid damage extend beyond the immediate molecular level. Accumulated mutations in DNA can lead to abnormal protein synthesis, potentially hindering vital cellular functions and impairing the plant's ability to adapt to environmental changes. ROS-induced DNA damage can also trigger cell cycle arrest and lead to senescence, significantly affecting the plant's growth and development.

3.4 Activation of Defense Mechanisms

When plants encounter stress from their environment, ROS play a pivotal role in activating their defense mechanisms. Though often considered harmful due to their potential to damage cellular components, ROS, at controlled levels, function as essential signaling molecules. These signals prompt the activation of various defense responses, such as the induction of antioxidant enzymes that neutralize excess ROS, thus preventing cellular damage. Additionally, ROS contribute to the fortification of plant cell walls by promoting the cross-linking of cell wall proteins, resulting in a more robust barrier against potential pathogens. ROS also stimulate the production of nonenzymatic antioxidants, including secondary metabolites, that further scavenge free radicals, disrupting harmful chain reactions and maintaining the cellular redox balance. These complex interactions allow plants to bolster their stress tolerance, supporting overall health and survival even under adverse environmental conditions.

3.4.1 Antioxidant Enzymes

The enzymatic antioxidants are the frontline soldiers that fight against ROS within plant cells. These enzymes transform harmful ROS into less reactive molecules, effectively shielding cells from oxidative damage. Among the key players is SOD, which converts $O_2^{\bullet-}$ into H_2O_2, a less reactive molecule. Subsequently, CAT breaks down H_2O_2 into water and oxygen, rendering it harmless to the cell (Nimse and Pal

2015). Also, PODs utilize H_2O_2 to oxidize various substrates, which further aids in the detoxification process. Gill and Tuteja (2010) have noted that SOD exists in multiple forms, each located in specific cellular compartments. They reported that the Cu/Zn-SOD variant is distributed across the cytosol, chloroplasts, mitochondria, and other sites such as peroxisomes, nuclei, and apoplasts. Fe-SOD predominantly occurs in chloroplasts, peroxisomes, and mitochondria, and Mn-SOD occurs mainly in peroxisomes, mitochondria, and vascular tissues (Ahmad et al. 2010; Zafra et al. 2018). Additionally, Chen et al. (2017) and Su et al. (2018) point out that other essential enzymes, including CAT, APX, and GR, operate in tandem to dismantle H_2O_2, thereby ensuring cellular protection. Regarding cellular compartmentation, CAT and APX are localized in peroxisomes and mitochondria. In addition, APX is also present in the cytosol and chloroplasts, where it plays a crucial role in the (Ascorbic acid-Glutathione (AsA–GSH) cycle (Anjum et al. 2016). On the other hand, GR converts oxidized glutathione back into its reduced form, which is essential for detoxifying H_2O_2 and preserving the cell's redox equilibrium (Gill and Tuteja 2010; Rao and Reddy 2008). Peroxiredoxins (Prxs) also perform a significant role in detoxifying various peroxides, including H_2O_2; reducing oxidative damage; and participating in ROS-related signaling processes (Dietz 2011; Kohli et al. 2019). Last, thioredoxins (Trxs) are small proteins that assist in regulating ROS levels by reducing disulfide bonds, which are vital for maintaining proper cellular function in processes involving ROS (Gelhaye et al. 2005; Dietz 2011).

3.4.2 Nonenzymatic Antioxidants

In addition to enzymatic defenses, plants also deploy a diverse array of nonenzymatic antioxidants to combat ROS (Ahmad et al. 2010). These include low-weight molecules like ascorbic acid (vitamin C), glutathione, tocopherols (vitamin E), and carotenoids, which neutralize ROS by donating electrons. Secondary metabolites such as flavonoids and alkaloids also play a crucial role, often working synergistically with enzymatic systems to enhance the plant's overall ROS-scavenging capacity (Hernandez et al. 1993; Petrov et al. 2015; Pandey et al. 2017). The GSH is a ubiquitous thiol tripeptide involved in the degradation of H_2O_2, a process catalyzed by glutathione peroxidase (Bartoli et al. 2017). GSH is a key component of the AsA–GSH pathway, where it assists dehydroascorbate reductase in scavenging H_2O_2 (Szalai et al. 2009). Ascorbic acid is another critical player in this pathway, donating electrons to APX to help break down H_2O_2. Additionally, ascorbic acid supports the production of other protective molecules like tocopherol and xanthophyll, which manage excess sunlight energy, preventing potential damage (Mittler 2002; Evans et al. 2004). Carotenoids, the pigments responsible for capturing light, contribute significantly to reducing oxidative stress by safely dissipating excess energy as heat (Hix et al. 2004; Uarrota et al. 2018). GABA is a special amino acid that accumulates in plants under stress. Not only does GABA act as an antioxidant, but it also regulates enzyme activity, helping plants adjust to challenging conditions like drought (Khan et al. 2021). GABA is metabolized via the GABA shunt pathway, involving several enzymes that maintain the energy balance sheet in plants (Jalil et al. 2017; Jalil et al. 2019; Ansari et al. 2021). Research has shown that plants lacking GABA become susceptible to stress, exhibiting reduced photosynthesis and premature aging (Jalil et al. 2017; Ansari et al. 2014).

The combined efforts of enzymatic and nonenzymatic antioxidants are crucial in controlling ROS within plant cells. By working in concert, these molecules protect plants from oxidative stress induced by harsh environmental conditions. Numerous studies have demonstrated that plants with robust antioxidant defenses are more adept at surviving challenges such as heat and heavy metal exposure (Thounaojam et al. 2012; Wilson et al. 2014). For instance, heat-tolerant varieties of mustard and wheat maintain elevated levels of antioxidants, helping them stay healthy under stress (Almeselmani et al. 2009; Wilson et al. 2014). Similarly, rice and lentils with strong antioxidant systems show enhanced resilience, highlighting the essential role of these protective molecules in plant survival under difficult conditions (Almeselmani et al. 2006; Chugh et al. 2017).

3.5 Conclusion

In summary, the generation of ROS within plant cells under stress conditions is a critical factor influencing plant health and survival. While ROS serve essential roles in signaling and defense, their excessive accumulation can lead to oxidative damage, disrupting cellular integrity and function. Plants have evolved sophisticated antioxidant defense systems to counteract these effects, maintaining redox balance and ensuring continued growth under stress. Understanding the intricate dynamics of ROS production, detoxification, and signaling offers valuable insights for developing strategies to enhance plant resilience, ultimately contributing to more sustainable and productive agricultural practices in challenging environments.

References

Ahmad P, Jaleel CA, Salem MA, Nabi G, Sharma S (2010) Roles of enzymatic and nonenzymatic antioxidants in plants during abiotic stress. Crit Rev Biotechnol 30:161–175

Ali MB, Hahn EJ, Paek KY (2005) Effects of temperature on oxidative stress defense systems, lipid peroxidation and lipoxygenase activity in *Phalaenopsis*. Plant Physiol Biochem 43:213–223. https://doi.org/10.1016/j.plaphy.2005.01.007

Almeselmani M, Deshmukh PS, Sairam RK, Kushwaha SR, Singh TP (2006) Protective role of antioxidant enzymes under high temperature stress. Plant Sci 171:382–388

Almeselmani M, Deshmukh P, Sairam R (2009) High temperature stress tolerance in wheat genotypes: role of antioxidant defence enzymes. Acta Agron Hung 57:1–14

Anjum NA, Sofo A, Scopa A, Roychoudhury A, Gill SS, Iqbal M, Lukatkin AS, Pereira E, Duarte AC, Ahmad I (2015) Lipids and proteins—major targets of oxidative modifications in abiotic stressed plants. Environ Sci Pollut Res 22:4099–4121. https://doi.org/10.1007/s11356-014-3917-1

Anjum NA, Sharma P, Gill SS, Hasanuzzaman M, Khan EA, Kachhap K, Mohamed AA, Thangavel P, Devi GD, Vasudhevan P et al (2016) Catalase and ascorbate peroxidase—representative H2O2-detoxifying heme enzymes in plants. Environ Sci Pollut Res 23:19002–19029

Ansari MI, Hasan S, Jalil SU (2014) Leaf senescence and GABA shunt. Bioinformation 10:734–736
Ansari MI, Jalil SU, Ansari SA et al (2021) GABA shunt: a key player in mitigation of ROS during stress. Plant Growth Regul 94:131–149
Apel K, Hirt H (2004) Reactive oxygen species: metabolism, oxidative stress, and signal transduction. Annu Rev Plant Biol 55:373–99. https://doi.org/10.1146/annurev.arplant.55.031903.141701. PMID: 15377225.
Baker A, Graham IA, Holdsworth M, Smith SM, Theodoulou FL (2006) Chewing the fat: β-oxidation in signalling and development. Trends Plant Sci 11:124–132. https://doi.org/10.1016/j.tplants.2006.01.005
Bartoli CG, Gómez F, Martínez DE, Guiamet JJ (2004) Mitochondria are the main target for oxidative damage in leaves of wheat (*Triticum aestivum* L.). J Exp Bot 55:1663–1669. https://doi.org/10.1093/jxb/erh199
Bartoli CG, Buet A, Grozeff GG, Galatro A, Simontacchi M (2017) Ascorbate-glutathione cycle and abiotic stress tolerance in plants. In: Hossain MA, Munne-Bosch S, Burritt DJ, Diaz-Vivancos P, Fujita M, Lorence A (eds) Ascorbic acid in plant growth, development and stress tolerance. Springer, Cham, pp 177–200
Benkő P, Gémes K, Fehér A (2022) Polyamine oxidase-generated reactive oxygen species in plant development and adaptation: the polyamine oxidase—NADPH oxidase nexus. Antioxidants 11:2488. https://doi.org/10.3390/antiox11122488
Berglund GI, Carlsson GH, Smith AT, Szöke H, Henriksen A, Hajdu J (2002) The catalytic pathway of horseradish peroxidase at high resolution. Nature 417:463–468. https://doi.org/10.1038/417463a
Bhuyan MB, Hasanuzzaman M, Parvin K, Mohsin SM, Al Mahmud J, Nahar K, Fujita M (2020) Nitric oxide and hydrogen sulfide: two intimate collaborators regulating plant defense against abiotic stress. Plant Growth Regul 90:409–424
Bindschedler LV et al (2006) Peroxidase-dependent apoplastic oxidative burst in *Arabidopsis* required for pathogen resistance. Plant J 47:851–863
Bose J, Rodrigo-Moreno A, Shabala S (2014) ROS homeostasis in halophytes in the context of salinity stress tolerance. J Exp Bot 65:1241–1257
Bolwell GP, Davies DR, Gerrish C, Auh CK, Murphy TM (1998) Comparative biochemistry of the oxidative burst produced by rose and french bean cells reveals two distinct mechanisms. Plant Phys 116:1379–1385
Bolwell GP, Blee KA, Butt VS, Davies DR, Gardner SL, Gerrish C, Minibayeva F, Rowntree EG, Wojtaszek P (1999) Recent advances in understanding the origin of the apoplastic oxidative burst in plant cells. Free Radic Res 31(Suppl):S137–S145
Chapman JM, Muhlemann JK, Gayomba SR, Muday GK (2019) RBOH-dependent ROS synthesis and ROS scavenging by plant specialized metabolites to modulate plant development and stress responses. Chem Res Toxicol 32:370–396
Chen N, Teng XL, Xiao XG (2017) Subcellular localization of a plant catalase-phenol oxidase, AcCATPO, from *Amaranthus* and identification of a non-canonical peroxisome targeting signal. Front Plant Sci 8:1345
Chen D, Shao Q, Yin L, Younis A, Zheng B (2018) Polyamine function in plants: metabolism, regulation on development, and roles in abiotic stress responses. Front Plant Sci 9:1945
Chmielowska-Bak J, Izbiańska K, Ekner-Grzyb A, Bayar M, Deckert J (2018) Cadmium stress leads to rapid increase in RNA oxidative modifications in soybean seedlings. Front Plant Sci 8:2219
Choi HW, Kim YJ, Lee SC, Hong JK, Hwang BK (2007) Hydrogen peroxide generation by the pepper extracellular peroxidase CaPO2 activates local and systemic cell death and defense response to bacterial pathogens. Plant Physiol 145:890–904
Chugh P, Kaur J, Grewal SK, Singh S, Agrawal SK (2017) Upregulation of superoxide dismutase and catalase along with proline accumulation mediates heat tolerance in lentil (*Lens culinaris* Medik.) genotypes during reproductive stage. Indian J Agric Biochem 30:195–199
Cilento G, Adam W (1995) From free radicals to electronically excited species. Free Radic Biol Med 19:103–114. https://doi.org/10.1016/0891-5849(95)00002-f

Corpas FJ, Del Río LA, Palma JM (2019) Plant peroxisomes at the crossroad of NO and H_2O_2 metabolism. J Integr Plant Biol 61:803–816

Cowley T, Walters DR (2002) Polyamine metabolism in an incompatible interaction between barley and the powdery mildew fungus, *Blumeria graminis* f. sp. *hordei*. J Phytopathol 150:581–586. https://doi.org/10.1046/j.1439-0434.2002.00816.x

Cruz DCM (2008) Drought stress and reactive oxygen species: production, scavenging and signaling. Plant Signal Behav 3:156–165. https://doi.org/10.4161/psb.3.3.5536

Das K, Roychoudhury A (2014) Reactive oxygen species (ROS) and response of antioxidants as ROS-scavengers during environmental stress in plants. Front Environ Sci 2:53. https://doi.org/10.3389/fenvs.2014.00053

Davies MJ (2005) The oxidative environment and protein damage. Biochim Biophys Acta Proteins Proteom 1703:93–109. https://doi.org/10.1016/j.bbapap.2004.08.007

de Dios Alché J (2019) A concise appraisal of lipid oxidation and lipoxidation in higher plants. Redox Biol 23:101136. https://doi.org/10.1016/j.redox.2019.101136

del Río LA, Lopez-Huertas E (2016) ROS generation in peroxisomes and its role in cell signaling. Plant Cell Physiol 57:1364–1376. https://doi.org/10.1093/pcp/pcw076

del Río LA, Pastori GM, Palma JM, Sandalio LM, Sevilla F, Corpas FJ, Jimenez A, Lopez-Huertas E, Hernández JA (1998) The activated oxygen role of peroxisomes in senescence. Plant Physiol 116:1195–1200. https://doi.org/10.1104/pp.116.4.1195

Dietz KJ (2011) Peroxiredoxins in plants and cyanobacteria. Antioxid Redox Signal 15:1129–1159

Dinakar C, Bartels D (2012) Light response, oxidative stress management and nucleic acid stability in closely related Linderniaceae species differing in desiccation tolerance. Planta 236:541–555

Evans MD, Dizdaroglu M, Cooke MS (2004) Oxidative DNA damage and disease: induction, repair and significance. Mutat Res Rev Mutat Res 567:1–61

Farvardin A, González-Hernández AI, Llorens E, García-Agustín P, Scalschi L, Vicedo B (2020) The apoplast: a key player in plant survival. Antioxidants 9:604

Fedorova GF, Trofimov AV, Vasil'ev RF, Veprintsev TL (2007) Peroxy-radical-mediated chemiluminescence: mechanistic diversity and fundamentals for antioxidant assay. ARKIVOC:163–215

Fedyaeva AV, Stepanov AV, Lyubushkina IV, Pobezhimova TP, Rikhvanov EG (2014) Heat shock induces production of reactive oxygen species and increases inner mitochondrial membrane potential in winter wheat cells. Biochemistry 79:1202–1210

Ferreira-Silva SL, Voigt EL, Silva EN, Maia JM, Aragão TCR, Silveira JAG (2012) Partial oxidative protection by enzymatic and non-enzymatic components in cashew leaves under high salinity. Biol Plant 56:172–176. https://doi.org/10.1007/s10535-012-0037-y

Fink SP, Reddy GR, Marnett LJ (1997) Mutagenicity in *Escherichia coli* of the major DNA adduct derived from the endogenous mutagen malondialdehyde. Proc Natl Acad Sci USA 94:8652–8657

Francoz E, Ranocha P, Nguyen-Kim H, Jamet E, Burlat V, Dunand C (2015) Roles of cell wall peroxidases in plant development. Phytochemistry 112:15–21. https://doi.org/10.1016/j.phytochem.2014.07.020

Gao C, Xing D, Li L, Zhang L (2008) Implication of reactive oxygen species and mitochondrial dysfunction in the early stages of plant programmed cell death induced by ultraviolet-C overexposure. Planta 227:755–767

Gelhaye E, Rouhier N, Navrot N, Jacquot JP (2005) The plant thioredoxin system. Cell Mol Life Sci 62:24–35

Gill SS, Tuteja N (2010) Reactive oxygen species and antioxidant machinery in abiotic stress tolerance in crop plants. Plant Physiol Biochem 48:909–930

Gui T, Gao D, Ding H, Yan X (2022) Identification of respiratory burst oxidase homolog (Rboh) family genes from *Pyropia yezoensis* and their correlation with archeospore release. Front Plant Sci 13:929299. https://doi.org/10.3389/fpls.2022.929299

Hameed A, Goher M, Iqbal N (2013) Drought induced programmed cell death and associated changes in antioxidants, proteases, and lipid peroxidation in wheat leaves. Biol Plant 57:370–374

Hasanuzzaman M, Bhuyan MHM, Zulfiqar F, Raza A, Mohsin SM, Mahmud JA, Fujita M, Fotopoulos V (2020) Reactive oxygen species and antioxidant defense in plants under abiotic stress: revisiting the crucial role of a universal defense regulator. Antioxidants 9:681

He M, He CQ, Ding NZ (2018) Abiotic stresses: general defenses of land plants and chances for engineering multi stress tolerance. Front Plant Sci 9:1771

Hernandez JA, Corpas FJ, Gomez M, del Rio LA, Sevilla F (1993) Salt-induced oxidative stress mediated by activated oxygen species in pea leaf mitochondria. Physiol Plant 89:103–110

Hix LM, Lockwood SF, Bertram JS (2004) Bioactive carotenoids: potent antioxidants and regulators of gene expression. Redox Rep 9:181–191

Hollosy F (2002) Effects of ultraviolet radiation on plant cells. Micron 33:179–197

Holzmeister C, Gaupels F, Geerlof A, Sarioglu H, Sattler M, Durner J, Lindermayr C (2015) Differential inhibition of *Arabidopsis* superoxide dismutases by peroxynitrite-mediated tyrosine nitration. J Exp Bot 66:989–999

Hu WH, Song XS, Shi K, Xia XJ, Zhou YH, Yu JQ (2008) Changes in electron transport, superoxide dismutase and ascorbate peroxidase isoenzymes in chloroplasts and mitochondria of cucumber leaves as influenced by chilling. Photosynthetica 46:581–588

Huang CH, Kuo WY, Weiss C, Jinn TL (2012) Copper chaperone-dependent and -independent activation of three copper-zinc superoxide dismutase homologs localized in different cellular compartments in *Arabidopsis*. Plant Physiol 158:737–746

Huang L, Liu Y, Wang X, Jiang C, Zhao Y, Lu M, Zhang J (2022) Peroxisome-mediated reactive oxygen species signals modulate programmed cell death in plants. Int J Mol Sci 23:10087. https://doi.org/10.3390/ijms231710087

Jalil SU, Ansari MI (2020) Physiological role of gamma-aminobutyric acid in salt stress tolerance. In: Hasanuzzaman M, Tanveer M (eds) Salt and drought stress tolerance in plants. Springer Nature, Cham, pp 337–350

Jalil SU, Ahmad I, Ansari MI (2017) Functional loss of GABA transaminase (GABA-T) expressed early leaf senescence under various stress conditions in *Arabidopsis thaliana*. Curr Plant Biol 9:11–22

Jalil SU, Khan MAA, Ansari MI (2019) Tremendous role of GABA shunt metabolic pathway in plant system. J Biol Chem Res 36:188–196

Janků M, Luhová L, Petřivalský M (2019) On the origin and fate of reactive oxygen species in plant cell compartments. Antioxidants 8:4. https://doi.org/10.3390/antiox8040105

Juan CA, Manuel J, Plou FJ (2021) The chemistry of reactive oxygen species (ROS) revisited: outlining their role in biological macromolecules (DNA, lipids and proteins) and induced pathologies. Int J Mol Sci 22:4642

Kadota Y, Shirasu K, Zipfel C (2015) Regulation of the NADPH oxidase RBOHD during plant immunity. Plant Cell Physiol 56:1472–1480

Kärkönen A, Kuchitsu K (2015) Reactive oxygen species in cell wall metabolism and development in plants. Phytochemistry 112:22–32. https://doi.org/10.1016/j.phytochem.2014.09.016

Karuppanapandian T, Kim W (2013) Cobalt-induced oxidative stress causes growth inhibition associated with enhanced lipid peroxidation and activates antioxidant responses in Indian mustard (*Brassica juncea* L.) leaves. Acta Physiol Plant 35:2429–2443. https://doi.org/10.1007/s11738-013-1277-y

Karuppanapandian T, Moon JC, Kim C, Manoharan K, Kim W (2011) Reactive oxygen species in plants: their generation, signal transduction, and scavenging mechanisms. Aust J Crop Sci 5:709

Khan MIR, Jalil SU, Chopra P, Chhillar H, Ferrante A, Khan NA, Ansari MI (2021) Role of GABA in plant growth, development and senescence. Plant Gene 26:100283

Khorobrykh S, Havurinne V, Mattila H, Tyystjärvi E (2020) Oxygen and ROS in photosynthesis. Plants 9:91. https://doi.org/10.3390/plants9010091

Kim MJ, Ciani S, Schachtman DP (2010) A peroxidase contributes to ROS production during *Arabidopsis* root response to potassium deficiency. Mol Plant 3:420–427

Kohli SK, Khanna K, Bhardwaj R, Abd_Allah EF, Ahmad P, Corpas FJ (2019) Assessment of subcellular ROS and NO metabolism in higher plants: multifunctional signaling molecules. Antioxidants 8:641

Kosova K, Vitamvas P, Urban MO, Prasil IT, Renaut J (2018) Plant abiotic stress proteomics: the major factors determining alterations in cellular proteome. Front Plant Sci 9:122. https://doi.org/10.3389/fpls.2018.00122

Krieger-Liszkay A, Shimakawa G (2022) Regulation of the generation of reactive oxygen species during photosynthetic electron transport. Biochem Soc Trans 50:1025–1034. https://doi.org/10.1042/BST20211246

Kumar V, Khare T, Sharma M, Wani SH (2017) ROS-induced signaling and gene expression in crops under salinity stress. In: Khan MIR, Khan NA (eds) Reactive oxygen species and antioxidant systems in plants: role and regulation under abiotic stress. Springer, Singapore, pp 159–184

Kumari S, Agrawal M, Singh A (2015) Effects of ambient and elevated CO2 and ozone on physiological characteristics, antioxidative defense system and metabolites of potato in relation to ozone flux. Environ Exp Bot 109:276–287. https://doi.org/10.1016/j.envexpbot.2014.06.015

Langebartels C, Wohlgemuth H, Kschieschan S, Grün S, Sandermann H (2002) Oxidative burst and cell death in ozone-exposed plants. Plant Physiol Biochem 40:567–576

Lazzarotto F, Turchetto-Zolet AC, Margis-Pinheiro M (2015) Revisiting the non-animal peroxidase superfamily. Trends Plant Sci 20:807–813. https://doi.org/10.1016/j.tplants.2015.08.005

Lin CC, Kao CH (2001) Cell wall peroxidase activity, hydrogen peroxide level and NaCl-inhibited root growth of rice seedlings. Plant Soil 230:135–143

Llorente F, Lopez-Cobollo RM, Catala R, Martinez-Zapater JM, Salinas J (2002) A novel cold-inducible gene from Arabidopsis, RCI3, encodes a peroxidase that constitutes a component for stress tolerance. Plant J 32:13–24

Lüthje S, Martinez-Cortes T (2018) Membrane-bound class III peroxidases: unexpected enzymes with exciting functions. Int J Mol Sci 19:2876. https://doi.org/10.3390/ijms19102876

Ma X, Wang W, Bittner F, Schmidt N, Berkey R, Zhang L, King H, Zhang Y, Feng J, Wen Y et al (2016) Dual and opposing roles of xanthine dehydrogenase in defense-associated reactive oxygen species metabolism in Arabidopsis. Plant Cell 28:1108–1126. https://doi.org/10.1105/tpc.15.00880

Marina M, Maiale SJ, Rossi FR, Romero MF, Rivas EI, Gárriz A, Ruiz OA, Pieckenstain FL (2008) Apoplastic polyamine oxidation plays different roles in local responses of tobacco to infection by the necrotrophic fungus Sclerotinia sclerotiorum and the biotrophic bacterium Pseudomonas viridiflava. Plant Physiol 147:2164–2178. https://doi.org/10.1104/pp.108.122614

Marino D, Dunand C, Puppo A, Pauly N (2012) A burst of plant NADPH oxidases. Trends Plant Sci 17:9–15. https://doi.org/10.1016/j.tplants.2011.10.001

Martinez C, Montillet JL, Bresson E, Agnel JP, Dai GH, Daniel JF, Geiger JP, Nicole M (1998) Apoplastic peroxidase generates superoxide anions in cells of cotton cotyledons undergoing the hypersensitive reaction to Xanthomonas campestris pv. malvacearum race 18. Mol Plant-Microbe Interact 11:1038–1047

Martinez V, Mestre TC, Rubio F, Girones-Vilaplana A, Moreno DA, Mittler R, Rivero RM (2016) Accumulation of flavonols over hydroxycinnamic acids favors oxidative damage protection under abiotic stress. Front Plant Sci 7:838. https://doi.org/10.3389/fpls.2016.00838

Maurya AK (2020) Oxidative stress in crop plants. In: Hasanuzzaman M (ed) Agronomic crops: stress responses and tolerance. Springer, Singapore, pp 349–380

Mellidou I, Karamanoli K, Beris D, Haralampidis K, Constantinidou HI, Roubelakis-Angelakis KA (2017) Underexpression of apoplastic polyamine oxidase improves thermotolerance in Nicotiana tabacum. J Plant Physiol 218:171–174

Meriga B, Reddy BK, Rao KR, Reddy LA, Kishor PK (2004) Aluminium-induced production of oxygen radicals, lipid peroxidation and DNA damage in seedlings of rice (*Oryza sativa*). J Plant Physiol 161:63–68. https://doi.org/10.1078/0176-1617-01156

Miller GAD, Suzuki N, Ciftci-Yilmaz S, Mittler R (2010) Reactive oxygen species homeostasis and signalling during drought and salinity stresses. Plant Cell Environ 33:453–467

Mitsuya Y, Takahashi Y, Berberich T, Miyazaki A, Matsumura H, Takahashi H, Terauchi R, Kusano T (2009) Spermine signaling plays a significant role in the defense response of Arabidopsis thaliana to cucumber mosaic virus. J Plant Physiol 166:626–643

Mittler R (2002) Oxidative stress, antioxidants and stress tolerance. Trends Plant Sci 7:405–410

Mittova V, Tal M, Volokita M, Guy M (2003) Up-regulation of the leaf mitochondrial and peroxisomal antioxidative systems in response to salt-induced oxidative stress in the wild salt-tolerant tomato species Lycopersicon pennellii. Plant Cell Environ 26:845–856. https://doi.org/10.1046/j.1365-3040.2003.01016.x

Miyamoto S, Martinez GR, Medeiros MHG, Di Mascio P (2014) Singlet molecular oxygen generated by biological hydroperoxides. J Photochem Photobiol B Biol 139:24–33

Moller IM, Jensen PE, Hansson A (2007) Oxidative modifications to cellular components in plants. Annu Rev Plant Biol 58:459–481. https://doi.org/10.1146/annurev.arplant.58.032806.103946

Moschou PN, Delis ID, Paschalidis KA, Roubelakis-Angelakis KA (2008a) Transgenic tobacco plants overexpressing polyamine oxidase are not able to cope with oxidative burst generated by abiotic factors. Physiol Plant 133:140–156

Moschou PN, Paschalidis KA, Delis ID, Andriopoulou AH, Lagiotis GD, Yakoumakis DI, Roubelakis-Angelakis KA (2008b) Spermidine exodus and oxidation in the apoplast induced by abiotic stress is responsible for H2O2 signatures that direct tolerance responses in tobacco. Plant Cell 20:1708–1724

Moschou PN, Sarris PF, Skandalis N, Andriopoulou AH, Paschalidis KA, Panopoulos NJ, Roubelakis-Angelakis KA (2009) Engineered polyamine catabolism preinduces tolerance of tobacco to bacteria and oomycetes. Plant Physiol 149:1970–1981. https://doi.org/10.1104/pp.108.134932

Moschou PN, Wu J, Cona A, Tavladoraki P, Angelini R, Roubelakis-Angelakis KA (2012) The polyamines and their catabolic products are significant players in the turnover of nitrogenous molecules in plants. J Exp Bot 63:695–709

Najafpour MM, Renger G, Hołynska M, Moghaddam AN, Aro EM, Carpentier R et al (2016) Manganese compounds as water-oxidizing catalysts: from the natural water-oxidizing complex to nanosized manganese oxide structures. Chem Rev 116:2886–2936. https://doi.org/10.1021/acs.chemrev.5b00340

Nelson N, Junge W (2015) Structure and energy transfer in photosystems of oxygenic photosynthesis. Annu Rev Biochem 84:659–683. https://doi.org/10.1146/annurev-biochem-092914-041942

Nimse SB, Pal D (2015) Free radicals, natural antioxidants, and their reaction mechanisms. RSC Adv 5:27986–28006

Nyathi Y, Baker A (2006) Plant peroxisomes as a source of signalling molecules. Biochim Biophys Acta, Mol Cell Res 1763:1478–1495. https://doi.org/10.1016/j.bbamcr.2006.08.031

O'Brien JA, Daudi A, Finch P, Butt VS, Whitelegge JP, Souda P, Ausubel FM, Bolwell GP (2012) A peroxidase-dependent apoplastic oxidative burst in cultured Arabidopsis cells functions in MAMP-elicited defense. Plant Physiol 158:2013–2027. https://doi.org/10.1104/pp.111.190140

Oracz K, Bouteau HEM, Farrant JM, Cooper K, Belghazi M, Job C, Job D, Corbineau F, Bailly C (2007) ROS production and protein oxidation as a novel mechanism for seed dormancy alleviation. Plant J 50:452–465. https://doi.org/10.1111/j.1365-313X.2007.03063.x

Ortega-Galisteo AP, Rodriguez-Serrano M, Pazmino DM, Gupta DK, Sandalio LM, Romero-Puertas MC (2012) S-Nitrosylated proteins in pea (Pisum sativum L.) leaf peroxisomes: changes under abiotic stress. J Exp Bot 63:2089–2103. https://doi.org/10.1093/jxb/err414

Pál M, Janda T (2017) Role of polyamine metabolism in plant pathogen interactions. J Plant Sci Phytopathol 1:95–100

Pandey P, Irulappan V, Bagavathiannan MV, Senthil-Kumar M (2017) Impact of combined abiotic and biotic stresses on plant growth and avenues for crop improvement by exploiting physio-morphological traits. Front Plant Sci 8:537

Petrov V, Hille J, Mueller-Roeber B, Gechev TS (2015) ROS-mediated abiotic stress-induced programmed cell death in plants. Front Plant Sci 6:69

Pnueli L, Liang H, Rozenberg M, Mittler R (2003) Growth suppression, altered stomatal responses, and augmented induction of heat shock proteins in cytosolic ascorbate peroxidase (Apx1)-deficient Arabidopsis plants. Plant J 34:187–203

Pospíšil P (2016) Production of reactive oxygen species by photosystem II as a response to light and temperature stress. Front Plant Sci 7:1950. https://doi.org/10.3389/fpls.2016.01950

Pospíšil P, Prasad A, Rác M (2014) Role of reactive oxygen species in ultra-weak photon emission in biological systems. J Photochem Photobiol B Biol 139:11–23

Prakash Sanyal R, Samant A, Prashar V, Sharan Misra H, Saini A (2018) Biochemical and functional characterization of OsCSD3, a novel CuZn superoxide dismutase from rice. Biochem J 475:3105–3121

Prasad A, Sedlářová M, Kale RS, Pospíšil P (2017) Lipoxygenase in singlet oxygen generation as a response to wounding: in vivo imaging in Arabidopsis thaliana. Sci Rep 7:9831. https://doi.org/10.1038/s41598-017-09758-1

Qiu QS, Huber JL, Booker FL, Jain V, Leakey AD, Fiscus EL, Yau PM, Donald RO, Huber SC (2008) Increased protein carbonylation in leaves of Arabidopsis and soybean in response to elevated [CO2]. Photosynth Res 97:155. https://doi.org/10.1007/s11120-008-9310-5

Rao AC, Reddy AR (2008) Glutathione reductase: a putative redox regulatory system in plant cells. In: Khan NA, Singh S, Umar S (eds) Sulfur assimilation and abiotic stress in plants. Springer, Berlin/Heidelberg, pp 111–147

Roldan-Arjona T, Ariza RR (2009) Repair and tolerance of oxidative DNA damage in plants. Mutat Res Rev Mutat Res 681:169–179

Romero-Puertas MC, McCarthy I, Sandalio LM, Palma JM, Corpas FJ, Gomez M, Del Rio LA (1999) Cadmium toxicity and oxidative metabolism of pea leaf peroxisomes. Free Radic Res 31:25–31. https://doi.org/10.1080/10715769900301281

Romero-Puertas MC, Palma JM, Gómez M, Del Rio LA, Sandalio LM (2002) Cadmium causes the oxidative modification of proteins in pea plants. Plant Cell Environ 25:677–686. https://doi.org/10.1046/j.1365-3040.2002.00850.x

Roychowdhury R, Khan MH, Choudhury S (2019) Physiological and molecular responses for metalloid stress in rice—a comprehensive overview. In: Hasanuzzaman M, Fujita M, Nahar K, Biswas JK (eds) Advances in rice research for abiotic stress tolerance. Woodhead Publishing, Cambridge, MA, pp 341–369

Sachdev S, Ansari SA, Ansari MI, Fujita M, Hasanuzzaman M (2021) Abiotic stress and reactive oxygen species: generation, signaling, and defense mechanisms. Antioxidants 10:277

Schertl P, Braun HP (2014) Respiratory electron transfer pathways in plant mitochondria. Front Plant Sci 5:163

Smirnoff N, Wheeler GL (2000) Ascorbic acid in plants: biosynthesis and function. Crit Rev Biochem Mol Biol 35(4):291–314. https://doi.org/10.1080/10409230008984166. PMID: 11005203.

Shah K, Chaturvedi V, Gupta S (2019) Climate change and abiotic stress-induced oxidative burst in rice. In: Hasanuzzaman M, Fujita M, Nahar K, Biswas JK (eds) Advances in rice research for abiotic stress tolerance. Woodhead Publishing, Cambridge, MA, pp 505–535

Sharma P, Jha AB, Dubey RS, Pessarakli M (2012) Reactive oxygen species, oxidative damage, and antioxidative defense mechanism in plants under stressful conditions. J Bot 2012:1. https://doi.org/10.1155/2012/217037

Shigeto J, Tsutsumi Y (2016) Diverse functions and reactions of class III peroxidases. New Phytol 209:1395–1402. https://doi.org/10.1111/nph.13738

Soares C, Carvalho ME, Azevedo RA, Fidalgo F (2019) Plants facing oxidative challenges—a little help from the antioxidant networks. Environ Exp Bot 161:4–25

Srivastava S, Brychkova G, Yarmolinsky D, Soltabayeva A, Samani T, Sagi M (2017) Aldehyde oxidase 4 plays a critical role in delaying silique senescence by catalyzing aldehyde detoxification. Plant Physiol 173:1977–1997. https://doi.org/10.1104/pp.16.01939

Su T, Wang P, Li H, Zhao Y, Lu Y, Dai P, Ren T, Wang X, Li X, Shao Q et al (2018) The Arabidopsis catalase triple mutant reveals important roles of catalases and peroxisome-derived signaling in plant development. J Integr Plant Biol 60:591–607

Suga M, Akita F, Hirata K, Ueno G, Murakami H, Nakajima Y et al (2015) Native structure of photosystem II at 1.95 angstrom resolution viewed by femtosecond X-ray pulses. Nature 517:99–103. https://doi.org/10.1038/nature13991

Suo J, Zhao Q, David L, Chen S, Dai S (2017) Salinity response in chloroplasts: insights from gene characterization. Int J Mol Sci 18:1011. https://doi.org/10.3390/ijms18051011

Survila M, Davidsson PR, Pennanen V, Kariola T, Broberg M, Sipari N, Heino P, Palva ET (2016) Peroxidase-generated apoplastic ROS impair cuticle integrity and contribute to DAMP-elicited defenses. Front Plant Sci 7:1945. https://doi.org/10.3389/fpls.2016.01945

Suzuki N, Miller G, Morales J, Shulaev V, Torres MA, Mittler R (2011) Respiratory burst oxidases: the engines of ROS signaling. Curr Opin Plant Biol 14:691–699. https://doi.org/10.1016/j.pbi.2011.07.014

Sweetlove LJ, Møller IM (2009) Oxidation of proteins in plants—mechanisms and consequences. Adv Bot Res 52:1–23

Szalai G, Kellős T, Galiba G, Kocsy G (2009) Glutathione as an antioxidant and regulatory molecule in plants under abiotic stress conditions. J Plant Growth Regul 28:66–80

Takács Z, Poór P, Tari I (2016) Comparison of polyamine metabolism in tomato plants exposed to different concentrations of salicylic acid under light or dark conditions. Plant Physiol Biochem 108:266–278

Talukdar D (2013) Arsenic-induced oxidative stress in the common bean legume, Phaseolus vulgaris L. seedlings and its amelioration by exogenous nitric oxide. Physiol Mol Biol Plants 19:69–79. https://doi.org/10.1007/s12298-012-0140-8

Tavladoraki P, Cona A, Federico R, Tempera G, Viceconte N, Saccoccio S, Battaglia V, Toninello A, Agostinelli E (2012) Polyamine catabolism: target for antiproliferative therapies in animals and stress tolerance strategies in plants. Amino Acids 42:411–426

Tavladoraki P, Cona A, Angelini R (2016) Copper-containing amine oxidases and FAD-dependent polyamine oxidases are key players in plant tissue differentiation and organ development. Front Plant Sci 7:824

Thounaojam TC, Panda P, Mazumdar P, Kumar D, Sharma GD, Sahoo L, Sanjib P (2012) Excess copper-induced oxidative stress and response of antioxidants in rice. Plant Physiol Biochem 53:33–39

Torres MA, Dangl JL (2005) Functions of the respiratory burst oxidase in biotic interactions, abiotic stress and development. Curr Opin Plant Biol 8:397–403. https://doi.org/10.1016/j.pbi.2005.05.014

Turrens JF (2003) Mitochondrial formation of reactive oxygen species. J Physiol 552:335–344

Uarrota VG, Stefen DLV, Leolato LS, Gindri DM, Nerling D (2018) Revisiting carotenoids and their role in plant stress responses: from biosynthesis to plant signaling mechanisms during stress. In: Gupta DK, Palma JM, Corpas FJ (eds) Antioxidants and antioxidant enzymes in higher plants. Springer Nature, Cham, pp 207–232

Veitch NC (2004) Structural determinants of plant peroxidase function. Phytochem Rev 3:3–18

Vogt L, Vinyard DJ, Khan S, Brudvig GW (2015) Oxygen-evolving complex of Photosystem II: an analysis of second-shell residues and hydrogen-bonding networks. Curr Opin Chem Biol 25:152–158. https://doi.org/10.1016/j.cbpa.2014.12.040

Voss I, Sunil B, Scheibe R, Raghavendra AS (2013) Emerging concept for the role of photorespiration as an important part of abiotic stress response. Plant Biol 15:713–722. https://doi.org/10.1111/j.1438-8677.2012.00710.x

Walters DR (2003) Polyamines and plant disease. Phytochemistry 64:97–107. https://doi.org/10.1016/S0031-9422(03)00329-7

Wang W, Liu JH (2016) CsPAO4 of Citrus sinensis functions in polyamine terminal catabolism and inhibits plant growth under salt stress. Sci Rep 6:31384. https://doi.org/10.1038/srep31384

Wang F, Chen ZH, Liu X, Colmer TD, Shabala L, Salih A, Zhao M, Shabala S (2017) Revealing the roles of GORK channels and NADPH oxidase in acclimation to hypoxia in Arabidopsis. J Exp Bot 68:3191–3204

Wang W, Paschalidis K, Feng JC, Song J, Liu JH (2019a) Polyamine catabolism in plants: a universal process with diverse functions. Front Plant Sci 10:561

Wang Y, Ye X, Yang K, Shi Z, Wang N, Yang L, Chen J (2019b) Characterization, expression, and functional analysis of polyamine oxidases and their role in selenium-induced hydrogen peroxide production in Brassica rapa. J Sci Food Agric 99:4082–4093

Wang RY, He F, Ning YS, Wang GL (2020) Fine-tuning of Rboh-mediated ROS signaling in plant immunity. Trends Plant Sci 25:1060–1062. https://doi.org/10.1016/j.tplants.2020.08.001

Welinder KG, Justesen AF, Kjaersgard IV, Jensen RB, Rasmussen SK, Jespersen HM, Duroux L (2002) Structural diversity and transcription of class III peroxidases from Arabidopsis thaliana. Eur J Biochem 269:6063–6081

Wilson RA, Sangha MK, Banga SS, Atwal AK, Gupta S (2014) Heat stress tolerance in relation to oxidative stress and antioxidants in Brassica juncea. J Environ Biol 35:383–387

Wongshaya P, Chayjarung P, Tothong C, Pilaisangsuree V, Somboon T, Kongbangkerd A, Limmongkon A (2020) Effect of light and mechanical stress in combination with chemical elicitors on the production of stilbene compounds and defensive responses in peanut hairy root culture. Plant Physiol Biochem 157:93–104

Yamane K, Taniguchi M, Miyake H (2012) Salinity-induced subcellular accumulation of H2O2 in leaves of rice. Protoplasma 249:301–308. https://doi.org/10.1007/s00709-011-0280-7

Yergaliyev TM, Nurbekova Z, Mukiyanova G, Akbassova A, Sutula M, Zhangazin S, Bari A, Tleukulova Z, Shamekova M, Masalimov ZK et al (2016) The involvement of ROS-producing aldehyde oxidase in plant response to Tombusvirus infection. Plant Physiol Biochem 109:36–44. https://doi.org/10.1016/j.plaphy.2016.09.001

Yesbergenova Z, Yang G, Oron E, Soffer D, Fluhr R, Sagi M (2005) The plant Mo-hydroxylases aldehyde oxidase and xanthine dehydrogenase have distinct reactive oxygen species signatures and are induced by drought and abscisic acid. Plant J 42:862–876. https://doi.org/10.1111/j.1365-313X.2005.02422.x

Yoda H, Yamaguchi Y, Sano H (2003) Induction of hypersensitive cell death by hydrogen peroxide produced through polyamine degradation in tobacco plants. Plant Physiol 132:1973–1981. https://doi.org/10.1104/pp.103.024737

Yoda H, Hiroi Y, Sano H (2006) Polyamine oxidase is one of the key elements for oxidative burst to induce programmed cell death in tobacco cultured cells. Plant Physiol 142:193–206

Yu Y, Zhou W, Zhou K, Liu W, Liang X, Chen Y, Sun D, Lin X (2018) Polyamines modulate aluminum-induced oxidative stress differently by inducing or reducing H2O2 production in wheat. Chemosphere 212:645–653. https://doi.org/10.1016/j.chemosphere.2018.08.133

Yu L, Fan J, Xu C (2019a) Peroxisomal fatty acid beta-oxidation negatively impacts plant survival under salt stress. Plant Signal Behav 14:1561121. https://doi.org/10.1080/15592324.2018.1561121

Yu Z, Jia D, Liu T (2019b) Polyamine oxidases play various roles in plant development and abiotic stress tolerance. Plants 8:184

Zafra A, Castro AJ, de Dios AJ (2018) Identification of novel superoxide dismutase isoenzymes in the olive (Olea europaea L.) pollen. BMC Plant Biol 18:114

GABA Transporters and Their Function in Stress Tolerance

4

Abstract

GABA transporters (GATs) help mediate plant stress tolerance. Exploring their mechanisms of action, regulation, and noteworthy role in various stress conditions assumes great significance, both from academic and crop augmentation points of view. GATs are specialized membrane-bound proteins that facilitate the uptake and release of GABA across cellular compartments, ensuring its proper distribution and availability for metabolic and signaling functions. Under abiotic stress conditions, such as drought, salinity, and temperature extremes, the regulation and activity of GATs are essential for maintaining cellular homeostasis. These transporters, encoded by specific gene families, are highly responsive to environmental cues, enabling the plant to adapt to adverse conditions by modulating GABA levels. The chapter highlights the molecular characteristics of GABA transporters, their proton-coupled transport mechanisms, and their localization in different cellular membranes. Furthermore, this chapter also covers recent advancements in understanding GABA transporter regulation, including posttranslational modifications and transcriptional control under stress conditions.

Keywords

GABA transporters · Membrane-bound proteins · Metabolic regulation · Stress tolerance

S. U. Jalil et al., *GABA Signaling System and Abiotic Stress Tolerance in Plants*,
https://doi.org/10.1007/978-981-95-0437-4_4

Highlights

- Mechanisms of GABA transport: proton-coupled transport and bidirectional movement across cellular compartments.
- Role of GABA transporters in abiotic stress tolerance (drought, salinity, heat, and cold) and their impact on cellular homeostasis.
- Gene families encoding GABA transporters (GATs) and their expression patterns under various stress conditions.
- Regulation of GABA transporter activity through environmental cues, transcriptional regulation, and posttranslational modifications.
- Cross talk between GABA transport and signaling pathways involved in plant defense and adaptation.
- Potential applications in crop improvement: enhancing stress resilience through manipulation of GABA transporter expression or function.

4.1 Introduction

GABA performs multifaceted roles across various biological systems (Jalil et al. 2016; Ansari et al. 2021). In plants, GABA is not merely a neurotransmitter as in animals but a pivotal component in regulating physiological processes, stress responses, and signaling (Bown and Shelp 1997; Fromm 2020; Jalil and Ansari 2020). It is synthesized through the decarboxylation of glutamate, and its production is tightly linked to plant stress adaptation mechanisms (Bown and Shelp 2020).

GABA's significance in plant physiology extends beyond its role in stress response; it influences growth, development, and metabolism. GABA acts as a signaling molecule, modulating various stress pathways and maintaining cellular homeostasis (Fait et al. 2008; Khan et al. 2021). The intricate balance of GABA within plant systems is crucial for coping with environmental challenges, such as drought, salinity, and extreme temperatures (Jalil et al. 2017, 2019a, b).

GABA transporters (GATs) facilitate the movement of GABA across cellular membranes, being integral to the regulation of GABA levels and its subsequent physiological effects (Yuan et al. 2023). These transporters help modulate intracellular GABA concentrations, thereby influencing stress tolerance, metabolic adjustments, and overall plant resilience (Breitkreuz et al. 1999; Batushansky et al. 2015). By managing the distribution and uptake of GABA, these transporters play a central role in how plants perceive and respond to stress.

The chapter aims to delve into the function of GATs in stress tolerance. We will explore the molecular mechanisms underlying their activity, their impact on stress-signaling pathways, and their contribution to enhancing plant resilience under adverse conditions. By examining recent advancements and current understanding in this field, we seek to provide a comprehensive overview of how GATs support plant adaptation to environmental stressors.

4.2 GABA Metabolism in Plants

4.2.1 Biosynthesis

GABA metabolism, encompassing both its biosynthesis and catabolism, is integral to plant growth, development, and stress response. GABA stands out as a versatile and dynamic player. The biosynthesis of GABA primarily occurs through the GABA shunt, a short pathway bypassing two steps of the Tricarboxylic Acid cycle (TCA) (Ansari et al. 2014). This pathway is a crucial part of plant metabolism, linking nitrogen and carbon metabolism and contributing to the plant's stress response mechanisms (Bouche and Fromm 2004).

The key enzyme involved in GABA biosynthesis is Glutamate Decarboxylase (GAD), which catalyzes the decarboxylation of glutamate to produce GABA and CO_2. This reaction occurs in the cytosol and is the first step of the GABA shunt. GAD activity is pivotal for maintaining GABA levels, particularly under stress conditions where rapid GABA synthesis is required (Baum et al. 1993). Besides the enzymatic pathway, GABA can also be synthesized from polyamines (PAs), such as spermidine and putrescine, and from proline, especially under oxidative stress conditions. This nonenzymatic route involves the conversion of these compounds into GABA through intermediate products like 4-aminobutanal and pyrroline (Signorelli et al. 2015).

GABA biosynthesis is tightly regulated by several mechanisms to ensure proper cellular function and response to environmental cues. GAD is regulated posttranslationally by Ca^{2+} through calmodulin (CaM). The binding of Ca^{2+}/CaM to GAD enhances its activity, thus increasing GABA synthesis (Baum et al. 1993). This regulation is crucial during stress conditions, where increased Ca^{2+} levels in the cytosol activate GAD, boosting GABA production to help the plant cope with stress. GAD activity is also modulated by cytosolic pH levels. Under stress conditions that cause cytosolic acidification, GAD activity elevates and accelerates GABA synthesis (Breitkreuz et al. 1999). Abiotic stresses, such as drought, salinity, and temperature extremes, can upregulate the expression of GAD genes, leading to increased GABA production. This helps the plant manage stress by providing a quick source of carbon and nitrogen for metabolic processes (Bouché et al. 2003).

4.2.2 Catabolism

GABA catabolism primarily occurs in the mitochondria through a series of well-coordinated enzymatic reactions. The first step in GABA catabolism involves the conversion of GABA to Succinic Semialdehyde (SSA) by the enzyme GABA Transaminase (GABA-T). This enzyme operates within the mitochondrial matrix and is crucial for maintaining GABA homeostasis by regulating its levels (Breitkreuz et al. 1999). SSA is then oxidized to succinate by Succinic Semialdehyde Dehydrogenase (SSADH), concurrently producing Nicotinamide Adenine Dinucleotide (NADH). Succinate enters the Krebs cycle, linking GABA catabolism to the central metabolic pathways and energy production of the cell (Breitkreuz et al. 1999).

SSA can also be converted into Gamma-Hydroxybutyrate (GHB) by the Succinic Semialdehyde Reductase (SSR) enzyme. This pathway provides an alternative route for SSA utilization, especially under conditions where SSADH activity is compromised (Hoover et al. 2007). GABA catabolism plays a significant role in maintaining cellular homeostasis, particularly under stress conditions. By feeding into the Krebs cycle, GABA catabolism provides essential intermediates like succinate, which are vital for energy production and metabolic balance (Ansari et al. 2005). The production of NADH during SSA oxidation by SSADH is crucial for maintaining the cellular redox balance. NADH is used in various metabolic reactions and helps in combating oxidative stress (Breitkreuz et al. 1999). Under stress conditions, GABA levels can rise significantly, providing a buffer against metabolic imbalances. The GABA shunt helps modulate carbon and nitrogen metabolism, supporting the overall stress resilience of plants (Bouche and Fromm 2004). GABA also acts as a signaling molecule, influencing the expression of stress-responsive genes and interacting with protein receptors to modulate cellular processes (Ramesh et al. 2015).

4.2.3 Transport

GABA transport across cellular membranes is essential for its function in different cellular compartments and tissues. Plants possess specific GATs that facilitate the uptake and release of GABA, ensuring its distribution and availability where needed. GATs are typically membrane-bound proteins that function through proton-coupled mechanisms, allowing for the active transport of GABA against concentration gradients (Ramesh et al. 2015). GATs are localized in various cellular membranes, including the plasma membrane, tonoplast (vacuolar membrane), and mitochondrial membrane. This distribution allows GABA to play a role in different cellular contexts, such as cytoplasmic pH regulation, vacuolar storage, and mitochondrial metabolism (Breitkreuz et al. 1999). These transporters are encoded by distinct gene families, and their expression can be highly regulated by developmental cues and environmental stresses.

4.3 Molecular Characterization of GATs

GABA stands out as a vital metabolite in plants, deeply embedded in various physiological *processes*, including stress response and development. The intricate movement of GABA across cellular membranes and into various organelles is orchestrated by specific transporters. Deciphering the molecular characteristics of these GATs sheds light on their functional roles and regulatory mechanisms within plant physiology.

Supplying GABA as the sole source of nitrogen in culture medium sustains *Arabidopsis thaliana* growth, demonstrating the presence of specific transporter(s) responsible for the uptake of exogenous GABA. The pivotal discovery of GATs in *Arabidopsis thaliana* has been seen as a significant milestone in plant biology (Breitkreuz et al. 1999). GABA transport within plants is a multifaceted process, encompassing its transfer between membranes and into various organelles. This intricate transport system is mediated by several transporters (Ramesh et al. 2015; Meyer et al. 2006; Dündar and Bush 2009; Snowden et al. 2015). These transporters, strategically positioned on the cell and organelle membranes, meticulously regulate the translocation of GABA into the intracellular space and across various organelles, underscoring their critical role in maintaining cellular homeostasis and responding to physiological demands (Meyer et al. 2006).

4.3.1 Mechanism of GABA Transport

GATs operate through several mechanisms to ensure the efficient movement of GABA within plant cells. These mechanisms include the following:

- *Proton-Coupled Transport*: Many GATs utilize proton gradients to drive the movement of GABA. This method involves coupling the transport of GABA with the movement of protons across the membrane. The electrochemical gradient created by proton pumps provides the necessary energy for GABA uptake or release (Meyer et al. 2006).
- *Anion Exchange*: Some transporters facilitate GABA transport through anion exchange mechanisms. For example, Aluminum-Activated Malate Transporter (ALMTs) can transport GABA by exchanging it with other anions, such as malate, across the plasma membrane (Ramesh et al. 2018).
- *Bidirectional Transport*: Certain transporters, like bidirectional amino acid transporters (BATs), can move GABA in both directions across membranes, allowing for the regulation of intracellular GABA levels in response to various physiological needs (Dündar and Bush 2009).

4.3.2 GATs: Localization and Tissue-Specific Expression

GATs are ubiquitously distributed in cells and perform locality-specific functions in signal transduction mechanisms.

4.3.2.1 Transmembrane GABA Transporters

Transmembrane GATs occur in plasma membranes that open bidirectional channels for various ions.

4.3.2.1.1 ALMTs

ALMTs comprise integral membrane proteins responsible for the bidirectional movement of ions and organic acids like malate, orchestrating key physiological processes within plants (Ramesh et al. 2018). The plant genome harbors a diverse array of 12 ALMT genes, each contributing uniquely to cellular functions (Ma et al. 2016). Among these, ALMT1 stands out for its significant role in the transport of GABA in plants (Fig. 4.1). Found across diverse species—including *Arabidopsis*, wheat, rice, and rape—ALMT1 has been most thoroughly examined in wheat (Duan et al. 2020). While its exact role in *Arabidopsis* remains a topic of ongoing research, ALMT1's ability to facilitate GABA transport is linked to its function as an anion channel. When activated by anions, ALMT1 generates an electrochemical gradient across the membrane, enhancing the efficiency of GABA transport (Gilliham and Tyerman 2016; Long et al. 2020). The interplay between ALMT1 and GABA is complex and gets influenced by various factors, including pH and the presence of

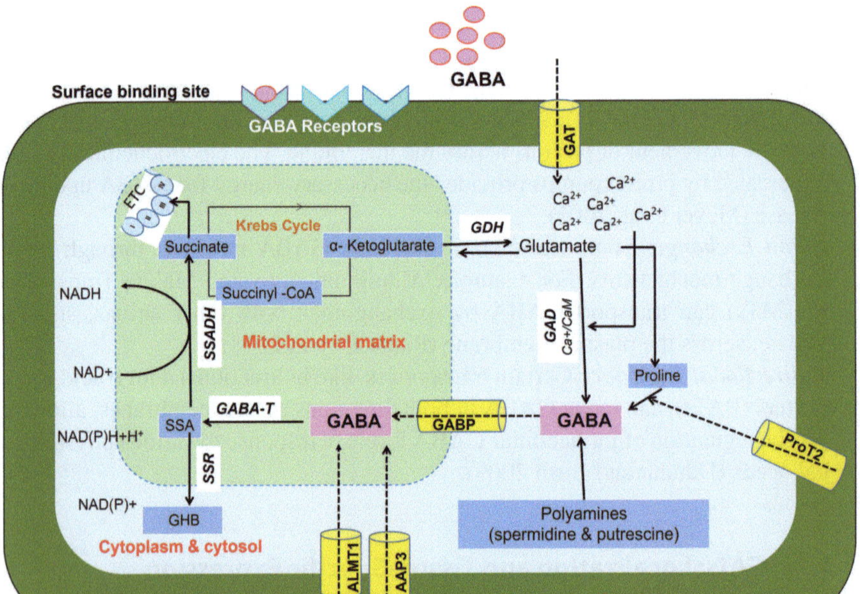

Fig. 4.1 Schematic representation of GABA transport and metabolism in plants. GABA is synthesized from glutamate via GAD under the influence of Ca^{2+}/CaM. GABA can be transported across the plasma membrane and tonoplast via various transporters, including GAT, GABA Permease (GABP), ALMT1, AAP3, and ProT2. Inside the mitochondria, GABA undergoes catabolism through the GABA shunt involving GABA-T, SSADH, and conversion to succinate, linking it to the Krebs cycle. Additionally, GABA interacts with PAs and contributes to the synthesis of proline and GHB, highlighting its central role in stress response and metabolic regulation

aluminum ions (Al^{3+}). In wheat under acidic conditions, aluminum enhances GABA release through TaALMT1, while an acidic apoplast further supports GABA uptake (Ramesh et al. 2018). However, the dynamics is not one-directional. Elevated GABA levels inhibit ALMT activity, suggesting that GABA may act as a negative regulator (Long et al. 2020). This inhibition is thought to occur through structural changes in ALMT1 induced by GABA, a mechanism reminiscent of GABA's effects on other proteins such as GabR and aspartate aminotransferase (Frezzini et al. 2019). In essence, the interaction between GABA and ALMTs, especially ALMT1, highlights their crucial role in regulating GABA transport and maintaining cellular ion homeostasis. This complex relationship underscores the sophisticated nature of plant cellular mechanisms and their adaptation to varying environmental conditions.

4.3.2.1.2 GATs

GATs are specialized proteins essential for the movement of GABA, a critical signaling molecule, across cell membranes (Meyer et al. 2006; Batushansky et al. 2015). These transporters belong to the amino acid–polyamine–organic acid gene family (Ma et al. 2016) and are represented in plants by four key members: GAT1, GAT2, GAT3, and GAT4. Among them, GAT1 is particularly notable for its role in transporting GABA from the apoplast (the extracellular space) into the cytoplasm (the interior of the cell) (Fig. 4.1). Distinct from ALMT1, another transporter involved in GABA movement, GAT1 operates independently of aluminum ions (Al^{3+}), which can inhibit ALMT1 activity (Long et al. 2020). GAT1 is a proton-coupled transporter, which means that it uses protons to facilitate the transport of GABA, and it exhibits a high affinity for GABA, enabling efficient transport even at low concentrations (Km $\approx$ 10 μM) (Meyer et al. 2006). GAT1 engages protons to effectively transport GABA in *Arabidopsis*, rice, and potato (Meyer et al. 2006). Fluorescent tagging/tracking experiment in tobacco confirms GAT1 localization in the cell membrane, which performs the transport of GABA across the membrane. Additionally, *Arabidopsis* plants with mutations in the GAT1 gene show impaired GABA uptake, further validating the importance of GAT1 in mediating GABA transport across the cell membrane. These findings underscore the crucial role of GAT1 in regulating GABA levels and its distinct functional characteristics compared to other transporters like ALMT1.

4.3.2.1.3 AAP3 and ProT2

AAP3 and ProT2 are integral membrane proteins involved in the transport of various amino acids, including GABA, across plant cell membranes, though they do so with less efficiency (Breitkreuz et al. 1999; Dinkeloo et al. 2018). In GAT1 mutants, other transporters can partially take over the role of GAT1. For instance, two transporters, namely AAP3 and ProT2, in yeast exist as alternatives to GAT1 for GABA transport (Young et al. 1999; Wipf et al. 2002).

AAP3, a member of the amino acid permease family, plays a significant role in the uptake of a range of amino acids into plant cells. While its primary function is to facilitate the transport of neutral and basic amino acids, AAP3 also mediates the uptake of GABA. In fact, AAP3 has a broad substrate range, which includes GABA, allowing it to contribute to the regulation of GABA levels within the plant cell (Breitkreuz et al. 1999). In *Arabidopsis*, AAP3 prefers other amino acids, like lysine, over GABA (Fischer et al. 2002). A proton gradient drives the process mediated by AAP3, which facilitates the active transport of amino acids, including GABA, into the cell.

ProT2, on the other hand, is specifically involved in the transport of proline and is known to influence proline homeostasis in plants. However, ProT2 also has the capability to transport GABA, though its primary role is in proline transport. ProT2 has a higher affinity for proline and glycine betaine than for GABA (Dinkeloo et al. 2018; Grallath et al. 2005). The dual function of ProT2 in transporting both proline and GABA suggests a degree of substrate versatility that is crucial for maintaining cellular amino acid balance, especially under stress conditions where both proline and GABA levels are known to fluctuate (Breitkreuz et al. 1999).

The interaction of AAP3 and ProT2 with GABA highlights the complex network of transporters responsible for regulating GABA homeostasis in plants. AAP3 facilitates GABA uptake by utilizing the proton gradient across the membrane, while ProT2, although primarily a proline transporter, contributes to GABA transport under certain physiological conditions (Fig. 4.1). Together, these transporters ensure that GABA levels are maintained appropriately, which is vital for various plant processes, including stress responses, signaling, and metabolism.

4.3.2.2 Cellular Organelle Membranes' GATs

4.3.2.2.1 BAT1

BAT1 is a critical protein found in plant mitochondria, known for its role in transporting various amino acids across the mitochondrial membrane. BAT1 operates bidirectionally, facilitating the movement of amino acids, such as arginine, glutamate, and lysine, in and out of mitochondria, thereby contributing to essential metabolic processes and cellular homeostasis (Wolf-Nicolas et al. 1998; Dündar and Bush 2009). Seven homologous BAT genes have been identified in *Arabidopsis*, potato, rice, and other crops (Ma et al. 2016; Tian et al. 2020), with BAT1 being known for transporting amino acids (Chen et al. 2001). Despite its broad amino acid transport capabilities, BAT1 itself does not directly transport GABA (Dündar and Bush 2009). Instead, a distinct mitochondrial protein, AtGABP, a splicing variant of BAT1, mediates the efficient transport of GABA (Michaeli et al. 2011). AtGABP distinguishes itself from other transporters like BAT1 that do not facilitate GABA movement. The role of BAT1 in GABA transport is thus indirect. While BAT1 manages a variety of amino acids essential for mitochondrial function and metabolism, AtGABP performs the dedicated task of GABA transport. The identification of AtGABP as a specialized GAT highlights a nuanced aspect of mitochondrial amino acid transport systems, where BAT1exhibits broader amino acid transport capabilities and also facilitates the more specific functions of other transporters like AtGABP.

4.3.2.2.2 CAATs

Cationic Amino acid Transporter (CAATs) are transport proteins located on the vacuolar membrane and are part of the APC gene family (Carter et al. 2004; Schmidt et al. 2007; Snowden et al. 2015). So far, nine homologous CAAT genes have been discovered in plants, with CAAT9 being primarily responsible for the bidirectional transport of GABA between the cytoplasm and the vacuole. CAAT9 has been identified in tomato, potato, *Arabidopsis*, and rice, and the role of its tomato counterpart, SlCAT9, in GABA transport has been experimentally confirmed (Whiteman et al. 2008; Snowden et al. 2015). GABA transport by SlCAT9 occurs via two main mechanisms: through the concentration gradient of the substrate and the driving force of the tonoplast proton pumps, which operate the charge exchange system. The vacuole, a specialized organelle, maintains its osmotic pressure, regardless of changes in amino acid content (Ma et al. 2013; Tsai et al. 2013). Consequently, transport processes are highly regulated. Research has also shown that SlCAT9 can transport glutamate and aspartate, suggesting its involvement in GABA metabolism and affecting the metabolic pathways of GABA in plants (Koike et al. 2013).

4.3.3 Role of GATs in Abiotic Stress Tolerance

GATs are essential for managing abiotic stress tolerance as they control the intracellular concentrations of GABA, which is a pivotal signaling molecule in stress responses. Under abiotic stress conditions, such as drought, salinity, and extreme temperatures, plants often experience an accumulation of GABA. This accumulation can be managed by GATs that facilitate the uptake and export of GABA, thus modulating its concentration within cells. By controlling GABA levels, these transporters help maintain cellular homeostasis, support stress-induced metabolic adjustments, and enhance the plant's resilience to withstand adverse conditions. Additionally, GATs are involved in the coordination of stress-signaling pathways, contributing to overall stress tolerance.

Amino acid transporters are pivotal in distributing essential nutrients across cellular membranes, thereby supporting plant growth, development, and defense mechanisms (Tegeder and Ward 2012). Particularly, membrane-bound GATs play a crucial role in facilitating the influx of Ca^{2+} into cells. GATs are activated by GABA, a metabolite often upregulated under stress conditions due to the expression of GABA shunt genes (Meyer et al. 2006). Serving as conduits for GABA, these transporters are integral to regulating carbon and nitrogen balance, pH homeostasis, nitrogen storage, and stress resilience (Shelp et al. 1999; Bouche and Fromm 2004). According to Bouche and Fromm (2004), the accretion of GABA impacts the activity of plant-specific anion transporters, thereby aiding root development and enhancing stress tolerance. In *Arabidopsis thaliana*, the GAT1 transporter exhibits a high affinity for GABA but is ineffective in transporting proline, thus highlighting its role in mediating carbon and nitrogen associations (Breitkreuz et al. 1999; Schwacke et al. 1999; Batushansky et al. 2015).

Overexpression of GATs can also upregulate the PeuGAT3 gene, which is involved in xylem tissue development in both *A. thaliana* and *Populus euphratica*. This upregulation enhances the lignin content in xylem tissues and increases proline accumulation in *P. euphratica* leaves, conferring improved tolerance to salt and drought stress (Bai et al. 2019). Further, proline transporters (ProTs) such as AtProT1 and AtProT2 also mediate GABA in *A. thaliana* and LeProT1 in *Solanum lycopersicum*, with AtProT2 being highly active under stress (Schwacke et al. 1999; Batushansky et al. 2015). GABA permease (AtGABP) in *A. thaliana* mediates the translocation of cytosolic GABA into mitochondria, where it is catabolized by GABA-T and SSADH to form succinate, which then enters the Krebs cycle and electron transport chain (Michaeli et al. 2011). In contrast, toxic SSA is transported from the mitochondria to the cytosol, where it is converted into GHB by SSR. Interestingly, In *A. thaliana*, knockout mutants for mitochondrial GAT exhibit reduced GABA consumption and elevated Krebs cycle activity (Michaeli et al. 2011). GABA also modulates the ALMT, which acts as a GABA receptor and influences membrane potentials, leading to immediate physiological responses in plants (Ramesh et al. 2015; Gilliham and Tyerman 2016). These findings suggest that GATs, either directly or indirectly, regulate GABA metabolism to mitigate Reactive Oxygen Species (ROS) and enhance stress tolerance.

GABA functions as a signaling molecule that orchestrates physiological and biochemical responses to improve drought resistance (Xu et al. 2021). It modulates stomatal aperture through interactions with ALMT proteins, such as ALMT9, which are crucial for stomatal movement. Xu et al. (2021) have demonstrated that ALMT9 is a key component of the GABA-signaling pathway, which reduces transpiration losses and improves water use efficiency, thereby enhancing drought resilience (Deng et al. 2022).

Under hypoxic conditions, GABA production involves proton consumption, which is essential for maintaining intracellular pH levels (Yang et al. 2015). GABA then joins a key metabolic process, helping keep important nutrient balance and reducing damage caused by low oxygen levels. Elevated GABA levels, induced by hypoxia, play a crucial role in restoring membrane potential and counteracting disturbances in cytoplasmic potassium (K^+) homeostasis and Ca^{2+} signaling, which are exacerbated by ROS (Wu Qi et al. 2021). This protective mechanism may involve the modulation by GABA of H^+-ATPase activity or its role in the GABA shunt and TCA cycle (Miyashita; Liao et al. 2017). GABA also alleviates excessive ROS accumulation and facilitates K^+ efflux by regulating espiratory Burst Oxidase Homolog (RBOH) genes and GABA-Activated K^+ channel (GOAK) channels, thus enhancing plant resilience to hypoxic stress (Wang et al. 2017).

GABA mediates plant responses to salt stress through a multifaceted approach. Increased GABA levels can suppress genes associated with salt stress responses, such as TIP2 (Sade et al. 2009). Ji et al. (2020) have reported that under salt stress, the activation of GADs and GABA-T facilitates GABA incorporation into the TCA cycle, enhancing salt tolerance. Furthermore, Su et al. (2019) reiterate that GABA maintains membrane potential and prevents potassium leakage by activating H^+-ATPase, thus mitigating salt stress effects. Cheng et al. (2018) have shown that GABA enhances seed germination under salt stress by improving Na^+/K^+ transport, increasing dehydration protectants, and regulating osmotic pressure. In consonance, the application of exogenous GABA significantly enhances the activity of several crucial antioxidant enzymes, including superoxide dismutase, peroxidase, catalase, ascorbic acid peroxidase, and glutathione peroxidase (Wu et al. 2020). This enzymatic activation leads to a reduction in ROS and Malondialdehyde (MDA) levels, thereby contributing to improved resilience against salt stress.

4.3.4 Role of GATs in Biotic Stress Tolerance

GATs play a crucial role in plants' resilience to biotic stress, including attacks by pathogens and pests, by managing the uptake and distribution of GABA within plant cells. These transporters enhance stress responses by facilitating GABA accumulation, which acts as a signaling molecule to activate defense mechanisms, such as the production of secondary metabolites and stress-related proteins. They also influence cellular signaling pathways and help maintain ion balance and pH levels during stress. This regulation supports cell wall reinforcement, improves antimicrobial defenses, and boosts overall plant resilience.

The initial defense response to microbial invasion typically involves a pattern-triggered immunity (PTI), followed by a more robust effector-triggered immunity (ETI) (Jones and Dangl 2006). Both immune responses are marked by an influx of Ca^{2+}, a critical early event following pathogen detection, which is directly linked to ligand–receptor interactions (Seybold et al. 2014). Second messengers, including cyclic Guanosine Monophosphate (GMP), mediate the activation of Ca^{2+} channels, such as those in the Cyclic Nucleotide-Gated Channel (CNGC) family (Ma et al. 2012; Zhao et al. 2013). The Ca^{2+} released into the cytosol binds to various proteins, including CaM, calcineurin B-like proteins (CBLs), and Ca^{2+}-dependent protein kinases (Schulz et al. 2013; Steinhorst and Kudla 2013).

The primary pathway for GABA synthesis in plants is the GABA shunt, where GABA arises as a product of glutamate decarboxylation by the action of GAD in the cytosol (Fait et al. 2008; Forde and Lea 2007). Notably, many plant GADs have a calmodulin-binding domain, which activates their decarboxylation activity in response to increased cytosolic Ca^{2+} levels (Shelp et al. 2006; Snedden et al. 1995). Thus, pathogen-induced GABA accumulation is regulated by Ca^{2+} influx resulting from the plant immune response. Additionally, GAD activity is stimulated by cytosolic acidification (Shelp et al. 2006; Snedden et al. 1995). Pathogen recognition, such as by bacteria and fungi, has been shown to induce GABA accumulation in various plants, including beans (*Phaseolus vulgaris*), tomatoes (*Solanum lycopersicum*), and soybeans (*Glycine max*) (Copley et al. 2017; O'Leary et al. 2016; Wang et al. 2019).

In *Solanum sitiens* mutants, a major source of cytosolic glutamate for the GABA shunt arises from the excessive activation of the Glutamine Synthetase/

Glutamine:2-oxoglutarate aminotransferase (GS/GOGAT) cycle (Seifi et al. 2013). However, this cycle can also act as a harmful electron sink when photosynthetic activity declines, a common occurrence during many plant–pathogen interactions (Liu et al. 2010). Regulation of the GS/GOGAT cycle is crucial for balancing carbon and nitrogen and is influenced by metabolic cues such as nitrogen availability and changes in carbohydrate and amino acid levels (Ishiyama et al. 2004). *Botrytis cinerea* infection in sunflower (*Helianthus annuus*) accelerates glutamate import from noninfected areas, advocating the existence of a mechanism to provide substrates for defense-related compounds (Dulermo et al. 2009).

GABA accumulation occurs in response to wounding (Wallace et al. 1984). Mechanical stimulation from insect feeding also triggers GABA accumulation in several investigations (Bown et al. 2002; Scholz et al. 2015). However, recent research suggests that GABA accumulation is not directly linked to Ca^{2+} increase from wounding or insect feeding but may instead be triggered by electrophysiological events (Scholz et al. 2017). This is consistent with the role of GABA in regulating membrane excitability via binding to ALMT anionic channels (Žárský 2015).

Insects affected by GABA often experience paralysis due to its inhibitory action at neuromuscular junctions (Irving et al. 1979), and several pesticides target GABA-responsive Cl^- channels to disrupt insect neuromuscular activity (Casida and Durkin 2015). Consequently, GABA accumulation induced by insects is likely a defensive strategy aimed at targeting phytophagous species with developed neuronal systems. Supporting this, experiments with GAD-overexpressing tobacco plants, which accumulate higher levels of GABA, demonstrate reduced nematode infestations compared to wild-type plants (McLean et al. 2003), suggesting a defensive role of GABA against both insects and nematodes (Shelp et al. 2006).

In sunflowers, the transport of glutamate to areas infected by *Botrytis cinerea* is believed to supply nitrogen to counteract necrotroph-induced senescence (Dulermo et al. 2009). Exogenous application of GABA has consistently been shown to alleviate symptoms caused by *Botrytis cinerea* in both tomato leaves and fruits (Seifi et al. 2013; Sun et al. 2019). Additionally, exogenous GABA in *Arabidopsis* increases the activity of enzymes associated with nitrogen metabolism, such as nitrate reductase and GOGAT (Barbosa et al. 2010), indicating that GABA may modulate metabolic adjustments in response to necrotrophic pathogens, functioning as a signaling molecule beyond its role as a mere metabolite (Mccraw et al. 2016), although this assumption requires further investigation.

4.4 Conclusion

GATs play a critical role in mediating plant responses to stress by regulating GABA homeostasis and facilitating its transport across cellular membranes. Their function is integral to plant adaptations to diverse environmental challenges, such as drought, salinity, and extreme temperatures. By influencing GABA levels, these transporters modulate stress-signaling pathways, support metabolic adjustments, and enhance overall stress resilience. Understanding the mechanisms of GATs provides valuable insights into their potential as targets for engineering stress-resistant crops. Advances in this area could lead to innovative strategies for improving crop performance under adverse conditions, ultimately contributing to global food security. Continued research into GATs and their interactions with other metabolic pathways will be essential for unraveling the full scope of their contributions to plant stress tolerance. The exploration of GATs is not only crucial for basic plant science but also holds promise for practical applications in agriculture. As our knowledge of these transporters deepens, it paves the way for developing novel approaches to enhance plant resilience, ensuring that crops can thrive in increasingly challenging environments.

References

Ansari MI, Lee R, Chen S (2005) A novel senescence-associated gene encoding γ-aminobutyric acid (GABA): pyruvate transaminase is upregulated during rice leaf senescence. Physiol Plant 123(1):1–8. https://doi.org/10.1111/j.1399-3054.2004.00430.x

Ansari MI, Hasan S, Jalil SU (2014) Leaf senescence and GABA shunt. Bioinformation 10:730–732

Ansari MI, Jalil SU, Ansari SA et al (2021) GABA shunt: a key-player in mitigation of ROS during stress. Plant Growth Regul 94:131–149

Bai X, Xu J, Shao X, Luo W, Niu Z, Gao C, Wan D (2019) A novel gene coding γ-aminobutyric acid transporter may improve the tolerance of Populus euphratica to adverse environments. Front Plant Sci 10:123–131

Barbosa JM, Singh NK, Cherry JH, Locy RD (2010) Nitrate uptake and utilization is modulated by exogenous γ-aminobutyric acid in Arabidopsis thaliana seedlings. Plant Physiol Biochem 48:443–450

Batushansky A, Kirma M, Grillich N, Pham PA, Rentsch D, Galili G, Fernie AR, Fait A (2015) The transporter GAT1 plays an important role in GABA-mediated carbon-nitrogen interactions in Arabidopsis. Front Plant Sci 6:785

Baum G, Chen Y, Arazi T, Takatsuji H, Fromm H (1993) A plant glutamate decarboxylase containing a calmodulin binding domain. Cloning, sequence, and functional analysis. J Biol Chem 268:19610–19617

Bouche N, Fromm H (2004) GABA in plants: just a metabolite? Trends Plant Sci 9:110–115

Bouché N, Lacombe B, Fromm H (2003) GABA signaling: a conserved and ubiquitous mechanism. Trends Cell Biol 13:607–610. https://doi.org/10.1016/j.tcb.2003.10.001

Bown AW, Shelp BJ (1997) The metabolism and functions of γ-aminobutyric acid. Plant Physiol 115:1–5

Bown AW, Shelp BJ (2020) Does the GABA shunt regulate cytosolic GABA? Trends Plant Sci 25:422–424

Bown AW, Hall DE, Macgregor KB (2002) Insect footsteps on leaves stimulate the accumulation of 4-aminobutyrate and can be visualized through increased chlorophyll fluorescence and superoxide. Plant Physiol 129:1430–1434

Breitkreuz KE, Shelp BJ, Fischer WN, Schwacke R, Rentsch D (1999) Identification and characterization of GABA, proline and quaternary ammonium compound transporters from Arabidopsis thaliana. FEBS Lett 450:280–284

Carter C, Pan S, Zouhar J, Avila EL, Girke T, Raikhel NV (2004) The vegetative vacuole proteome of Arabidopsis thaliana reveals predicted and unexpected proteins. Plant Cell 16:3285–3303

Casida JE, Durkin KA (2015) Novel GABA receptor pesticide targets. Pestic Biochem Physiol 121:22–30

Chen L, Ortiz-Lopez A, Jung A, Bush DR (2001) ANT1, an aromatic and neutral amino acid transporter in Arabidopsis. Plant Physiol 125:1813–1820

Cheng B, Li Z, Liang L, Cao Y, Zeng W, Zhang X, Ma X, Huang L, Nie G, Liu W et al (2018) The γ-aminobutyric acid (GABA) alleviates salt stress damage during seeds germination of White Clover associated with Na+/K+ transportation, dehydrins accumulation, and stress-related genes expression in White Clover. Int J Mol Sci 19:2520

Copley TR, Aliferis KA, Kliebenstein DJ, Jabaji SH (2017) An integrated RNAseq-1H NMR metabolomics approach to understand soybean primary metabolism regulation in response to Rhizoctonia foliar blight disease. BMC Plant Biol 17:84

Deng J, Kong L, Zhu Y, Pei D, Chen X, Wang Y, Qi J, Song C, Yang S, Gong Z (2022) BAK1 plays contrasting roles in regulating abscisic acid-induced stomatal closure and abscisic acid-inhibited primary root growth in Arabidopsis. J Integr Plant Biol 64:17

Dinkeloo K, Boyd S, Pilot G (2018) Update on amino acid transporter functions and on possible amino acid sensing mechanisms in plants. Semin Cell Dev Biol 74:105–113

Duan Y, Zhu X, Shen J, Xing H, Zou Z, Ma Y, Wang Y, Fang W (2020) Genome-wide identification, characterization and expression analysis of the amino acid permease gene family in tea plants (Camellia sinensis). Genomics 112:2866–2874

Dulermo T, Rascle C, Chinnici G, Gout E, Bligny R, Cotton P (2009) Dynamic carbon transfer during pathogenesis of sunflower by the necrotrophic fungus Botrytis cinerea: from plant hexoses to mannitol. New Phytol 183:1149–1162

Dündar E, Bush DR (2009) BAT1, a bidirectional amino acid transporter in Arabidopsis. Planta 229:1047–1056

Fait A, Fromm H, Walter D, Galili G, Fernie AR (2008) Highway or byway: the metabolic role of the GABA shunt in plants. Trends Plant Sci 13:14–19

Fischer WN, Loo DD, Koch W, Ludewig U, Boorer KJ, Tegeder M, Rentsch D, Wright EM, Frommer WB (2002) Low and high affinity amino acid H+-cotransporters for cellular import of neutral and charged amino acids. Plant J 29:717–731

Forde BG, Lea PJ (2007) Glutamate in plants: metabolism, regulation, and signalling. J Exp Bot 58:2339–2358

Frezzini M, Guidoni L, Pascarella S (2019) Conformational transitions induced by γ-aminobutyrate binding in GabR, a bacterial transcriptional regulator. Sci Rep 9:19319

Fromm H (2020) GABA signaling in plants: targeting the missing pieces of the puzzle. J Exp Bot 71:6238–6245

Gilliham M, Tyerman SD (2016) Linking metabolism to membrane signaling: the GABA–malate connection. Trends Plant Sci 21:295–301

Grallath S, Weimar T, Meyer A, Gumy C, Suter-Grotemeyer M, Neuhaus JM, Rentsch D (2005) The AtProT family: compatible solute transporters with similar substrate specificity but differential expression patterns. Plant Physiol 137:117–126

Hoover GJ, Van Cauwenberghe OR, Breitkreuz KE, Clark SM, Merrill AR, Shelp BJ (2007) Characteristics of an Arabidopsis glyoxylate reductase: general biochemical properties and substrate specificity for the recombinant protein, and developmental expression and implications for glyoxylate and succinic semialdehyde metabolism in planta. Can J Bot 85:883–895

Irving SN, Osborne MP, Wilson RG (1979) Studies on L-glutamate in insect hemolymph. I. Effect of injected L-glutamate. Physiol Entomol 4:139–146

Ishiyama K, Inoue E, Watanabe-Takahashi A, Obara M, Yamaya T, Takahashi H (2004) Kinetic properties and ammonium-dependent regulation of cytosolic isoenzymes of glutamine synthetase in Arabidopsis. J Biol Chem 279:16598–16605

Jalil SU, Ansari MI (2020) Physiological role of gamma aminobutyric acid in salt stress tolerance. In: Hasanuzzaman M, Tanveer M (eds) Salt and drought stress tolerance in plants, signaling and communication in plants. Springer, Cham, pp 337–350

Jalil SU, Ahmad I, Ansari MI (2016) Phenotypic characterization of GABA-transaminase mutants of *Arabidopsis thaliana*. Adv Life Sci 5:10005–10008

Jalil SU, Ahmad I, Ansari MI (2017) Functional loss of GABA transaminase (GABA-T) expressed early leaf senescence under various stress conditions in Arabidopsis thaliana. Curr Plant Biol 9–10:11–22

Jalil SU, Khan MAA, Ansari MI (2019a) Tremendous role of GABA shunt metabolic pathway in plant system. J Biol Chem Res 36:188–196

Jalil SU, Khan MQR, Ansari MI (2019b) Role of GABA transaminase in the regulation of development and senescence in Arabidopsis thaliana. Curr Plant Biol 19:100119

Ji J, Shi Z, Xie T, Zhang X, Chen W, Du C, Sun J, Yue J, Zhao X, Jiang Z, Shi S (2020) Responses of GABA shunt coupled with carbon and nitrogen metabolism in poplar under NaCl and CdCl2 stresses. Ecotoxicol Environ Saf 193:110322. https://doi.org/10.1016/j.ecoenv.2020.110322

Jones JDG, Dangl JL (2006) The plant immune system. Nature 444:323–329

Khan M, Jalil SU, Chopra P, Chhillar H, Ansari MI (2021) Role of GABA in plant growth, development and senescence. Plant Gene 26:100283

Koike S, Matsukura C, Takayama M, Asamizu E, Ezura H (2013) Suppression of gamma-aminobutyric acid (GABA) transaminases induces prominent GABA accumulation, dwarfism and infertility in the tomato (Solanum lycopersicum L.). Plant Cell Physiol 54:793–807

Liao J, Wu X, Xing Z, Li Q, Duan Y, Fang W, Zhu X (2017) γ-Aminobutyric acid (GABA) accumulation in Tea (Camellia sinensis L.) through the GABA shunt and polyamine degradation pathways under anoxia. J Agric Food Chem 65:3013–3018

Liu G, Ji Y, Bhuiyan NH, Pilot G, Selvaraj G, Zou J, Wei Y (2010) Amino acid homeostasis modulates salicylic acid-associated redox status and defense responses in Arabidopsis. Plant Cell 22:3845–3863

Long Y, Tyerman SD, Gilliham M (2020) Cytosolic GABA inhibits anion transport by wheat ALMT1. New Phytol 225:671–678

Ma Y, Walker RK, Zhao Y, Berkowitz GA (2012) Linking ligand perception by PEPR pattern recognition receptors to cytosolic Ca2+ elevation and downstream immune signaling in plants. Proc Natl Acad Sci USA 109:19852–19857

Ma D, Lu P, Shi Y (2013) Substrate selectivity of the acid-activated glutamate/γ-aminobutyric acid (GABA) antiporter GadC from *Escherichia coli*. J Biol Chem 288:15148–15153

Ma H, Cao X, Shi S, Li S, Gao J, Ma Y, Zhao Q, Chen Q (2016) Genome-wide survey and expression analysis of the amino acid transporter superfamily in potato (*Solanum tuberosum* L.). Plant Physiol Biochem 107:164–177

McCraw SL, Park DH, Jones R, Bentley MA, Rico A, Ratcliffe RG, Preston GM (2016) GABA (γ-aminobutyric acid) uptake via the GABA permease GabP represses virulence gene expression in *Pseudomonas syringae* pv. *tomato* DC3000. Mol Plant-Microbe Interact 29:938–949

McLean MD, Yevtushenko DP, Deschene A, Van OR, Makhmoudova A, Potter JW, Barry J (2003) Overexpression of glutamate decarboxylase in transgenic tobacco plants confers resistance to the northern root-knot nematode. Mol Breed 11:277–285

Meyer A, Eskandari S, Grallath S, Rentsch D (2006) AtGAT1, a high affinity transporter for gamma-aminobutyric acid in *Arabidopsis thaliana*. J Biol Chem 281:7197–7204

Michaeli S, Fait A, Lagor K, Nunes-Nesi A, Grillich N, Yellin A, Bar D, Khan M, Fernie AR, Turano FJ et al (2011) A mitochondrial GABA permease connects the GABA shunt and the TCA cycle, and is essential for normal carbon metabolism. Plant J 67:485–498

O'Leary BM, Neale HC, Geilfus CM, Jackson RW, Arnold DL, Preston GM (2016) Early changes in apoplast composition associated with defence and disease in interactions between *Phaseolus vulgaris* and the halo blight pathogen *Pseudomonas syringae* pv. *phaseolicola*. Plant Cell Environ 39:2172–2184

Ramesh SA, Tyerman SD, Xu B, Bose J, Kaur S, Conn V, Domingos P, Ullah S, Wege S, Shabala S et al (2015) GABA signalling modulates plant growth by directly regulating the activity of plant-specific anion transporters. Nat Commun 6:7879

Ramesh SA, Kamran M, Sullivan W, Chirkova L, Okamoto M, Degryse F, McLaughlin M, Gilliham M, Tyerman SD (2018) Aluminum-activated malate transporters can facilitate GABA transport. Plant Cell 30:1147–1164

Sade N, Vinocur BJ, Diber A, Shatil A, Ronen G, Nissan H, Wallach R, Karchi H, Moshelion M (2009) Improving plant stress tolerance and yield production: is the tonoplast aquaporin SlTIP2;2 a key to isohydric to anisohydric conversion? New Phytol 181:651–661

Schmidt UG, Endler A, Schelbert S, Brunner A, Schnell M, Neuhaus HE, Marty-Mazars D, Marty F, Baginsky S, Martinoia E (2007) Novel tonoplast transporters identified using a proteomic approach with vacuoles isolated from cauliflower buds. Plant Physiol 145:216–229

Scholz SS, Reichelt M, Mekonnen DW, Ludewig F, Mithöfer A (2015) Insect herbivory-elicited GABA accumulation in plants is a wound-induced, direct, systemic, and jasmonate-independent defense response. Front Plant Sci 6:388

Scholz SS, Malabarba J, Reichelt M, Heyer M, Ludewig F, Mithöfer A (2017) Evidence for GABA-induced systemic GABA accumulation in *Arabidopsis* upon wounding. Front Plant Sci 8:1–9

Schulz P, Herde M, Romeis T (2013) Calcium-dependent protein kinases: Hubs in plant stress signaling and development. Plant Physiol 163:523–530

Schwacke R, Grallath S, Breitkreuz KE, Stransky E, Stransky H, Frommer WB, Rentsch D (1999) LeProT1, a transporter for proline, glycine betaine, and gamma-aminobutyric acid in tomato pollen. Plant Cell 11:377–391

Seifi HS, Curvers K, De Vleesschauwer D, Delaere I, Aziz A, Höfte M (2013) Concurrent overactivation of the cytosolic glutamine synthetase and the GABA shunt in the ABA-deficient sitiens mutant of tomato leads to resistance against *Botrytis cinerea*. New Phytol 199:490–504

Seybold H, Trempel F, Ranf S, Scheel D, Romeis T, Lee J (2014) Ca2+ signalling in plant immune response: from pattern recognition receptors to Ca2+ decoding mechanisms. New Phytol 204:782–790

Shelp BJ, Bown AW, McLean MD (1999) Metabolism and function of gamma aminobutyric acid. Trends Plant Sci 4:446–452

Shelp BJ, Bown AW, Faure D (2006) Extracellular γ-aminobutyrate mediates communication between plants and other organisms. Plant Physiol 142:1350–1352

Signorelli S, Dans PD, Coitiño EL, Borsani O, Monza J (2015) Connecting proline and γ-aminobutyric acid in stressed plants through non-enzymatic reactions. PLoS One 10:e0115349. https://doi.org/10.1371/journal.pone.0115349

Snedden WA, Arazi T, Fromm H, Shelp BJ (1995) Calcium/calmodulin activation of soybean glutamate decarboxylase. Plant Physiol 108:543–549

Snowden CJ, Thomas B, Baxter CJ, Smith JA, Sweetlove LJ (2015) A tonoplast Glu/Asp/GABA exchanger that affects tomato fruit amino acid composition. Plant J 81:651–660

Steinhorst L, Kudla J (2013) Calcium and reactive oxygen species rule the waves of signaling. Plant Physiol 163:471–485

Su N, Wu Q, Chen J, Shabala L, Mithöfer A, Wang H, Qu M, Yu M, Cui J, Shabala S (2019) GABA operates upstream of H+-ATPase and improves salinity tolerance in *Arabidopsis* by enabling cytosolic K+ retention and Na+ exclusion. J Exp Bot 70:6349–6361

Sun C, Jin L, Cai Y, Huang Y, Zheng X, Yu T (2019) L-Glutamate treatment enhances disease resistance of tomato fruit by inducing the expression of glutamate receptors and the accumulation of amino acids. Food Chem 293:263–270

Tegeder M, Ward JM (2012) Molecular evolution of plant AAP and LHT amino acid transporters. Front Plant Physiol 3:1–11

Tian R, Yang Y, Chen M (2020) Genome-wide survey of the amino acid transporter gene family in wheat (*Triticum aestivum* L.): identification, expression analysis and response to abiotic stress. Int J Biol Macromol 162:1372–1387

Tsai MF, McCarthy P, Miller C (2013) Substrate selectivity in glutamate-dependent acid resistance in enteric bacteria. Proc Natl Acad Sci USA 110:5898–5902

Wallace W, Secor J, Schrader LE (1984) Rapid accumulation of γ-aminobutyric acid and alanine in soybean leaves in response to an abrupt transfer to lower temperature, darkness, or mechanical manipulation. Plant Physiol 75:170–175

Wang F, Chen ZH, Liu X, Colmer TD, Shabala L, Salih A, Zhou M, Shabala S (2017) Revealing the roles of GORK channels and NADPH oxidase in acclimation to hypoxia in *Arabidopsis*. J Exp Bot 68:3191–3204

Wang G, Kong J, Cui D, Zhao H, Niu Y, Xu M, Wang W (2019) Resistance against *Ralstonia solanacearum* in tomato depends on the methionine cycle and the γ-aminobutyric acid metabolic pathway. Plant J 97:1032–1047

Whiteman SA, Serazetdinova L, Jones AM, Sanders D, Rathjen J, Peck SC, Maathuis FJ (2008) Identification of novel proteins and phosphorylation sites in a tonoplast-enriched membrane fraction of *Arabidopsis thaliana*. Proteomics 8:3536–3547

Wipf D, Ludewig U, Tegeder M, Rentsch D, Koch W, Frommer WB (2002) Conservation of amino acid transporters in fungi, plants and animals. Trends Biochem Sci 27:139–147

Wolf-Nicolas F, Bruno A, Doris R, Sylvia K, Mechthild T, Kevin B, Wolf BF (1998) Amino acid transport in plants. Trends Plant Sci 3:188–195

Wu X, Jia Q, Ji S, Gong B, Li J, Lü G, Gao H (2020) Gamma-aminobutyric acid (GABA) alleviates salt damage in tomato by modulating Na+ uptake, the GAD gene, amino acid synthesis and reactive oxygen species metabolism. BMC Plant Biol 20:465

Wu Qi SN, Huang X, Cui J, Shabala L, Zhou M, Yu M, Shabala S (2021) Hypoxia-induced increase in GABA content is essential for restoration of membrane potential and preventing ROS-induced disturbance to ion homeostasis. Plant. Communications 2:100188. https://doi.org/10.1016/j.xplc.2021.100188

Xu B, Sai N, Gilliham M (2021) The emerging role of GABA as a transport regulator and physiological signal. Plant Physiol 187:2005–2016

Yang R, Guo Y, Wang S, Gu Z (2015) Ca2+ and aminoguanidine on gamma-aminobutyric acid accumulation in germinating soybean under hypoxia-NaCl stress. J Food Drug Anal 23:287–293

Young GB, Jack DL, Smith DW, Saier MH (1999) The amino acid–auxin–proton symport permease family. Biochim Biophys Acta 1415:306–322

Yuan D, Wu X, Gong B et al (2023) GABA metabolism, transport and their roles and mechanisms in the regulation of abiotic stress (hypoxia, salt, drought) resistance in plants. Meta 13:347. https://doi.org/10.3390/metabo13030347

Žárský V (2015) Signal transduction: GABA receptor found in plants. Nat Plants 1:15115

Zhao Y, Qi Z, Berkowitz GA (2013) Teaching an old hormone new tricks: cytosolic Ca2+ elevation involvement in plant brassinosteroid signal transduction cascades. Plant Physiol 163:555–565

GABA as a Modulator of the Defense System Under Oxidative Stress

5

Abstract

GABA plays a multifaceted role in modulating defense mechanisms, particularly under oxidative stress. As plants encounter environmental stressors, such as drought, salinity, and heavy metals, oxidative stress is often induced, characterized by the accumulation of Reactive oxygen species (ROS). GABA modulates metabolic pathways linked to stress resilience that influence antioxidant systems via the synthesis of antioxidant enzymes, balance ROS levels, and maintain cellular redox homeostasis. GABA signaling also intersects with other metabolic and hormonal pathways to ameliorate tolerance to oxidative stress, boosting the overall defense capacity. Emphasis is placed on recent advances in understanding GABA's involvement in ROS scavenging, signaling pathways, and its regulatory impact on plant growth and survival under adverse conditions.

Keywords

Antioxidants · GABA · Oxidative stress · Redox homeostasis · ROS

Highlights

- GABA modulates plant defense against oxidative stress, acting as both a Reactive Oxygen Species (ROS) scavenger and a regulator of stress-signaling pathways.
- It upregulates key antioxidant enzymes, contributing to effective ROS detoxification and maintaining redox balance.
- GABA interacts with hormonal signaling molecules, modulating their effects to enhance stress adaptation and defense.
- GABA offers potential benefits for the development of crop stress resilience and sustainable agriculture.

S. U. Jalil et al., *GABA Signaling System and Abiotic Stress Tolerance in Plants*,
https://doi.org/10.1007/978-981-95-0437-4_5

5.1 Introduction

Plants constantly face environmental challenges that disrupt their cellular homeostasis, particularly in the form of oxidative stress (Grene 2002). This stress arises when the generation of Reactive oxygen species (ROS) exceeds the plant's antioxidant defenses, leading to an accumulation of these highly reactive molecules (Hasanuzzaman et al. 2020; Sachdev et al. 2021). ROS, including superoxide anions ($O_2{\bullet}^-$), hydrogen peroxide (H_2O_2), and hydroxyl radicals (•OH), are unavoidable by-products of normal metabolic processes like photosynthesis and respiration. Under optimal conditions, ROS function as signaling molecules, mediating processes like growth, development, and defense. However, during environmental stress—such as drought, salinity, heat and cold stress, and pathogen invasion—ROS levels surge, causing widespread damage to cellular structures. Lipid peroxidation, protein denaturation, and DNA fragmentation are hallmarks of oxidative stress, ultimately impairing vital cellular functions and threatening plant survival (Huang et al. 2019; Dumanović et al. 2021).

To combat this, plants have evolved a complex network of defense mechanisms, including both enzymatic and nonenzymatic antioxidant systems (Rahman et al. 2024). Among these defenses, an often-overlooked but increasingly recognized player is GABA, which rapidly accumulates in response to stress and is critical to the modulation of plant responses to oxidative stress (Al-Quraan et al. 2013; Jalil et al. 2017). Consequently, GABA acts not only as a metabolic buffer but also as a signaling molecule that influences a wide range of physiological processes (Song et al. 2010; Salah et al. 2019; Jin et al. 2023; Guo et al. 2023).

During oxidative stress, GABA biosynthesis is rapidly upregulated, primarily through the Glutamate decarboxylase (GAD) activity that accomplishes the decarboxylation of glutamate to produce GABA (Mei et al. 2016). This surge in GABA accumulation serves multiple protective roles (Khan et al. 2021). First, GABA acts as a buffer to stabilize cytosolic pH, which often fluctuates under stress conditions (Allan et al. 2009; Shelp et al. 2017). Additionally, GABA directly contributes to ROS detoxification by interacting with the plant's antioxidant defense machinery (Ansari et al. 2021). It enhances the activity of key antioxidant enzymes such as Superoxide dismutase (SOD), Catalase (CAT), and Ascorbate peroxidase (APX), which are crucial for neutralizing excess ROS and preventing oxidative damage (Ansari et al. 2021; Abdullah et al. 2024).

Beyond its role in direct ROS scavenging, GABA is deeply integrated into the plant's broader defense networks. It interacts with other metabolic pathways and stress-signaling molecules, including phytohormones like abscisic acid (ABA) and ethylene, which are central to the plant's stress response (Jalil and Ansari, 2020; Islam et al. 2024). GABA also participates in redox signaling, helping maintain cellular homeostasis by regulating the balance between oxidative and reductive reactions (Allan et al. 2009; Shelp et al. 2017). This cross talk between GABA and other signaling pathways allows the plant to fine-tune its defense mechanisms in response to fluctuating environmental conditions (Ansari et al. 2021).

The modulation of GABA under oxidative stress is particularly significant as it directly influences the antioxidant defense system. By enhancing the activity of antioxidant enzymes, GABA contributes to reducing ROS levels, thereby protecting plants from oxidative damage (Jalil et al. 2019a, b; Li et al. 2021; Abdel Razik et al. 2021). Furthermore, recent studies have highlighted GABA's involvement in mitigating oxidative stress through its role in the regulation of the Ascorbate–Glutathione (ASC-GSH) cycle, a key pathway for ROS detoxification. By modulating this cycle, GABA helps maintain the redox balance within cells, ensuring that ROS levels remain under control without compromising the plant's ability to signal stress responses (Yang et al. 2011; Abdel Razik et al. 2021). GABA also extends its role to coordinate the cross talk between metabolic and hormonal pathways, making it an integral component in maintaining redox homeostasis. Understanding how GABA modulates the plant defense system during oxidative stress provides new insights into improving stress tolerance, which is crucial for developing resilient crop varieties in the face of climate change.

5.2 Oxidative Stress in Plants: Mechanisms and Consequences

Oxidative stress is a major physiological challenge that plants encounter when exposed to both abiotic and biotic stress factors. The central cause of oxidative stress is the excessive generation of ROS, which are highly reactive molecules derived from molecular oxygen. Under normal conditions, ROS are by-products of essential processes such as photosynthesis, respiration, and photorespiration. Under stress conditions, the balance between the production and removal of ROS is disturbed, resulting in the accumulation of ROS and subsequent oxidative damage (Fig. 5.1).

5.2.1 ROS Generation Under Abiotic and Biotic Stress

Abiotic stressors, such as drought, salinity, extreme temperatures, heavy metal toxicity, and UV radiation, as well as biotic factors, like pathogen infection, herbivore attack, and wounding, significantly elevate ROS production in plant cells (Nawaz et al. 2023). For instance, during drought and salt stress, water shortage disrupts photosynthesis and respiration, causing electrons to leak from the electron transport chains in chloroplasts and mitochondria, which promotes the overproduction of ROS, such as $O_2^{\bullet -}$, H_2O_2, and •OH (Ma et al. 2019; dos Santos et al. 2022). Similarly, pathogen invasion triggers a rapid ROS burst, commonly known as the oxidative burst, which acts as a defense mechanism against pathogens by strengthening cell walls, inducing cell death in infected tissues, and activating further defense pathways (Wojtaszek 1997; Torres et al. 2006).

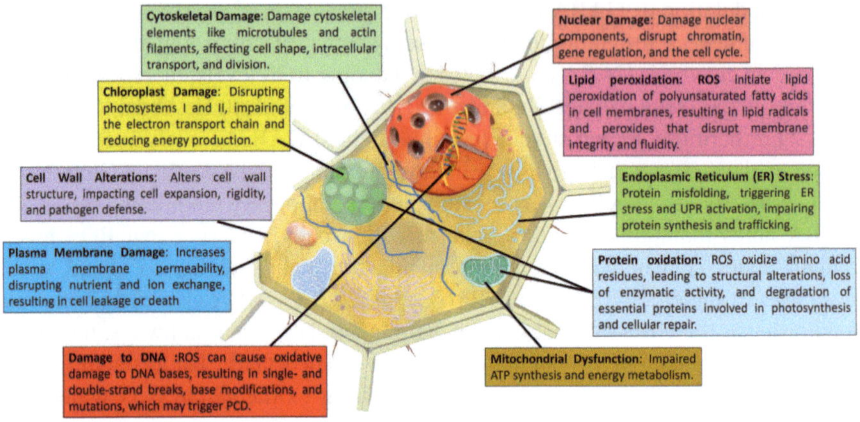

Fig. 5.1 Effects of oxidative stress on plant cells. ROS accumulation disrupts nuclear integrity, leading to chromatin damage, impaired gene regulation, and cell cycle arrest. Lipid peroxidation affects membrane fluidity by generating lipid radicals and peroxides, compromising cellular homeostasis. In the endoplasmic reticulum, oxidative stress induces protein misfolding and unfolded protein response (UPR) activation, impairing protein synthesis and trafficking. Protein oxidation alters enzymatic activity and degrades essential proteins involved in photosynthesis and cellular repair. Mitochondrial dysfunction reduces Adenosine Triphosphate (ATP) synthesis and energy metabolism, further exacerbating cellular stress. Chloroplast damage disrupts photosystems I and II, impairing the electron transport chain and reducing energy production. Cytoskeletal damage affects microtubules and actin filaments, compromising cell shape, intracellular transport, and division. Structural alterations in the cell wall impact expansion, rigidity, and pathogen defense, while increased plasma membrane permeability leads to disrupted ion exchange, nutrient loss, and potential cell death. DNA damage, characterized by oxidative modifications, strand breaks, and mutations, can trigger Programmed Cell Death (PCD). This comprehensive depiction emphasizes the widespread effects of oxidative stress on plant cellular functions and survival

5.2.2 Oxidative Stress Effect on Cellular Components

The excessive accumulation of ROS under oxidative stress conditions has negative effects on various cellular components, including lipids, proteins, and DNA (Jakubczyk et al. 2020). Lipids, especially polyunsaturated fatty acids in membrane structures, are particularly vulnerable to ROS attack, leading to lipid peroxidation (Ayala et al. 2014). This process not only damages membrane integrity but also disrupts cellular compartmentalization, ion transport, and cell signaling (Repetto et al. 2012). Proteins are similarly affected, with ROS causing the oxidation of amino acid residues, leading to enzyme inactivation, protein denaturation, and aggregation (Zhang et al. 2013). Such oxidative modifications compromise protein functionality and disrupt key metabolic pathways (Demidchik 2015). ROS can also induce breaks and mutations in DNA, impairing genetic integrity and potentially leading to PCD (Avery 2011).

5.2.3 Oxidative Stress Triggering Cellular Signaling Pathways

While ROS are often viewed as harmful, they also play a dual role as essential signaling molecules in plants. ROS serve as second messengers that activate a range of signaling pathways in response to stress. When ROS levels rise, they trigger specific signaling cascades that regulate gene expression involved in defense, repair, and adaptation mechanisms (D'Autréaux and Toledano, 2007). Central to this signaling network are the mitogen-activated protein kinase (MAPK) cascades that mediate the expression of antioxidant and stress-responsive genes during oxidative stress (Plotnikov et al. 2011). Moreover, ROS signaling is tightly linked to hormonal pathways involving ABA, Jasmonis acid (JA), and Salicyclic acid (SA), which coordinate systemic defense responses across plant tissues (Devireddy et al. 2021). In addition to activating antioxidant defenses, ROS modulate redox homeostasis by interacting with other redox-active molecules such as GSH and ASC, which are components of the ASC-GSH cycle (Foyer and Kunert 2024). This cycle plays a vital role in detoxifying ROS and maintaining the cellular redox environment. ROS-driven signaling also triggers autophagy and PCD under severe oxidative stress conditions, thereby eliminating damaged cells and preserving overall plant integrity (Foyer and Kunert 2024).

5.3 Regulation and Metabolism of GABA Under Oxidative Stress

GABA has emerged as a critical metabolite in plant responses to various environmental stresses, particularly oxidative stress, playing multiple roles in stress signaling, metabolic regulation, and cellular protection. Its rapid accumulation under oxidative stress suggests that GABA metabolism is tightly regulated and integrated into the plant's defense network (Fig. 5.2).

5.3.1 GABA Biosynthesis Regulation

The primary biosynthetic pathway for GABA in plants operates through GAD, which converts glutamate into GABA via decarboxylation (Jalil et al. 2017; Ansari et al. 2021; Khan et al. 2021). This reaction is pivotal, as glutamate is a key amino acid involved in nitrogen metabolism, and its conversion to GABA acts as a crucial step in modulating cellular responses to stress (Khan et al. 2021). GAD is activated under various stress conditions, particularly those that elevate cytosolic calcium ion (Ca^{2+}) levels, such as oxidative stress, drought, salinity, and pathogen attack. The increase in cytosolic Ca^{2+} stimulates the calmodulin-binding domains on GAD, triggering its activation and enhancing GABA production.

Further, the cellular redox state intimately regulates GABA biosynthesis. Under oxidative stress, ROS such as H_2O_2 not only damage cellular components but also serve as signaling molecules that upregulate stress-responsive genes, including

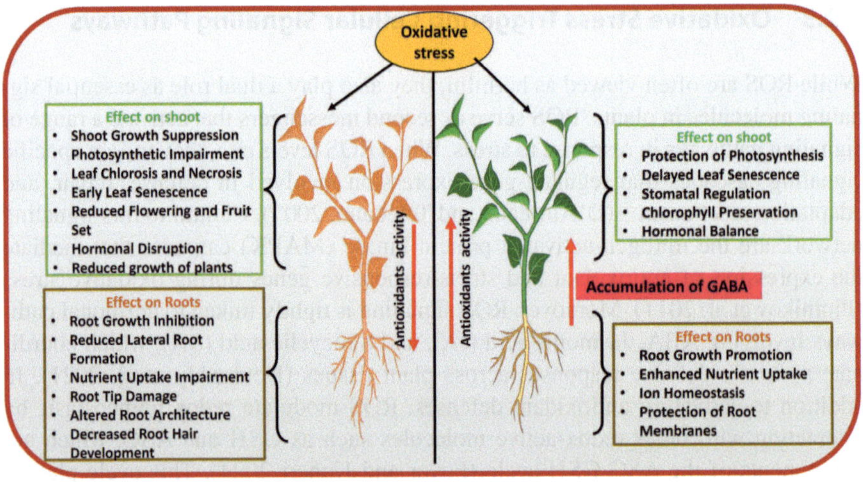

Fig. 5.2 GABA accumulation arising from oxidative stress counteracts and improves plant growth. On the left, oxidative stress leads to shoot growth suppression, photosynthetic impairment, leaf chlorosis and necrosis, early senescence, reduced flowering, hormonal disruption, and overall stunted plant growth. Similarly, root growth is inhibited, and lateral root/root hair formation is suppressed, altering root architecture and impairing nutrient uptake. On the right, GABA accumulation enhances antioxidant activity, protecting plants from oxidative damage. Shoots exhibit improved photosynthesis, delayed senescence, regulated stomatal function, chlorophyll preservation, and hormonal balance. In roots, GABA promotes growth, enhances nutrient uptake, maintains ion homeostasis, and protects root membranes. Thus, GABA accumulation highlights protection from oxidative stress and recovery of plant growth and productivity

those involved in GABA metabolism (Kabała and Janicka 2024). This regulation ensures that GABA biosynthesis is fine-tuned to respond swiftly to oxidative challenges, enabling plants to rapidly accumulate GABA as part of their defense strategy.

5.3.2 GABA Accumulation and Oxidative Stress

GABA accumulation is one of the most notable metabolic shifts observed in plants during oxidative stress. The rapid increase in GABA levels is not merely a by-product of stress but rather an adaptive response aimed at mitigating the damaging effects of ROS. GABA accumulation serves several protective roles. First, it acts as a buffer to maintain cytosolic pH, which often fluctuates under oxidative stress. This pH stabilization is crucial for preserving enzyme function and metabolic homeostasis in stressed cells.

GABA interacts with the antioxidant defense system and comprehensively detoxifies ROS, preventing oxidative damage to cellular structures. Its accumulation under stress also provides a carbon and nitrogen reserve to support essential metabolic processes when normal metabolism is disrupted. Furthermore, the osmotic adjustment by GABA accumulation helps plants cope with water deficit conditions, which often accompany oxidative stress. The accumulation of GABA in higher quantities remarkably performs a positive role in stress alleviation. The exogenous treatment of plants with GABA increases the content of endogenous GABA (Ramos-Ruiz et al. 2019). When stressed plants are treated with GABA, it helps improve their photosynthesis, raises the endogenous levels of GABA and antioxidants, and lowers the level of ROS and malondialdehyde (MDA) to maintain cell membrane integrity (Ramos-Ruiz et al. 2019; Sita and Kumar 2020). Also, the higher levels of GABA in the plant help protect it from both water loss and damage caused by stress (Jalil et al. 2017; Ansari et al. 2021). Treating different plants species and tissues, such as banana peels, tomato seedlings, tea plants, and anthurium cut flowers, with exogenous GABA helps them tolerate oxidative stress caused by low temperatures, triggering several protective mechanisms, such as increased endogenous GABA content, fortification of the antioxidant system, reduction in ROS content, reduced electrolyte leakage and MDA level, and/or increased accumulation of proline-induced osmoregulation (Malekzadeh et al. 2014; Wang et al. 2014; Aghdam et al. 2016; Zhu et al. 2019). GABA efficiently and significantly offsets the production of O_2•- and H_2O_2 in tobacco leaves under water stress (Liu et al. 2021), promotes the activity of antioxidant enzymes, lowers the levels of MDA and O2•-, and helps protect against oxidative damage in corn seedlings exposed to salt stress (Wang et al. 2017). Additionally, GABA enhances the activity of enzymes involved in the AsA-GSH cycle, such as Ascorbic acid (APX), Dehydroascorbate reductase (DHAR), and Monodehydroascorbate reductase (MDHAR). When peaches are exposed to chilling stress, GABA treatment increases the activities of APX, DHAR, and MDHAR, as well as Glutathione peroxidase (GPX), Glutathione S-transferase (GST), CAT, and SOD (Yang et al. 2011). Additionally, when GABA is added to sunflowers under drought and heat stress, it helps the plants better manage stress by improving their metabolism, boosting protective proteins, and increasing the activity of antioxidants. This helps the plants fight off the negative effects of both drought and heat (Abdel Razik et al. 2021).

5.3.3 GABA Shunt Components as Regulators of GABA Metabolism

The GABA shunt is a vital metabolic pathway that helps plants manage stress, particularly oxidative stress, by regulating GABA levels. At the heart of this pathway is GAD, responsible for GABA synthesis. GAD activity is tightly controlled during stress, especially when calcium levels in the cell rise. This regulation is further influenced by processes like phosphorylation and calmodulin binding, which link GAD to cellular signaling pathways activated under stress conditions (Mazzucotelli et al. 2006; Al-Quraan et al. 2013; Al-Quraan and Al-Share 2016).

GABA-T regulates GABA accumulation by converting GABA into succinic semialdehyde, which Succinic semialdehyde dehydrogenase (SSADH) catalytically converts to succinate. Together, these steps form the GABA shunt, a pathway that bypasses part of the TCA cycle. This bypass allows plants to maintain energy production even when the TCA cycle is disrupted by oxidative stress, ensuring that cellular respiration continues efficiently (Al-Quraan and Al-Share 2016; Jalil et al. 2017; Ansari et al. 2021).

Under stress conditions, such as hypoxia (low oxygen), GABA levels rise, helping stabilize the cell membrane and regulate H_2O_2 levels. This prevents damage from ROS, which can accumulate during stress. GABA also acts as a signaling molecule, influencing the activity of potassium channels and other cellular processes. For example, GABA can directly modulate the opening of potassium efflux channels, such as Guard cell Outward Rectifying K^+ channel (GORK), which helps maintain cellular ion balance (Wu et al. 2021).

The importance of the GABA shunt in stress tolerance is evident from studies on *Arabidopsis* mutants. For instance, the *pop2–5* mutant, which accumulates higher levels of GABA, shows enhanced resistance to salt stress. This resistance is linked to increased activity of proteins like PM-H^+-ATPase, higher expression of sodium-proton antiporters (SOS1 and NHX1), and reduced H_2O_2 levels in tissues (Su et al. 2019). In contrast, mutants with reduced GABA levels, such as *pop2–1* and *pop2–3*, exhibit premature leaf aging, reduced photosynthesis, and increased membrane damage under stress (Jalil et al. 2017). Similarly, *Arabidopsis* mutants lacking SSADH (*ssadh* mutants) accumulate a by-product called γ-hydroxybutyrate and are more sensitive to heat and UV stress. These mutants develop necrotic lesions and show elevated H_2O_2 levels when exposed to stress (Bouché et al. 2003). This sensitivity highlights the role of the GABA shunt in preventing ROS accumulation and maintaining cellular health during stress.

The GABA shunt plays a critical role in maintaining ROS homeostasis, particularly in guard cells, which regulate stomatal openings. GABA synthesis disruption, as seen in *gad2* or *gad1/2/4/5* mutants, leads to higher ROS levels, increased calcium concentrations in guard cells, and greater susceptibility to drought. Treating these mutants with ethylene glycol tetraacetic acid, a calcium-binding agent, reduces ROS accumulation and improves stomatal function, demonstrating the interplay between GABA, calcium, and ROS regulation (Xu et al. 2024). Overall, the coordination between GABA production (via GAD) and breakdown (via GABA-T and SSADH) is essential for maintaining GABA homeostasis during stress. This dynamic balance ensures metabolic flexibility and stress resilience, allowing plants to respond swiftly to oxidative stress and recover effectively. The GABA shunt not only supports energy production but also acts as a critical regulator of ROS and calcium signaling, safeguarding cellular survival under adverse conditions.

5.4 GABA as an ROS Scavenger

GABA is increasingly recognized for its multifaceted role in plant defense, particularly as a modulator of oxidative stress. In addition to its function as a signaling molecule and metabolic regulator, GABA plays a crucial role in scavenging ROS, thereby reducing oxidative damage and enhancing plant resilience under stress. Its involvement in both enzymatic and nonenzymatic antioxidant systems highlights GABA's integrative function in maintaining cellular redox balance.

5.4.1 GABA Mechanisms for Oxidative Damage Reduction

Under oxidative stress, excessive ROS (O_2•-, •OH, and H_2O_2) can cause significant damage to cellular components, including lipids, proteins, and DNA. GABA, through its accumulation during stress, acts as a critical buffer against this oxidative damage. One proposed mechanism is its ability to neutralize ROS, either directly or through its interaction with antioxidant defense pathways. GABA's role in pH regulation, particularly in stabilizing cytoplasmic pH, further protects the cell from the deleterious effects of ROS, as acidic shifts can exacerbate oxidative damage.

Moreover, GABA's integration into the GABA shunt—an alternative metabolic pathway connected to the TCA cycle—provides a means to bypass ROS-induced inhibition of central metabolism. By contributing to cellular energy metabolism and detoxifying ROS, the GABA shunt supports both energy production and ROS mitigation during stress. This dual role of GABA in oxidative stress management positions it as a key player in maintaining cellular homeostasis under adverse conditions.

5.4.2 GABA Interaction with Enzymatic and Nonenzymatic Antioxidant Systems

GABA confers plant resilience to oxidative stress by interacting with both enzymatic and nonenzymatic antioxidant systems. These interactions help plants neutralize ROS, which are harmful by-products of stress conditions, such as salinity, low temperatures, and environmental pollutants. Enzymatic antioxidants, such as SOD, CAT, and Peroxidase (POD), form the first line of defense against ROS, rapidly converting superoxide radicals and hydrogen peroxide into less reactive molecules. GABA indirectly enhances the activity of these enzymes by regulating redox-sensitive signaling pathways, which activate the expression of antioxidant genes. For example, under salt stress, GABA increases the activity of CAT, APX, and SOD, thereby improving the plant's ability to cope with oxidative stress (Kalhor et al. 2018; Aljuaid and Ashour 2022; Ullah et al. 2023). This upregulation of enzymatic antioxidants ensures that plants can swiftly respond to stress-induced ROS accumulation.

In addition to enzymatic antioxidants, GABA supports nonenzymatic antioxidant systems by promoting the synthesis and regeneration of key redox molecules such as ASC and GSH. ASC acts as a direct ROS scavenger, while GSH plays a crucial role in redox regulation and detoxification. GABA's involvement in maintaining the pools of these antioxidants enhances the plant's overall capacity to mitigate oxidative damage. For instance, in plants exposed to phenanthrene (a polycyclic aromatic hydrocarbon), GABA treatment promotes GSH levels by stimulating the expression of its biosynthetic genes (GSH1 and GSH2). The GSH-dependent mechanism is essential for reducing ROS production and alleviating stress (Guo et al. 2021).

Stress conditions often offset the balance between ROS production and antioxidant defense systems, leading to excessive ROS accumulation and oxidative damage. GABA helps restore this balance by stimulating antioxidant systems, thereby improving plant survival under unfavorable conditions (Li et al. 2017a, b; Ramos-Ruiz et al. 2019; Sita and Kumar 2020; AbdElbar et al. 2021). For example, GABA application improves photosynthetic parameters and electron transport rates while reducing non-photochemical quenching in lettuce plants exposed to salt stress. Additionally, GABA augments the activity of CAT, APX, and SOD, mitigating the negative effects of salt stress (Kalhor et al. 2018). Similar protective effects of GABA have been observed in other plant species under various stress conditions. In maize seedlings and mung bean plants subjected to salt stress, GABA treatment increases the activity of antioxidant enzymes, reducing oxidative damage and improving growth (Aljuaid and Ashour 2022; Ullah et al. 2023). In anthurium cut flowers stored at low temperatures, GABA treatment delays flower browning, increases the ratio of unsaturated to saturated fatty acids, and reduces H_2O_2 accumulation by enhancing the activity of SOD, CAT, APX, and Glutathione Reductase (GR) (Aghdam et al. 2016). Similarly, in tomato plants exposed to low temperatures, GABA attenuates oxidative damage by increasing the activity of phenylalanine ammonia-lyase, CAT, SOD, and APX, thereby protecting cell membrane integrity and reducing MDA and H_2O_2 levels (AbdElbar et al. 2021).

GABA performs a dual role in ROS regulation, contributing to both the production and elimination of H_2O_2. In the initial phase of stress, GABA promotes the production of H_2O_2, which acts as a signaling molecule to trigger defense responses. For example, in *Caragana intermedia* under salt stress, GABA treatment facilitates the expression of gene-encoding proteins such as respiratory burst oxidase homolog (RBOH) and Diamine oxidase (DAO) involved in H_2O_2 production (Shi et al. 2010). Later, GABA helps eliminate the excess ROS by activating antioxidant systems, ensuring that ROS levels do not reach damaging concentrations. This dual role is particularly evident in hypoxia-induced stress, where plants experience a significant increase in GABA levels. Hypoxia leads to H_2O_2 accumulation, which activates ROS-induced Ca^{2+} uptake channels and exacerbates K^+ loss through nonselective cation channels. GABA counteracts these effects by restoring membrane potential through pH-dependent regulation of the plasma membrane H^+-ATPase and generating energy via the GABA shunt pathway and the TCA cycle (Wu et al. 2021).

Overall, GABA's interaction with both enzymatic and nonenzymatic antioxidant systems is crucial for enhancing plants' stress tolerance. By modulating the activity of antioxidant enzymes and supporting the synthesis of redox molecules like ascorbate and glutathione, GABA ensures that plants can effectively neutralize ROS and maintain cellular function under stress. This dual role of GABA in ROS regulation, combined with its ability to restore membrane potential and energy production, highlights its importance as a key metabolite and signaling molecule in plant stress responses.

5.4.3 GABA Modulation of ASC-GSH Cycle

The ASC-GSH cycle is a pivotal antioxidant system in plants, responsible for detoxifying ROS, particularly H_2O_2. GABA contributes to the modulation of this cycle by enhancing the availability and recycling of ASC and GSH. During oxidative stress, the ASC-GSH cycle works to reduce H_2O_2 to water, with ASC acting as an electron donor and GSH regenerating ascorbate from its oxidized form. GABA, through its influence on redox-sensitive pathways, stimulates the expression of key enzymes in the ASC-GSH cycle, including APX, GR, and DHAR.

By modulating these enzymes, GABA ensures efficient recycling of ASC and GSH, maintaining their reduced forms, which are essential for continuous ROS detoxification. Moreover, GABA's contribution to the regeneration of Nicotinamide Adenine Dinucleotide Phosphate (NADPH), a cofactor required for the activity of GR, further supports the ASC-GSH cycle. This coordinated action of GABA in regulating both the synthesis and recycling of ASC and GSH amplifies the plant's antioxidant capacity, enabling it to cope with prolonged or severe oxidative stress. In addition to activating antioxidant defenses, ROS modulate redox homeostasis by interacting with other redox-active molecules such as GSH and ASC, which are also components of the ASC-GSH cycle (Foyer and Kunert 2024). This cycle plays a vital role in detoxifying ROS and maintaining the cellular redox environment. ROS-driven signaling also triggers autophagy and PCD under severe oxidative stress conditions, thereby eliminating damaged cells and preserving overall plant integrity (Foyer and Kunert 2024).

5.4.4 GABA Cross Talk with Other Signaling Molecules in Defense Modulation

The regulatory role of GABA extends beyond direct enzyme modulation, as it interacts with other key signaling molecules involved in stress defense. This cross talk between GABA and other metabolites enhances the plant's ability to manage oxidative stress more effectively.

GABA cross talks with other signaling molecules, including Ca^{2+}, ROS, hormones, and nitric oxide (NO), in orchestrating plant defense responses. This cross talk between GABA and other signaling molecules represents an integrated stress response, ensuring that multiple defense systems are activated simultaneously to cope with the challenges of oxidative stress.

5.4.4.1 GABA and Calcium Signaling

Calcium ions (Ca^{2+}) are essential signaling molecules in plants, helping them respond to various environmental stresses, such as pathogen attacks, drought, and salinity (Tuteja and Mahajan 2007). GABA plays a significant role in regulating Ca^{2+} dynamics in plants. It does this by influencing the activity of calcium channels and transporters (Ma et al. 2019). GABA can bind to glutamate receptor–like (GLR) channels, which are special channels that allow Ca^{2+} to enter cells. When GABA binds to these channels, it triggers a flow of Ca^{2+} into the cell, activating a series of downstream signals that strengthen the plant's defense mechanisms. This interaction between GABA and Ca^{2+} is especially important during pathogen attacks, where rapid changes in Ca^{2+} levels are needed to activate genes and proteins involved in defense.

GABA plays a protective role in plants during salinity stress, and its supplementation in barley greatly increases phenolic compounds related to stress defense involving Ca^{2+} signaling (Ma et al. 2019). GABA also triggers an increase in calmodulin (CaM) protein that binds Ca^{2+} and enhances the activity of Ca^{2+}-ATPase, an enzyme that helps regulate Ca^{2+} levels in cells. Additionally, GABA upregulates the expression of calcium-binding proteins, such as calcium-dependent protein kinase (CDPK) and calmodulin-dependent protein kinase (CaMK), which are crucial for stress responses. These changes in Ca^{2+} signaling and distribution, particularly in root tips, highlight the importance of Ca^{2+} in GABA-mediated stress tolerance. The accumulation of phenolic compounds under salinity stress further emphasizes the role of Ca^{2+} in GABA's signaling pathway, helping plants cope with salt stress (Ma et al. 2019).

5.4.4.2 GABA and Nitric Oxide

Nitric oxide is a key signaling molecule involved in plant immunity, often acting in concert with GABA. The application of NO and GABA promotes plant growth and stress tolerance under various adverse conditions (Jiang 2023), such as salinity, water deficit, and metal toxicity. Both NO donors (e.g., sodium nitroprusside) and GABA improve physiological parameters like photosynthetic efficiency, antioxidant defense, and ion homeostasis, while reducing oxidative damage and sodium accumulation in turfgrass and white clover (Ji et al. 2019; Jiang 2023). Metabolomic analyses reveal that NO and GABA activate overlapping and distinct metabolic pathways, increasing the levels of amino acids, sugars, organic acids, and sugar alcohols, with specific compounds like citric acid linked to NO and glycine, methionine, and aconitic acid associated with GABA (Li et al. 2020). GABA-induced NO generation plays a critical role in stress tolerance, with NO production mediated by pathways involving nitrate reductase (NR) and nitric oxide synthase (NOS)–like activity. GABA enhances NO accumulation under stress conditions, which, in turn, promotes phenolic compound biosynthesis and antioxidant responses. This interplay between GABA and NO is crucial for mitigating stress-induced damage, as evidenced by studies on creeping bentgrass (Tang et al. 2020), muskmelon (Xu et al. 2021), soybean (Xie et al. 2021), and wheat (Khanna et al. 2021). Additionally, NO can act upstream of GABA, modulating GABA synthesis and catabolic enzymes to improve cold tolerance in bamboo (Wang et al. 2017a, b). The relationship between NO and GABA also extends to metal toxicity, where NO is essential for GABA-dependent stress tolerance, but GABA is not required for NO-mediated improvements in tomatoes and brinjals (Suhel et al. 2023; Suhel et al. 2024). Furthermore, NO and GABA interact in pathogen defense, with GABA regulating NO levels through enzymes like S-nitrosoglutathione reductase (GSNOR) to maintain resistance in cherry tomatoes (Wang et al. 2022). NO signaling involves second messengers like cGMP and protein kinase G, which mediate GABA accumulation under UV-B stress in soybean sprouts (Jiao et al. 2019). Posttranslational modifications, such as S-nitrosylation, also play a role in NO–GABA interactions, influencing enzyme activity and stress responses. Overall, the synergistic and independent actions of NO and GABA highlight their importance in enhancing plant resilience to abiotic and biotic stresses through complex signaling and metabolic networks.

5.4.4.3 GABA and ROS

GABA plays a pivotal role in regulating oxidative stress by influencing the cellular redox balance, which is essential for maintaining the equilibrium between the production of ROS and the cell's antioxidant capacity. This balance prevents oxidative damage while allowing ROS to function as signaling molecules. GABA contributes to this delicate equilibrium by modulating ROS generation and enhancing antioxidant defense mechanisms. For instance, GABA enhances the activity of key enzymes involved in redox regulation, such as GR, which maintains the reduced state of GSH, a major cellular antioxidant. By promoting GSH recycling, GABA ensures that the antioxidant pool remains replenished, enabling continuous detoxification of ROS. In addition, the GABA shunt supports the production of NADH and NADPH, which are essential cofactors for antioxidant enzymes like GR. This metabolic flexibility allows plants to simultaneously maintain energy production and redox balance, enhancing their ability to cope with oxidative stress (Gill and Tuteja 2010; Mittler et al. 2011).

Beyond its direct role in antioxidant defense, GABA also modulates ROS-signaling pathways. At low concentrations, ROS such as H_2O_2 act as second messengers in stress-responsive signaling cascades. GABA helps regulate ROS levels, ensuring that these signaling functions are preserved without causing excessive oxidative damage to the cell. This dual role of GABA highlights its importance in maintaining cellular homeostasis under stress conditions.

ROS are oxygen-containing molecules with unpaired electrons (radicals) or O–O bonds, making them highly reactive. Radicals like the superoxide anion ($O_2^{\bullet -}$), hydroperoxide radical ($HO_2^{\bullet}$), and hydroxyl radical ($HO^{\bullet}$) are typically short-lived and remain localized to their site of generation. Non-radical ROS, such as singlet oxygen ($^1O_2$), ozone (O_3), and H_2O_2, are less reactive and can diffuse across cellular membranes, allowing them to participate in signaling processes (Shields et al. 2021). While ROS are often considered harmful due to their ability to damage lipids, proteins, and nucleic acids, they also play essential roles in cellular signaling. For example, ROS can oxidize amino acids, fragment polypeptide chains, and induce protein cross-linking or aggregation, leading to increased protein degradation (Anjum et al. 2015; Romero-Puertas et al. 2002). In DNA, ROS cause strand breaks, base modifications, and chromatin fragmentation, disrupting genetic integrity (Juan et al. 2021). Lipid peroxidation, initiated by hydroxyl radicals, generates reactive intermediates like lipid hydroperoxides, which can further degrade into aldehydes and alcohols, perpetuating oxidative damage (Mohammadi and Karr 2001; Juan et al. 2021).

Despite their destructive potential, ROS like H_2O_2 also serve as critical signaling molecules. At low concentrations, H_2O_2 activates defense mechanisms in plants under stress, while at high concentrations, it can cause significant cellular damage (Gill and Tuteja 2010; Mittler et al. 2011). This dual nature of H_2O_2 underscores the importance of maintaining a balance between its beneficial and harmful effects. H_2O_2 is relatively stable compared to other ROS and can be transported across cell membranes via aquaporins, facilitating its role in signaling (Miller et al. 2010).

H_2O_2 production occurs in various cellular compartments, including chloroplasts, mitochondria, peroxisomes, the endoplasmic reticulum, and the apoplast. Key sources of H_2O_2 include disruptions in the electron transport chain in chloroplasts and mitochondria, plasma membrane NADPH oxidases (RBOH), and cell wall–associated peroxidases (Foyer and Noctor 2003; Gupta et al. 2016). NADPH oxidases generate superoxide anions, which are rapidly converted to H_2O_2 by SOD. Additionally, Polyamines (PAs) oxidases, e.g., DAO, contribute to H_2O_2 production during PA degradation (Moschou et al. 2008).

Plants exhibit remarkable tolerance to high levels of H_2O_2, which would be lethal to animal cells. This tolerance is attributed to their highly efficient antioxidant systems, which enable them to survive under stressful conditions (Huang et al. 2019). The ability to regulate ROS production and maintain redox balance is critical for plant survival and adaptation to environmental challenges. GABA, through its modulation of ROS levels and enhancement of antioxidant defenses, ensures that plants can effectively manage oxidative stress while preserving essential cellular functions.

5.4.4.4 GABA and PAs

PAs and GABA work together to regulate oxidative stress responses in plants, maintaining a balance in ROS levels (Ansari et al. 2021). Under stress conditions, the catabolism of PAs, driven by enzymes like DAO, leads to the production of GABA, a process that also generates H_2O_2. This H_2O_2, as a signaling molecule, activates antioxidant defenses to mitigate oxidative damage (Shelp et al. 2012; Gupta et al. 2016). Stressful conditions induce the activities of GAD and DAO, promoting GABA synthesis and H_2O_2 production, which together help plants cope with stress (Bouche and Fromm 2004). The close relationship between PAs and GABA is evident as their levels fluctuate in response to environmental stresses, such as hypoxia, and return to normal once the stress is removed (Song et al. 2012). Major PAs like spermidine, spermine, and putrescine act as signaling molecules that mediate plant responses to various stresses, while GABA plays a protective role by regulating PA metabolism and enhancing stress tolerance (Hu et al. 2015; Wang et al. 2017a, b).

PA degradation through DAO and Polyamine Oxidase (PAO) produces GABA, which is further metabolized into succinate and enters the Krebs cycle, contributing to energy production (Duan et al. 2008). However, only a portion of GABA is derived from PA degradation, as seen in germinating fava beans under hypoxia (Yang et al. 2013). Exogenous GABA application can suppress the expression and activity of DAO and PAO, altering PA degradation more than biosynthesis, suggesting a feedback mechanism where high GABA levels inhibit its own production (Wang et al. 2014). GABA also helps maintain PA levels, reducing damage under stress conditions like hypoxia (Wang et al. 2014). In drought-stressed plants, exogenous GABA enhances endogenous GABA content, improving tolerance by regulating GABA shunt, PA, and proline metabolism, while reducing oxidative damage markers like electrolyte leakage and lipid peroxidation (Yong et al. 2017). GABA treatment also increases PA biosynthesis and suppresses PA catabolism, leading to higher levels of putrescine and spermidine, as well as enhanced proline accumulation, which further aids stress adaptation (Yong et al. 2017).

PAs directly regulate antioxidant enzymes, such as catalase, and are involved in ROS signaling during plant–pathogen interactions (Sung et al. 2011; Lou et al. 2016). For example, spermine catabolism triggers a hypersensitive response in tobacco and cotton, while exogenous spermine induces H_2O_2 production, aiding signal transduction in *Arabidopsis* during viral infections (Mo et al. 2015; Yoda et al. 2003). Conversely, inhibiting putrescine catabolism reduces H_2O_2 production, suppressing immune responses in *Arabidopsis* (Lou et al. 2016). GABA interplay with PAs provides additional regulation of ROS production (Seifi and Shelp 2019; Gupta et al. 2016). The strong connection between GABA and PA metabolism under both abiotic and biotic stress highlights the potential for strategically modulating these pathways to control ROS signaling and enhance stress tolerance (Hatmi et al. 2015; Podlešáková et al. 2019).

5.4.4.5 GABA and Phytohormones

GABA does not act in isolation but interacts with various hormonal signaling pathways that are crucial for stress adaptation (Podlešáková et al. 2019; Kabała and Janicka 2024). One of the key interactions is with ABA, a hormone central to stress responses. ABA signaling is known to regulate a wide range of stress-responsive genes and physiological processes, including stomatal closure, osmotic adjustment, and antioxidant defenses. GABA modulates ABA signaling by influencing the synthesis and degradation of ABA and affecting the expression of ABA-responsive genes.

Similarly, GABA interacts with ethylene signaling, which is involved in regulating plant responses to stress, including oxidative stress. Ethylene-signaling pathways control various stress responses, such as cell wall reinforcement and pathogen defense. GABA modulates ethylene biosynthesis and signaling by interacting with key ethylene-responsive transcription factors and enzymes. This interaction ensures that the plant can integrate signals from both GABA and ethylene pathways to mount a coordinated defense response (Hijaz et al. 2018).

During oxidative stress, GABA and phytohormones work synergistically to enhance plant resilience. In *Pinus radiata*, elevated levels of Indole-3-acetic acid (IAA), ABA, SA, JA, and zeatin riboside provide protection from stress (De Diego et al. 2015). In creeping bentgrass (*Agrostis stolonifera*), GABA, ABA, and SA collectively mitigate oxidative damage by stabilizing membranes, enhancing antioxidant enzyme activities, and maintaining water balance (Li et al. 2017a, b). GABA specifically stimulates amino acid and organic acid biosynthesis, which plays a vital role in scavenging ROS. On the other hand, ABA enhances the TCA cycle intermediates, and SA promotes carbohydrate accumulation. Furthermore, the transgenic creeping bentgrass with elevated cytokinin levels increases GABA production, indicating a link between cytokinin signaling and oxidative stress tolerance (Li et al. 2017a, b).

Exogenous GABA application confers oxidative stress resilience in apple seedlings by enhancing ABA signaling, reducing transpiration, and increasing stomatal closure (Liu et al. 2021). In *Arabidopsis*, BRL3 overexpression (a brassinosteroid receptor) results in high GABA, proline, and tyrosine levels under oxidative stress, highlighting a connection between GABA and brassinosteroid signaling (Fàbregas et al. 2018).

Under salinity-induced oxidative stress, GABA helps regulate hormonal equilibrium. In *Cassia italica*, GABA priming ushers in high auxin and gibberellin levels and reduces ABA and JA concentrations, promoting oxidative stress tolerance (Alqarawi et al. 2016). In *Caragana intermedia*, GABA promotes ABA receptor genes and ethylene synthesis in response to salt stress. Poplar seedlings treated with GABA exhibit an altered expression of ethylene biosynthesis and signaling genes, reinforcing GABA's role in modulating hormonal responses during oxidative stress. In *Cucurbita pepo*, epibrassinolide (EBL) pretreatment results in GABA accumulation that inhibits lipid peroxidation and boosts antioxidant defenses (Shi et al. 2010). Similarly, in grapevine, exogenous GABA, brassinolide, and SA improve growth and fruit quality under salinity by alleviating oxidative stress (Zhao et al. 2023).

GABA also plays a key role in temperature-induced oxidative stress adaptation. In *Origanum vulgare*, GABA provides superior protection under salt stress, while SA becomes more effective against heat stress (Garoosi et al. 2023). In *Capsicum annuum*, SA increases GAD activity under osmotic stress but does not directly enhance GABA accumulation. In chilling-sensitive *Capsicum annuum* fruit, GABA levels correlate with reduced seed browning, suggesting its role in auxin-mediated cold stress responses (Çekiç 2017).

Overall, GABA acts as a crucial regulator of oxidative stress tolerance by interacting with multiple phytohormones. Through its role in modulating metabolism, enhancing antioxidant enzyme activities, and maintaining hormonal balance, GABA contributes to plant resilience against oxidative stress, reinforcing its importance as a key adaptive mechanism.

5.5 Conclusion

GABA's multifaceted role in modulating defense mechanisms during oxidative stress underscores its importance as a versatile regulator in plant stress responses. By acting as a key player in ROS scavenging, influencing antioxidant enzyme activities, and integrating with hormonal and MAPK-signaling pathways, GABA enhances the plant's ability to manage oxidative damage and maintain cellular homeostasis. This comprehensive involvement in stress signaling and defense mechanisms positions GABA as a critical factor in enhancing plant resilience. The implications of GABA research for sustainable agriculture are significant, offering potential strategies for developing crops with improved stress tolerance and resilience. By leveraging GABA's role in stress modulation, agricultural practices can be optimized to improve crop protection and productivity, contributing to more sustainable and resilient agricultural systems.

References

Abdel Razik ES, Alharbi BM, Pirzadah TB, Alnusairi GS, Soliman MH, Hakeem KR (2021) γ-Aminobutyric acid (GABA) mitigates drought and heat stress in sunflower (Helianthus annuus L.) by regulating its physiological, biochemical and molecular pathways. Physiol Plant 172:505–527

AbdElbar OH, Elkelish A, Niedbała G, Farag R, Wojciechowski T, Mukherjee S, Abou-Hadid AF, El-Hennawy HM, Abou El-Yazied A, Abd El-Gawad HG et al (2021) Protective effect of γ-aminobutyric acid against chilling stress during reproductive stage in tomato plants through modulation of sugar metabolism, chloroplast integrity, and antioxidative defence systems. Front Plant Sci 12:663750

Abdullah, Wani KI, Naeem M et al (2024) From neurotransmitter to plant protector: the intricate world of GABA signaling and its diverse functions in stress mitigation. J Plant Growth Regul. https://doi.org/10.1007/s00344-024-11470-0

Aghdam SM, Naderi R, Jannatizadeh A, Sarcheshmeh MAA, Babalar M (2016) Enhancement of postharvest chilling tolerance of Anthurium cut flowers by γ-aminobutyric acid (GABA) treatments. Sci Hortic 198:52–60

Aljuaid BS, Ashour H (2022) Exogenous γ-aminobutyric acid (GABA) application mitigates salinity stress in maize plants. Life 12:1860

Allan WL, Clark SM, Hoover GJ, Shelp BJ (2009) Role of plant glyoxylate reductases during stress: a hypothesis. Biochem J 423:15–22

Alqarawi AA, Hashem A, Abd_Allah EF, Al-Huqail AA, Alshahrani TS, Alshalawi SR, Egamberdieva D (2016) Protective role of γ-aminobutyric acid on Cassia italica Mill under salt stress. Legum Res 39:396–404

Al-Quraan NA, Al-Share AT (2016) Characterization of the γ-aminobutyric acid shunt pathway and oxidative damage in Arabidopsis thaliana pop2 mutants under various abiotic stresses. Biol Plant 60:132–138

Al-Quraan NA, Sartawe FAB, Qaryouti MM (2013) Characterization of γ-aminobutyric acid metabolism and oxidative damage in wheat (Triticum aestivum L.) seedlings under salt and osmotic stress. J Plant Physiol 170:1003–1009

Anjum NA, Sofo A, Scopa A, Roychoudhury A, Gill SS, Iqbal M, Lukatkin AS, Pereira E, Duarte AC, Ahmad I (2015) Lipids and proteins—major targets of oxidative modifications in abiotic stressed plants. Environ Sci Pollut Res 22:4099–4121

Ansari MI, Jalil SU, Ansari SA, Hasanuzzaman M (2021) GABA shunt: a key-player in mitigation of ROS during stress. Plant Growth Regul 94:131–149

Avery SV (2011) Molecular targets of oxidative stress. Biochem J 434(2):201–210

Ayala A, Muñoz MF, Argüelles S (2014) Lipid peroxidation: production, metabolism, and signaling mechanisms of malondialdehyde and 4-hydroxy-2-nonenal. Oxidative Med Cell Longev 2014:360438. https://doi.org/10.1155/2014/360438. Epub 2014 May 8. PMID: 24999379; PMCID: PMC4066722

Bouche N, Fromm H (2004) GABA in plants: just a metabolite? Trends Plant Sci 9:110–115

Bouché N, Fait A, Bouchez D, Møller SG, Fromm H (2003) Mitochondrial succinic-semialdehyde dehydrogenase of the γ-aminobutyrate shunt is required to restrict levels of reactive oxygen intermediates in plants. Proc Natl Acad Sci USA 100:6843–6848

Çekiç FÖ (2017) Effects of salicylic acid on GABA metabolism in pepper plants under stress conditions. Hacet J Biol Chem 3:359–364

De Diego N, Saiz-Fernández I, Rodríguez JL, Pérez-Alfocea P, Sampedro MC, Barrio RJ, Lacuesta M, Moncaleán P (2015) Metabolites and hormones are involved in the intraspecific variability of drought hardening in Radiata pine. J Plant Physiol 188:64–71
Demidchik V (2015) Mechanisms of oxidative stress in plants: from classical chemistry to cell biology. Environ Exp Bot 109:212–228
Devireddy AR, Tschaplinski TJ, Tuskan GA, Muchero W, Chen JG (2021) Role of reactive oxygen species and hormones in plant responses to temperature changes. Int J Mol Sci 22(16):8843. https://doi.org/10.3390/ijms22168843. PMID: 34445546; PMCID: PMC8396215
D'Autréaux B, Toledano MB (2007) ROS as signalling molecules: mechanisms that generate specificity in ROS homeostasis. Nat Rev Mol Cell Biol 8(10):813–824. https://doi.org/10.1038/nrm2256. PMID: 17848967
Duan JJ, Li J, Guo SR, Kang YY (2008) Exogenous spermidine affects polyamine metabolism in salinity-stressed Cucumis sativus roots and enhances short-term salinity tolerance. J Plant Physiol 165:1620–1635
Dumanović J, Nepovimova E, Natić M, Kuča K, Jaćević V (2021) The significance of reactive oxygen species and antioxidant defence system in plants: a concise overview. Front Plant Sci 11:552969
Fàbregas N, Lozano-Elena F, Blasco-Escámez D, Tohge T, Martínez-Andújar C, Albacete A, Osorio S, Bustamante M, Riechmann JL, Nomura T et al (2018) Overexpression of the vascular brassinosteroid receptor BRL3 confers drought resistance without penalizing plant growth. Nat Commun 9:4680
Foyer CH, Noctor G (2003) Redox sensing and signalling associated with reactive oxygen in chloroplasts, peroxisomes and mitochondria. Physiol Plant 119:355–364
Foyer CH, Kunert K (2024) The ascorbate-glutathione cycle coming of age. J Exp Bot 75(9): 2682–2699. https://doi.org/10.1093/jxb/erae023. PMID: 38243395; PMCID: PMC11066808
Garoosi MK, Sanjarian F, Chaichi M (2023) The role of γ-aminobutyric acid and salicylic acid in heat stress tolerance under salinity conditions in *Origanum vulgare* L. PLoS One 18:e0288169
Gill SS, Tuteja N (2010) Polyamines and abiotic stress tolerance in plants. Plant Signal Behav 5:26–33
Grene R (2002) Oxidative stress and acclimation mechanisms in plants. Arabidopsis Book 1:e0036
Guo Z, Lv J, Dong X, Du N, Piao F (2021) Gamma-aminobutyric acid improves phenanthrene phytotoxicity tolerance in cucumber through the glutathione-dependent system of antioxidant defence. Ecotoxicol Environ Saf 217:112254
Guo Z, Gong J, Luo S, Zuo Y, Shen Y (2023) Role of gamma-aminobutyric acid in plant defence response. Metabolites 13:741
Gupta K, Sengupta A, Chakraborty M, Gupta B (2016) Hydrogen peroxide and polyamines act as double-edged swords in plant abiotic stress responses. Front Plant Sci 7:1343
Hasanuzzaman M, Bhuyan MHMB, Zulfiqar F, Raza A, Mohsin SM, Mahmud JA, Fujita M, Fotopoulos V (2020) Reactive oxygen species and antioxidant defence in plants under abiotic stress: revisiting the crucial role of a universal defence regulator. Antioxidants 9:681
Hatmi S, Gruau C, Trotel-Aziz P, Villaume S, Rabenoelina F, Baillieul F, Aziz A (2015) Drought stress tolerance in grapevine involves activation of polyamine oxidation contributing to improved immune response and low susceptibility to *Botrytis cinerea*. J Exp Bot 66:775–787
Hijaz F, Nehela Y, Killiny N (2018) Application of gamma-aminobutyric acid increased the level of phytohormones in *Citrus sinensis*. Planta 248:909–918
Hu X, Xu Z, Xu W, Li J, Zhao N, Zhou Y (2015) Application of γ-aminobutyric acid demonstrates a protective role of polyamine and GABA metabolism in muskmelon seedlings under $Ca(NO_3)_2$ stress. Plant Physiol Biochem 92:1–10
Huang H, Ullah F, Zhou D-X, Yi M, Zhao Y (2019) Mechanisms of ROS regulation of plant development and stress responses. Front Plant Sci 10:800
Islam SNU, Kouser S, Hassan P et al (2024) Gamma-aminobutyric acid interactions with phytohormones and its role in modulating abiotic and biotic stress in plants. Stress Biol 4:36
Jakubczyk K, Drużga A, Katarzyna J, Skonieczna-Żydecka K (2020) Antioxidant potential of curcumin-a meta-analysis of randomized clinical trials. Antioxidants (Basel) 9(11):1092. https://doi.org/10.3390/antiox9111092. PMID: 33172016; PMCID: PMC7694612

Jalil SU, Ansari MI (2020) Physiological role of gamma-aminobutyric acid in salt stress tolerance. In: Salt and drought stress tolerance in plants. Springer, Cham, pp 337–350

Jalil SU, Ahmad I, Ansari MI (2017) Functional loss of GABA transaminase (GABA-T) expressed early leaf senescence under various stress conditions in *Arabidopsis thaliana*. Curr Plant Biol 9:11–22

Jalil SU, Khan MAA, Ansari MI (2019a) Tremendous role of GABA shunt metabolic pathway in plant system. J Biol Chem Res 36:188–196

Jalil SU, Khan MIR, Ansari MI (2019b) Role of GABA transaminase in the regulation of development and senescence in *Arabidopsis thaliana*. Curr Plant Biol 19:100119

Ji Z, Camberato JJ, Zhang C, Jiang Y (2019) Effects of 6-benzyladenine, γ-aminobutyric acid, and nitric oxide on plant growth, photochemical efficiency, and ion accumulation of perennial ryegrass cultivars to salinity stress. HortScience 54:1418–1422

Jiang Y (2023) Application of gamma-aminobutyric acid and nitric oxide on turfgrass stress resistance: current knowledge and perspectives. Grass Res 3:1–6

Jiao C, Duan Y, Lin Q (2019) MAPK mediates NO/cGMP-induced GABA accumulation in soybean sprouts. LWT 100:253–262

Jin Y, Zhi L, Tang X, Chen Y, Hancock JT, Hu X (2023) The function of GABA in plant cell growth, development and stress response. Phyton Int J Exp Bot 92:2211–2225

Juan CA, Pérez de la Lastra JM, Plou FJ, Pérez-Lebeña E (2021) The chemistry of reactive oxygen species (ROS) revisited: outlining their role in biological macromolecules (DNA, lipids and proteins) and induced pathologies. Int J Mol Sci 22:4642

Kabała K, Janicka M (2024) Relationship between the GABA pathway and signaling of other regulatory molecules. Int J Mol Sci 25:10749. https://doi.org/10.3390/ijms251910749

Kalhor MS, Aliniaeifard S, Seif M, Asayesh EJ, Bernard F, Hassani B, Li T (2018) Enhanced salt tolerance and photosynthetic performance: implication of 7-amino butyric acid application in salt-exposed lettuce (*Lactuca sativa* L.) plants. Plant Physiol Biochem 130:157–172

Khan MIR, Jalil SU, Chopra P, Chhillar H, Ferrante A, Khan NA, Ansari MI (2021) Role of GABA in plant growth, development and senescence. Plant Gene 26:100283

Khanna RR, Jahan B, Iqbal N, Khan NA, AlAjmi MF, Tabish Rehman M, Khan MIR (2021) GABA reverses salt-inhibited photosynthetic and growth responses through its influence on NO-mediated nitrogen-sulfur assimilation and antioxidant system in wheat. J Biotechnol 325:73–82

Li Y, Fan Y, Ma Y, Zhang Z, Yue H, Wang L, Li J, Jiao Y (2017a) Effects of exogenous γ-aminobutyric acid (GABA) on photosynthesis and antioxidant system in pepper (*Capsicum annuum* L.) seedlings under low light stress. J Plant Growth Regul 36:436–449

Li Z, Yu J, Peng Y, Huang B (2017b) Metabolic pathways regulated by abscisic acid, salicylic acid and γ-aminobutyric acid in association with improved drought tolerance in creeping bentgrass (*Agrostis stolonifera*). Physiol Plant 159:42–58

Li L, Wei S, Shen W (2020) The role of methane in plant physiology: a review. Plant Cell Rep 39:171–179

Li L, Dou N, Zhang H, Wu C (2021) The versatile GABA in plants. Plant Signal Behav 16(3):1862565

Liu C, Wang H, Zhang X, Ma F, Guo T, Li C (2021) Activation of the ABA signal pathway mediated by GABA improves the drought resistance of apple seedlings. Int J Mol Sci 22:12676

Lou YR, Bor M, Yan J, Preuss AS, Jander G (2016) *Arabidopsis* NATA1 acetylates putrescine and decreases defence related hydrogen peroxide accumulation. Plant Physiol 171(2):1443–1455

Ma Y, Wang P, Gu Z, Tao Y, Shen C, Zhou Y, Han Y, Yang R (2019) Ca involved in GABA signal transduction for phenolics accumulation in germinated hulless barley under NaCl stress. Food Chem X 2:100023

Malekzadeh P, Khara J, Heydari R (2014) Alleviating effects of exogenous Gamma-aminobutiric acid on tomato seedling under chilling stress. Physiol Mol Biol Plants 20(1):133–137. https://doi.org/10.1007/s12298-013-0203-5

Mazzucotelli E, Tartari A, Cattivelli L, Forlani G (2006) Metabolism of γ-aminobutyric acid during cold acclimation and freezing and its relationship to frost tolerance in barley and wheat. J Exp Bot 57(14):3755–3766

Mei X, Chen Y, Zhang L et al (2016) Dual mechanisms regulating glutamate decarboxylases and accumulation of gamma-aminobutyric acid in tea (*Camellia sinensis*) leaves exposed to multiple stresses. Sci Rep 6:23685. https://doi.org/10.1038/srep23685

Miller EW, Dickinson BC, Chang CJ (2010) Aquaporin-3 mediates hydrogen peroxide uptake to regulate downstream intracellular signaling. Proc Natl Acad Sci USA 107:15681–15686

Mittler R, Vanderauwera S, Suzuki N, Miller G, Tognetti VB, Vandepoele K, Gollery M, Shulaev V, Van Breusegem F (2011) ROS signaling: the new wave? Trends Plant Sci 16:300–309

Mo H, Wang X, Zhang Y, Zhang G, Zhang J, Ma Z (2015) Cotton polyamine oxidase is required for spermine and camalexin signalling in the defence response to *Verticillium dahliae*. Plant J 83:962–975

Mohammadi M, Karr AL (2001) Membrane lipid peroxidation, nitrogen fixation and leghemoglobin content in soybean root nodules. J Plant Physiol 158:9–19

Moschou PN, Paschalidis KA, Roubelakis-Angelakis KA (2008) Plant polyamine catabolism. Plant Signal Behav 3:1061–1066

Nawaz M, Sun J, Shabbir S, Khattak WA, Ren G, Nie X, Bo Y, Javed Q, Du D, Sonne C (2023) A review of plants strategies to resist biotic and abiotic environmental stressors. Sci Total Environ 900:165832. https://doi.org/10.1016/j.scitotenv.2023.165832. Epub 2023 Jul 29. PMID: 37524179

Plotnikov A, Zehorai E, Procaccia S, Seger R (2011) The MAPK cascades: signaling components, nuclear roles and mechanisms of nuclear translocation. Biochim Biophys Acta 1813(9):1619–1633. https://doi.org/10.1016/j.bbamcr.2010.12.012. Epub 2010 Dec 16. PMID: 21167873

Podlešáková K, Ugena L, Spíchal L, Doležal K, De Diego N (2019) Phytohormones and polyamines regulate plant stress responses by altering GABA pathway. New Biotechnol 48:53–65

Rahman M, Ghosh PK, Akter M, Al Noor MM, Rahman MA et al (2024) Green vanguards: harnessing the power of plant antioxidants, signal catalysts, and genetic engineering to combat reactive oxygen species under multiple abiotic stresses. Plant Stress 13:100547. https://doi.org/10.1016/j.stress.2024.100547

Ramos-Ruiz R, Martinez F, Knauf-Beiter G (2019) The effects of GABA in plants. Cogent Food Agric 5:1670553

Repetto M, Semprine J, Boveris A (2012) Lipid peroxidation: chemical mechanism, biological implications and analytical determination. In: Catala A (ed) Lipid peroxidation. IntechOpen, London, pp 4–30. https://doi.org/10.5772/45943

Romero-Puertas MC, Palma JM, Gómez M, Del Río LA, Sandalio LM (2002) Cadmium causes the oxidative modification of proteins in pea plants. Plant Cell Environ 25:677–686

Sachdev S, Ansari SA, Ansari MI, Fujita M, Hasanuzzaman M (2021) Abiotic stress and reactive oxygen species: generation, signaling, and defence mechanisms. Antioxidants (Basel) 10(2):277. https://doi.org/10.3390/antiox10020277

Salah A, Zhan M, Cao C et al (2019) γ-Aminobutyric acid promotes chloroplast ultrastructure, antioxidant capacity, and growth of waterlogged maize seedlings. Sci Rep 9:484. https://doi.org/10.1038/s41598-018-36334-y

Santos A, Xiao L, Labandeira C, Néraudeau D, Dépré É et al (2022) Plant–insect interactions from the mid-Cretaceous at Puy-Puy (Aquitaine Basin, western France) indicates pref erential herbivory for angiosperms amid a forest of ferns, gymnosperms, and angiosperms. Bot Lett 169(4):568–587. https://doi.org/10.1080/23818107.2022.2092772.insu-03713015

Seifi HS, Shelp BJ (2019) Spermine differentially refines plant defence responses against biotic and abiotic stresses. Front Plant Sci 10:117. https://doi.org/10.3389/fpls.2019.00117

Shelp BJ, Bozzo GG, Trobacher CP, Zarei A, Deyman KL, Brikis CJ (2012) Hypothesis/review: contribution of putrescine to 4-aminobutyrate (GABA) production in response to abiotic stress. Plant Sci 193–194:130–135

Shelp BJ, Bown AW, Zarei A (2017) 4-Aminobutyrate (GABA): a metabolite and signal with practical significance. Botany 95:1015–1032

Shi S, Shi Z, Jiang Z, Qi L, Sun X, Li C, Liu J, Xiao W, Zhang S (2010) Effects of exogenous GABA on gene expression of *Caragana intermedia* roots under NaCl stress: regulatory roles for H2O2 and ethylene production. Plant Cell Environ 33:149–162

Shields HJ, Traa A, Van Raamsdonk JM (2021) Beneficial and detrimental effects of reactive oxygen species on lifespan: a comprehensive review of comparative and experimental studies. Front Cell Dev Biol 9:628157

Sita K, Kumar V (2020) Role of gamma amino butyric acid (GABA) against abiotic stress tolerance in legumes: a review. Plant Physiol Rep 25:654–663

Song H, Xu X, Wang H, Wang H, Tao Y (2010) Exogenous gamma-aminobutyric acid alleviates oxidative damage caused by aluminium and proton stresses on barley seedlings. J Sci Food Agric 90:1410–1416

Song S, Li J, Gao H, Li QY, Yang LW, Gong RJ (2012) Effect of exogenous γ-aminobutyric acid on inorganic nitrogen metabolism and mineral elements contents of melon seedling under hypoxia stress. Acta Hort Sin 39:695–704

Su N, Wu Q, Chen J, Shabala L, Mithöfer A, Wang H, Qu M, Yu M, Cui J, Shabala S (2019) GABA operates upstream of H+-ATPase and improves salinity tolerance in *Arabidopsis* by enabling cytosolic K+ retention and Na+ exclusion. J Exp Bot 70:6349–6361

Suhel M, Husain T, Prasad SM, Singh VP (2023) GABA requires nitric oxide for alleviating arsenate stress in tomato and brinjal seedlings. J Plant Growth Regul 42:670–683

Suhel M, Husain T, Prasad SM, Singh VP (2024) GABA and nitric oxide reduce AsV toxicity in tomato and brinjal seedlings by regulating sulphur assimilation and proline metabolism. Plant Stress 11:100338

Sung MS, Chow TJ, Lee TM (2011) Polyamine acclimation alleviates hypersalinity-induced oxidative stress in a marine green microalga, *Ulva fasciata*, by modulation of antioxidative enzyme gene expression. J Phycol 47:538–547

Tang M, Li Z, Luo L, Cheng B, Zhang Y, Zeng W, Peng Y (2020) Nitric oxide signal, nitrogen metabolism, and water balance affected by γ-aminobutyric acid (GABA) in relation to enhanced tolerance to water stress in creeping bentgrass. Int J Mol Sci 21:7460

Torres MA, Jones JD, Dangl JL (2006) Reactive oxygen species signaling in response to pathogens. Plant Physiol 141(2):373–378. https://doi.org/10.1104/pp.106.079467. PMID: 16760490; PMCID: PMC1475467

Tuteja N, Mahajan S (2007) Calcium signaling network in plants: an overview. Plant Signal Behav 2:79–85

Ullah A, Ali I, Noor J, Zeng F, Bawazeer S, Eldin SM, Asghar MA, Javed HH, Saleem K, Ullah S et al (2023) Exogenous γ-aminobutyric acid (GABA) mitigated salinity-induced impairments in mungbean plants by regulating their nitrogen metabolism and antioxidant potential. Front Plant Sci 13:1081188

Wang CY, Li JR, Xia QP, Wu XL, Gao HB (2014) Influence of exogenous γ-aminobutyric acid (GABA) on GABA metabolism and amino acid contents in roots of melon seedling under hypoxia stress. J Appl Ecol 25:2011–2018

Wang F, Chen ZH, Liu X, Colmer TD, Shabala L, Salih A, Zhao M, Shabala S (2017) Revealing the roles of GORK channels and NADPH oxidase in acclimation to hypoxia in *Arabidopsis*. J Exp Bot 68(12):3191–3204

Wang D, Li L, Xu Y, Limwachiranon J, Li D, Ban Z, Luo Z (2017a) Effect of exogenous nitro oxide on chilling tolerance, polyamine, proline, and γ-aminobutyric acid in bamboo shoots (*Phyllostachys praecox* f. *Prevernalis*). J Agric Food Chem 65:5607–5613

Wang Y, Gu W, Meng Y, Xie T, Li L, Li J (2017b) γ-Aminobutyric acid imparts partial protection from salt stress injury to maize seedlings by improving photosynthesis and upregulating osmoprotectants and antioxidants. Sci Rep 7:43609. https://doi.org/10.1038/srep43609

Wang X, Cao J, Qiao J, Pan J, Zhang S, Li Q, Wang Q, Gong B, Shi J (2022) GABA keeps nitric oxide in balance by regulating GSNOR to enhance disease resistance of harvested tomato against *Botrytis cinerea*. Food Chem 392:133299

Wojtaszek P (1997) Oxidative burst: an early plant response to pathogen infection. Biochem J 322(Pt3):681–692. https://doi.org/10.1042/bj3220681. PMID: 9148737; PMCID: PMC1218243

Wu Q, Su N, Huang X, Cui J, Shabala L et al (2021) Hypoxia-induced increase in GABA content is essential for restoration of membrane potential and preventing ROS-induced disturbance to ion homeostasis. Plant Commun 2(3):100188. https://doi.org/10.1016/j.xplc.2021.100188

Xie C, Wang P, Sun M, Gu Z, Yang R (2021) Nitric oxide mediates γ-aminobutyric acid signaling to regulate phenolic compounds biosynthesis in soybean sprouts under NaCl stress. Food Biosci 44:101356

Xu J, Liu T, Qu F, Jin X, Huang N, Wang J, Hu X (2021) Nitric oxide mediates γ-aminobutyric acid-enhanced muskmelon tolerance to salinity–alkalinity stress conditions. Sci Hortic 286:110229

Xu B, Feng X, Piechatzek A, Zhang S, Konrad KR, Kromdijk J, Hedrich R, Gilliham M (2024) The GABA shunt contributes to ROS homeostasis in guard cells of Arabidopsis. New Phytol 241:73–81

Yang A, Cao S, Yang Z, Cai Y, Zheng Y (2011) γ-Aminobutyric acid treatment reduces chilling injury and activates the defence response of peach fruit. Food Chem 129(4):1619–1622

Yang R, Guo Q, Gu Z (2013) GABA shunt and polyamine degradation pathway on γ-aminobutyric acid accumulation in germinating fava bean (*Vicia faba* L.) under hypoxia. Food Chem 136:152–159. https://doi.org/10.1016/j.foodchem.2012.08.008

Yoda H, Yamaguchi Y, Sano H (2003) Induction of hypersensitive cell death by hydrogen peroxide produced through polyamine degradation. Plant Physiol 132(4):1973–1198

Yong B, Xie H, Li Z, Li YP, Zhang Y, Nie G, Zhang XQ, Ma X, Huang LK, Yan YH (2017) Exogenous application of GABA improves PEG-induced drought tolerance positively associated with GABA-shunt, polyamines, and proline metabolism in white clover. Front Physiol 8:1107. https://doi.org/10.3389/fphys.2017.01107

Zhang W, Xiao S, Ahn DU (2013) Protein oxidation: basic principles and implications for meat quality. Crit Rev Food Sci Nutr 53(11):1191–1120

Zhao M, Li J, Shi X, Malik MS, Quan Y, Guo D, Wang L, Wang S (2023) Effects of exogenous plant regulators on growth and development of "Kyoho" grape under salt alkali stress. Front Plant Sci 14:1274684

Zhu X, Liao J, Xia X, Xiong F, Li Y, Shen J, Wen B, Ma Y, Wang Y, Fang W (2019) Physiological and iTRAQ-based proteomic analyses reveal the function of exogenous γ-aminobutyric acid (GABA)inimproving tea plant (Camellia sinensis L.) tolerance at cold temperature. BMC Plant Biol 19(1):1–20

GABA Against Nutrient Starvation

6

Abstract

Under nutrient-deficient conditions, plants reprogram their metabolic pathways to sustain growth and survival. GABA modulates nutrient uptake, carbon/nitrogen balance, and stress-related gene expression. It is rapidly synthesized via the decarboxylation of glutamate and accumulates in response to deficiencies in essential nutrients, such as nitrogen, phosphorus, and potassium. GABA not only serves as a carbon and nitrogen reservoir but also regulates ion transporters, maintains cellular pH homeostasis, and interacts with polyamines and proline to enhance stress resilience. Moreover, GABA influences root architecture, improves nutrient use efficiency, and supports energy production through its integration into the Tricarboxylic Acid (TCA) cycle via the GABA shunt. Recent findings highlight the role of GABA transporters and receptors in facilitating long-distance signaling under nutrient stress. This chapter discusses the biosynthesis, transport, and signaling mechanisms of GABA in plants during nutrient starvation, along with its interplay with metabolic and hormonal pathways. Understanding GABA's multifaceted role provides insights into developing nutrient-efficient crops capable of withstanding harsh environmental conditions.

Keywords

Cellular homeostasis · Crop resilience · GABA · Nutrient starvation · Plant stress response

Highlights Nutrient scarcity in soil poses significant challenges to plant growth and productivity.

S. U. Jalil et al., *GABA Signaling System and Abiotic Stress Tolerance in Plants*,
https://doi.org/10.1007/978-981-95-0437-4_6

- GABA plays a crucial role in plant response to nutrient starvation.
- GABA functions as a key player in nitrogen metabolism, stress response, and signaling.
- This chapter explores GABA's molecular mechanisms in plants under nutrient-deprived conditions.
- It highlights how GABA regulates physiological processes to maintain cellular homeostasis.
- GABA's potential applications in enhancing crop resilience and improving agricultural sustainability in nutrient-limited environments are discussed.
- This chapter provides insights into leveraging GABA for crop improvement strategies in agriculture.

6.1 Introduction

Mineral nutrients play an indispensable role in the growth and development of plants (Chen et al. 2023). Recent research has unveiled the intricate nature of nutrient uptake in plants, which is influenced by numerous factors, including external environmental conditions and inherent plant characteristics (Goldstein et al. 2013; Nath and Tuteja 2016). Plants possess a sophisticated array of mechanisms to deal with nutrient exploration, uptake, and assimilation in order to survive in diverse environmental conditions. At the core of this survival strategy lies the concept of nutrient starvation, wherein plants are deprived of essential elements, such as nitrogen, phosphorus, potassium, or other essential nutrients, crucial for their growth and vitality. The scarcity of these resources can trigger cascading physiological and biochemical responses, affecting plant health and yield. GABA is a four-carbon, non-protein amino acid compound that is widely distributed in animals, plants, and microorganisms (Wang et al. 2020). In recent years, GABA has emerged as a pivotal player in the dynamic orchestra of plant responses to nutrient starvation (Yang et al. 2023). This versatile molecule, originally recognized for its role as an inhibitory neurotransmitter in animals, has since been unveiled as a multifaceted mediator in the intricate signaling networks of plants (Fromm 2020). GABA's presence and influence extend far beyond its initial identification, extending into the realms of stress tolerance (Kaspal et al. 2021), defense against pathogens, and, as we shall explore in this chapter, protection against the perils of nutrient scarcity. Studies have revealed that applying exogenous GABA can promote plant growth and increase the contents of endogenous GABA, amino acids, and endogenous hormones by upregulating the expression of several critical genes for phytohormone synthesis (Ali et al. 2010, Hijaz et al. 2018). In terms of nutrient uptake, GABA may influence the metabolism of free amino acids by regulating Aluminum-Activated Malate Transporter (ALMT) activity, thereby affecting nitrogen metabolism and absorption in *Arabidopsis* (Batushansky et al. 2015).

This chapter navigates the scientific background that highlights GABA's profound influence on nutrient-starved plants, shedding light on its intricate biochemical pathways and signaling networks. GABA employs mechanisms that regulate nutrient allocation, modify root architecture, and communicate stress signals to the rest of the plant. Moreover, we will explore the broader implications of GABA-mediated responses for sustainable agriculture and food security in our rapidly changing world.

6.2 GABA Shunt: Metabolic Reprogramming During Starvation

During nutrient starvation, plants undergo significant metabolic reprogramming to sustain essential physiological processes. One critical adaptation is the activation of the GABA shunt bypass, which provides an alternative route for energy production when traditional pathways are compromised. Metabolic and transcriptional analyses of *Arabidopsis thaliana* have shed light on this mechanism, particularly under dark-induced starvation conditions.

GABA displays a tight correlation with succinate levels under stress conditions (Caldana et al. 2011). Succinate, an important intermediate of the TCA cycle, plays a pivotal role in respiration and energy production. Under starvation, GABA shunt activity ensures that succinate continues to be produced even when conventional carbon sources are depleted, allowing for the continuation of mitochondrial respiration. This metabolic shift becomes crucial in maintaining cellular energy balance.

On a transcriptional level, starving plants exhibit a notable decrease in the expression of SCOAS genes, which are essential for the TCA cycle function, alongside a simultaneous increase in the expression of Succinic Semialdehyde Dehydrogenase (SSADH), a key enzyme in the GABA shunt (Caldana et al. 2011). This gene expression pattern highlights the plant's strategic shift toward utilizing the GABA shunt as a bypass to maintain energy production in the face of carbon starvation. The upregulation of SSADH emphasizes the enhanced conversion of GABA to succinate, bypassing the TCA cycle steps that are energy-intensive and providing a more direct route to support respiration.

During nutrient scarcity, the metabolic adjustments highlight GABA's important role in maintaining energy production and point to its broader function as a signaling molecule. By influencing both metabolic pathways and gene expression under low-energy conditions, GABA acts as a key signal that coordinates the plant's adaptive responses, helping it survive prolonged periods of starvation.

6.3 Nutrient Starvation in Plants

Nutrient starvation in plants refers to the condition in which essential nutrients required for plant growth and metabolic functions are limited or unavailable in the environment. Nutrients essential for plant growth are classified into two categories: macronutrients (nitrogen, phosphorus, potassium, calcium, magnesium, and sulfur) and micronutrients (iron, manganese, zinc, copper, molybdenum, boron, chlorine, and nickel). Nutrient starvation can manifest as either single-nutrient deficiency, where only one nutrient is lacking, or multiple-nutrient deficiency, where plants are deprived of more than one essential element. Both forms affect critical processes such as photosynthesis, respiration, and protein synthesis, leading to diminished plant growth and productivity.

6.3.1 Global Significance of Nutrient Limitation in Agriculture

Nutrient limitation is a significant challenge in global agriculture, particularly in regions where soil nutrient content is naturally low or depleted due to over-farming and inadequate fertilization practices. Nutrient-poor soils are common in many developing countries, where resource-limited farmers cannot afford costly fertilizers. As a result, crop yields suffer, leading to food insecurity and economic instability. Nitrogen and phosphorus are the most commonly limiting nutrients in agricultural systems, and their deficiencies are closely linked to reduced crop productivity. To address this, sustainable agricultural practices such as balanced fertilization, crop rotation, and the use of organic amendments are increasingly being adopted. The global push for food security under the pressures of climate change further amplifies the need for efficient nutrient management in farming, highlighting the critical role of research and innovation in nutrient use efficiency and plant adaptation to nutrient-limited conditions.

6.3.2 Implications for Plant Growth and Development

Nutrient starvation has profound effects on plant physiology and development. When deprived of essential nutrients, plants experience stunted growth, chlorosis, necrosis, and poor root development, all of which compromise their ability to carry out photosynthesis and respiration. Nitrogen deficiency, for instance, leads to reduced chlorophyll content and leaf yellowing, limiting photosynthetic efficiency and reducing biomass production. Phosphorus deficiency affects energy transfer and root architecture, hampering plant resilience. Prolonged nutrient starvation can trigger adaptive mechanisms such as root elongation, altered hormonal responses, and enhanced nutrient remobilization from older tissues (Fig. 6.1).

6.3.3 GABA Accumulation in Response to Nutrient Starvation

GABA functions as both a metabolite and a signaling molecule, especially in response to nutrient starvation. When plants face limited nutrient availability, such as carbon or nitrogen starvation, they engage alternative metabolic pathways to sustain essential physiological functions. One such pathway is the GABA shunt, which becomes particularly prominent during these stress conditions (Fig. 6.1).

Studies on *Arabidopsis thaliana* mutants impaired in enzymes essential for respiration under extended dark-induced carbon starvation reveal a notable increase in

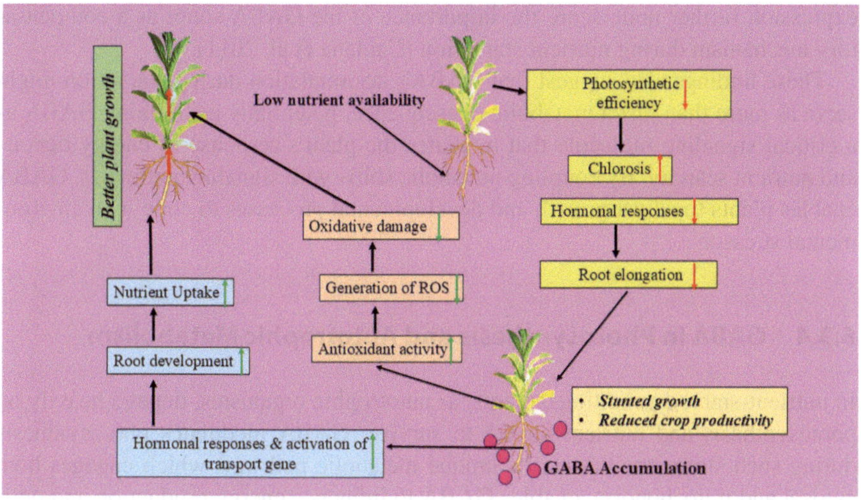

Fig. 6.1 Effect of low nutrient availability on plants and the role of GABA in nutrient starvation tolerance. Nutrient-deficient conditions induce chlorosis, reduced photosynthetic efficiency, and impaired hormonal responses that manifest as stunted root elongation and diminished growth and crop productivity. Additionally, nutrient shortages trigger reactive oxygen species (ROS) generation and induce oxidative damage. However, GABA accumulation under nutrient stress activates antioxidant mechanisms, reducing ROS levels and oxidative damage. GABA also promotes root development by influencing hormonal responses and activating transport genes, which, in turn, enhance nutrient uptake. Thus, GABA supports plant growth and resilience during nutrient stress, mitigating the adverse effects of nutrient deficiency

amino acids typically associated with starvation-induced protein degradation. Alongside these amino acids, both GABA and the TCA cycle intermediate, succinate, significantly increase their levels (Araújo et al. 2010). Interestingly, neither GABA nor succinate is a direct product of protein degradation, suggesting that their accumulation during nutrient starvation points to a critical adaptive role. This elevated GABA concentration suggests that GABA shunt activity increases to produce mitochondrial succinate, maintaining respiration when traditional carbon sources are scarce. The conversion of GABA into succinate through the action of GABA-transaminase (GABA-T) and SSADH bypasses several steps of the TCA cycle, enabling plants to preserve energy production and cellular homeostasis under nutrient-limited conditions.

Further metabolic and transcriptional analyses conducted under dark-induced starvation have confirmed the centrality of GABA in responding to such stress. In these studies, GABA emerges as the dominant non-proteinogenic amino acid, closely correlated with succinate levels (Caldana et al. 2011). This metabolic adjustment allows plants to meet their energy needs and manage stress more effectively, supporting the notion that GABA's role extends beyond basic metabolism to include signaling functions that regulate stress response pathways. The reduced expression of TCA cycle–associated SCOAS genes and the simultaneous increase in SSADH expression further underscore the importance of the GABA shunt as a compensatory mechanism during nutrient starvation (Caldana et al. 2011).

These findings also suggest that GABA's accumulation during starvation might serve as more than just a metabolic consequence, potentially positioning GABA as a critical signaling molecule that mediates the plant's response to energy deficits and nutrient scarcity. By coupling metabolic shifts with signaling pathways, GABA enables plants to adjust growth and developmental processes to cope with environmental stressors.

6.3.4 GABA in Photosynthesis and Autotrophic Metabolism

In nutrient-starved conditions, plants, as autotrophic organisms, depend heavily on photosynthesis and nutrient uptake to survive. GABA modulates photosynthesis during such stress conditions. Its unique metabolic pathway, which engages both cytosol and mitochondria, enables GABA to influence energy production and optimize photosynthetic efficiency under nutrient-limiting conditions.

In silico studies on cyanobacteria have highlighted the significance of the GABA shunt in optimizing photosynthesis during autotrophic metabolism, demonstrating the importance of GABA in maintaining energy flow when carbon supplies are limited (Nogales et al. 2012). These findings suggest that even in prokaryotic organisms like cyanobacteria, GABA plays a pivotal role in ensuring efficient energy production via photosynthesis, particularly during periods of nutrient scarcity.

In higher plants such as rice and tomato, GABA's involvement in photosynthesis regulation extends beyond its mitochondrial role. Certain GABA-T isoforms have been identified in plastids, implying GABA's participation in plastidial functions, which are essential for the photosynthetic machinery (Clark et al. 2009; Shimajiri et al. 2013). The plastid localization of these GABA-T isoforms suggests a more direct role of GABA in modulating processes like carbon fixation and energy conversion, particularly under conditions where nutrients are limited.

One study demonstrates the practical implications of GABA's role in photosynthesis regulation, where exogenous GABA application augments photosynthesis parameters in muskmelon seedlings under hypoxic stress (Xia et al. 2011). This suggests that GABA not only supports energy production via the GABA shunt but also plays a regulatory role in maintaining photosynthetic efficiency during environmental stress, thereby helping plants cope with nutrient starvation and ensuring their survival.

6.3.5 GABA's Role in Nutrient Uptake

Beyond its metabolic functions, GABA exerts a significant influence on nutrient uptake, especially nitrogen. Its role in regulating nutrient absorption has been demonstrated in multiple plant species. For example, GABA controls nitrate uptake in *Brassica napus* (Beuve et al. 2004) and *Arabidopsis thaliana* (Barbosa et al. 2011).

These studies reveal that GABA-signaling pathways are crucial in modulating root function and architecture, directly influencing how plants absorb and utilize essential nutrients.

Internally, GABA's function as a metabolic safeguard comes to the forefront when plants face nutrient deprivation. Under conditions of carbon, hydrogen, nitrogen, and sulfur shortages, GABA application helps retain these essential elements, ensuring that critical physiological processes such as photosynthesis and mitochondrial respiration continue uninterrupted (Khan et al. 2023). GABA not only supports these life-sustaining processes but also acts as a key player in Ca^{2+} signaling. This signaling mechanism is crucial for balancing Na^+ and K^+ ions, especially under salt stress. By bolstering Ca^{2+} availability, GABA enables plants to maintain K^+/Na^+ selectivity, which prevents Na^+ toxicity by inhibiting its uptake through nonselective cation channels (Grenzi et al. 2021).

Moreover, as demonstrated in wheat plants subjected to both salt and endoplasmic reticulum stresses, GABA significantly impacts ionic homeostasis by reducing Na^+ accumulation while supporting the retention of K^+ and Ca^{2+} levels. This balance helps prevent photoinhibition and sustain energy production in stressed plants (Kumari et al. 2024). Furthermore, GABA's ability to enhance mineral consumption is particularly evident in *Lemna minor*, where its application significantly promotes growth and commands functions of GABA receptors remarkably similar to those found in the animal nervous system (Kinnersley and Lin 2000). These findings emphasize GABA's dual role both as a metabolite and a signaling molecule, helping plants optimize nutrient uptake, even under stress conditions like nutrient starvation. Through these signaling pathways, GABA can modify root architecture and function, ensuring efficient nutrient absorption and promoting overall plant growth.

6.3.6 GABA in Symbiotic Nitrogen Fixation

GABA has emerged as a significant player in the regulation of symbiotic nitrogen fixation in legumes, a process that is essential for plant growth in nitrogen-deficient environments. Legumes rely on specialized root structures called nodules, where symbiotic bacteria, known as *Rhizobia,* convert atmospheric nitrogen (N_2) into a form usable by plants. This biological nitrogen fixation is crucial for plant development, especially under nutrient-limiting conditions. GABA's role in enhancing this symbiotic relationship has been increasingly recognized, particularly for its regulatory influence over nodule activity and nitrogen fixation efficiency.

For instance, GABA significantly upregulates the activity of nodules in legume plants, contributing to more efficient nitrogen fixation through its interaction with nitrogen metabolism and signaling pathways within the plant (Sulieman and Schulze 2010). GABA mediates root nodule activity and enhances nitrogen fixation as and when legumes encounter nitrogen-starved conditions. Consequently, GABA levels within the plant tissues tend to increase, which appears to be a mechanism to enhance nitrogen fixation when external nitrogen availability is low. This adaptive response ensures that the plants can still acquire sufficient nitrogen for their growth and development despite environmental limitations.

In this context, GABA functions as more than just a metabolite; it also acts as a signaling molecule that modulates the interaction between the plant and *Rhizobia*. The increase in GABA levels in legumes triggers an upregulation of nodule activity, leading to a higher fixation rate of atmospheric nitrogen. This finding points to GABA as a key factor in the energy-intensive process of nitrogen fixation, where the symbiotic bacteria convert N_2 into ammonia (NH_3), which is then assimilated into amino acids and other nitrogen-containing compounds by the plant.

Sulieman (2011) has demonstrated that GABA not only promotes nodule activity but also directly influences the efficiency of nitrogen fixation in legumes. The accumulation of GABA during nitrogen-limiting conditions helps optimize the nodule function, ensuring that legumes can continue to fix nitrogen even when soil nitrogen levels are insufficient. This regulatory role of GABA is critical because nitrogen fixation is an energy-demanding process, and plants must allocate resources efficiently to sustain this function during nutrient stress.

GABA appears to play a dual role in this process. First, it acts as a metabolic intermediary, contributing to cellular respiration and energy production via the GABA shunt. Second, it serves as a signaling molecule, influencing gene expression and enzymatic activity associated with nitrogen fixation. Thus, GABA strikes a balance between plant growth and nutrient acquisition under stress conditions.

6.3.7 GABA and Soil Nutrient Availability

GABA also stimulates plants to overcome nutrient starvation by acting as both a metabolic modulator and a signaling molecule (Yang et al. 2023). Beyond its well-known internal regulatory functions, GABA extends its influence to the soil environment, where it actively shapes nutrient availability through its impact on root secretions. For instance, GABA application in loquat seedlings alters the composition of root exudates and lowers soil pH in such a fashion that it elevates the bioavailability of soil nitrogen and phosphorus (Yang et al. 2023). This modification of the soil environment is not an isolated phenomenon; GABA brings similar effects across different plant species, highlighting its dynamic role in optimizing nutrient uptake under stress conditions (Zhou 2004; Song et al. 2012).

The application of exogenous GABA or the overexpression of *MdGAD1* in apple seedling roots enhances not only the accumulation of endogenous indole-3-acetic acid (IAA) but also their tolerance to abiotic stress arising from low soil phosphorus. Elevated GABA levels contribute to improved root system architecture and the expansion of the mature root zone, which, in turn, activates phosphate starvation–induced (PSI) genes, leading to more efficient phosphorus uptake (Chen et al. 2023).

Hence, GABA plays a critical role in shaping root architecture, enabling plants to adapt to changes in water and nutrient availability. By maintaining key root traits such as seminal root angle, root area, diameter, and length, GABA ensures efficient nutrient acquisition even under extreme stress conditions (Kumari et al. 2024). This capacity to alter soil conditions likely stems from GABA's role in regulating root secretions, promoting organic acids that influence nutrient availability (Renault et al. 2013; Li et al. 2018).

Through its dual role in both internal metabolic processes and external nutrient optimization, GABA emerges as a central figure in the plant's response to nutrient starvation, ensuring resilience and adaptability in ever-changing environmental conditions. Its influence stretches from the root–soil interface to the intricate cellular machinery, weaving a narrative of survival and endurance against the odds of nutrient scarcity.

6.4 Conclusion

Thus, there exists an intricate relationship between GABA and nutrient starvation in plants. GABA's multifaceted roles as a protective agent, signaling molecule, and regulator of stress responses have far-reaching implications for crop improvement and sustainable agriculture. By elucidating the molecular mechanisms and physiological responses underlying GABA's actions, the quest for global food security and the optimization of agricultural practices in nutrient-limited environments appears to be highly feasible.

References

Ali MMB, Ashrafuzzaman MUPM, Ismail MRUP, Shahidullah MSUP, Prodhan AKMA (2010) Influence of foliar applied GABA on growth and yield contributing characters of white gourd (*Benincasa hispida*). Int J Agric Biol 12:373–376

Araújo WL, Ishizaki K, Nunes-Nesi A et al (2010) Identification of the 2-hydroxyglutarate and isovaleryl-CoA dehydrogenases as alternative electron donors linking lysine catabolism to the electron transport chain of *Arabidopsis* mitochondria. Plant Cell 22:1549–1563. https://doi.org/10.1105/tpc.110.075630

Barbosa JM, Singh NK, Cherry JH, Locy RD (2011) Nitrate uptake and utilization is modulated by exogenous γ-aminobutyric acid in *Arabidopsis thaliana* seedlings. Plant Physiol Biochem 48:443–450. https://doi.org/10.1016/j.plaphy.2010.01.020

Batushansky A, Kirma M, Grillich N et al (2015) The transporter GAT1 plays an important role in GABA-mediated carbon-nitrogen interactions in *Arabidopsis*. Front Plant Sci 6:785

Beuvé N, Rispail N, Laine P, Cliquet J-B, Ourry A, Le Deunff E (2004) Putative role of g-aminobutyric acid as a long distance signal in up-regulation of nitrate uptake in Brassica napus L. Plant Cell Environ 27(8):1035–1046. https://doi.org/10.1111/j.13653040.2004.01208.x

Caldana C, Degenkolbe T, Cuadros-Inostroza A et al (2011) High-density kinetic analysis of the metabolomic and transcriptomic response of *Arabidopsis* to eight environmental conditions. Plant J 67:869–884. https://doi.org/10.1111/j.1365-313X.2011.04640.x

Chen X, Tian Y, Xie Y et al (2023) GABA responds to low phosphorus stress by interfering with endogenous auxin levels in apple. Plant Soil 490:435–450. https://doi.org/10.1007/s11104-023-06088-2

Clark SM, Di Leo R, Van Cauwenberghe OR et al (2009) Subcellular localization and expression of multiple tomato γ-aminobutyrate transaminases that utilize both pyruvate and glyoxylate. J Exp Bot 60:3255–3267. https://doi.org/10.1093/jxb/erp161

Fromm H (2020) GABA signaling in plants: targeting the missing pieces of the puzzle. J Exp Bot 71:6238–6245

Goldstein G, Bucci SJ, Scholz FG (2013) Why do trees adjust water relations and hydraulic architecture in response to nutrient availability? Tree Physiol 33:238–240

Grenzi M, Resentini F, Vanneste S et al (2021) Illuminating the hidden world of calcium ions in plants with a universe of indicators. Plant Physiol 187:550–571

Hijaz F, Nehela Y, Killiny N (2018) Application of gamma-aminobutyric acid increased the level of phytohormones in *Citrus sinensis*. Planta 248:909–918

Kaspal M, Kanapaddalagamage MH, Ramesh SA (2021) Emerging roles of γ-aminobutyric acid (GABA) gated channels in plant stress tolerance. Plants 10:2178

Khan MIR, Nazir F, Maheshwari C et al (2023) Mineral nutrients in plants under changing environments: a road to future food and nutrition security. Plant Genome 21:e20362

Kinnersley AM, Lin F (2000) Receptor modifiers indicate that 4-aminobutyric acid (GABA) is a potential modulator of ion transport in plants. Plant Growth Regul 32:65–76. https://doi.org/10.1023/a:1006305120202

Kumari S, Nazir F, Singh A et al (2024) γ-Aminobutyric acid (GABA) strengthened nutrient accumulation, defense metabolism, growth and yield traits against salt and endoplasmic reticulum stress conditions in wheat plants. Plant Soil 498:409–429. https://doi.org/10.1007/s11104-023-06444-2

Li J, Tian Z, Wu X, Lv G, Ma W, Zhang Y, Gao H (2018) Gamma-aminobutyric acid (GABA) modulates nitrate concentrations and metabolism in the leaves of pakchoi (Brassica campestris ssp. chinensis Makino) treated with a nitrogen-rich solution. Plant Mol Biol Report 36:530–542

Nath M, Tuteja N (2016) NPKS uptake, sensing, and signaling and miRNAs in plant nutrient stress. Protoplasma 253:767–786

Nogales J, Gudmundsson S, Knight EM, Palsson BO, Thiele I (2012) Detailing the optimality of photosynthesis in cyanobacteria through systems biology analysis. Proc Natl Acad Sci USA 109:2678–2683. https://doi.org/10.1073/pnas.1117907109

Renault H, El AA, Berger A, Mouille G, Soubigou-Taconnat L, Bouchereau A, Deleu C (2013) Gamma-aminobutyric acid transaminase deficiency impairs central carbon metabolism and leads to cell wall defects during salt stress in Arabidopsis roots. Plant Cell Environ 36:1009–1018

Shimajiri Y, Ozaki K, Kainou K, Akama K (2013) Differential subcellular localization, enzymatic properties and expression patterns of γ-aminobutyric acid transaminases (GABA-Ts) in rice (Oryza sativa). J Plant Physiol 170:196–201. https://doi.org/10.1016/j.jplph.2012.09.007

Song SL, Li JR, Gao HB, Li QY, Yang LW, Gong RJ (2012) Effects of exogenous γ-aminobutyric acid on inorganic nitrogen metabolism and mineral elements contents of melon seedling under hypoxia stress. Acta Hortic Sin 39:695–704

Sulieman S (2011) Does GABA increase the efficiency of symbiotic N2 fixation in legumes? Plant Signal Behav 6:32–36. https://doi.org/10.4161/psb.6.1.14318

Sulieman S, Schulze J (2010) Phloem-derived γ-aminobutyric acid (GABA) is involved in upregulating nodule N2 fixation efficiency in the model legume Medicago truncatula. Plant Cell Environ 33:2162–2172. https://doi.org/10.1111/j.1365-3040.2010.02214.x

Wang SS, Liu XJ, Hu Y, Ji YL, Bai T, Zhu MX, Zhang YH (2020) Metabolism and enrichment mechanism of γ-aminobutyric acid in plants. J Anhui Agric Sci 48:9–12

Xia QP, Gao HB, Li JR (2011) Effects of γ-aminobutyric acid on the photosynthesis and chlorophyll fluorescence parameters of muskmelon seedlings under hypoxia stress. Chin J Appl Ecol 22:999–1006

Yang Z, Xu Y, Song P, Li X, Zhou J, Lin L, Xia H, Liang D, Luo X, Zhang H et al (2023) Effects of gamma amino butyric acid (GABA) on nutrient uptake of loquat [Eriobotrya japonica (Thunb.) Lindl.] seedlings. Horticulturae 9:196

Zhou X (2004) Physiological effects of salt stress-induced GABA accumulation in maize seedlings. China Agricultural University, Beijing

GABA as a Protectant Against Salinity Stress

7

Abstract

Salinity represents a major environmental hurdle that restricts plant growth and crop productivity on a global scale. Its detrimental effects stem from osmotic pressure, ion toxicity, oxidative stress, and metabolic disturbances. GABA has emerged as a vital molecule that contributes to salt tolerance in plants by facilitating osmotic regulation, maintaining ion equilibrium, activating antioxidant defenses, and influencing hormone signaling. This chapter offers an in-depth exploration of GABA's role in enhancing plant resilience under saline conditions. It discusses the synthesis and metabolic pathways of GABA, outlines the signaling mechanisms involved in GABA-mediated stress responses, and explains how GABA supports physiological and molecular adjustments in plants facing salinity stress. The chapter also highlights recent progress in identifying GABA receptors and their interactions with other stress-responsive networks. The insight distilled from the current knowledge and research findings would provide valuable inputs to engage GABA against salinity stress, ensuring sustainable crop productivity.

Keywords

GABA · Homeostasis · Plants · ROS · Salinity stress

Highlights

- Salinity stress disrupts plant growth and productivity by causing osmotic imbalance, ionic disturbances, oxidative stress, and metabolic disarray.
- GABA helps in osmotic adjustment, ion homeostasis, antioxidant defense, and hormonal regulation.

S. U. Jalil et al., *GABA Signaling System and Abiotic Stress Tolerance in Plants*,
https://doi.org/10.1007/978-981-95-0437-4_7

- GABA receptor/signaling cross talks with diverse stress-responsive pathways.
- Acting as a crucial signaling compound, GABA helps counteract the harmful effects of salinity in plants.
- GABA-based strategies are expected to enhance crop salinity tolerance, offering sustainable solutions for agriculture and food security.

7.1 Introduction

Soil salinity currently affects nearly 20% of the world's arable land and around 60% of land under irrigation, posing a severe challenge to global agriculture (Machado and Serralheiro 2017). This increasing salinization, driven by the overuse of saline irrigation water, inadequate land management, and the broader impacts of climate change, has emerged as a key factor threatening food security. Salinity-induced stress in plants affects their morphology, physiology, and biochemical functions, ultimately impairing growth and productivity (Ashraf 1993). When excess salts accumulate in the soil, they disturb the osmotic and ionic balance within plant cells, resulting in dehydration, ionic toxicity, and oxidative stress, which can cause irreversible cellular damage or even plant death (Fu and Yang 2023).

Recent studies have focused on understanding the mechanisms through which plants perceive and adapt to salinity stress. Such insights are vital for the development of resilient crop varieties capable of tolerating elevated salt levels. Among the various biochemical mediators and signaling molecules identified in plant stress responses, GABA has attracted significant interest for its role in alleviating salinity stress.

GABA is a small, nonprotein amino acid found across biological kingdoms, including plants, animals, and microbes. In plants, it is increasingly recognized for its multifunctional roles, including growth regulation and stress mitigation. Its involvement in salinity tolerance is supported by studies indicating that GABA functions as an osmolyte to protect cells from desiccation and as an antioxidant agent that neutralizes reactive oxygen species (ROS) during stressful conditions (Carillo 2018). The GABA shunt pathway plays a crucial role in modulating GABA metabolism, particularly under stress conditions (Al-Quraan et al. 2016; Jalil et al. 2017).

Exogenous application of GABA has been shown to improve plant development and bolster antioxidant defense by inducing the expression of genes associated with antioxidative enzymes in a concentration-dependent manner (Shi et al. 2010; Vijayakumari et al. 2016; Li et al. 2017). It also aids in maintaining osmotic balance, thereby increasing stress tolerance (Yu et al. 2014; Vijayakumari and Puthur 2015). GABA mitigates oxidative damage by controlling the transcription of genes responsible for producing both harmful free radicals and their scavengers (Shi et al. 2010). Despite promising results across various abiotic stress conditions, the specific contribution of GABA to salinity tolerance is still under-explored.

Elucidating the ways in which GABA enhances plant resistance to salinity stress is critical for advancing strategies to breed or engineer salt-tolerant crops. Leveraging

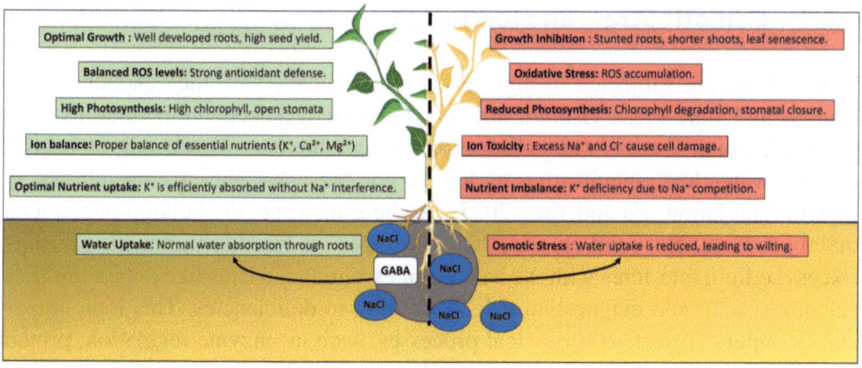

Fig. 7.1 Illustration of GABA's role in alleviating salinity stress in plants
The left side represents a healthy plant under optimal conditions, exhibiting well-developed roots, high seed yield, balanced ROS levels, efficient photosynthesis, proper ion balance, and optimal nutrient and water uptake. The right side illustrates the negative impacts of salinity stress, including growth inhibition, oxidative stress, reduced photosynthesis, ion toxicity, nutrient imbalance, and osmotic disruption, all contributing to poor development and premature aging. GABA application in the rhizosphere helps alleviate salinity stress by improving ion homeostasis, reducing oxidative stress, enhancing water, and supporting better nutrient and water absorption, thereby fostering healthier plant growth

GABA's potential could pave the way for more sustainable agricultural systems capable of withstanding salinity-induced challenges. This chapter seeks to synthesize the current understanding of GABA's protective functions in plants under salinity stress, offering a foundation for future research in this field. GABA-mediated salinity tolerance is largely attributed to its roles in activating antioxidative mechanisms, supporting osmotic balance, and regulating gene expression during salt exposure (Fig. 7.1). Ultimately, gaining deeper insights into GABA's functions could help shape innovative approaches to improve crop resilience, ensuring both productivity and sustainability in increasingly saline agricultural landscapes.

7.2 Salinity Stress Impact on Plant Growth

Salinity entails severe implications for plant growth, development, and agricultural productivity. Soil salinity, resulting from excessive salt concentrations, significantly hinders plant survival in both natural and agricultural ecosystems. The presence of excessive salts disrupts essential plant physiological processes, impairs water uptake, causes nutrient imbalance and oxidative damage, and alters metabolism. Gaining a clear understanding of these physiological disruptions is fundamental for designing effective strategies to increase salt tolerance in crops and ensure better performance under a saline environment.

7.2.1 Osmotic Stress and Ion Toxicity

When plants are subjected to high levels of salt in the soil, they encounter both osmotic and ionic stress. Osmotic stress is triggered by a decrease in the soil's water potential, which restricts water absorption through the roots. This condition leads to cellular water loss, manifesting as low turgor pressure, wilting, and ultimately, growth retardation (Fu and Yang 2023). At the same time, the buildup of salt ions, mainly sodium (Na^+) and chloride (Cl^-), disrupts ion equilibrium inside the cells. Excess sodium interferes with the balance of essential nutrients like potassium (K^+), calcium (Ca^{2+}), and magnesium (Mg^{2+}), leading to deficiencies. This ionic imbalance hampers critical physiological processes, such as enzyme regulation, protein synthesis, and photosynthetic efficiency (Sun et al. 2024).

7.2.2 Oxidative Stress

Salinity not only disturbs osmotic and ionic balance but also promotes the generation of ROS, including $O_2^{\bullet -}$, hydrogen peroxide (H_2O_2), and •OH, which damage plant cells (Hasanuzzaman et al. 2021). These ROS attack vital biomolecules, lipids, proteins, and DNA, causing oxidative damage and the disruption of cellular integrity. Additionally, ROS accumulation activates complex signaling networks that induce defensive responses (Schieber and Chandel 2014). To mitigate this stress, plants boost their antioxidant defense systems by increasing the activity of enzymes such as Superoxide Dismutase (SOD), Catalase (CAT), and Peroxidase (POD), which help neutralize ROS and restore redox balance (Rajput et al. 2021).

7.2.3 Nutritional Imbalance

Salinity significantly alters nutrient availability and transport in plants, leading to deficiencies that restrict their growth and development. High salt levels disrupt mineral nutrient uptake due to the competitive interaction between harmful ions like Na^+ and Cl^- and beneficial nutrients such as K^+, Ca^{2+}, and NO_3^-. This results in an imbalance that affects various structural and metabolic functions essential for plant vitality (Grattan and Grieves 1999; Jalil and Ansari 2020). Studies have shown that elevated NaCl levels lead to a sharp rise in sodium and chloride concentrations, accompanied by a decline in nitrogen, phosphorus, potassium, magnesium, and calcium contents in species like *Foeniculum vulgare* (Abd El-Wahab 2006), ajwain (Ashraf and Orooj 2006), peppermint and lemon verbena (*Mentha piperita* and *Aloysia citrodora*) (Tabatabaie and Nazari 2007), chamomile (*Matricaria recutita*) (Baghalian et al. 2008), and *Achillea* (Abd EL-Azim and Ahmed 2009).

7.2.4 Hyperosmotic Conditions and Water Uptake Disruption

Hyperosmotic stress caused by salinity limits water absorption by roots and increases water loss from leaves, mimicking drought conditions (Arzani and Ashraf 2016). The elevated salt concentration in the soil modifies the osmotic gradient between the soil and plant roots, resulting in a higher osmotic potential outside the root cells. This shift hinders the passive flow of water into the roots via osmosis (Atta et al. 2023). Typically, water moves from the soil (low solute concentration) into root cells (higher solute concentration). However, in saline soils, the increased external solute concentration lowers the water potential of the soil, reducing water availability to plant cells (Fu and Yang 2023). Consequently, plants experience internal water deficits and must adapt through physiological changes, such as reducing leaf surface area, closing stomata to conserve water, and adjusting their metabolism to cope with limited moisture availability (Lu and Fricke 2023).

7.2.5 Cellular Enzyme Inhibition and Ion Homeostasis

When plants encounter high salt concentrations in the soil or water, they experience multiple physiological and biochemical disturbances that impair normal cellular operations. A key outcome of salinity stress is the disturbance of ion homeostasis, which often results from the inhibition of critical cellular enzymes (Zhao et al. 2021). Under optimal conditions, plants sustain a fine balance of ions like Na^+, K^+, Ca^{2+}, and others within their cells to support essential biological functions. This balance is crucial for proper cellular functions, including enzyme activities, osmotic regulation, and signal transduction. However, salinity stress disrupts this delicate balance by directly inhibiting cellular enzymes involved in ion transport and regulation. High salt concentrations in the surrounding environment alter the electrochemical gradients across cell membranes that translate into the inhibition or malfunctioning of these enzymes. This disruption impairs the transport of essential ions, such as K^+ and Ca^{2+}, and promotes the accumulation of toxic ions within the plant cells. The imbalance between Na^+ and K^+ ions also disorients cellular signaling processes (Sun et al. 2024). Potassium is a crucial regulator of enzyme activities and is involved in numerous cellular functions, including photosynthesis, respiration, and protein synthesis. The presence of high Na^+ levels drastically alters the K^+/Na^+ ratio and impairs vital processes, with diminishing overall plant performance. The primary process responsible for energy acquisition and its conversion into carbon compounds for plant growth and maintenance is photosynthesis. Under salinity stress, photosynthesis is severely hindered, leading to reduced energy acquisition. Moreover, a portion of the energy is redirected toward tolerance and defense mechanisms. Additionally, excessive salinity causes harm to chloroplast proteins, reduces stomatal conductance, alters respiration rates, and promotes the accumulation of harmful ROS (Hameed et al. 2021). Consequently, ROS trigger augmented lipid peroxidation on chloroplast and mitochondrial membranes, resulting in the production of detrimental aldehydes. As a result, salinity not only diminishes energy production but also amplifies energy consumption. Along with this come the impairment in the transport of essential ions, such as K^+ and Ca^{2+}, and promotion of toxic ions' accumulation within the plant cells.

7.3 GABA in Salinity Stress Response

GABA is a vital component in plant responses to salinity stress, influencing a wide range of physiological activities. It contributes to plant resilience under saline conditions through various mechanisms. Below are some of the primary ways in which GABA aids plants in coping with salinity stress:

7.3.1 GABA Accumulation and Salinity Stress

Plants tend to accumulate higher levels of GABA when exposed to adverse environmental stimuli, including salinity (Kinnersley and Turano 2000). This increase is largely attributed to the activity of glutamate decarboxylase (GAD), the enzyme chiefly responsible for GABA biosynthesis. GAD operates in a calcium/calmodulin (CaM)-dependent manner, containing a specific calmodulin-binding domain (Kumar et al. 2022). The enzyme's activation is enhanced by elevated levels of H^+ or Ca^{2+}, both in vitro and in vivo (Chun et al. 2019; González-Orenga et al. 2019). When plants encounter salinity stress, there is a rapid and significant increase in cytosolic Ca^{2+} levels (Dell'Aversana et al. 2021). For instance, tomato cells subjected to NaCl show a noticeable rise in calmodulin messenger RNA (mRNA) within 15 minutes of treatment (Thouvenot et al. 2015). Similarly, salinity-stressed seedlings of mung beans (*Vigna radiata*) have shown upregulation of the calmodulin gene MBCaM-1 (de Oliveira et al. 2020). These findings strongly indicate that salt-induced GABA accumulation in plants is driven by the activation of GAD through Ca^{2+}–calmodulin signaling.

Interestingly, salinity stress results in decreased levels of polyamines (PAs) vis-à-vis an increase in GABA content and diamine oxidase (DAO) activity. Presumably, PAs act through the formation of GABA as an oxidative product, explaining an inverse relationship between their levels in plants. However, very limited information is available on the relationship between PA degradation and GABA accumulation under salinity stress. In a very detailed and extended investigation, the soybean plants treated with graded NaCl concentrations exhibited GABA accumulation to the extent of 11–17-fold, accompanied by a corresponding decrease in PAs (putrescine (Put), cadaverine (Cad), and spermidine (Spd) in their roots. However, the experiment incorporating aminoguanidine (AG), a specific DAO inhibitor, along with NaCl concentrations, exhibits the recovery of PA levels and a simultaneous decrease in GABA levels within 6 days of treatment in soybean roots. Therefore, it becomes very clear that salinity stress–induced DAO activity performs oxidative degradation of PAs to GABA and confers salinity stress resilience, establishing the role of PAs through GABA in the acquisition process of salt tolerance (Xing et al. 2007).

7.3.2 GABA Signaling Pathways and Salt Tolerance

Under salinity stress, plants produce excessive ROS, which cause oxidative injury to cellular components. GABA signaling modulates ROS levels and redox homeostasis, primarily through the regulation of antioxidant gene expression (Bouché et al. 2003; Che-Othman et al. 2020). GABA-mediated ROS signaling helps mitigate oxidative stress and prevents cellular damage under high salt concentrations (Jalil and Ansari 2020). The signaling pathways involved in GABA-mediated salt tolerance encompass multiple mechanisms that collectively enhance plant adaptation and survival under salinity stress. The involvement of GABA receptors, ion flux regulation, osmotic adjustment, ROS signaling, and hormonal cross talk highlights the complexity of GABA-mediated responses. Stress-induced signaling begins with increased cytosolic Ca^{2+}, which activates the CaM-dependent GAD enzyme. This, along with other components of the GABA shunt, assists in scavenging ROS and reducing oxidative damage during salt exposure (Jalil and Ansari 2020).

GABA signaling under salinity stress involves plasma membrane–bound receptors, whose activation initiates downstream signaling cascades affecting ion channels and transporters. These cascades adjust ion distribution, particularly Na^+ and K^+, to preserve ionic homeostasis under stress. In barley, both GABA and glutamate have been linked to long-distance electrical signaling (Felle and Zimmermann 2007). Additionally, glutamate receptor–like (GLR) proteins modulate intracellular Ca^{2+} signals, which, in turn, influence GAD activity and GABA-related signals (Toyota et al. 2018; Shao et al. 2020; Xu et al. 2021a). GABA has also been shown to interact with the guard cell outward-rectifying K^+ (GORK) channels, which are involved in stress-induced K^+ efflux (Adem et al. 2020; Xu et al. 2021b). Potassium serves both nutritional and signaling roles in plants (Wu et al. 2018; Isayenkov et al. 2019), and its loss through GORK and non-selective cation (NSCC) channels is a typical response to salinity stress.

Cytosolic K^+ loss may function as a "metabolic switch," halting energy-intensive biosynthesis and reallocating resources for stress survival and life cycle completion (Demidchik et al. 2014; Adem et al. 2020). Exogenous GABA applications have been shown to improve Na^+/K^+ ratios in salinity-stressed white clover (*Trifolium repens* cv. Haifa) by upregulating genes associated with ion transport and detoxification, such as *SOS1*, *H^+-ATPase*, *HKT1*, *HKT8*, and *HAL2* (Cheng et al. 2018). In tomatoes, GABA reduces Na^+ accumulation in roots, preventing its transport to leaves and minimizing further ionic damage (Wu et al. 2020). Proteomic studies in creeping bentgrass under salinity stress support GABA's role in ion regulation, showing increased expression of *SOS1*, *NHX2/4/6*, *ATPaB2*, and *PPa2* in roots, along with decreased *HKT1/4* in leaves, helping maintain a favorable K^+/Na^+ ratio and compartmentalize Na^+ into vacuoles (Li et al. 2020).

In corn, GABA-treated plants show elevated expression of the *ZmHKT1* gene and suppressed *ZmSOS1* and *ZmNHX1* gene expression under salinity stress. This expression pattern suggests that GABA aids in preventing Na^+ toxicity by promoting its exclusion from the cytosol and maintaining cellular function (Aljuaid and Ashour 2022). Moreover, putrescine (a GABA precursor) improves salt tolerance in white clover by enhancing the expression of ion transporters (*HKT1*, *HKT8*, *SOS1*, *SKOR*, and *NHX6*) and detoxification-related genes like *HAL2* (Cheng et al. 2022). Similarly, spermine, a related GABA precursor, elicits comparable responses in bentgrass, indicating a shared mode of action among GABA-related compounds (Li et al. 2022). These findings collectively underscore GABA's multifaceted role in stress signaling and ionic regulation in plants under salinity.

GABA also interacts closely with hormonal signaling pathways involved in abiotic stress responses, particularly abscisic acid (ABA), jasmonic acid (JA), and ethylene. It promotes the accumulation of these stress-responsive hormones, which, in turn, activate downstream genes involved in stress adaptation (Ansari et al. 2014). GABA has been observed to enhance ABA-mediated responses such as stomatal closure, thus reducing transpirational water loss. It also modulates ethylene production and signaling, influencing both growth and defense mechanisms. Experimental evidence further suggests that GABA treatment leads to the upregulation of several metabolites, including amino acids, sugars, and endogenous GABA itself, along with elevated expression of genes involved in the biosynthesis of key phytohormones. For example, GABA application in citrus (*Citrus* × *sinensis*) increases the expression of GABA-transaminase (*GABA-T*), succinate semialdehyde dehydrogenase (*SSADH*), and tricarboxylic acid (TCA) cycle enzymes (malate dehydrogenase and succinic dehydrogenase) and amino acids such as glycine, L-alanine, L-proline, L-glutamine, and L-asparagine. Additionally, phytohormone-related genes are also induced, raising the levels of compounds like indole acetic acid, trans-jasmonic acid, benzoic acid, indole propionic acid, cinnamic acid, ABA, and salicylic acid (Hijaz et al. 2018).

Transcriptomic analyses in *Populus* × *tomentosa* treated with GABA under salinity stress further reveal the involvement of the genes related to ABA metabolism (*ABAH2*, *ABAH4*, and *ABAG*) and receptors (*PYL1/2/4/6*), along with the genes linked to ethylene biosynthesis (*ACO1/2/5* and *ACS1/7*) (Ji et al. 2018). This hormonal cross talk enhances GABA's role in mediating a robust tolerance response to salinity.

7.3.3 GABA-Mediated Physiological and Molecular Mechanisms

GABA contributes to plant salt tolerance through a multifaceted network of physiological and molecular mechanisms. These mechanisms function together to help plants maintain homeostasis and survive under salinity stress. Below are some of the key processes through which GABA confers this tolerance:

7.3.3.1 Osmotic Adjustment and Ion Homeostasis

Under saline conditions, GABA accumulation plays a crucial role in supporting osmotic balance by raising the concentration of solutes within plant cells. This increase helps maintain water uptake and reduces the risk of cellular dehydration. GABA also assists in sustaining ion balance by influencing the activity of ion transporters and channels, limiting the absorption of harmful ions like sodium, and promoting their compartmentalization into vacuoles. This mechanism allows plants to alleviate the primary effects of osmotic stress and the secondary effects of ion toxicity commonly triggered by salt exposure.

Osmotic adjustment acts as an adaptive response that enables plant cells to resist drought-like conditions induced by salinity. GABA either functions directly as an osmoprotectant or promotes the synthesis of other osmolytes during stress (Shelp et al. 1999; Ansari et al. 2021). In *Zea mays* seedlings, salinity stress leads to an increase in key osmolytes such as proline and glycine betaine (Mansour et al. 2005), with proline levels particularly rising in response to salinity (Kaya et al. 2010). A dose-dependent pattern is observed; plants treated with 150 mM NaCl show elevated levels of proline and soluble sugars, whereas these levels decline significantly at 300 mM NaCl, suggesting threshold limits to the stress response (Wang et al. 2017).

Salinity initially induces osmotic stress, followed by ionic imbalance as a secondary stressor. In this context, GABA likely acts as a compatible solute, helping plant cells endure dehydration without inflicting cellular damage (Carillo 2018). Transporters like AtProT2 in *Arabidopsis thaliana* and LeProT1 in tomato are known to facilitate the transport of both proline and GABA. These transporters become more active under water or salinity stress, enabling the coordinated movement of these osmoprotective compounds within plant tissues (Breitkreuz and Shelp 1995; Schwacke et al. 1999). Exogenously applied GABA has also been shown to enhance proline and sugar accumulation in maize plants exposed to both 150 mM and 300 mM NaCl. However, plants exposed to the lower salt concentration exhibit greater accumulation, indicating more efficient stress mitigation at moderate salinity levels (Wang et al. 2017). Therefore, GABA treatment contributes to the buildup of osmolytes, aiding plant survival under harsh environmental conditions (Chen et al. 2007).

Additionally, GABA feeds into the Krebs cycle, a vital metabolic pathway that extracts energy through the breakdown of carbohydrates, fatty acids, and proteins. The oxidation of GABA generates succinate and α-ketoglutarate, intermediates of the Krebs cycle, which supply the energy needed to support stress-related cellular activities. Thus, the link between GABA metabolism and enhanced levels of proline and sugars highlights GABA's indirect role in providing the energy required to counteract salinity-induced damage.

7.3.3.2 GABA as an ROS Scavenger

Under salinity stress, plants experience an overproduction of ROS, which, in turn, elevates malondialdehyde (MDA) levels, a well-known indicator of oxidative stress and lipid peroxidation that causes membrane damage (Mittler et al. 2004). The plant's antioxidant defense system, which includes enzymes like SOD, ascorbate peroxidase (APX), CAT, and POD, plays a crucial role in controlling ROS levels (Qiu et al. 2013). Some studies report that salinity can suppress CAT activity while simultaneously enhancing SOD activity, indicating a complex response pattern depending on species and stress levels (Zhu et al. 2004; Li et al. 2010). In maize seedlings, antioxidant responses are often enhanced to counteract salt-induced oxidative damage (Wang et al. 2017). The rise in MDA and O_2^- under salinity stress suggests a direct correlation between salt exposure and oxidative stress. To manage this, plants employ several molecular strategies, including regulated gene expression of antioxidants (Zorb et al. 2004).

Salinity stress activates the transcription of gene coding for antioxidant enzymes, thereby strengthening the plant's defense against oxidative harm. In *Zea mays*, increased CAT activity under salinity has been linked to elevated mRNA levels in response to ROS accumulation (Menezes-Benavente et al. 2004). Similarly, in *Solanum lycopersicum*, salinity stress upregulates genes associated with SOD, APX, and POD, although CAT expression is suppressed (Luo et al. 2011). Application of GABA at appropriate doses promotes plant growth and stimulates the antioxidant system at both the enzymatic and transcript levels, resulting in reduced oxidative injury under salinity stress (Li et al. 2017). For example, *Zea mays* seedlings exposed to 200 mM NaCl exhibit higher expression of aspartate (ASP) and POD genes than those treated with 100 mM NaCl, although extreme salinity can suppress antioxidant gene expression. Importantly, GABA supplementation in plants under both 150 mM and 300 mM NaCl stress reduces MDA and O_2^- levels while boosting antioxidant enzyme activity (Wang et al. 2017).

Though the exact mechanisms by which GABA enhances antioxidant responses remain under investigation, existing evidence indicates that GABA acts as both a local and a systemic signaling molecule. In *Brassica napus*, GABA functions as a shoot-to-root signal regulating nitrate uptake (Beuve et al. 2004). Moreover, GABA application mitigates oxidative stress in *Hordeum vulgare* roots under aluminum and proton stress (Song et al. 2010). In *Caragana intermedia*, GABA represses the expression of genes responsible for hydrogen peroxide production, thereby limiting ROS accumulation and oxidative injury (Shi et al. 2010). Studies also confirm that exogenous GABA significantly enhances antioxidant enzyme activity in various plants, such as *Solanum lycopersicum* (Luo et al. 2011), *Lolium perenne* (Krishnan et al. 2013), *Oryza sativa* (Nayyar et al. 2014), and *Piper nigrum* (Vijayakumari and Puthur 2015).

Pre-sowing seed treatment with GABA (1–2 mM) in *Zea mays* has been shown to elevate antioxidant activity and reduce the negative effects of salinity on germination (Tian et al. 2014). GABA also helps mitigate low-light-induced oxidative stress by promoting antioxidant defenses (Li et al. 2017). Furthermore, GABA treatment increases endogenous levels of proline and other compounds that enhance antioxidant capacity, as observed in germinating barley under salinity stress (Maa et al. 2018). Collectively, these findings suggest that GABA fortifies the plant's antioxidant machinery, enabling efficient ROS scavenging and providing cellular protection under adverse environmental conditions (Carillo 2018).

7.3.3.3 GABA and Stabilization of Cellular Membranes

Salinity stress can displace calcium ions from their binding sites on the plasma membrane, which leads to increased membrane permeability. This disturbance results in greater ion leakage and elevated leaf-relative conductivity, indicating membrane destabilization (Pitann et al. 2013). Exposure to high levels of NaCl can severely damage cellular organelles, interfere with vital cellular activities, cause chlorophyll degradation, and eventually, lead to plant death (Davenport et al. 2005; Quintero et al. 2007).

GABA plays a protective role in maintaining the membrane structure and function under saline conditions. It helps minimize lipid peroxidation, ensures membrane fluidity, and restricts the loss of electrolytes from plant cells. Interestingly, studies on the salt-tolerant cyanobacterium *Aphanothece halophytica* show GABA accumulation under salinity stress, suggesting that genetic engineering strategies to boost GABA biosynthesis in plants could be a promising approach to enhancing stress tolerance (Boonburapong et al. 2016).

GABA treatments have been shown to improve germination rates and reduce salt-induced cellular damage in crops like wheat and maize (Luo et al. 2011). External application of GABA has also been linked to improved osmotic balance, which aids in plant resistance against environmental stresses (Vijayakumari and Puthur 2015; Yu et al. 2014; Malekzadeh et al. 2014). In *Zea mays* seedlings treated with NaCl, salinity causes noticeable membrane damage and decreases relative leaf water content. However, GABA supplementation in such cases promotes plant growth by increasing shoot and root length, as well as fresh weight (Wang et al. 2017).

In rice, GABA application helps stabilize cell membranes, preserve leaf color, and support mitochondrial function during episodes of high temperature or hypoxia (Nayyar et al. 2014). Research has shown that mutants of *Arabidopsis thaliana* lacking succinate semialdehyde dehydrogenase (SSADH) are highly sensitive to stress due to impaired hydrogen peroxide detoxification. The GABA shunt pathway contributes intermediates like succinate and nicotinamide adenine dinucleotide (NADH) to the electron transport chain, which supports energy production and helps limit ROS generation during stress. Deficiencies in this pathway can impair TCA cycle activity, as seen in yeast mutants lacking GABA shunt genes, which are more vulnerable to oxidative damage from hydrogen peroxide (Coleman et al. 2001; Bouché et al. 2003).

Although it is evident that GABA enhances membrane stability and protects cellular structures from oxidative stress, the precise mechanisms behind its regulation of water retention and suppression of ROS-induced damage are not yet fully understood. Further investigation is needed to uncover the molecular basis of GABA's protective functions in salinity-stressed plants.

7.3.3.4 Modulation of Stress-Responsive Gene Expression

GABA plays an important role in regulating stress-responsive genes in plants under salinity stress. When plants are exposed to high salt concentrations, they activate a variety of defense mechanisms. Among these, GABA accumulation, signaling interactions, and metabolic cross talk are critical elements. Although the full understanding of how GABA is metabolized and transported in plant systems is still evolving, it is evident that GABA influences gene expression through specific signaling pathways. Acting as a signaling molecule, GABA initiates intracellular responses by binding to receptors, which subsequently activate downstream components involved in transcriptional regulation. This ultimately leads to the modulation of genes associated with osmotic balance, ion regulation, antioxidant defense, and stress signaling.

Evidence from studies on tomato plants reveals that silencing GAD and GABA-T genes increases sensitivity to salinity stress and raises ROS levels. Notably, GABA content rises in GABA-T mutants and drops in GAD mutants, yet both mutants show a reduction in succinate, a key intermediate. The suppression of SSADH leads to phenotypic abnormalities such as curled leaves, dwarfism, and elevated ROS levels, even in the absence of stress. This confirms the GABA shunt's function in controlling ROS via an alternative route that converts succinic semialdehyde into γ-hydroxybutyrate, and its disruption results in ROS accumulation under saline conditions (Bao et al. 2015).

Introducing the GABA-T gene from *Morus* (mulberry) into *Arabidopsis* and mulberry hairy roots leads to lower GABA levels and higher ROS content, making the plants more sensitive to salinity stress. However, applying GABA externally helps reduce ROS accumulation, boosts antioxidant enzyme activity (SOD, POD, and CAT), and enhances salinity stress tolerance (Zhang et al. 2022). Further genetic studies on *pop2* (GABA-T mutant) and *gad1/gad2* (GAD double mutant) lines show that *pop2* mutants with elevated GABA levels are more salt-tolerant, while *gad1/gad2* mutants lacking GABA production from glutamate exhibit greater salt sensitivity.

GABA-mediated salt tolerance is also linked to the activity and expression of membrane transporters and ion channels such as H^+-ATPase, SOS1, NHX1, and GORK. These proteins are essential for maintaining cellular ion homeostasis, particularly in regulating Na^+ and K^+ levels, and membrane potential and suppressing hydrogen peroxide formation (Su et al. 2019). However, the exact mechanisms by which GABA influences the transcription and function of these transporters remain to be elucidated.

Beyond these roles, GABA interacts with other signaling compounds and plant hormones to modulate stress-related gene expression. In *Arabidopsis* cell suspensions, polyamines contribute to GABA biosynthesis via the activity of aldehyde dehydrogenase enzymes AtALDH10A8 and AtALDH10A9, which convert 4-aminobutanal and 3-aminopropanal into GABA and β-alanine, respectively. Mutations that impair polyamine metabolism result in reduced GABA accumulation and increased sensitivity to salinity stress (Zarei et al. 2016).

GABA transporters (GATs) are also involved in stress regulation. Overexpression of GAT genes from *Populus euphratica* in both *Arabidopsis* and poplar plants enhances tolerance to salinity stress, likely due to increased lignin biosynthesis in xylem tissues and higher proline levels in leaves (Bai et al. 2019). Additionally, the GABA shunt supports GABA accumulation, integrating with other metabolic and signaling pathways to mitigate stress effects. Enhancing the levels of GABA or its precursors holds potential as a strategy to strengthen plant resistance to salinity.

7.4 Conclusion

GABA has emerged as a valuable protective molecule in enhancing plant tolerance to salinity stress, showing strong potential to improve crop resilience and yield under challenging environmental conditions (Table 7.1). Acting both as a signaling compound and a key metabolic player, GABA contributes to a wide range of physiological, biochemical, and molecular adaptations that enable plants to cope with salt-induced damage. It plays a central role in sustaining ion homeostasis, regulating osmotic pressure, and detoxifying harmful ROS. Additionally, GABA influences the expression of stress-related genes and participates in hormonal signaling pathways, reinforcing the plant's overall defense system. Its positive effects extend to promoting photosynthesis and fostering robust root growth, both of which are critical for plant survival in saline environments. Applying GABA externally or enhancing its biosynthesis within plants presents a promising strategy for developing salt-tolerant crops and advancing sustainable agricultural practices. However, further research is needed to decode the detailed mechanisms behind GABA's protective effects, explore its interactions with other signaling systems, and tailor its application across diverse crops and environmental scenarios. A deeper understanding of GABA's role in salinity stress management could play a pivotal part in mitigating soil salinization impacts and securing food production in an era of global climate change.

Table 7.1 Role of GABA in managing salinity stress tolerance across various crop species

S. No.	Plants	Mechanism	References
1	Apple	Sustaining ion homeostasis	Shi et al. (2024)
2	*Glycine max* L. (soybean)	Modulating nutrient uptake, osmolyte accumulation, and the ascorbate–glutathione antioxidant pathway	Qian et al. (2024)
3	*Triticum aestivum* (wheat)	Minimizing oxidative stress, controlling stomatal function, enhancing photosynthetic efficiency, stimulating antioxidant defenses, and increasing the expression of stress-related genes.	Badr et al. (2024)
4	*Cyperus esculentus* L. (chufa)	Managing nitrogen utilization and enhancing antioxidant capacity	Ullah et al. (2023)
5	*Zea mays* (maize)	Improving nitrogen metabolism under salinity stress and supporting the TCA cycle's anaplerotic reactions through the modulation of the GABA shunt pathway	Wang et al. (2023)
6	*Zea mays* (maize)	GABA promotes chlorophyll production under stress conditions	Aljuaid and Ashour (2022)
7	*Oryza sativa* (rice)	Modulating antioxidant systems and carbohydrate metabolism under saline–alkaline stress conditions	Kaur and Zhawar (2021)

(continued)

Table 7.1 (continued)

S. No.	Plants	Mechanism	References
8	*Lycopersicon esculentum* (tomato)	Adjusting sodium ion uptake, regulating GAD gene expression, enhancing amino acid biosynthesis, and managing reactive oxygen species metabolism	Wu et al. (2020)
9	*Arabidopsis thaliana*	Controlling H^+-ATPase activity, supporting cytosolic potassium retention, and aiding in sodium ion exclusion	Su et al. (2019)
10	*Cucumis melon* (muskmelon)	Maintaining redox homeostasis and enhancing chlorophyll production to strengthen tolerance against salinity–alkaline stress	Jin et al. (2019)
11	Hulless barley	Improving physiological and biochemical processes along with strengthening antioxidant defenses	Ma et al. (2018)
12	*Lactuca sativa* (lettuce)	Improved tolerance to salinity and increased efficiency of photosynthesis	Kalhora et al. (2018)
13	*Lycopersicon esculentum* (tomato)	Enhancing stress tolerance by increasing internal GABA levels and phenolic acid accumulation	Cekic (2018)
14	*Trifolium repens* (white clover)	Enhancing salinity-stressed seed germination by regulating Na+/K+ transport, enhancing dehydrin accumulation, and upregulating stress-related genes	Cheng et al. (2018)
15	*Zea mays* (maize)	Alleviating salt-induced damage by enhancing photosynthetic activity and upregulating osmoprotectants and antioxidants	Wang et al. (2017)
16	*Cassia italica* (cassia)	Mitigating salt toxicity by boosting the activity of antioxidant enzymes	Alqarawi et al. (2016)
17	*Piper nigrum* (black pepper)	Improving resistance to osmotic stress induced by polyethylene glycol (PEG)	Vijayakumari and Puthur (2015)
18	*Cucumis melon* (muskmelon)	Promoting polyamine synthesis, facilitating the conversion of free putrescine into its bound form, and limiting PA breakdown	Hu et al. (2015)
19	*Triticum aestivum* (wheat)	Minimizing oxidative harm during salinity stress conditions	Al-Quraan et al. (2013)
20	*Lycopersicon esculentum* (tomato)	Regulation of oxidative metabolism and chlorophyll fluorescence characteristics	Luo et al. (2011)
21	*Lycopersicon esculentum* (tomato)	Salinity stress enhances the GABA content in tomato fruit	Zushi and Matsuzoe (2007)
22	*Zea mays* (maize)	Enhanced resistance through modulation of antioxidant enzyme activity	Tian et al. (2005)

References

Abd EL-Azim WM, Ahmed STH (2009) Effect of salinity and cutting date on growth and chemical constituents of *Achillea fragratissima* Forssk, under Ras Sudr conditions. Res J Agric Biol Sci 5:1121–1129

Abd El-Wahab MA (2006) The efficiency of using saline and fresh water irrigation as alternating methods of irrigation on the productivity of Foeniculum vulgare Mill subsp. vulgare var. vulgare under North Sinai conditions. Res J Agric Biol Sci 2:571

Adem GD, Chen G, Shabala L, Chen Z-H, Shabala S (2020) GORK channel: a master switch of plant metabolism? Trends Plant Sci 25:434–445

Aljuaid BS, Ashour H (2022) Exogenous-aminobutyric acid (GABA) application mitigates salinity stress in maize plants. Life 12:1860

Abdulaziz AA, Abeer H, Elsayed A-AF, Asma A-HA, Thobayet AS, Sa'ad AR, Dilfuza E (2016) Protective role of gamma amminobutyric acid onCassia italica Mill under salt stress. Legum Res 39(3):396–404. https://doi.org/10.18805/lr.v0iOF.9561

Al-Quraan NA, Al-Share AT (2016) Characterization of the γ-aminobutyric acid shunt pathway and oxidative damage in Arabidopsis thaliana pop2 mutants under various abiotic stresses. Biol Plantarum 60:132–138

Al-Quraan NA, Sartawe FAB, Qaryouti MM (2013) Characterization of γ-aminobutyric acid metabolism and oxidative damage in wheat (Triticum aestivum L.) seedlings under salt and osmotic stress. J Plant Physiol 170(11):1003–1009

Ansari MI, Hasan S, Jalil SU (2014) Leaf senescence and GABA shunt. Bioinformation 10:730–732

Ansari MI, Jalil SU, Ansari SA et al (2021) GABA shunt: a key-player in mitigation of ROS during stress. Plant Growth Regul 94:131–149

Arzani A, Ashraf M (2016) Smart engineering of genetic resources for enhanced salinity tolerance in crop plants. Crit Rev Plant Sci 35:146–189

Ashraf M (1993) Effect of sodium chloride on water relations and some organic osmotica in arid zone plant species Melilotus indica (L.) All. Tropenlandwirt 94:95–102

Ashraf M, Orooj A (2006) Salt stress effects on growth, ion accumulation and seed oil concentration in an arid zone traditional medicinal plant ajwain (Trachyspermum ammi [L.] Sprague). J Arid Environ 64:209–220

Atta K, Mondal S, Gorai S et al (2023) Impacts of salinity stress on crop plants: improving salt tolerance through genetic and molecular dissection. Front Plant Sci 14:1241736. https://doi.org/10.3389/fpls.2023.1241736

Badr A, Basuoni MM, Ibrahim M et al (2024) Ameliorative impacts of gamma-aminobutyric acid (GABA) on seedling growth, physiological biomarkers, and gene expression in eight wheat (Triticum aestivum L.) cultivars under salt stress. BMC Plant Biol 24:605. https://doi.org/10.1186/s12870-024-05264-5

Baghalian K, Haghiry A, Naghavi MR, Mohammadi A (2008) Effect of saline irrigation water on agronomical and phytochemical characters of chamomile (Matricaria recutita L.). Sci Hortic 116:437–441

Bai X, Xu J, Shao X et al (2019) A novel gene coding γ-aminobutyric acid transporter may improve the tolerance of Populus euphratica to adverse environments. Front Plant Sci 10:1083

Bao H, Chen X, Lv S et al (2015) Virus-induced gene silencing reveals control of reactive oxygen species accumulation and salt tolerance in tomato by γ-aminobutyric acid metabolic pathway: effects of GABA shunt on tomato salt tolerance. Plant Cell Environ 38:600–613

Beuve N, Rispail N, Laine P et al (2004) Putative role of γ-aminobutyric acid (GABA) as a long distance signal in upregulation of nitrate uptake in Brassica napus L. Plant Cell Environ 27:1035–1046

Boonburapong B, Laloknam S, Incharoensakdi A (2016) Accumulation of gamma aminobutyric acid in the halotolerant cyanobacterium Aphanothece halophytica under salt and acid stress. J Appl Phycol 28:141–148

Bouché N, Fait A, Bouché D et al (2003) Mitochondrial succinate-semialdehyde dehydrogenase of the gamma-aminobutyrate shunt is required to restrict levels of reactive oxygen intermediates in plants. Proc Natl Acad Sci USA 100:6843–6848

Breitbreuz KE, Shelp BJ (1995) Subcellular compartmentation of the 4-aminobutyric shunt in protoplasts from developing soybean cotyledons. Plant Physiol 108:99–103

Carillo P (2018) GABA shunt in durum wheat. Front Plant Sci 9:100

Çekiç F (2018) Exogenous GABA stimulates endogenous GABA and phenolic acid contents in tomato plants under salt stress. Celal Bayar Univ J Sci 14(1):61–64. https://doi.org/10.18466/cbayarfbe.348935

Chen Z et al (2007) Compatible solute accumulation and stress-mitigating effects in barley genotypes contrasting in their salt tolerance. J Exp Bot 58:4245–4255

Cheng B, Li Z, Liang L, Cao Y, Zeng W, Zhang X, Ma X, Huang L, Nie G, Liu W et al (2018) The-aminobutyric acid (GABA) alleviates salt stress damage during seed germination of white clover associated with Na^+/K^+ transportation, dehydrins accumulation, and stress related gene expression. Int J Mol Sci 19:2520

Cheng B, Hassan MJ, Feng G, Zhao J, Liu W, Peng Y, Li Z (2022) Metabolites reprogramming and Na^+/K^+ transportation associated with putrescine-regulated white clover seed germination and seedling tolerance to salt toxicity. Front Plant Sci 13:856007

Che-Othman MH, Jacoby RP, Millar AH, Taylor NL (2020) Wheat mitochondrial respiration shifts from the tricarboxylic acid cycle to the GABA shunt under salt stress. New Phytol 225:1166–1180

Chun HJ, Baek D, Cho HM, Jung HS, Jeong MS, Jung WH, Choi CW, Lee SH, Jin BJ, Park MS et al (2019) Metabolic adjustment of Arabidopsis root suspension cells during adaptation to salt stress and mitotic stress memory. Plant Cell Physiol 60:612–625

Coleman ST, Fang TK, Rovinsky SA et al (2001) Expression of a glutamate decarboxylase homologue is required for normal oxidative stress tolerance in Saccharomyces cerevisiae. J Biol Chem 276:244–250

Davenport R, James RA, Zakrisson-Plogander A, Tester M, Munns R (2005) Control of sodium transport in durum wheat. Plant Physiol 137:807–818

Dell'Aversana E, Hessini K, Ferchichi S, Fusco GM, Woodrow P, Ciarmiello LF, Abdelly C, Carillo P (2021) Salinity duration differently modulates physiological parameters and metabolites profile in roots of two contrasting barley genotypes. Plants 10:307

Demidchik V, Straltsova D, Medvedev SS, Pozhvanov GA, Sokolik A, Yurin V (2014) Stress induced electrolyte leakage: the role of K^+-permeable channels and involvement in programmed cell death and metabolic adjustment. J Exp Bot 65:1259–1270

Felle HH, Zimmermann MR (2007) Systemic signalling in barley through action potentials. Planta 226:203

Fu H, Yang Y (2023) How plants tolerate salt stress. Curr Issues Mol Biol 45:5914–5934

González-Orenga S, Ferrer-Gallego PP, Laguna E, López-Gresa MP, Donat-Torres MP, Verdeguer M, Vicente O, Boscaiu M (2019) Insights on salt tolerance of two endemic Limonium species from Spain. Metabolites 9:294

Grattan SR, Grieves CM (1999) Salinity-mineral nutrient relations in horticultural crops. Sci Hortic 78:127–157

Hameed A, Ahmed MZ, Hussain T, Aziz I, Ahmad N, Gul B, Nielsen BL (2021) Effects of salinity stress on chloroplast structure and function. Cells 10:2023

Hasanuzzaman M, Raihan MRH, Masud AAC, Rahman K, Nowroz F, Rahman M, Nahar K, Fujita M (2021) Regulation of reactive oxygen species and antioxidant defense in plants under salinity. Int J Mol Sci 22:9326

Hijaz F, Nehela Y, Killiny N (2018) Application of gamma-aminobutyric acid increased the level of phytohormones in Citrus sinensis. Planta 248:909–918

Hu X, Xu Z, Xu W, Li J, Zhao N, Zhou Y (2015) Application of γ-aminobutyric acid demonstrates a protective role of polyamine and GABA metabolism in muskmelon seedlings under Ca(NO3)2 stress. Plant Physiol Biochem 92:1–10. https://doi.org/10.1016/j.plaphy.2015.04.006

Isayenkov SV, Maathuis FJM (2019) Plant salinity stress: many unanswered questions remain. Front Plant Sci 10:80

Jalil SU, Ansari MI (2020) Physiological role of gamma-aminobutyric acid in salt stress tolerance. In: Salt and drought stress tolerance in plants. Springer, Cham, pp 337–350

Jalil SU, Ahmad I, Ansari MI (2017) Functional loss of GABA transaminase (GABA-T) expressed early leaf senescence under various stress conditions in Arabidopsis thaliana. Curr Plant Biol 9–10:11–22

Ji J, Yue J, Xie T, Chen W, Du C, Chang E, Chen L, Jiang Z, Shi S (2018) Roles of gamma aminobutyric acid on salinity-responsive genes at transcriptomic level in poplar: involving in abscisic acid and ethylene-signalling pathways. Planta 248:675–690

Jin X, Liu T, Xu J, Gao Z, Hu X (2019) Exogenous GABA enhances muskmelon tolerance to salinity-alkalinity stress by regulating redox balance and chlorophyll biosynthesis. BMC Plant Biol 19(1):48. https://doi.org/10.1186/s12870-019-1660-y. PMID: 30709373; PMCID: PMC6359809

Kalhor MS, Aliniaeifard S, Seif M et al (2018) Enhanced salt tolerance and photosynthetic performance: implication of γ-amino butyric acid application in salt-exposed lettuce (*Lactuca sativa* L.) plants. Plant Physiol Biochem 130:157–172

Kaur R, Zhawar VK (2021) Regulation of secondary antioxidants and carbohydrates by gamma-aminobutyric acid under salinity-alkalinity stress in rice (Oryza sativa L.). Biol Futur 72(2):229–239. https://doi.org/10.1007/s42977-020-00055-z. Epub 2021 Feb 5. PMID: 34554480

Kaya C, Tuna AL, Okant AM (2010) Effect of foliar applied kinetin and indole acetic acid on maize plants grown under saline conditions. Turk J Agric For 34:529–538

Kinnersley AM, Turano FJ (2000) Gamma-aminobutyric acid (GABA) and plant responses to stress. Crit Rev Plant Sci 19:479–509

Krishnan S, Laskowski K, Shukla V, Merewitz EB (2013) Mitigation of drought stress damage by exogenous application of a non-protein amino acid gamma aminobutyric acid on perennial ryegrass. J Am Soc Hortic Sci 138:358–366

Kumar P, Choudhary M, Halder T, Prakash NR, Singh V, Vineeth VT, Sheoran S, Ravikiran TK, Longmei N, Rakshit S et al (2022) Salinity stress tolerance and omics approaches: revisiting the progress and achievements in major cereal crops. Heredity 128:497–518

Li Y, Bai Q, Jin X, Wen H, Gu Z (2010) Effects of cultivar and culture conditions on γ aminobutyric acid accumulation in germinated fava beans (Vicia faba L.). J Sci Food Agric 90:52–57

Li Y, Fan Y, Ma Y, Zhang Z, Yue H, Wang L, Li J, Jiao Y (2017) Effects of exogenous γ aminobutyric acid (GABA) on photosynthesis and antioxidant system in pepper (Capsicum annuum L.) seedlings under low light stress. J Plant Growth Regul 36:1–14

Li Z, Cheng B, Zeng W, Zhang X, Peng Y (2020) Proteomic and metabolomic profilings reveal crucial functions of γ-aminobutyric acid in regulating ionic, water, and metabolic homeostasis in creeping bentgrass under salt stress. J Proteome Res 19:769–780

Li Z, Cheng B, Liu W, Feng G, Zhao J, Zhang L, Peng Y (2022) Global metabolites reprogramming induced by spermine contributing to salt tolerance in creeping bentgrass. Int J Mol Sci 23:4472

Lu Y, Fricke W (2023) Salt stress-regulation of root water uptake in a whole-plant and diurnal context. Int J Mol Sci 24:8070

Luo HY, Yang LW, Gao HB, Wu XL, Liu HH (2011) Physiological mechanism of GABA soaking to tomato seed germination and seedling development under NaCl stress. Acta Bot Boreal-Occident Sin 31:2235–2242

Ma Y, Wang P, Chen Z, Gu Z, Yang R (2018) GABA enhances physio-biochemical metabolism and antioxidant capacity of germinated hulless barley under NaCl stress. J Plant Physiol 231:192–201

Machado R, Serralheiro R (2017) Soil salinity: effect on vegetable crop growth. Management practices to prevent and mitigate soil salinization. Horticulturae 3:30

Malekzadeh P, Khara J, Heydari R (2014) Alleviating effects of exogenous gamma aminobutyric acid on tomato seedling under chilling stress. Physiol Mol Biol Plants 20:133–137

Mansour MMF, Salama KHA, Ali FZM, Abou HAF (2005) Cell and plant responses to NaCl in Zea mays cultivars differing in salt tolerance. Gen Appl Plant Physiol 31:29–41

Menezes-Benavente L, Kernodle SP, Margis-Pinheiro M, Scandalios JG (2004) Salt-induced antioxidant metabolism defenses in maize (Zea mays L.) seedlings. Redox Rep 9:29–36

Mittler R, Vanderauwera S, Gollery M, Van Breusegem F (2004) Reactive oxygen gene network of plants. Trends Plant Sci 9:490–498

Nayyar H, Kaur R, Kaur S, Singh R (2014) γ-Aminobutyric acid (GABA) imparts partial protection from heat stress injury to rice seedlings by improving leaf turgor and upregulating osmoprotectants and antioxidants. J Plant Growth Regul 33:408–419

de Oliveira DF, de Sousa Lopes L, Gomes-Filho E (2020) Metabolic changes associated with differential salt tolerance in sorghum genotypes. Planta 252:34

Pitann B, Mohamed AK, Neubert AB, Schubert S (2013) Tonoplast Na^+/H^+ antiporters of newly developed maize (Zea mays) hybrids contribute to salt resistance during the second phase of salt stress. J Plant Nutr Soil Sci 176:148–156

Qian Z, Lu L, Zihan W et al (2024) Gamma-aminobutyric acid (GABA) improves salinity stress tolerance in soybean seedlings by modulating their mineral nutrition, osmolyte contents, and ascorbate-glutathione cycle. BMC Plant Biol 24:365

Qiu ZB, Li JT, Zhang MM, Bi ZZ, Li ZL (2013) He-Ne laser pretreatment protects wheat seedlings against cadmium-induced oxidative stress. Ecotoxicol Environ Saf 88:135–141

Quintero JM, Fournier JM, Benlloch M (2007) Na^+ accumulation in shoot is related to water transport in K^+-starved sunflower plants but not in plants with a normal K^+ status. J Plant Physiol 164:60–67

Rajput VD, Harish, Singh RK, Verma KK, Sharma L, Quiroz-Figueroa FR, Meena M, Gour VS, Minkina T, Sushkova S, Mandzhieva S (2021) Recent developments in enzymatic antioxidant defence mechanism in plants with special reference to abiotic stress. Biology (Basel) 10:267

Schieber M, Chandel NS (2014) ROS function in redox signaling and oxidative stress. Curr Biol 24:R453–R462

Schwacke R et al (1999) LeProT1, a transporter for proline, glycine betaine, and gamma aminobutyric acid in tomato pollen. Plant Cell 11:377–392

Shao Q, Gao Q, Lhamo D, Zhang H, Luan S (2020) Two glutamate- and pH-regulated Ca^{2+} channels are required for systemic wound signaling in Arabidopsis. Sci Signal 13:eaba1453

Shelp BJ, Bown AW et al (1999) Metabolism and function of gamma aminobutyric acid. Trends Plant Sci 4:446–452

Shi SQ, Zheng S, Jiang ZP, Qi LW, Sun XM, Li CX, Liu JF, Xiao WF, Zhang SG (2010) Effects of exogenous GABA on gene expression of Caragana intermedia roots under NaCl stress: regulatory roles for H_2O_2 and ethylene production. Plant Cell Environ 33:149–162

Shi Y, Li Y, Liu T, Guo C, Liang W, Ma F, Li C (2024) γ-Aminobutyric acid enhances salt tolerance by sustaining ion homeostasis in apples. Plant Physiol Biochem 206:108306

Song H, Xu X, Wang H, Wang H, Tao Y (2010) Exogenous gamma-aminobutyric acid alleviates oxidative damage caused by aluminium and proton stresses on barley seedlings. J Sci Food Agric 90:1410–1416

Su N, Wu Q, Chen J, Shabala L, Mithöfer A, Wang H, Qu M, Yu M, Cui J, Shabala S (2019) GABA operates upstream of H^+-ATPase and improves salinity tolerance in Arabidopsis by enabling cytosolic K^+ retention and Na^+ exclusion. J Exp Bot 70:6349–6361

Sun L, Cao X, Du J, Wang Y, Zhang F (2024) Canola (Brassica napus) enhances sodium chloride and sodium ion tolerance by maintaining ion homeostasis, higher antioxidant enzyme activity and photosynthetic capacity fluorescence parameters. Funct Plant Biol 51:FP23089

Tabatabaie SJ, Nazari J (2007) Influence of nutrient concentration and NaCl salinity on growth, photosynthesis and essential oil content of peppermint and lemon verbena. Turk J Agric 31:245–253

Thouvenot L, Deleu C, Berardocco S, Haury J, Thiébaut G (2015) Characterization of the salt stress vulnerability of three invasive freshwater plant species using a metabolic profiling approach. J Plant Physiol 175:113–121

Tian L, Greenberg SA, Kong SW, Altschuler J, Kohane IS, Park PJ (2005) Discovering statistically significant pathways inexpression profiling studies. Proc Natl Acad Sci U S A 102(38):13544–13549. https://doi.org/10.1073/pnas.0506577102. Epub 2005 Sep 8. PMID: 16174746; PMCID: PMC1200092

Tian Y, Guan B, Zhou D, Yu J, Li G, Lou Y (2014) Responses of seed germination, seedling growth, and seed yield traits to seed pretreatment in maize (Zea mays L.). Sci World J 2014:8

Toyota M, Spencer D, Sawai-Toyota S, Jiaqi W, Zhang T, Koo AJ, Howe GA, Gilroy S (2018) Glutamate triggers long-distance, calcium-based plant defense signaling. Science 361:1112–1115

Ullah A, Tariq A, Zeng F, Noor J, Sardans J, Asghar MA, Zhang Z, Peñuelas J (2023) Application of GABA (γ-aminobutyric acid) to improve saline stress tolerance of chufa (Cyperus esculentus L. var. sativus Boeck) plants by regulating their antioxidant potential and nitrogen assimilation. S Afr J Bot 157:540–552

Vijayakumari K, Puthur JT (2015) γ-Aminobutyric acid (GABA) priming enhances the osmotic stress tolerance in Piper nigrum Linn. plants subjected to PEG-induced stress. Plant Growth Regul 78:1–11

Vijayakumari K, Jisha C, Puthur JT (2016) GABA/BABA priming: a means for enhancing abiotic stress tolerance potential of plants with less energy investments on defence cache. Acta Physiol Plant 38:230

Wang Y, Gu W, Yao M, Xie T, Li L, Jing L, Shi W (2017) γ-Aminobutyric acid imparts partial protection from salt stress injury to maize seedlings by improving photosynthesis and upregulating osmoprotectants and antioxidants. Sci Rep 7:43609

Wang Y, Cao H, Wang S, Guo J, Dou H, Qiao J, Yang Q, Shao R, Wang H (2023) Exogenous γ-aminobutyric acid (GABA) improves salt-inhibited nitrogen metabolism and the anaplerotic reaction of the tricarboxylic acid cycle by regulating GABA-shunt metabolism in maize seedlings. Ecotoxicol Environ Saf 254:114756

Wu H, Zhang X, Giraldo JP, Shabala S (2018) It is not all about sodium: revealing tissue specificity and signalling roles of potassium in plant responses to salt stress. Plant Soil 431:1–17

Wu X, Jia Q, Ji S, Gong B, Li J, Lü G, Gao H (2020) Gamma-aminobutyric acid (GABA) alleviates salt damage in tomato by modulating Na^+ uptake, the GAD gene, amino acid synthesis and reactive oxygen species metabolism. BMC Plant Biol 20:465

Xing SG, Jun YB, Hau ZW, Liang LY (2007) Higher accumulation of gamma-aminobutyric acid induced by salt stress through stimulating the activity of diamine oxidases in Glycine max (L.) Merr. roots. Plant Physiol Biochem 45:560–566

Xu B, Long Y, Feng X, Zhu X, Sai N, Chirkova L, Betts A, Herrmann J, Edwards EJ, Okamoto M et al (2021a) GABA signalling modulates stomatal opening to enhance plant water use efficiency and drought resilience. Nat Commun 12:1952

Xu B, Sai N, Gilliham M (2021b) The emerging role of GABA as a transport regulator and physiological signal. Plant Physiol 187:2005–2016

Yu C, Zeng L, Sheng K, Chen F, Zhou T, Zheng X, Yu T (2014) γ-Aminobutyric acid induces resistance against Penicillium expansum by priming of defence responses in pear fruit. Food Chem 159:29–37

Zarei A, Trobacher CP, Shelp BJ (2016) Arabidopsis aldehyde dehydrogenase 10 family members confer salt tolerance through putrescine-derived 4-aminobutyrate (GABA) production. Sci Rep 6:35115

Zhang M, Liu Z, Fan Y, Liu C, Wang H, Li Y, Xin Y, Gai Y, Ji X (2022) Characterization of GABA-transaminase gene from mulberry (Morus multicaulis) and its role in salt stress tolerance. Genes 13:501

Zhao S, Zhang Q, Liu M, Zhou H, Ma C, Wang P (2021) Regulation of plant responses to salt stress. Int J Mol Sci 22:4609

Zhu ZJ, Wei GQ, Li J, Qian QQ, Yu JQ (2004) Silicon alleviates salt stress and increases antioxidant enzymes activity in leaves of salt-stressed cucumber (Cucumis sativus L.). Plant Sci 167:527–533

Zörb C et al (2004) The biochemical reaction of maize (Zea mays L.) to salt stress is characterized by a mitigation of symptoms and not by a specific adaptation. Plant Sci 167:91–100

Zushi K, Matsuzoe N (2007) Seasonal and cultivar differences in salt-induced change in ascorbic acid and dehydroascorbic acid contents of tomato fruit. Environ Control Biol 45(3):165–171. https://doi.org/10.2525/ecb.45.165

GABA as Protectant Against Cold and Chilling Stress

8

Abstract

Cold stress is a significant abiotic factor that negatively impacts plant growth and productivity by disrupting physiological and biochemical processes. Gamma-aminobutyric acid (GABA), a key signaling molecule, is implicated in various stress responses, including cold stress. GABA modulates stress-responsive mechanisms that stabilize cellular structures, maintain metabolic balance, and enforce antioxidant defense systems. Cold stress accelerates GABA accumulation, which regulates expression of key stress-responsive genes and confers cold resilience in crops. Further, GABA also interacts with several metabolic pathways responsible for improved cold tolerance. The chapter highlights the multifaceted roles of GABA and its potential in developing cold-tolerant crop varieties.

Keywords

Antioxidant defense · Cold stress tolerance · Crop resilience GABA · Metabolic adjustments · Osmotic regulation · Stress-responsive genes

Highlights Gamma-aminobutyric acid (GABA) as a key signaling molecule enhances plant resilience under cold stress conditions.

- Cold stress induces GABA accumulation, triggering protective responses in plants.
- GABA-mediated regulation of stress-responsive gene expression occurs to support cold tolerance.
- GABA stabilizes cellular structures and maintains metabolic homeostasis during low temperatures.

S. U. Jalil et al., *GABA Signaling System and Abiotic Stress Tolerance in Plants*,
https://doi.org/10.1007/978-981-95-0437-4_8

- GABA invokes antioxidant defense systems and mitigates cold-induced oxidative stress.
- GABA and other metabolic pathways synergize to improve cold stress tolerance in crops.

8.1 Introduction

Cold stress, encompassing both chilling (0–15 °C) and freezing (<0 °C) temperatures, is a significant abiotic stress that adversely affects plant growth, development, and survival. Plants, being sessile organisms, cannot escape adverse environmental conditions and must develop intricate mechanisms to perceive, respond, and adapt to cold stress (Satyakam et al. 2022). Cold stress can lead to cellular damage, metabolic imbalances, and physiological disruptions, severely impairing photosynthesis, nutrient uptake, and overall plant productivity (Banerjee and Roychoudhury 2019). Understanding the mechanisms of cold stress response is crucial for developing strategies to enhance plant resilience and ensure agricultural sustainability in the face of climate variability.

The ability of plants to tolerate cold stress is of paramount importance for agriculture and ecosystems, particularly in temperate and boreal regions. Cold stress can significantly reduce crop yield and quality, posing a threat to food security and economic stability. With the increasing unpredictability of weather patterns due to climate change, enhancing cold stress tolerance in crops has become a critical goal for plant breeders and biotechnologists. Developing cold-tolerant varieties can help mitigate yield losses, extend growing seasons, and improve the geographical range of crop cultivation.

Gamma-aminobutyric acid (GABA) is a nonprotein amino acid that plays diverse roles in plant physiology and stress responses. Initially discovered as a neurotransmitter in animals (Xu et al. 2008), GABA has gained recognition as a crucial metabolite in plants, involved in multiple cellular processes (Abdullah et al. 2025). GABA is synthesized from glutamate by the enzyme glutamate decarboxylase (GAD) and is catabolized through the GABA shunt, involving GABA-T and succinate semialdehyde dehydrogenase (SSADH). This shunt integrates with the tricarboxylic acid (TCA) cycle, linking GABA metabolism to primary energy production pathways (Jalil et al. 2017; Ansari et al. 2021). Recent research has highlighted GABA's role in abiotic stress tolerance, including drought, salinity, and cold stress, making it a focal point for studies on plant resilience.

GABA also plays a multifaceted critical role in enhancing cold stress tolerance in plants. We will explore the metabolic pathways of GABA, its accumulation and protective functions under cold stress, and the genetic and molecular mechanisms underlying its action. We aim to provide a comprehensive understanding of GABA's potential in improving cold stress tolerance and contributing to sustainable agricultural practices.

8.2 Effect of Cold/Freezing Stress

8.2.1 Plant Morphology and Productivity

Cold stress triggers a series of adverse effects in plants, starting with visible symptoms in the leaves and extending to critical stages like germination and reproduction. Initially, leaves show chlorosis and wilting, which can escalate to necrosis and stunted growth if the cold persists in wheat and maize (Equiza et al. 2001; Janowiak et al. 2002). In wheat, cold stress delays and reduces germination, leading to lower overall productivity (Jame and Cutforth 2004). During the reproductive stage, cold stress causes premature flower shedding, deforms pollen tubes (Chakrabarti et al. 2011), induces pollen sterility (Ji et al. 2017), and disrupts grain development (Barton et al. 2014). These factors collectively result in a marked reduction in final productivity (Hussain et al. 2018). Root growth and surface area are significantly restricted, diminishing the plant's ability to absorb essential nutrients and water (Hussain et al. 2018).

8.2.2 Physiological and Biochemical Alterations

Cold stress acts as a formidable adversary to plant growth and productivity. When temperatures drop, plants experience a cascade of physiological and biochemical changes, severely impairing their development. Significant adverse impacts on plant physiology caused by cold stress arise from the disruption of membrane stability and protein structures (Chen et al. 2018). The stress induces the formation of ice crystals within plant tissues, leading to a decrease in apoplastic water potential and causing water to move out of the cells. (Puhakainen 2004).

This cellular-level stress often triggers drought conditions, which further hinder growth and productivity (Hassan et al. 2021). Drought stress results in diminished root growth, reduced root hydraulic conductivity, and decreased leaf turgidity (Siddique et al. 2000). Further, low temperature extremes interfere with soil microbial activity, adversely affecting plant nutrient relationships (Massenssini et al. 2015). The situation severely impairs plants' ability to uptake water and nutrients, disrupting carbohydrate metabolism and leading to substantial yield losses (Hassan et al. 2021).

The formation of ice crystals also manifests as electrolyte leakage and lipid peroxidation (Hassan et al. 2021). Cell puncturing may occur due to these crystals, resulting in cytosol leakage and cell death (Zhang et al. 2011a, b; Demidchik et al. 2014; Sun et al. 2019). Therefore, the prevention of ice crystal formation becomes indispensable for acquiring plant cold tolerance. To mitigate the damage caused by cold stress, plants often accumulate cryoprotective polypeptides and osmolytes (Ritonga and Chen 2020).

8.2.3 Photosynthetic and Respiratory Responses

Cold stress significantly impairs photosynthesis, which is essential for economic yield, including grain production. One major effect is the overexcitation of PSII, which incurs energy loss through nonradiative reactions. This inefficiency in energy use can severely limit the photosynthetic capacity of plants under cold conditions (Venzhik et al. 2011; Cvetkovic et al. 2017).

Additionally, cold stress reduces the synthesis of chlorophyll, the pigment crucial for capturing light energy. This reduction in chlorophyll content directly decreases photosynthetic efficiency (Rinalducci et al. 2011; Khan et al. 2017). Furthermore, the activity of Rubisco, the enzyme responsible for carbon fixation, is diminished under cold stress, further limiting photosynthesis (Yamori et al. 2009). Cold temperatures also impair electron transport within the chloroplasts and reduce stomatal conductivity, both of which are vital for efficient photosynthesis and gas exchange (Bota et al. 2004; Hussain et al. 2018). These combined effects decrease the production of assimilates, which are necessary for plant growth and yield, resulting in significant yield losses (Hussain et al. 2018).

Cold stress damages the structure of mitochondria and impairs plant respiration. This structural damage disrupts enzymatic activities within the mitochondria, slowing down the flow of kinetic energy and thereby diminishing the overall respiration rate (Ikkonen et al. 2020). This reduction in respiration rate limits the energy available for essential cellular processes, compounding the stress experienced by the plant and further affecting growth and productivity.

8.2.4 Changes in Cellular Structure

Cold stress significantly impacts the cellular structure of plants, leading to various harmful morphological and physiological changes that can hinder growth and development (Adhikari et al. 2022). One of the main effects of cold stress is the damage to cell membranes, which can result from alterations in membrane fluidity, low water potential, changes in lipid composition, reduced ATP supply, accumulation of toxic substances, imbalanced ion levels, and solute leakage (Theocharis et al. 2012).

The composition of lipids and their metabolism during cold stress play a crucial role in determining the extent of cold damage in plants. For instance, an increase in unsaturated fatty acids in the plasma membrane of acclimated plants enhances the fluidity that confers membrane stabilization. Research using fatty acid desaturase (fad2) mutants has shown that polyunsaturated lipids are vital for surviving cold temperatures in *Arabidopsis* (Miquel et al. 1993). For instance, the cold-acclimated rye (*Secale cereale* L.) displays high levels of di-unsaturated lipids like phosphatidylcholine and phosphatidylethanolamine and maintains the membrane stability in the face of adversity (Lynch and Steponkus 1987).

The cell wall serves as the first barrier against environmental stress and is essential for maintaining cell stability and support. Its dynamic structure allows for growth and adaptation under stress conditions (Yan et al. 2019; Liu et al. 2020). The expansion and remodeling of the cell wall are vital for enhancing leaf area and boosting photosynthesis (Clemente-Moreno et al. 2019). Important regulators of this process include xyloglucan endotransglucosylase/hydrolases, glycoside hydrolases, and expansins, which are crucial for preserving structural integrity under stress conditions (Tenhaken 2015). Additionally, signaling pathways that maintain cell wall integrity convey important signals, including reactive oxygen species (ROS) and phytohormones like jasmonate and ethylene, contributing to the antioxidative response and redox balance (Wolf et al. 2012; Liu et al. 2019a, b). A balance between reactive oxygen species (ROS) levels and antioxidant activity is vital for cell wall expansion and plant adaptation to stress (Tenhaken 2015).

Freezing stress imposes the formation of intra- and intercellular ice, which can lead to cell shrinkage, deformation, and even cell death (Uemura et al. 1995). When temperatures drop below freezing, cells may contract due to the osmotic gradient created between the ice-filled spaces and the fluid-filled cytoplasm (Li and Palva 2012; Fujikawa and Kuroda 2000). For example, in non-acclimated rye, the protoplast contraction at 0–5 °C temperature leads to cell lysis upon thawing (Li and Palva 2012), a phenomenon not seen in acclimated plants. Other cellular effects of cold stress include increased oxidative stress and ROS production, slowed metabolic processes, dehydration, reduced nutrient absorption, and diminished photosynthesis (Dreyer and Dietz 2018).

8.2.5 Disturbance in ROS and Antioxidant Defense System

Cold stress significantly offsets ROS balance and the antioxidant defense system in plants. Under cold stress, there is an increase in ROS, particularly in plant chloroplasts, which enhances the oxygenation response and leads to the accumulation of glycolate. This glycolate is then converted to glyoxylate in the peroxisomes, resulting in the accumulation of hydrogen peroxide (Hassan et al. 2021). Despite the harmful effects of increased ROS, plants possess a robust antioxidant defense system designed to scavenge these reactive molecules (Ritonga and Chen 2020). The response to cold stress involves several steps, including stress detection, signal perception, transduction, and the upregulation of stress-responsive genes (Ganeshan et al. 2008). Numerous genes have been identified that trigger a cascade of transcriptional, biochemical, and physiological processes essential for chilling tolerance (Kosová et al. 2008).

Plants counteract the effects of cold stress by accumulating various osmolytes and soluble sugars, which help stabilize cellular membranes and reduce water content (Thakur and Nayyar 2013; Turk et al. 2014; Yokota et al. 2015). These compounds also play a critical role in decreasing ROS and malondialdehyde (MDA) levels, a marker of lipid peroxidation. This reduction is achieved through the enhancement of antioxidant enzyme activities, including catalase (CAT), peroxidase (POD), and superoxide dismutase (SOD) (Sun et al. 2019).

8.3 Cold Stress Perception/Response Mechanisms

8.3.1 Signaling Pathways

8.3.1.1 Calcium Signaling

Cold stress triggers complex cellular and molecular responses in plants, prominently mediated by Ca^{2+} signaling pathways (Iqbal et al. 2022). Upon exposure to low temperatures, Ca^{2+} serves as a crucial secondary messenger, orchestrating a cascade of events that regulate gene expression essential for cold acclimation (Yuan et al. 2018).

Key components of Ca^{2+} signaling in cold stress responses include Ca^{2+} channels, such as mechanosensitive channels MCA1 and MCA2 (Fig. 8.1), which facilitate cold-induced increases in cytosolic Ca^{2+}. Mutants lacking MCA1 and MCA2 exhibit reduced cold-induced cytosolic Ca^{2+} levels and heightened sensitivity to chilling and freezing stresses (Mori et al. 2018). Besides, vesicle membrane Ca^{2+}/H^+ antiporter AtCAX1 is subtly involved in regulating cold acclimation responses by modulating the induction of C-repeat binding factor/Dehydration-responsive

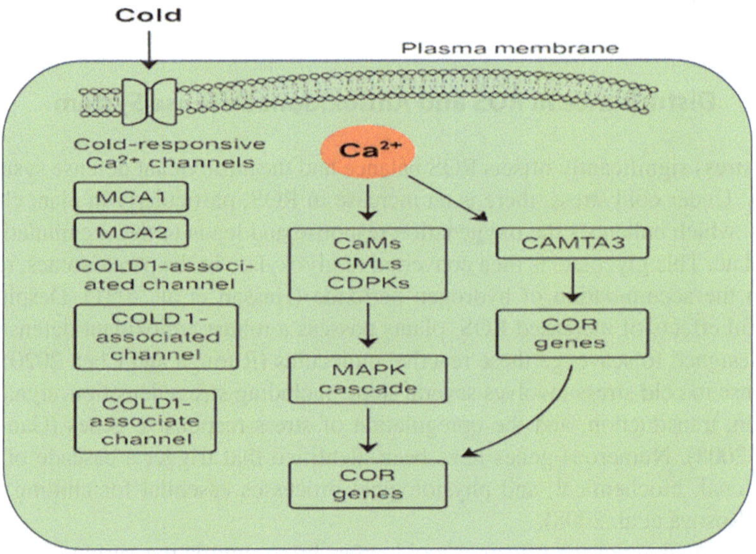

Fig. 8.1 Ca^{2+}-mediated signaling pathways involved in plant cold stress tolerance. Cold stress is perceived by the plasma membrane through changes in fluidity, activating Ca^{2+}-permeable channels such as MCA1, MCA2, and COLD1-associated channels. The influx of Ca^{2+} into the cytosol generates specific calcium signatures decoded by calcium-binding proteins like CaMs, CMLs, and calcium-dependent protein kinases (CDPKs). These sensors initiate downstream signaling cascades including MAPK activation, ROS production, and transcription factor modulation (e.g., CAMTA), which collectively lead to the expression of COR genes and enhanced cold tolerance mechanisms

element-binding protein 1 (CBF/DREB1) and downstream cold-responsive (COR) genes (Catalá et al. 2003).

The Ca^{2+}/cation antiporter superfamily proteins, identified in *Saccharum* spp., play pivotal roles in various environmental stresses, including cold stress (Su et al. 2021). Similarly, cyclic nucleotide-gated channels (CNGCs), localized to the plasma membrane, are critical for thermal sensing and thermotolerance. In rice, OsCNGC9 transcriptionally activated by OsDREB1A enhances chilling tolerance by facilitating cold-induced Ca^{2+} influx and elevation of cytosolic Ca^{2+} (Wang et al. 2021). These channels, like OsCNGC9, are implicated across diverse plant species, such as Chinese jujube (*Ziziphus jujuba* Mill.) and *Brassica oleracea*, in regulating cold stress responses via Ca^{2+} signals (Wang et al. 2020; Kakar et al. 2017).

The G-protein regulator CHILLING TOLERANCE DIVERGENCE1 (COLD1), in association with RICE G-PROTEIN α SUBUNIT1 (RGA1), facilitates cold stress signaling by promoting intracellular Ca^{2+} influx in *Oryza sativa* (Ma et al. 2015). Another plasma membrane-bound Ca^{2+} channel, GLR (glutamate receptor-like channel), responds to cold stress in an AtGLR3.4-dependent manner, positively regulating cold tolerance in Arabidopsis by modulating jasmonate signaling (Zheng et al. 2018). Annexins, Ca^{2+}-permeable transporters, also contribute to cold stress responses. In Arabidopsis, ANNEXIN1 (AtANN1) regulates cold-induced Ca^{2+} influx and freezing tolerance downstream of OST1, a key regulator in cold stress signaling (Liu et al. 2021). Similarly, the organellar Ca^{2+} channel GhCAX3 from *Gossypium hirsutum* influences cold and ABA signal transduction, affecting the expression of COR genes (Xu et al. 2013a, b). Collectively, these findings underscore the importance of calcium signaling in cold stress responses, where both plasma membrane and organellar Ca^{2+} channels coordinate with hormonal and transcriptional networks to promote plant survival under chilling conditions.

8.3.1.2 ROS Signaling

Cold stress presents a formidable challenge to plants, threatening their growth and productivity by disrupting cellular functions essential for survival. When exposed to cold temperatures, plants respond by triggering various defense mechanisms to combat these adverse conditions. One crucial aspect of this response revolves around ROS, once considered harmful but now recognized as pivotal signaling molecules in stress adaptation.

Under cold stress, plant cells experience oxidative stress, leading to the generation of ROS such as superoxide, hydrogen peroxide, and hydroxyl radicals (Dreyer and Dietz 2018). The primary source of ROS production is the damage inflicted on chloroplasts, particularly on the thylakoid membranes of PSII, which disrupts photosynthetic processes (Miura and Furumoto 2013; Zhang et al. 2020). This disruption initiates a cascade of signaling events where ROS act as key messengers, transmitting signals of stress to activate protective responses within the plant.

Moderate levels of ROS serve as signals to induce protective mechanisms against stress. However, excessive ROS accumulation poses a threat, causing oxidative damage to cellular components and potentially leading to cell death (Gechev et al. 2006). To counteract these detrimental effects, plants mobilize an arsenal of

ROS-scavenging enzymes such as CAT, SOD, POD, and GSTs (An et al. 2012). For instance, in cassava (*Manihot esculenta*), genes encoding CAT2, POD 72, and GST1 are upregulated in response to cold stress, enhancing the detoxification of ROS and thereby bolstering stress tolerance (An et al. 2012) (Fig. 8.2).

Moreover, cold stress induces redox changes that play a critical role in activating transcription factors crucial for cold tolerance, such as C-repeat binding factors and other redox-sensitive proteins. Antioxidant proteins like thioredoxins, glutaredoxins, and peroxiredoxins regulate the redox state of these transcription factors through reversible thiol-disulfide exchange reactions (Chi et al. 2013). When plants encounter stress, these antioxidant proteins facilitate the reduction of disulfide bonds in inactive forms of transcription factors, triggering their transformation into active monomeric or dimeric forms (Lee et al. 2021; Shaikhali et al. 2008, 2012). These activated transcription factors then migrate to the nucleus, where they bind to specific DNA sequences, initiating the expression of genes that confer stress tolerance.

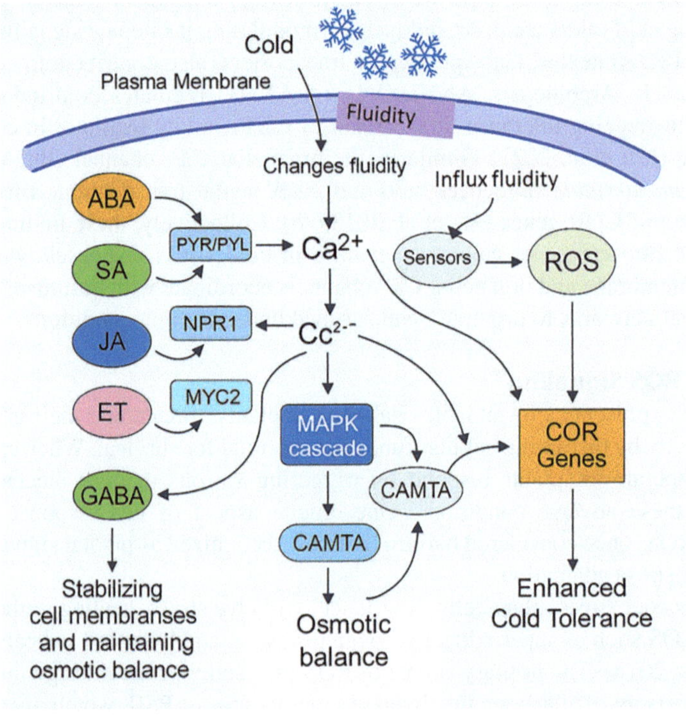

Fig. 8.2 Integrated signaling network showing interactions among molecules conferring cold stress resilience in plants. Cold stress activates calcium-permeable channels (e.g., MCA1, COLD1), triggering cytosolic Ca^{2+} influx. Calcium sensors (e.g., CaMs, CDPKs, CBLs) relay signals via protein phosphorylation and secondary messengers. These signals interact with ROS, ABA, and MAPK pathways, activating transcription factors like CBFs, CAMTAs, and ICE1 to induce cold-responsive gene expression. Crosstalk among Ca^{2+}, ROS, hormones (ABA, JA, SA), and gene expression mediates cellular responses such as osmotic adjustment, antioxidant defense, and membrane stabilization, enhancing cold stress resilience

8.3.1.3 Hormonal Signaling

Phytohormones play pivotal roles in mediating plant responses to cold stress, orchestrating various adaptive mechanisms to enhance cold tolerance. Upon exposure to cold stress, phytohormone levels fluctuate, with increased levels of ABA and JA and decreased levels of cytokinins, ethylene, and gibberellins (Deng et al. 2018; Pottosin et al. 2014). These hormones not only regulate the transcription of COR genes but also participate in ROS scavenging and the metabolism of biomolecules like sugars and starch (An et al. 2012).

Cold stress perception by plants involves the transduction of signals through secondary messengers such as hydrogen peroxide and calcium ions mediated by these phytohormones (Theocharis et al. 2012). The ABA-dependent signaling pathway is well-studied in response to cold stress, where increased ABA levels induce antioxidant defenses to scavenge ROS and mitigate oxidative damage (Zhang et al. 2006). Most phytohormones act via the CBF-independent pathway to regulate cold stress responses (Theocharis et al. 2012). For instance, in cassava, genes like SAUR, AREB3, GAI, ERF2, and JAZ7 are induced by cold stress and regulated by various phytohormones (An et al. 2012).

Under cold conditions, cytokinin and GA signaling are generally suppressed, which contributes to the upregulation of cold-responsive genes such as *CBF1* and *CBF3*. This hormonal suppression enhances freezing tolerance by promoting the expression of genes involved in cold acclimation (Novillo et al. 2007; Thomashow 2001). Interestingly, not all components of the CBF regulatory pathway have a positive effect. For instance, *CBF2* acts as a negative regulator of *CBF1* and *CBF3*, and its overexpression leads to increased plant sensitivity to freezing, highlighting the fine-tuned balance within the CBF regulatory network. On the other hand, ethylene signaling has also been implicated in modulating cold stress responses. Proteins such as ETHYLENE-INSENSITIVE3 (EIN3) have been shown to negatively regulate CBF-mediated pathways, thereby reducing cold tolerance when ethylene signaling is activated (Robison et al. 2019).

ABA activates ABFs to regulate COR gene transcription, CK stimulates CBF expression, and GA activates DELLA proteins that repress the CBF transcriptional pathway (Shi et al. 2018). JA, on the other hand, enhances cold tolerance by suppressing ICE1/2 genes, thereby upregulating downstream COR genes. JA also interacts with other hormone signaling pathways such as auxin, ethylene, and gibberellin to synergistically enhance cold tolerance (Hu et al. 2017; Wang et al. 2020; Yang et al. 2019).

8.3.2 COR Genes and Their Expression

Cold stress significantly impacts plant growth and productivity (Guan et al. 2023). Therefore, scientists intend to explore various mechanisms that plants use to tolerate cold stress and efficiently withstand low temperatures. There are two primary pathways that plants use to manage cold stress: C-repeat/DREB binding factors (CBF)-dependent and CBF-independent (Park et al. 2018; Qian et al. 2024).

Understanding these pathways can lead to developing crops that can survive and thrive in cold climates.

In the CBF-dependent pathway, cold stress is perceived by sensors in the plasma membrane, leading to an influx of Ca^{2+} into the cytosol. This calcium signaling activates the Inducer of CBF Expression 1 (ICE1), a MYC-type basic helix-loop-helix transcription factor. ICE1 binds to the promoters of CBF genes, inducing their expression. The CBF proteins then bind to the C-repeat/dehydration responsive element (CRT/DRE) motifs in the promoters of COR genes, activating their transcription. The products of COR genes contribute to freezing tolerance by stabilizing cellular structures and maintaining osmotic balance (Shi et al. 2018).

The CBF-independent pathway involves alternative mechanisms that confer cold tolerance without relying on CBF transcription factors. For instance, the ESKIMO1 (ESK1) gene in Arabidopsis encodes a protein that regulates freezing tolerance through cell wall modifications, independently of the CBF pathway (Xin et al. 2007). Similarly, the GIGANTEA (GI) gene integrates circadian and stress signals, modulating cold responses possibly through epigenetic regulation (Cao et al. 2005). Another gene, HOS9, encodes a homeodomain transcription factor that influences chromatin remodeling and stress gene expression, contributing to cold tolerance without involving CBFs (Zhu et al. 2004). The genes act as an emergency response system, helping the plant survive when temperatures drop without relying on the usual CBF pathway.

Lipid transfer proteins also modulate the response of plants to cold stress. These proteins help stabilize cell membranes under cold stress. The overexpression of lipid transfer protein 3 (LTP3) in Arabidopsis plants enhances their freeze tolerance. Essentially, LTP3 helps the plant cells remain intact and functional during freezing temperatures. This improvement occurs without changing the CBF-regulated COR genes, thanks to regulation by the MYB96 transcription factor (Guo et al. 2013). Transcription factors are proteins that help turn specific genes on or off. They play a significant role in managing how plants respond to cold. For instance, BZR1 is a transcription factor that positively influences freezing tolerance by regulating genes through both CBF-dependent and CBF-independent pathways. BZR1 impacts several genes, including PYL6, WRKY6, SAG21, SOC1, JMT, and ESM 1 (Li et al. 2017). These genes contribute to various functions, such as stress responses and developmental processes, helping the plant adapt to cold conditions.

Cold stress affects gene expression in plants, leading to changes in their metabolism and overall function. In maize, cold stress strongly downregulates genes involved in crucial processes like DNA replication, translation, and photosynthesis (Avila et al. 2018). This downregulation is a way for the plant to conserve energy and resources during stressful conditions. In another study, *Lotus japonicus,* having been exposed to cold stress for 24 h, reveals 713 upregulated genes and 364 downregulated genes based on RNA sequencing. The upregulated genes primarily exhibit association with lipid, cell wall, sugar, phenylpropanoid, and proline metabolism. On the contrary, the downregulated genes mainly relate to photosynthesis and chloroplast development (Calzadilla et al. 2016). This change in the pattern of gene expression helps the plant adapt its metabolic processes to better survive the cold.

MicroRNAs (miRNAs) are small molecules that help regulate gene expression. Cold stress can affect the expression of specific miRNAs. For example, miR389a.1 expression is downregulated under cold stress (Chinnusamy et al. 2007). This change can impact various physiological processes in the plant, contributing to its overall stress response.

Plants also produce certain proteins and compounds that help them tolerate cold. In Arabidopsis, the accumulation of cold shock domain proteins (AtCSP3) enhances freezing tolerance (Kim et al. 2009). These proteins act like a protective shield, helping the plant cells survive the cold. In temperate cereals such as wheat and barley, a gene called VERNALIZATION1 (VRN1) is crucial for low-temperature tolerance (Mickelbart et al. 2015). The gene prompts the plants to get ready for winter by adjusting their growth cycle to the changing seasons. Exogenous application of melatonin, a plant growth-boosting hormone, stimulates antioxidant capacity and enhances cold tolerance (Li et al. 2017). In maize, combining potassium chloride, gibberellin, and a chemical called DCPTA improves cold tolerance during germination (Wang et al. 2018). These treatments help the plants cope with cold stress by supporting various physiological processes.

Plants often face multiple stresses simultaneously, such as cold, drought, and salinity. There is a significant interplay among various responses to these different stresses, involving molecules like kinases, transcription factors, enzymes, and hormones. For example, in alfalfa, cold-inducible K3 dehydrin genes exhibit association with the desiccation or drying-out response (Dubé et al. 2013). DREB transcription factors regulate responses to cold, drought, and salinity, often functioning independently of ABA (Lata and Prasad 2011). The cold-induced DREB1 genes in black poplar show variable expression under dehydration stress (Chu et al. 2014). Additionally, treatments with salicylic acid can mimic the effects of cold acclimation, reducing freezing injury in winter wheat leaves (Taşgın et al. 2003). Glycine betaine, occurring in many plants, enhances freezing tolerance during cold acclimation, ABA application, and water stress in Arabidopsis (Xing and Rajashekar 2001).

Plants dispatch a complex and multifaceted response to cold stress, involving various genes, proteins, transcription factors, and chemical compounds. Understanding these mechanisms can help develop crops that are more resilient to cold and other environmental stresses, ultimately improving agricultural productivity and sustainability.

8.4 GABA Protection and Cold Stress Tolerance

As mentioned in the beginning, GABA serves a crucial protective role in enhancing cold stress tolerance in plants, offering a multifaceted defense mechanism against the adverse effects of low temperatures (Table 8.1). As a non-proteinogenic amino acid, GABA accumulates in response to cold stress, functioning both as a signaling molecule and as a key regulator of physiological processes that promote survival under cold conditions.

Table 8.1 GABA-mediated mechanisms of cold stress tolerance

Plant	Mechanism	References
Triticum aestivum (wheat) and barley	Advancement in cold acclimation and frost resistance mechanisms	Mazzucotelli et al. (2006)
Prunus persica (peach fruits)	Enhanced proline accumulation associated with mitigation of chilling injury	Shang et al. (2011)
Prunus persica (peach fruits)	Boosting antioxidant enzyme activity and maintaining energy status	Yang et al. (2011)
Lycopersicon esculentum (tomato)	Mitigation of cold stress through upregulation of antioxidant enzymes and suppression of lipid peroxidation	Malekzadeh et al. (2014)
Musa spp. (Banana)	Alleviation of chilling injury and upregulation of antioxidant capacity in banana peel	Wang et al. (2014)
Anthurium andraeanum (Anthurium)	Improved postharvest chilling tolerance through activation of antioxidant defense and suppression of ROS accumulation	Aghdam et al. (2016)
Oryza sativa (Rice)	Mitigated cold stress by restoring membrane integrity	Jia et al. (2017)
Cucumis sativus (cucumber)	Modulated proline, endogenous GABA, antioxidants (CAT, SOD), and membrane-associated enzymes (PLC, PLD, LOX)	Malekzadeh et al. (2017)
Morus alba (white mulberry)	Adaptation and tolerance to different cold storage condition	Li et al. (2018)
Cucurbita pepo (zucchini)	Conferring cold tolerance in postharvest storage conditions	Palma et al. (2019)
Camellia sinensis (tea plant)	Cold temperature tolerance mechanism	Zhu et al. (2019)
Lycopersicon esculentum (tomato)	Chilling stress protection via sugar metabolism and antioxidant defense	Abd Elbar et al. (2021)
Medicago ruthenica	GABA-induced cold tolerance via metabolic regulation, membrane stability, and leaf structure enhancement	Li et al. (2022b)

8.4.1 Chlorophyll Accumulation

GABA plays a pivotal role in enhancing chlorophyll biosynthesis under cold stress, thereby aiding in maintaining photosynthetic efficiency and overall plant health (Li et al. 2022a, b). Cold stress typically disrupts chlorophyll biosynthesis and accelerates its degradation, but GABA mitigates these effects through several mechanisms. When plants are exposed to cold stress, the biosynthesis of chlorophyll can be severely affected. This disruption leads to a decline in photosynthetic efficiency, which in turn impacts plant growth and productivity. GABA intervenes by modulating the chlorophyll biosynthesis pathway. It enhances the expression of genes encoding key enzymes involved in chlorophyll synthesis, ensuring a steady supply of precursors for chlorophyll synthesis even under stress conditions. By upregulating these critical biosynthetic enzymes, GABA helps maintain chlorophyll levels that are essential for efficient photosynthesis (Xiang et al. 2016; Malekzadeh et al. 2017; Palma et al. 2019; Li et al. 2022a, b).

Further, GABA also downregulates chlorophyll-degrading enzymes such as chlorophyllase and pheophorbide a oxygenase (Rossi et al. 2020). These enzymes are responsible for breaking down chlorophyll, and their inhibition by GABA helps preserve chlorophyll content in the plant tissues (Hörtensteiner 2013). This preservation is crucial for maintaining the green color and photosynthetic capacity of leaves during periods of cold stress. Furthermore, GABA enhances the plant's antioxidant defense systems. Cold stress often leads to the production of reactive oxygen species (ROS), which can damage cellular components, including chlorophyll. By boosting the antioxidant defense, GABA helps neutralize ROS and protect chlorophyll from oxidative damage. This protective effect ensures that the chlorophyll remains intact and functional, thereby supporting continued photosynthesis under adverse conditions. GABA also functions as an osmoprotectant, stabilizing cell membranes and maintaining cellular osmotic balance (Li et al. 2022a, b; Liu et al. 2020). Cold stress can lead to cellular dehydration and membrane instability, which in turn can affect chloroplast structure and function (Guan et al. 2023). By maintaining osmotic balance, GABA ensures that the chloroplast structure remains intact, facilitating efficient chlorophyll synthesis and accumulation.

The overall impact of GABA on chlorophyll and carotenoid contents under cold stress is significant. Chlorophyll content directly reflects the adaptation potential of plants to environmental changes (Wan et al. 2008). Carotenoids play a crucial role in light absorption and protection against photooxidative damage (Fracheboud et al. 2004). Both chlorophyll and carotenoid contents decline at low temperatures, indicating that stress may stimulate chlorophyll degradation, inhibit chlorophyll synthesis, or disorganize chloroplast tissue via photooxidation (Camejo et al. 2006). On the other hand, GABA application enhances chlorophyll fluorescence in the leaves of the tea plant (*Camellia sinensis* L.) under cold stress (Zhu et al. 2019) and prevents decreases in chlorophyll a and carotenoid contents in *Medicago ruthenica* at low temperatures (Li et al. 2022a, b). Hence, GABA alleviates the rapid degradation of photosynthetic pigments and maintains photosynthetic efficiency under cold stress.

8.4.2 Photosynthesis Resilience

Cold stress typically impairs photosynthetic efficiency due to the disruption of chloroplast function, resulting in reduced chlorophyll content and increased production of reactive oxygen species (ROS). GABA helps mitigate the sensitivity of photosynthesis to cold stress, thereby maintaining plant growth and productivity. GABA also counteracts the cold damage to PSII, a crucial component in the carbon metabolic process (Nunes-Nesi et al. 2010).

Cold stress disrupts various metabolic pathways in plants, including those related to signaling molecules like nitrate, ammonium, sugars, amino acids, and organic acids. These molecules interact within the broader network of carbon and nitrogen metabolism (Zhang et al. 2018). The integration of nitrogen uptake and metabolism with carbon metabolism is essential since nitrogen metabolism requires carbon skeletons, reducing power, ATP, and cotransporters (Nunes-Nesi et al. 2010; Kruger and von Schaewen 2003; Sahrawy et al. 2004; Gao et al. 2008). This coordination is likely to operate at the level of gene expression (Escobar et al. 2006). In response to cold stress, the upregulation occurs for only certain enzymes like aminomethyl transferase and fructose 1,6-bisphosphate aldolases that participate in the TCA cycle, glyoxylate and dicarboxylate metabolism, and carbon fixation in photosynthetic organisms. Specifically, fructose 1,6-bisphosphate aldolases, which are crucial for the Calvin-Benson cycle, show significant alterations in tomato seedlings subjected to heat or cold stress (Cai et al. 2018).

Exogenous GABA application enhances photosynthesis in tea plants by regulating low-temperature responses through the expression and activity of FBA enzymes (Zhu et al. 2019). The regulation is vital for maintaining photosynthetic efficiency and ensuring plant growth under cold stress conditions. By modulating these metabolic pathways, GABA not only helps in alleviating cold-induced damage but also supports the overall metabolic balance necessary for plant survival and productivity.

8.4.3 Osmotic Balance and Maintenance of Cellular Structures

GABA regulates intracellular redox balance and osmotic balance and maintains cellular structures, which plants effectively employ to survive in the cold. When the temperature drops, plants generate harmful ROS like superoxide and hydrogen peroxide (Liu et al. 2020). If left unchecked, these molecules can wreak havoc on cellular components, leading to oxidative stress and cell death. However, GABA steps in as a protagonist, enhancing the activity of antioxidant enzymes like CAT and SOD. These enzymes neutralize ROS, safeguarding the cellular machinery of the plants (Boudsocq and Sheen 2013; Lv et al. 2018).

When cold stress strikes, plants risk losing water and becoming dehydrated due to cellular shrinkage, but the accumulation of GABA serves as a crucial osmoprotectant. GABA helps maintain cellular osmotic balance, stabilizes cell membranes, and preserves cell turgor, acting like an internal hydration system to keep the plant resilient against the cold (Cheng et al. 2016). The plant cell wall, providing structural support, undergoes expansion and remodeling to protect intracellular functions, playing a critical role in sensing and responding to cold stress (Tenhaken et al. 2015; Wolf et al. 2012; Yan et al. 2019). Low-temperature stress can cause lipid membrane damage, physiological/metabolic disorder, and destruction to cell structure (Liu et al. 2018). Indicators of oxidative stress, such as increased electrical conductivity and MDA content, show reduced membrane stability. However, GABA inhibits MDA formation during lipid oxidation, effectively protecting the membrane system (Deng et al. 2010; Li et al. 2016). Studies on tomato and *Medicago ruthenica* reveal that GABA delays an increase in relative conductivity and MDA content under low temperatures, highlighting its protective role in maintaining membrane integrity (Abd El-Gawad et al. 2021; Li et al. 2022a, b).

There are intriguing connections between the structural components of the cell wall and ROS metabolism. Some structural proteins within the cell wall can cross-link with antioxidative enzymes, enhancing their ability to remove excess hydrogen peroxide (Tenhaken 2015). As plants adjust their cell walls, they also regulate intracellular redox balance by monitoring these changes. This process involves modulating the expression of antioxidant enzyme genes, thereby boosting their activities, enhancing cell membrane stability, and fortifying defense against cold-induced damage (Kärkönen and Kuchitsu 2015).

8.5 Antioxidant Defense and ROS Scavenging

GABA steps into action by activating ROS-scavenging enzymes like CAT and SOD, which are frontline defenders, swiftly neutralizing ROS in plants. This, in turn, prevents oxidative damage that could otherwise lead to cell death (Xu et al. 2022). By reducing ROS accumulation, GABA ensures the integrity of cellular membranes and proteins, which is crucial for sustaining plant health under cold temperatures.

In several studies, the exogenous application of GABA amplifies its protective effects. In tomato seedlings, for instance, exposure to chilling conditions prompts a natural increase in GABA levels. Exogenous application of GABA not only enhances its cellular titers but also promotes antioxidant enzyme activity and the accumulation of beneficial compounds like proline and sugars, further fortifying the plant against cold damage (Malekzadeh et al. 2014). To investigate further, scientists apply exogenous GABA to some of the seedlings of tomatoes. What they find remarkable is that the treated seedlings not only enhance their own GABA levels even more but also show reduced levels of MDA, a marker of oxidative stress. The decrease in MDA level hints at the curbing role of GABA in oxidative damage under cold stress conditions. Moreover, GABA tremendously elevates the activities of antioxidant enzymes like CAT, SOD, and APX (Abd Elbar et al. 2021). These enzymes act as frontline defenders to neutralize harmful ROS generated during cold stress.

GABA influences gene expression that is partly responsible for its protective prowess; exogenous GABA application upregulates the GAD gene and suppresses GABA-T, an enzyme that breaks down GABA (Shekari et al. 2021). This dual action ensures that GABA levels remain elevated, reinforcing the plant's defense against cold-induced oxidative stress. In banana fruit, exogenous GABA application alleviates chilling injury by promoting the accumulation of proline, another important osmoprotectant, and enhancing overall antioxidant capacity (Wang et al. 2014; Hayat et al. 2023). The finding underscores the broad applicability of GABA in mitigating cold stress across different plant species and organs. ROS, while essential for various cellular processes, can also pose a threat when produced excessively under stress conditions (Waszczak et al. 2018; Hasanuzzaman et al. 2020). Here, GABA steps in as a key regulator, orchestrating a defense strategy that balances ROS production and scavenging.

The GABA shunt pathway, bypassing vulnerable enzymes in the TCA cycle, emerges as a crucial route for maintaining ROS homeostasis (Janse van Rensburg and Van den Ende 2020; Xu et al. 2022). Moreover, many GABA-related components play vital roles in ROS scavenging and detoxification under stress conditions (Jalil et al. 2017; Ansari et al. 2021; Xu et al. 2022). Within this pathway, enzymes like SSADH that are integral to the GABA shunt play indispensable roles. SSADH ensures the conversion of succinic semialdehyde to succinate in mitochondria, effectively contributing to ROS detoxification (Bao et al. 2015; Xu et al. 2022). This illustrates how GABA not only directly scavenges ROS but also influences downstream pathways essential for oxidative stress management in plants.

8.6 GABA-Mediated Signaling in Cold Stress

Plants operate intricate systems to cope with low temperatures. When plants sense cold, they trigger protective mechanisms like reducing malondialdehyde (MDA) content, synthesizing cryoprotectant molecules, and enhancing the scavenging activity of ROS (Hare et al. 1998; Gusta et al. 2004; Wang et al. 2003). Chlorophyll fluorescence imaging is also used to assess cold tolerance in various plants (Li et al. 2017). These changes help plants maintain a balance of substance and energy metabolism in cold conditions.

GABA accumulates quickly in response to cold stress, aiding in adaptation (Mazzucotelli et al. 2006; Yang et al. 2011). However, endogenous GABA levels decrease under cold stress, for the metabolite degrades to promote the synthesis of other stress-response substances and feeds back into the TCA cycle, as seen by the upregulation of amino acids and differentially abundant proteins (DAPs) in the TCA cycle at low temperatures (Zhu et al. 2019).

GABA metabolism in tea plants also regulates the GABA precursor glutamate and polyamine contents to alleviate anoxia stress (Liao et al. 2017). Exogenous GABA application promotes the carbon and nitrogen cycles via feedback to the TCA cycle. Under cold conditions, the contents of most amino acids, except for alanine and lysine, decline (Zhu et al. 2019). Alanine aminotransferase catalyzes the reversible transformation of glutamic acid to pyruvic acid, while lysine, synthesized from aspartic acid, enters the TCA cycle via acetyl coenzyme A (Arruda et al. 2000).

Protein analysis using iTRAQ is a valuable tool for identifying stress-induced dynamic proteins and key metabolic pathways involved in cold resistance in tea plants with exogenous GABA application. According to Zhu et al. (2019), notable alterations were observed in several metabolic pathways, such as amino acid and flavonoid biosynthesis, glyoxylate and dicarboxylate metabolism, carbon fixation, and the pentose phosphate pathway. Flavonoids, such as anthocyanins, play decisive roles in plant functions, including antioxidant activities and defense against pathogens (Grotewold 2006). Low temperatures induce anthocyanin synthesis, enhancing drought tolerance by removing ROS (Chalker-Scott 1999; Fiorentino et al. 2007; Mohammadi et al. 2017). Cold treatment upregulates genes like CsBAN involved in anthocyanin biosynthesis, while CsCHS and CsF3H are downregulated (Wang et al.

2018). Exogenous GABA affects the expression of these genes differently, impacting mRNA and protein expression in flavonoid metabolism (Zhu et al. 2019).

The most significant changes in amino acids like GABA, proline, tryptophan, histidine, asparagine, and alanine occur in response to GABA application at low temperatures (Zhu et al. 2019). After exogenous GABA treatment at low temperatures, the contents of most amino acids in leaves increase significantly (Zhu et al. 2019). Glutamine synthetase, which participates in nitrogen assimilation, is downregulated and thus is involved in GABA signaling under cold stress (Zhu et al. 2019). Cysteine synthase, which catalyzes cysteine synthesis, also shows increased activity, suggesting a regulatory mechanism involving transcriptional and posttranslational levels (Liu et al. 2016; Kasukabe et al. 2004).

The TCA cycle, vital for amino acid synthesis and energy supply, is impacted by low temperatures and GABA application. Malate dehydrogenase levels decrease under stress, but exogenous GABA mitigates this reduction, suggesting its role in malate metabolism during cold stress (Musrat et al. 1998; Xu and Huang 2010a; Ramírez-Benítez et al. 2008). GABA promotes alanine accumulation, enhancing the TCA cycle and proline metabolism, crucial for carbon and nitrogen metabolism (Fait et al. 2008; Merewitz et al. 2012; Xu et al. 2013a, b; Shi et al. 2015).

The glyoxylate cycle, important for biosynthetic processes and respiration, also responds to exogenous GABA under cold stress. Proteins involved in this cycle are generally downregulated, but GABA helps protect PSII, which is essential for carbon metabolism (Nunes-Nesi et al. 2010). Signaling molecules like nitrate, ammonium, sugars, and amino acids interact in carbon and nitrogen metabolism at the gene expression level (Kruger and Von 2003; Sahrawy et al. 2004; Gao et al. 2008; Escobar and Geisler 2006).

Exogenous GABA impacts not only the pentose phosphate pathway, which is crucial for energy and metabolic intermediates, but also the energy budget, influencing purine metabolism and enhancing cold resistance in tea plants (Nunes-Nesi et al. 2010; Zhu et al. 2019).

8.7 Conclusion

In conclusion, GABA emerges as a crucial player in the plant response to cold stress, functioning not only as a signaling molecule but also as a key component in various physiological and biochemical pathways. Its ability to enhance osmotic regulation, boost antioxidant defenses, and influence gene expression underscores its potential in developing cold-tolerant plant varieties. As climate change continues to challenge agricultural productivity, harnessing GABA's properties could lead to innovative strategies for improving crop resilience against cold stress. Future research should focus on elucidating the precise mechanisms of GABA action and exploring its application in agricultural practices to enhance cold tolerance in economically important crops.

References

Abd Elbar OH, Elkelish A, Niedbała G, Farag R, Wojciechowski T, Mukherjee S et al (2021) Protective effect of γ-aminobutyric acid against chilling stress during reproductive stage in tomato plants through modulation of sugar metabolism, chloroplast integrity, and antioxidative defense systems. Front Plant Sci 12. https://doi.org/10.3389/fpls.2021.663750

Abdullah, Wani KI, Naeem M et al (2025) From neurotransmitter to plant protector: the intricate world of GABA signaling and its diverse functions in stress mitigation. J Plant Growth Regul 44:403–418. https://doi.org/10.1007/s00344-024-11470-0

Adhikari L, Baral R, Paudel D, Min D, Makaju SO, Poudel HP, Acharya JP, Missaoui AM (2022) Cold stress in plants: strategies to improve cold tolerance in forage species. Plant Stress 4:100081. https://doi.org/10.1016/j.stress.2022.100081

Aghdam MS, Naderi R, Jannatizadeh A, Babalar M, Sarcheshmeh MA, Faradonbe MZ (2016) Impact of exogenous GABA treatments on endogenous GABA metabolism in anthurium cut flowers in response to postharvest chilling temperature. Plant Physiol Biochem 106:11–15. https://doi.org/10.1016/j.plaphy.2016.04.045. Epub 2016 Apr 25. PMID: 27135813

An D, Yang J, Zhang P (2012) Transcriptome profiling of low temperature-treated cassava apical shoots showed dynamic responses of tropical plant to cold stress. BMC Genomics 13:64. https://doi.org/10.1186/1471-2164-13-64. PMID: 22321773; PMCID: PMC3339519

Ansari MI, Jalil SU, Ansari SA, Hasanuzzaman M (2021) GABA shunt: a key-player in mitigation of ROS during stress. Plant Growth Regul 94(2):131–149. https://doi.org/10.1007/s10725-021-00710-y

Arruda P, Kemper EL, Papes F, Leite A (2000) Regulation of lysine catabolism in higher plants. Trends Plant Sci 5:324–330

Avila LM, Obeidat W, Earl H et al (2018) Shared and genetically distinct Zea mays transcriptome responses to ongoing and past low temperature exposure. BMC Genomics 19:761–761

Banerjee A, Roychoudhury A (2019) Cold stress and photosynthesis. In: Photosynthesis, productivity and environmental stress. Wiley, Hoboken, pp 27–37

Bao H, Chen X, Lv S, Jiang P, Feng J, Fan P et al (2015) Virus-induced gene silencing reveals control of reactive oxygen species accumulation and salt tolerance in tomato by γ-aminobutyric acid metabolic pathway. Plant Cell Environ 38:600–613. https://doi.org/10.1111/pce.12419

Barton DA, Cantrill LC, Law AMK, Phillips CG, Sutton BG, Overall RL (2014) Chilling to zero degrees disrupts pollen formation but not meiotic microtubule arrays in Triticum aestivum L. Plant Cell Environ 37:2781–2794. https://doi.org/10.1111/pce.12358

Bota J, Medrano H, Flexas J (2004) Is photosynthesis limited by decreased Rubisco activity and RuBP content under progressive water stress? New Phytol 162:671–681. https://doi.org/10.1111/j.1469-8137.2004.01056.x

Boudsocq M, Sheen J (2013) CDPKs in immune and stress signaling. Trends Plant Sci 18:30–40

Cai B, Li Q, Liu F, Bi H, Ai X (2018) Decreasing fructose 1,6-bisphosphate aldolase activity reduces plant growth and tolerance to chilling stress in tomato seedlings. Physiol Plant 163:247–258

Calzadilla PI, Signorelli S, Escaray FJ, Menéndez AB, Monza J, Ruiz OA, Maiale SJ (2016) Photosynthetic responses mediate the adaptation of two Lotus japonicus ecotypes to low temperature. Plant Sci 250:59–68

Camejo D, Ana J, Juan Jose A, Walfredo T, Juana María G, Francisca S (2006) Changes in photosynthetic parameters and antioxidant activities following heat-shock treatment in tomato plants. Funct Plant Biol 33:177–187. https://doi.org/10.1071/FP05067

Cao S, Ye M, Jiang S (2005) Involvement of GIGANTEA gene in the regulation of the cold stress response in Arabidopsis. Plant Cell Rep 24:683–690. https://doi.org/10.1007/s00299-005-0061-x

Catalá R, Santos E, Alonso JM, Ecker JR, Martínez-Zapater JM, Salinas J (2003) Mutations in the Ca2+/H+ transporter CAX1 increase CBF/DREB1 expression and the cold-acclimation response in Arabidopsis. Plant Cell 15:2940–2951. https://doi.org/10.1105/tpc.015248

Chakrabarti B, Singh S, Nagarajan S, Aggarwal P (2011) Impact of temperature on phenology and pollen sterility of wheat varieties. Proceedings of Southern Cross Publishing, Brisbane

Chalker-Scott L (1999) Environmental significance of anthocyanins in plant stress responses. Photochem Photobiol 70:1–9

Chen L, Zhao Y, Xu S, Zhang Z, Xu Y, Zhang J et al (2018) OsMADS57 together with OsTB1 coordinates transcription of its target OsWRKY94 and D14 to switch its organogenesis to defense for cold adaptation in rice. New Phytol 218:219–231. https://doi.org/10.1111/nph.14977

Cheng F, Lu JY, Gao M, Shi K, Kong QS, Huang Y, Bie ZL (2016) Redox signaling and CBF-responsive pathway are involved in salicylic acid-improved photosynthesis and growth under chilling stress in watermelon. Front Plant Sci 7:1519

Chi Y, Yang Y, Zhou Y, Zhou J, Fan B, Yu JQ, Chen Z (2013) Protein-protein interactions in the regulation of WRKYtranscription factors. Mol Plant 6(2):287–300. https://doi.org/10.1093/mp/sst026. Epub 2013 Mar 2. PMID: 23455420

Chinnusamy V, Zhu J, Zhu JK (2007) Cold stress regulation of gene expression in plants. Trends Plant Sci 12(10):444–451. https://doi.org/10.1016/j.tplants.2007.07.002. Epub 2007 Sep 12. PMID: 17855156

Chu Y, Huang Q, Zhang B et al (2014) Expression and molecular evolution of two DREB1 genes in black poplar (Populus nigra). PLoS One 9:e98334

Clemente-Moreno MJ, Gago J, Díaz-Vivancos P, Bernal A, Miedes E, Bresta P, Liakopoulos G, Fernie AR, Hernández JA, Flexas J (2019) The apoplastic antioxidant system and altered cell wall dynamics influence mesophyll conductance and the rate of photosynthesis. Plant J 99:1031–1046

Cvetkovic J, Müller K, Baier M (2017) The effect of cold priming on the fitness of Arabidopsis thaliana accessions under natural and controlled conditions. Sci Rep 7:44055. https://doi.org/10.1038/srep44055

Demidchik V, Straltsova D, Medvedev SS, Pozhvanov GA, Sokolik A, Yurin V (2014) Stress-induced electrolyte leakage: the role of K+^++-permeable channels and involvement in programmed cell death and metabolic adjustment. J Exp Bot 65:1259–1270

Deng X, Wang J, Li Y et al (2018) Comparative transcriptome analysis reveals phytohormone signalings, heat shock module and ROS scavenger mediate the cold- tolerance of rubber tree. Sci Rep 8:1–16

Deng Y, Xu LJ, Zeng X, Li Z, Qin B, He N (2010) New perspective of GABA as an inhibitor of formation of advanced lipoxidation end-products: its interaction with malondialdehyde. J Biomed Nanotechnol 6:318–324

Dreyer A, Dietz KJ (2018) Reactive oxygen species and the redox-regulatory network in cold stress acclimation. Antioxidants 7:169

Dubé MP, Castonguay Y, Cloutier J et al (2013) Characterization of two novel cold- inducible K3 dehydrin genes from alfalfa (Medicago sativa spp. sativa L.). Theor Appl Genet 126:823–835

Equiza MA, Miravé JP, Tognetti JA (2001) Morphological, anatomical and physiological responses related to differential shoot vs. root growth inhibition at low temperature in spring and winter wheat. Ann Bot 87:67–76

Escobar MA, Geisler DA, Rasmusson AG (2006) Reorganization of the alternative pathways of the Arabidopsis respiratory chain by nitrogen supply: opposing effects of ammonium and nitrate. Plant J 45:775–788

Fait A, Fromm H, Walter D, Galili G, Fernie AR (2008) Highway or byway: the metabolic role of the GABA shunt in plants. Trends Plant Sci 13:14–19

Fiorentino A, D'Abrosca B, Pacifico S, Golino A, Mastellone C, Oriano P, Monaco P (2007) Reactive oxygen species scavenging activity of flavone glycosides from *Melilotus neapolitana*. Molecules 12:263–270

Fracheboud Y, Jompuk C, Ribaut JM, Stamp P, Leipner J (2004) Genetic analysis of cold-tolerance of photosynthesis in maize. Plant Mol Biol 56:241–253

Fujikawa S, Kuroda K (2000) Cryo-scanning electron microscopic study on freezing behavior of xylem ray parenchyma cells in hardwood species. Micron 31:669–686

Ganeshan S, Vitamvas P, Fowler DB, Chibbar RN (2008) Quantitative expression analysis of selected COR genes reveals their differential expression in leaf and crown tissues of wheat (*Triticum aestivum* L.) during an extended low temperature acclimation regimen. J Exp Bot 59:2393–2402

Gao P, Xin Z, Zheng ZL (2008) The OSU1/QUA2/TSD2-encoded putative methyltransferase is a critical modulator of carbon and nitrogen nutrient balance response in Arabidopsis. PLoS One 3:e106

Gechev TS, Van Breusegem F, Stone JM, Denev I, Laloi C (2006) Reactive oxygen species as signals that modulate plant stress responses and programmed cell death. BioEssays 28(11):1091–1101. https://doi.org/10.1002/bies.20493. PMID: 17041898

Grotewold E (2006) The science of flavonoids. Springer, New York

Guan Y, Hwarari D, Korboe HM, Ahmad B, Cao Y, Movahedi A, Yang L (2023) Low temperature stress-induced perception and molecular signaling pathways in plants. Environ Exp Bot 207:105190

Guo L, Yang H, Zhang X et al (2013) Lipid transfer protein 3 as a target of MYB96 mediates freezing and drought stress in Arabidopsis. J Exp Bot 64:1755–1767

Gusta LV, Wisniewski M, Nesbitt NT, Gusta ML (2004) The effect of water, sugars, and proteins on the pattern of ice nucleation and propagation in acclimated and nonacclimated canola leaves. Plant Physiol 135:1642–1653

Hare PD, Cress WA, Van Staden J (1998) Dissecting the roles of osmolyte accumulation during stress. Plant Cell Environ 21(6):535–553

Hasanuzzaman M, Bhuyan MHMA, Zulfiqar F, Raza A, Mohsin SM, Mahmud JA et al (2020) Reactive oxygen species and antioxidant defense in plants under abiotic stress: revisiting the crucial role of a universal defense regulator. Antioxidants 9(8):681. https://doi.org/10.3390/antiox9080681

Hassan MU, Chattha MU, Khan I, Chattha MB, Barbanti L, Aamer M et al (2021) Heat stress in cultivated plants: nature, impact, mechanisms, and mitigation strategies. Plant Biosyst 155:211–234

Hayat F, Khan U, Li J, Ahmed N, Khanum F, Iqbal S, Altaf MA, Ahmad J, Javed HU, Peng Y et al (2023) γ aminobutyric acid (GABA): a key player in alleviating abiotic stress resistance in horticultural crops: current insights and future directions. Horticulturae 9:647. https://doi.org/10.3390/horticulturae9060647

Hörtensteiner S (2013) Update on the biochemistry of chlorophyll breakdown. Plant Mol Biol 82:505–517. https://doi.org/10.1007/s11103-012-9940-z

Hu Y, Jiang Y, Han X et al (2017) Jasmonate regulates leaf senescence and tolerance to cold stress: crosstalk with other phytohormones. J Exp Bot 68:1361–1369

Hussain HA, Hussain S, Khaliq A, Ashraf U, Anjum SA, Men S et al (2018) Chilling and drought stresses in crop plants: implications, cross talk, and potential management opportunities. Front Plant Sci 9:393. https://doi.org/10.3389/fpls.2018.00393

Ikkonen EN, Shibaeva TG, Sherudilo EG, Titov AF (2020) Response of winter wheat seedlings respiration to long-term cold exposure and short-term daily temperature drops. Russ J Plant Physiol 67:538–544

Iqbal Z, Memon AG, Ahmad A, Iqbal MS (2022) Calcium mediated cold acclimation in plants: underlying signaling and molecular mechanisms. Front Plant Sci 13:855559

Jalil SU, Ahmad I, Ansari MI (2017) Functional loss of GABA transaminase (GABA-T) expressed early leaf senescence under various stress conditions in Arabidopsis thaliana. Curr Plant Biol 9:11–22. https://doi.org/10.1016/j.cpb.2017.02001

Jame YW, Cutforth HW (2004) Simulating the effects of temperature and seeding depth on germination and emergence of spring wheat. Agric For Meteorol 124:207–218. https://doi.org/10.1016/j.agrformet.2004.01.012

Janowiak F, Maas B, Dörffling K (2002) Importance of abscisic acid for chilling tolerance of maize seedlings. J Plant Physiol 159:635–643

Janse van Rensburg HC, Van den Ende W (2020) Priming with γ-aminobutyric acid against Botrytis cinerea reshuffles metabolism and reactive oxygen species: dissecting signalling and metabolism. Antioxidants 9(12):1174. https://doi.org/10.3390/antiox9121174

Ji H, Xiao L, Xia Y, Song H, Liu B, Tang L et al (2017) Effects of jointing and booting low temperature stresses on grain yield and yield components in wheat. Agric For Meteorol 243:33–42. https://doi.org/10.1016/j.agrformet.2017.04.016

Jia Y, Zou D, Wang J et al (2017) Effects of γ-aminobutyric acid, glutamic acid, and calcium chloride on rice (Oryza sativa L.) under cold stress during the early vegetative stage. J Plant Growth Regul 36:240–253. https://doi.org/10.1007/s00344-016-9634-x

Kakar KU, Nawaz Z, Kakar K, Ali E, Almoneafy AA, Ullah R et al (2017) Comprehensive genomic analysis of the CNGC gene family in Brassica oleracea: novel insights into synteny, structures, and transcript profiles. BMC Genomics 18:1–18. https://doi.org/10.1186/s12864-017-4244-y

Kärkönen A, Kuchitsu K (2015) Reactive oxygen species in cell wall metabolism and development in plants. Phytochemistry 112:22–32

Kasukabe Y, He L, Nada K, Misawa S, Ihara I, Tachibana S (2004) Overexpression of spermidine synthase enhances tolerance to multiple environmental stresses and up-regulates the expression of various stress-regulated genes in transgenic Arabidopsis thaliana. Plant Cell Physiol 45(6):712

Khan TA, Fariduddin Q, Yusuf M (2017) Low-temperature stress: is phytohormone application a remedy? Environ Sci Pollut Res 24:21574–21590. https://doi.org/10.1007/s11356-017-9948-7

Kim M-H, Sasaki K, Imai R (2009) Cold shock domain protein 3 regulates freezing tolerance in Arabidopsis thaliana. J Biol Chem 284:23454–23460

Kosová K, Prášil IT, Vítámvás P (2008) The relationship between vernalization- and photoperiodically-regulated genes and the development of frost tolerance in wheat and barley. Biol Plant 52:601–615. https://doi.org/10.1007/s10535-008-0120-6

Kruger NJ, von Schaewen A (2003) The oxidative pentose phosphate pathway: structure and organisation. Curr Opin Plant Biol 6(3):236–246

Lata C, Prasad M (2011) Role of DREBs in regulation of abiotic stress responses in plants. J Exp Bot 62:4731–4748

Lee ES, Park JH, Wi SD et al (2021) Redox-dependent structural switch and CBF activation confer freezing tolerance in plants. Nat Plants 7:914–922

Li PH, Palva ET (2012) Plant cold hardiness: gene regulation and genetic engineering. Springer Science & Business Media

Li Z, Yu JJ, Peng Y, Huang BR (2016) Metabolic pathways regulated by γ-aminobutyric acid (GABA) contributing to heat tolerance in creeping bentgrass (Agrostis stolonifera). Sci Rep-UK 6:30338. https://doi.org/10.1038/srep30338

Li X, Brestic M, Tan DX, Zivcak M, Zhu X, Liu S, Song F, Reiter RJ, Liu F (2017) Melatonin alleviates low PS I limited carbon assimilation under elevated CO2 and enhances the cold tolerance of offspring in chlorophyll b-deficient mutant wheat. J Pineal Res 64(1):e12453. https://doi.org/10.1111/jpi.12453

Li E, Luo X, Liao S, Shen W, Li Q, Liu F, Zou Y (2018) Accumulation of γ-aminobutyric acid during cold storage in mulberry leaves. Int J Food Sci Technol 53(12):2664–2672. https://doi.org/10.1111/ijfs.13875

Li M, Hwarari D, Li Y, Ahmad B, Min T, Zhang W, Wang J, Yang L (2022a) The bZIP transcription factors in *Liriodendron chinense*: genome-wide recognition, characteristics and cold stress response. Front Plant Sci 13:1035627. https://doi.org/10.3389/fpls.2022.1035627

Li Y, Yu X, Ma K (2022b) Physiological effects of γ-aminobutyric acid application on cold tolerance in *Medicago ruthenica*. Front Plant Sci 13:958029. https://doi.org/10.3389/fpls.2022.958029

Liao J, Wu X, Xing Z, Li Q, Duan Y, Fang W, Zhu X (2017) Gamma-aminobutyric acid accumulation in tea (Camellia sinensis L.) through the GABA shunt and polyamine degradation pathways under anoxia. J Agric Food Chem 65:3013–3018

Liu X, Yu W, Wang G, Cao F, Cai J, Wang H (2016) Comparative proteomic and physiological analysis reveals the variation mechanisms of leaf coloration and carbon fixation in a Xantha mutant of *Ginkgo biloba* L. Int J Mol Sci 17(11):1794

Liu XM, Zhou YL, Xiao JM, Bao F (2018) Effects of chilling on the structure, function and development of chloroplasts. Front Plant Sci 9:1715. https://doi.org/10.3389/fpls.2018.01715

Liu T, Du QJ, Li SZ, Yang JY, Li XJ, Xu JJ et al (2019a) GSTU43 gene involved in ALA-regulated redox homeostasis to maintain coordinated chlorophyll synthesis of tomato at low temperature. BMC Plant Biol 19:323

Liu T, Xu JJ, Li JM, Hu XH (2019b) NO is involved in JA- and H2O2-mediated ALA-induced oxidative stress tolerance at low temperatures in tomato. Environ Exp Bot 161:334–343

Liu T, Jiao X, Yang S, Zhang Z, Ye X, Li J, Qi H, Hu X (2020) Crosstalk between GABA and ALA to improve antioxidation and cell expansion of tomato seedling under cold stress. Environ Exp Bot 180:104228. https://doi.org/10.1016/j.envexpbot.2020.104228

Liu Q, Ding Y, Shi Y, Ma L, Wang Y, Song C et al (2021) The calcium transporter ANNEXIN1 mediates cold-induced calcium signaling and freezing tolerance in plants. EMBO J 40:e104559. https://doi.org/10.15252/embj.2020104559

Lv XZ, Li HZ, Chen XX, Xiang X, Guo ZX, Yu JQ, Zhou YH (2018) The role of calcium-dependent protein kinase in hydrogen peroxide, nitric oxide and ABA dependent cold acclimation. J Exp Bot 69:4127–4139

Lynch DV, Steponkus PL (1987) Plasma membrane lipid alterations associated with cold acclimation of winter rye seedlings (*Secale cereale* L. cv Puma). Plant Physiol 83:761–767

Ma Y, Dai X, Xu Y, Luo W, Zheng X, Zeng D et al (2015) COLD1 confers chilling tolerance in rice. Cell 160:1209–1221. https://doi.org/10.1016/j.cell.2015.01.046

Malekzadeh P, Khara J, Heydari R (2014) Alleviating effects of exogenous gamma-aminobutyric acid on tomato seedling under chilling stress. Physiol Mol Biol Plants 20(1):133–137. https://doi.org/10.1007/s12298-013-0203-5

Malekzadeh P, Khosravi-Nejad F, Hatamnia AA, Mehr RS (2017) Impact of postharvest exogenous γ-aminobutyric acid treatment on cucumber fruit in response to chilling tolerance. Physiol Mol Biol Plants 23:827–836

Massenssini AM, Bonduki VHA, Melo CAD, Tótola MR, Ferreira FA, Costa MD (2015) Relative importance of soil physico-chemical characteristics and plant species identity to the determination of soil microbial community structure. Appl Soil Ecol 91:8–15. https://doi.org/10.1016/j.apsoil.2015.02.009

Mazzucotelli E, Tartari A, Cattivelli L, Forlani G (2006) Metabolism of γ-aminobutyric acid during cold acclimation and freezing and its relationship to frost tolerance in barley and wheat. J Exp Bot 57(14):3755–3766

Merewitz EB, Du H, Yu W, Liu Y, Gianfagna T, Huang B (2012) Elevated cytokinin content in *ipt* transgenic creeping bentgrass promotes drought tolerance through regulating metabolite accumulation. J Exp Bot 63(3):1315–1328

Mickelbart MV, Hasegawa PM, Bailey-Serres J (2015) Genetic mechanisms of abiotic stress tolerance that translate to crop yield stability. Nat Rev Genet 16:237–251

Miquel M, James D, Dooner H et al (1993) Arabidopsis requires polyunsaturated lipids for low-temperature survival. Proc Natl Acad Sci USA 90:6208–6212

Miura K, Furumoto T (2013) Cold signaling and cold response in plants. Int J Mol Sci 14(3):5312–5337. https://doi.org/10.3390/ijms14035312. PMID: 23466881; PMCID: PMC3634503

Mohammadi MHS, Etemadi N, Arab MM, Aalifar M, Arab M, Pessarakli M (2017) Molecular and physiological responses of Iranian perennial ryegrass as affected by Trinexapac ethyl, Paclobutrazol and abscisic acid under drought stress. Plant Physiol Biochem 111:129–143

Mori K, Renhu N, Naito M, Nakamura A, Shiba H, Yamamoto T et al (2018) Ca^{2+}-permeable mechanosensitive channels MCA1 and MCA2 mediate cold-induced cytosolic Ca^{2+} increase and cold tolerance in Arabidopsis. Sci Rep 8:1–10. https://doi.org/10.1038/s41598-017-17483-y

Musrati RA, Kollárová M, Mernik N, Mikulásová D (1998) Malate dehydrogenase: distribution, function and properties. Gen Physiol Biophys 17(3):193

Novillo F, Medina J, Salinas J (2007) Arabidopsis CBF1 and CBF3 have a different function than CBF2 in cold acclimation and define different gene classes in the CBF regulon. Proc Natl Acad Sci 104:21002–21007

Nunes-Nesi A, Fernie AR, Stitt M (2010) Metabolic and signaling aspects underpinning the regulation of plant carbon nitrogen interactions. Mol Plant 3(6):973–996

Palma F, Carvajal F, Jimenez-Muñoz R, Pulido A, Jamilena M, Garrido D (2019) Exogenous γ-aminobutyric acid treatment improves the cold tolerance of zucchini fruit during postharvest storage. Plant Physiol Biochem 136:188–195. https://doi.org/10.1016/j.plaphy.2019.01.023

Park S, Gilmour SJ, Grumet R, Thomashow MF (2018) CBF-dependent and CBF-independent regulatory pathways contribute to the differences in freezing tolerance and cold-regulated gene expression of two Arabidopsis ecotypes locally adapted to sites in Sweden and Italy. PLoS One 13(12):e0207723. https://doi.org/10.1371/journal.pone.0207723

Pottosin I, Velarde-Buendía AM, Bose J et al (2014) Cross-talk between reactive oxygen species and polyamines in regulation of ion transport across the plasma membrane: implications for plant adaptive responses. J Exp Bot 65:1271–1283

Puhakainen T (2004) Short-day potentiation of low temperature-induced gene expression of a C-repeat-binding factor-controlled gene during cold acclimation in silver birch. Plant Physiol 136:4299–4307. https://doi.org/10.1104/pp.104.047258

Qian Z, He L, Li F (2024) Understanding cold stress response mechanisms in plants: an overview. Front Plant Sci 15:1443317. https://doi.org/10.3389/fpls.2024.1443317

Ramírez-Benítez JE, Chee-González L, Hernandez-Sotomayor SM (2008) Aluminium induces changes in organic acids metabolism in Coffea arabica suspension cells with differential Al-tolerance. J Inorg Biochem 102(8):1631–1637

Rinalducci S, Egidi MG, Karimzadeh G, Jazii FR, Zolla L (2011) Proteomic analysis of a spring wheat cultivar in response to prolonged cold stress. Electrophoresis 32:1807–1818. https://doi.org/10.1002/elps.201000663

Ritonga FN, Chen S (2020) Physiological and molecular mechanism involved in cold stress tolerance in plants. Plants 9:560. https://doi.org/10.3390/plants9050560

Robison JD, Yamasaki Y, Randall SK (2019) The ethylene signaling pathway negatively impacts CBF/DREBregulated cold response in Soybean (Glycine max). Front Plant Sci 10:121–121

Rossi S, Chapman C, Huang B (2020) Suppression of heat-induced leaf senescence by γ-aminobutyric acid, proline, and ammonium nitrate through regulation of chlorophyll degradation in creeping bentgrass. Environ Exp Bot 177:104116. https://doi.org/10.1016/j.envexpbot.2020.104116

Sahrawy M, Ávila C, Chueca A, Cánovas FM, López-Gorgé J (2004) Increased sucrose level and altered nitrogen metabolism in Arabidopsis thaliana transgenic plants expressing antisense chloroplastic fructose-1,6-bisphosphatase. J Exp Bot 55(408):2495–2503

Satyakam Z, Zinta G, Singh RK, Kumar R (2022) Cold adaptation strategies in plants—an emerging role of epigenetics and antifreeze proteins to engineer cold resilient plants. Front Genet 13:909007. https://doi.org/10.3389/fgene.2022.909007

Shaikhali J, Heiber I, Seidel T et al (2008) The redox-sensitive transcription factor Rap2. 4a controls nuclear expression of 2-Cys peroxiredoxin A and other chloroplast antioxidant enzymes. BMC Plant Biol 8:1–14

Shaikhali J, Noren L, De Dios Barajas-Lopez J et al (2012) Redox-mediated mechanisms regulate DNA binding activity of the G-group of basic region leucine zipper (bZIP) transcription factors in Arabidopsis. J Biol Chem 287:27510–27525

Shang H, Shifeng C, Zhenfeng Y, Yuting C, Yonghua Z (2011) Effect of exogenous γ-aminobutyric acid treatment on proline accumulation and chilling injury in peach fruit after long-term cold storage. J Agric Food Chem 59:1264–1268

Shekari A, Hassani RN, Aghdam MS (2021) Exogenous application of GABA retards cap browning in Agaricus bisporus and its possible mechanism. Postharvest Biol Technol 174:111434. https://doi.org/10.1016/j.postharvbio.2020.111434

Shi H, Jiang C, Ye T, Tan DX, Reiter RJ, Zhang H, Liu R, Chan Z (2015) Comparative physiological, metabolomic, and transcriptomic analyses reveal mechanisms of improved abiotic stress resistance in bermudagrass [Cynodon dactylon (L). Pers.] by exogenous melatonin. J Exp Bot 66(3):681–694

Shi Y, Ding Y, Yang S (2018) Molecular regulation of CBF signaling in cold acclimation. Trends Plant Sci 23(7):623–637. https://doi.org/10.1016/j.tplants.2018.04.002

Siddique MRB, Hamid A, Islam MS (2000) Drought stress effects on water relations of wheat. Bot Bull Acad Sin 41:35–39

Su W, Zhang C, Wang D, Ren Y, Sun T, Feng J et al (2021) The CaCA superfamily genes in Saccharum: comparative analysis and their functional implications in response to biotic and abiotic stress. BMC Genomics 22:1–19. https://doi.org/10.1186/s12864-021-07828-3

Sun X, Zhu Z, Zhang L, Fang L, Zhang J, Wang Q et al (2019) Overexpression of ethylene response factors VaERF080 and VaERF087 from Vitis amurensis enhances cold tolerance in Arabidopsis. Sci Hortic 243:320–326. https://doi.org/10.1111/tpj.14378

Taşgın E, Atící O, Nalbantoglu B (2003) Effects of salicylic acid and cold on freezing tolerance in winter wheat leaves. Plant Growth Regul 41:231–236

Tenhaken R (2015) Cell wall remodeling under abiotic stress. Front Plant Sci 5:771

Thakur P, Nayyar H (2013) Facing the cold stress by plants in the changing environment: sensing, signaling, and defending mechanisms. In: Tuteja N, Gill SS (eds) Plant acclimation to environmental stress. Springer Science+Business Media, New York, pp 29–69

Theocharis A, Clément C, Barka EA (2012) Physiological and molecular changes in plants grown at low temperatures. Planta 235:1091–1105. https://doi.org/10.1007/s00425-012-1641-y

Thomashow MF (2001) So what's new in the field of plant cold acclimation? Lots! Plant Physiol 125:89–93

Turk H, Erdal S, Genisel M, Atici O, Demir Y, Yanmis D (2014) The regulatory effect of melatonin on physiological, biochemical and molecular parameters in cold-stressed wheat seedlings. Plant Growth Regul 74:139–152. https://doi.org/10.1007/s10725-014-9905-0

Uemura M, Joseph RA, Steponkus PL (1995) Cold acclimation of Arabidopsis thaliana: effect on plasma membrane lipid composition and freeze-induced lesions. Plant Physiol 109:15

Venzhik YV, Titov AF, Talanova VV, Frolova SA, Talanov AV, Nazarkina YA (2011) Influence of lowered temperature on the resistance and functional activity of the photosynthetic apparatus of wheat plants. Biol Bull 38:132–137

Wan HW, Yang Y, Bai SQ, Xu YH, Bai YF (2008) Variations in leaf functional traits of six species along a nitrogen addition gradient in Leymus chinensis steppe in Inner Mongolia. Chin J Plant Ecol 32:611–621. https://doi.org/10.3773/j.issn.1005-264x.2008.03.010

Wang Y, Zhang Z, He X, Zhou H, Wen Y, Dai J, Zhang J, Chen S (2003) A rice transcription factor OsbHLH1 is involved in cold stress response. Theor Appl Genet 107:1402–1409

Wang Y, Luo Z, Huang X, Yang K, Gao S, Du R (2014) Effect of exogenous γ-aminobutyric acid (GABA) treatment on chilling injury and antioxidant capacity in banana peel. Sci Hortic 168:132–137. https://doi.org/10.1016/j.scienta.2014.01.022

Wang J, Guo M, Li Y, Wu R, Zhang K (2018) High-throughput transcriptome sequencing reveals the role of anthocyanin metabolism in Begonia semperflorens under high light stress. Photochem Photobiol 94:105–114

Wang L, Li M, Liu Z, Dai L, Zhang M, Wang L et al (2020) Genome-wide identification of CNGC genes in Chinese jujube (Ziziphus jujuba Mill.) and ZjCNGC2 mediated signalling cascades in response to cold stress. BMC Genomics 21:1–16. https://doi.org/10.1186/s12864-020-6601-5

Wang J, Ren Y, Liu X, Luo S, Zhang X, Liu X et al (2021) Transcriptional activation and phosphorylation of OsCNGC9 confer enhanced chilling tolerance in rice. Mol Plant 14:315–329. https://doi.org/10.1016/j.molp.2020.11.022

Waszczak C, Carmody M, Kangasjärvi J (2018) Reactive oxygen species in plant signaling. Annu Rev Plant Biol 69:209–236. https://doi.org/10.1146/annurev-arplant-042817-040322

Wolf S, Hématy K, Höfte H (2012) Growth control and cell wall signaling in plants. Annu Rev Plant Biol 63:381–407

Xiang LX, Hu LP, Xu WN, Zhen A, Zhang L, Hu XH et al (2016) Exogenous γ-aminobutyric acid improves the structure and function of photosystem II in muskmelon seedlings exposed to salinity-alkalinity stress. PLoS One 11:e0164847. https://doi.org/10.1371/journal.pone.0164847

Xin Z, Mandaokar A, Chen J, Last RL, Browse J (2007) Arabidopsis ESK1 encodes a novel regulator of freezing tolerance. Plant J 49:786–799. https://doi.org/10.1111/j.1365-313X.2006.02994.x

Xing W, Rajashekar C (2001) Glycine betaine involvement in freezing tolerance and water stress in Arabidopsis thaliana. Environ Exp Bot 46:21–28

Xu C, Huang B (2010a) Differential proteomic response to heat stress in thermal Agrostis scabra and heat-sensitive Agrostis stolonifera. Physiol Plant 139:192–204

Xu Y, Du H, Huang B (2013a) Identification of metabolites associated with superior heat tolerance in thermal bentgrass through metabolic profiling. Crop Sci 53:1626–1635

Xu L, Zahid KR, He L, Zhang W, He X, Zhang X et al (2013b) GhCAX3 gene, a novel Ca2+/H+ exchanger from cotton, confers regulation of cold response and ABA induced signal transduction. PLoS One 8:e66303. https://doi.org/10.1371/journal.pone.0066303

Xu M, Yang Q, Bai G, Li P, Yan J (2022) Polyamine pathways interconnect with GABA metabolic processes to mediate the low-temperature response in plants. Front Plant Sci 13:1035414. https://doi.org/10.3389/fpls.2022.1035414

Yamori W, Noguchi K, Hikosaka K, Terashima I (2009) Cold-tolerant crop species have greater temperature homeostasis of leaf respiration and photosynthesis than cold-sensitive species. Plant Cell Physiol 50:203–215. https://doi.org/10.1093/pcp/pcn189

Yan JW, Huang Y, He H, Han T, Di PC, Sechet J et al (2019) Xyloglucan endotransglucosylase-hydrolase30 negatively affects salt tolerance in Arabidopsis. J Exp Bot 70:5495–5506

Yang J, Duan G, Li C et al (2019) The crosstalks between jasmonic acid and other plant hormone signaling highlight the involvement of jasmonic acid as a core component in plant response to biotic and abiotic stresses. Front Plant Sci 10:1349

Yang A, Cao S, Yang Z, Cai Y, Zheng Y (2011) γ-Aminobutyric acid treatment reduces chilling injury and activates the defence response of peach fruit. Food Chem 129:1619–1622

Yokota H, Iehisa JCM, Shimosaka E, Takumi S (2015) Line differences in Cor/Lea and fructan biosynthesis-related gene transcript accumulation are related to distinct freezing tolerance levels in synthetic wheat hexaploids. J Plant Physiol 176:78–88. https://doi.org/10.1016/j.jplph.2014.12.007

Yuan P, Yang T, Poovaiah B (2018) Calcium signaling-mediated plant response to cold stress. Int J Mol Sci 19:3896. https://doi.org/10.3390/ijms19123896

Zhang A, Jiang M, Zhang J et al (2006) Mitogen-activated protein kinase is involved in abscisic acid-induced antioxidant defense and acts downstream of reactive oxygen species production in leaves of maize plants. Plant Physiol 141:475–487

Zhang F, Jiang Y, Bai LP, Zhang L, Chen LJ, Li HG et al (2011a) The ICE-CBF-COR pathway in cold acclimation and AFPs in plants. Middle East J Sci Res 8:493–498

Zhang J, Zhang Y, Du Y, Chen S, Tang H (2011b) Dynamic metabonomic responses of tobacco (Nicotiana tabacum) plants to salt stress. J Proteome Res 10:1904–1914

Zhang X, Misra A, Nargund S, Coleman GD, Sriram G (2018) Concurrent isotope assisted metabolic flux analysis and transcriptome profiling reveal responses of poplar cells to altered nitrogen and carbon supply. Plant J 93:472–488

Zhang F, Lu K, Gu Y et al (2020) Effects of low-temperature stress and brassinolide application on the photosynthesis and leaf structure of tung tree seedlings. Front Plant Sci 10:1767

Zheng Y, Luo L, Wei J, Chen Q, Yang Y, Hu X et al (2018) The glutamate receptors AtGLR1.2 and AtGLR1.3 increase cold tolerance by regulating jasmonate signaling in Arabidopsis thaliana. Biochem Biophys Res Commun 506:895–900. https://doi.org/10.1016/j.bbrc.2018.10.153

Zhu J, Shi H, Lee BH, Damsz B, Cheng S, Stirm V et al (2004) An Arabidopsis homeodomain transcription factor gene, HOS9, mediates cold tolerance through a CBF-independent pathway. Proc Natl Acad Sci USA 101:9873–9878. https://doi.org/10.1073/pnas.0403166101

Zhu X, Liao J, Xia X et al (2019) Physiological and iTRAQ-based proteomic analyses reveal the function of exogenous γ-aminobutyric acid (GABA) in improving tea plant (Camellia sinensis L.) tolerance at cold temperature. BMC Plant Biol 19:43. https://doi.org/10.1186/s12870-019-1646-9

GABA Against Drought Stress

9

Abstract

Drought stress represents a significant challenge to global agriculture, threatening crop yield and food security. Understanding the intricate mechanisms underlying plant responses to water scarcity is essential for developing resilient agricultural systems. Gamma-aminobutyric acid (GABA) is involved in osmotic regulation, antioxidant defense, and signal transduction pathways, conferring drought stress resistance in plants. Furthermore, GABA regulates the intricate network of molecular processes and metabolic pathways that ensures plant survival during drought stress. We delve into its role as a compatible solute, its contribution to the maintenance of cellular turgor, and its ability to mitigate oxidative stress. Moreover, GABA is involved in modulating stress-related gene expression, ion homeostasis, and stomatal regulation. Understanding the molecular and physiological mechanisms by which GABA enhances plant resilience to drought stress is pivotal for the development of novel strategies to combat the adverse effects of water scarcity on crop productivity and global food security.

Keywords

Antioxidant defense · Osmoprotectant · Osmotic regulation · Plant resilience · Water deficit

Highlights

Gamma-aminobutyric acid (GABA) is responsible for:

- Ion homeostasis, osmotic adjustment, and cellular turgor maintenance and protection against oxidative damage under drought stress.

S. U. Jalil et al., *GABA Signaling System and Abiotic Stress Tolerance in Plants*,
https://doi.org/10.1007/978-981-95-0437-4_9

- Modulation of stress-responsive gene expression, drought-induced signal transduction pathways, and metabolic reprogramming under drought conditions.
- Improvement in drought resilience, adaptability, and stomatal regulation in crops, ensuring sustainable food production.

9.1 Introduction

Drought stress poses a significant threat to global agriculture, limiting crop productivity and food security. As the frequency and severity of drought events increase due to climate change, understanding the mechanisms that plants employ to withstand water scarcity becomes paramount. In recent years, gamma-aminobutyric acid (GABA) has emerged as a central player in conferring advantages to plants in combating water deficit conditions.

GABA primarily acts as an osmoprotectant to offset drought stress in plants. GABA accumulation in plant cells helps maintain cellular turgor pressure, stabilizes membranes, and prevents protein denaturation under drought conditions. Stomatal regulation is critical for water conservation during drought stress, eliciting a discussion on GABA as a signaling molecule to induce stomatal closure, reducing water loss through transpiration, and improving water use efficiency in plants. Under drought stress, plants often produce excessive reactive oxygen species (ROS), leading to oxidative damage, which GABA attenuates to protect plant cells from oxidative stress (Fig. 9.1).

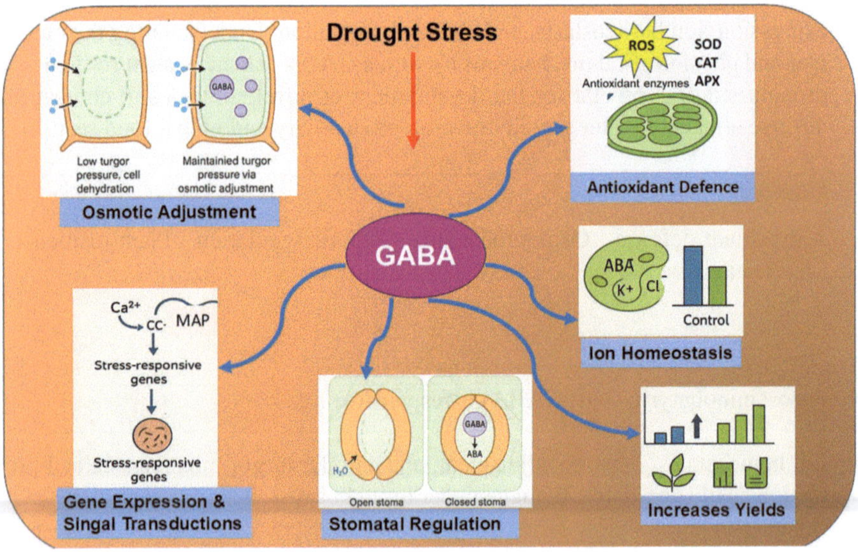

Fig. 9.1 GABA and drought stress tolerance in plants. Under drought stress, GABA accumulates and modulates various physiological and molecular processes

GABA does not act in isolation but interacts with other signaling molecules and phytohormones involved in drought stress responses. This section discusses the crosstalk between GABA and hormones such as abscisic acid, jasmonic acid, and salicylic acid, highlighting the synergistic effects on drought tolerance. As our understanding of the functions of GABA continues to grow, its potential increasingly becomes spectacular to revolutionize agriculture and mitigate the impact of drought on global food security.

9.2 Drought Stress Impact

Drought stress, marked by limited water availability, adversely affects plants at various levels, from molecular to physiological, ultimately impacting their survival and productivity. Reduced water availability can hinder germination and early plant development (Yigit et al. 2016). As plants grow, drought disrupts water relations, leading to significant metabolic disturbances dependent on the severity and duration of the water stress (Osakabe et al. 2014). One major effect is the reduction in photosynthetic activity (Bota et al. 2004; Flexas et al. 2006), which, along with oxidative stress-induced cell membrane damage, causes metabolic disorders and lowers plant yield (Vurayai et al. 2011; Mafakheri et al. 2010; Wang et al. 2019a,b; Sharma and Zheng 2019). Plants respond to drought stress through various physiological, biochemical, and molecular mechanisms, including antioxidative responses to diminish damage from ROS (Fahad et al. 2017). Before delving into the specific role of GABA, it is essential to establish the physiological and molecular responses of plants to drought stress. This section provides an overview of the effects of drought stress on plant water relations, photosynthesis, stomatal regulation, and hormonal responses.

9.2.1 Stomatal Closure

As first observed by Kramer (1969), drought stress reduces leaf area, cell size, and intercellular volume. Drought stress, characterized by water deficits, is one of the most significant environmental challenges impacting agricultural productivity worldwide, often leading to considerable yield reductions (Deka et al. 2018). When plants experience water deficit stress, their growth processes are notably affected, particularly elongation and expansion growth (Anjum et al. 2003; Kusaka et al. 2005; Shao et al. 2008).

Imagine a crop variety that can resist the flow of water from its stomata into the atmosphere; this trait is considered beneficial for drought tolerance. When soil moisture decreases, leaf water content drops, causing guard cells to lose turgor pressure. As a result, the size of stomatal pores reduces (Tezara et al. 2002), leading to stomatal closure. Increased stomatal resistance, a consequence of this closure, further reduces water transport within leaves, thereby decreasing stomatal conductance. This reduction in stomatal conductance has a crucial survival benefit. By closing their stomata, plants reduce transpiration, conserving essential soil water reserves in the root zone, which can prolong plants' survival during extended dry periods. Stomatal closure also helps maintain high leaf water content and leaf water potential, which is vital for sustaining plant health. However, this conservation strategy comes at a cost: a reduction in photosynthetic activity (Hsiao 1973).

Stomatal closure restricts transpiration and CO_2 intake. Nevertheless, plants maintain higher photosynthetic efficiency with partial stomatal closure that sustains greater biomass accumulation and improved crop yields under certain favorable conditions. Thus, the balance between water conservation and carbon dioxide intake is a delicate one. Plants often close their stomata under drought conditions to conserve water. While this reduces water loss through transpiration, it simultaneously restricts carbon dioxide intake necessary for photosynthesis.

9.2.2 Impaired Photosynthesis

Drought stress profoundly affects photosynthesis, the fundamental process by which plants convert light energy into chemical energy. A deficit in water availability, i.e., water stress, significantly hampers photosynthetic activity in plants. Nezhadahmadi et al. (2013) emphasize that one of the primary consequences of drought is the suppression of photosynthesis. The intricate balance of plant physiology gets disrupted, leading to several detrimental effects.

Reduced leaf area and increased stomatal closure are among the first responses to water scarcity. These changes hinder the transpiration process, escalating osmotic stress and causing damage to the photosynthetic apparatus (Zare et al. 2011; Bhargava and Sawant 2013). The primary reason behind the decline in photosynthesis during drought is the reduced CO_2 conductance through the stomata and mesophyll limitations (Singh and Thakur 2018). Stomata close up to conserve water, but this also limits the intake of CO_2 necessary for photosynthesis (Chaves et al. 2009; Marcińska et al. 2013). The drop in CO_2 intake sets off a cascade of adverse effects. Deepak et al. (2019) highlight that this reduction impacts Rubisco activity, an essential enzyme in the Calvin cycle. It also affects the function of NR and sucrose phosphate synthase, which are crucial for sugar production and energy storage. Interestingly, Sicher and Barnaby (2012) have demonstrated that CO_2 enrichment could mitigate many early metabolic and transcriptomic responses to drought stress in maize, suggesting potential avenues for alleviation.

Drought stress does not just stop at photosynthesis; it also modifies plant morphology. Reduced leaf area per shoot changes the canopy structure, affecting gas exchange, water relations, vegetative growth, and the development of sinks like fruits and grains (Rahmati et al. 2018). For example, water stress alters berry sugar concentration in grapevines (Zsófi et al. 2011), while prolonged water stress reduces kernel number and weight in maize (Ge et al. 2012a,b).

Reduced chlorophyll content is a clear indicator of photo-oxidation and chlorophyll degradation during drought stress (Anjum et al. 2011). In soybeans, drought stress alters chlorophyll synthesis and the chlorophyll a/b ratio (Chowdhury et al.

2017). The decline in photosynthetic activity, chlorophyll content, and efficiency of Photosystem II leads to reduced plant productivity (Xiang et al. 2013). Drought-induced ROS, like O_2^- and H_2O_2, cause lipid peroxidation, further degrading chlorophyll (Karimpour 2019). Water deficit impacts not only photosynthesis but also carbon allocation and metabolism. This disruption results in energy dissipation and reduced yield (Cuellar-Ortiz et al. 2008). For instance, drought stress suppresses photosynthetic product production and disrupts carbon balance in soybeans (Du et al. 2020) and reduces the abundance of several Calvin cycle proteins, including Rubisco, in olives (Abdallah et al. 2018). Severe drought conditions cause damage to photosynthetic enzymes, including Rubisco, coupled with a loss of photosynthetic pigment content (Brito et al. 2019). Thus, drought stress profoundly affects photosynthesis through various mechanisms, including reduced CO_2 conductance, stomatal limitations, and chlorophyll degradation. These effects cascade into broader impacts on plant growth, architecture, and yield.

9.2.3 Oxidative Stress

Drought stress is a significant challenge for plants, leading to the overproduction of ROS such as O_2^-, OH^-, ${}_1O^2$, and H_2O_2. These ROS cause oxidative stress, damaging cellular structures, proteins, and DNA, thus compromising plant integrity and function (Impa et al. 2012). The balance between ROS generation and antioxidant enzyme activities is crucial, as it regulates the cellular redox state. When ROS levels exceed physiological limits, they result in cellular injury and death (Møller et al. 2007).

It is reiterated that the drought condition triggers the activities of key antioxidant enzymes like catalase (CAT), superoxide dismutase (SOD), peroxidase (POD), and ascorbate peroxidase (APX) (Cao et al. 2017). Tolerant plant species or varieties typically exhibit higher antioxidant enzyme activities than those sensitive to drought stress, highlighting their essential role in controlling the production and diffusion of H_2O_2 and O_2^- in leaf tissues (Moussa and Abdel-Aziz 2008). For example, drought-tolerant potato genotypes show increased activities of SOD, POD, and CAT, which help manage ROS levels (Shi et al. 2015). APX, which uses ascorbate to convert H_2O_2 to water, also plays a crucial role in scavenging excess ROS, particularly in the chloroplasts and cytoplasm (Hossain et al. 2012; Caverzan et al. 2016).

The activities of SOD, POD, CAT, and APX are vital in protecting various plants, including peony, from acute water deficit (Wang et al. 2019a,b). Plants like *Coleus plectranthus* exhibit gains in nonenzymatic antioxidants such as ascorbic acid, reduced glutathione, and α-tocopherol, along with antioxidant enzymes under drought stress (Prathyusha and Chaitanya 2019). In *Vicia faba*, limited water availability stimulates the activities of SOD, CAT, and POD, which are higher in drought-tolerant genotypes than in the drought-sensitive ones (Abid et al. 2017; Siddiqui et al. 2015). This improvement in both enzymatic and nonenzymatic antioxidants under mild and moderate water deficit conditions supports drought tolerance (Flexas et al. 2012). Similarly, *Glycyrrhiza glabra* L. shows increased CAT, POD, and SOD levels in response to drought, aimed at counteracting H_2O_2 spread (Hosseini et al. 2018). In *Vigna mungo*, water deficit elevates SOD and POD levels, which enhances drought tolerance (Gurumurthy et al. 2019). In addition, plants like *Adonis amurensis* and *Adonis pseudoamurensis* elevate the activities of CAT, SOD, POD, and APX under drought conditions that reduce ROS levels and mitigate oxidative stress (Gao et al. 2020).

Overall, the severity of drought stress directly correlates with the production of ROS, leading to the peroxidation of membrane lipids, degradation of nucleic acids, and damage to both structural and functional proteins. Chloroplasts, mitochondria, and peroxisomes are the primary sites and initial targets of ROS generated under drought stress. The enhanced activity of antioxidant enzymes is a critical adaptive mechanism that plants employ to mitigate the detrimental effects of drought-induced oxidative stress.

9.2.4 Nutrient Imbalances

Decreasing water availability during drought typically leads to reduced nutrient uptake and lower nutrient concentrations in crop plant tissues (Farooq et al. 2009). One significant impact of water deficit is on the nutrient acquisition by roots and their subsequent transport to shoots. The roots face difficulties in absorbing inorganic nutrients, a process that is further complicated by the decrease in transpiration rate. This flow is crucial as it helps move nutrients from the roots to other parts of the plant. Under water deficit conditions, this system starts to break down, disrupting the water and nutrient balance in plants (Garg 2003).

Not all plants react to drought in the same way. Some species show remarkable resilience, while others falter. For instance, plants generally experience reduced potassium uptake due to its low mobility arising from the weakened action of root membrane transporters due to a drop in transpiration rate under water deficit conditions (Hu and Schmidhalter 2005). The drought-stressed *Malus hupehensis* plants also experience similar diminutive K levels (Qi et al. 2019). Interestingly, some plant genotypes have evolved to handle drought better than others. For instance, it was discovered that the drought-resistant genotypes of *Triticum durum* maintain high potassium levels, whereas its susceptible genotypes accumulate more sodium (Karimpour 2019). This genetic variance shows how some plants can adapt to harsh conditions by tweaking their nutrient uptake mechanisms.

As a matter of fact, drought stress represses genes encoding potassium transporters (Li et al. 2009), blocking potassium channels in grapevine roots (Cuéllar et al. 2010). It may be reiterated that the drought stress exhibits site-specific ill effects, for the same potassium channels remain activated in leaves of the grapevine being stimulated by a protein kinase (CIPK23) in conjunction with calcium sensors (Cuéllar et al. 2010). This complex regulation underscores how plants try to strike an internal nutrient balance under drought stress. Apart from affecting potassium dwindling, drought stress variably offsets nitrogen levels in some plants. For example, the nitrogen level remains unaffected in *Mentha piperita* and *Salvia lavandulifolia* but drops significantly in *Lavandula latifolia*, with a reduction in photosynthesis and an acceleration in leaf senescence (Da Silva et al. 2011). Phosphorus levels also generally decrease under drought conditions, impacting various physiological functions.

Different plants exhibit unique responses to drought conditions. Limited water availability decreases potassium in *Ocimum basilicum* and *Ocimum americanum* (Khalid 2006). Similarly, *Thymus daenensis* also reduces its potassium levels under drought stress (Bahreininejad et al. 2013). The common thread is that drought disrupts the delicate balance of nutrient uptake and distribution, with significant impacts on plant health and productivity. Interestingly, drought stress can sometimes lead to the accumulation of certain nutrients. Samarah et al. (2004) have noted increased levels of manganese, molybdenum, phosphorus, potassium, copper, calcium, and zinc in soybean plants under drought conditions. This suggests that while some nutrient pathways are hindered, others may be upregulated as a compensatory mechanism.

In essence, drought stress reshapes the nutrient landscape for plants. It can severely disrupt the uptake and transport of essential minerals, leading to nutrient imbalances and deficiencies. Understanding these dynamics helps us appreciate the resilience and adaptability of plants and guides us in developing strategies to enhance crop tolerance to water scarcity, ensuring better productivity even under challenging conditions.

9.2.5 Crop Growth and Yield

Drought stress is widely recognized as a critical limiting factor that affects multiple aspects of plant growth and development. Imagine a young plant starting its journey from a tiny seed; the very first challenge it faces is seed germination, which is highly sensitive to drought stress. Sourour et al. (2017) have noted significant changes in the germination process of various plant species, including major crops like maize (Queiroz et al. 2019), sorghum (Patanè et al. 2013), and wheat (Qayyum et al. 2011), when subjected to drought conditions.

As the plant progresses beyond germination, the early vegetative stages can exhibit visible symptoms of water scarcity. Leaves may begin to wilt, plant height may decrease, and the establishment of buds and flowers may be interrupted (Bhatt and Rao 2005). Limited soil moisture also hampers nutrient uptake, leading to shorter stem lengths, as observed in different plant species, such as *Lathyrus sativus* (Razmjoo et al. 2008; Gheidary et al. 2017). Interestingly, when water is scarce, plants adapt by enhancing their root systems to seek water from deeper soil layers (Asadi et al. 2012). Roots, being the primary sensors of water availability, regulate their growth characteristics, including root length, spread, and the number of lateral roots (Salazar et al. 2015). This adaptive mechanism is crucial as roots are involved in various biological activities, such as nutrient accumulation and water absorption. For instance, *Crocus sativus* shows increased root length under drought stress, highlighting the importance of a robust root system in sustaining plant growth during dry periods (Maleki et al. 2011).

However, this adaptation comes at a cost. As the root-to-shoot ratio increases, the overall plant biomass tends to decrease (Akhtar and Nazir 2013). Leaves, the main sites of photosynthesis, also suffer from water stress. The number of leaves and their area decrease, leading to reduced photosynthesis and, consequently, lower crop yields. Plants like *Andrographis paniculata* and *Petroselinum crispum* exhibit significant reductions in leaf area when subjected to drought, which is a survival strategy to minimize water loss through transpiration (Bhargavi et al. 2017; Najla et al. 2012; Srivastava and Srivastava 2014). The reduction in leaf area can be attributed to the inhibition of cell division and the resulting loss of cell turgidity (Bangar et al. 2019). As soil moisture declines, so does leaf water content, causing a drop in turgor pressure and leading to stomata closure (Deka et al. 2018). This not only reduces water loss but also limits the photosynthetic capacity of plants. Moreover, drought conditions accelerate the rate of premature leaf senescence, further compromising plant growth and productivity (Farooq et al. 2009).

Drought stress triggers a complex array of responses in plants, from altered seed germination and reduced vegetative growth to enhanced root development and decreased photosynthetic activity. Understanding these responses helps us appreciate the resilience and adaptability of plants in the face of environmental challenges and underscores the importance of developing strategies to mitigate the impacts of drought on agriculture.

9.3 GABA Protection in Plant Drought Tolerance

Imagine a world where plants constantly battle against an invisible yet frightening enemy—drought. This environmental stressor, exacerbated by climate change, threatens plant growth and development globally, causing a cascade of challenges that plants must overcome to survive and thrive (Buragohain et al. 2024). One of the primary weapons for this adversary is the overproduction of ROS, which can wreak havoc on plant cellular structures, leading to significant damage (Cruz de Carvalho 2008).

Yet, not all is as bleak as it seems to be in the intricate web of plant life. Recent studies have uncovered a fascinating dual role of ROS—not only as destructive agents but also as crucial signaling molecules that trigger protective responses within plants (Hasan et al. 2020). These responses are part of an elaborate defense mechanism that plants deploy when under stress.

One of the heroes in this tale of resilience is proline, a compound whose synthesis is triggered by ROS during stressful conditions (Khan et al. 2017; Hasan et al. 2018). Alongside proline, other vital metabolites such as alanine, glutamate, and the spotlight of our story, GABA, accumulate in the plant as part of the GABA shunt pathway, orchestrating a defense against ROS (Zhang and Becker 2015; Suzuki and Mittler 2006; Bouche and Fromm 2004). Within the plant cell, certain ROS, like H_2O_2, act as key messengers. They activate signal transduction pathways that aim at ensuring plant survival. For example, H_2O_2 facilitates the reduction of NADPH in the nucleus, protecting cellular processes from oxidative toxicity. In wheat plants, a decrease in H_2O_2 levels is vital for regulating redox reactions and gene expression during seed development, illustrating the delicate balance plants maintain with ROS (Pulido et al. 2009).

The role of mitochondrial glutamate dehydrogenase, which ROS stimulate to provide glutamate for the synthesis of ornithine and GABA, triggers a cascade of events leading to plant defense mechanisms against stressors (Skopelitis et al. 2006). GABA particularly provides a protective shield against oxidative damage induced by drought stress (Table 9.1) in common beans. Exogenous GABA application to leaves enhances drought tolerance by increasing fresh and dry shoot weights and leaf area through osmotic adjustment (Abd El-Gawad et al. 2020). This demonstrates the critical role of GABA in fortifying plants against drought. Further research highlights the efficacy of exogenous GABA in reducing drought-induced leaf damage. GABA-treated plants exhibit higher relative water content along with lower electrolyte leakage and lipid peroxidation, indicating better drought resistance (Yong et al. 2017; Abdullah et al. 2025). GABA also promotes the accumulation of amino acids, organic acids, and other osmotic compounds, all of which are crucial for secondary metabolism and overall plant health (Li et al. 2016a,b,c). In white clover leaves, exogenous GABA treatment leads to increased activity of GABA transaminase and alpha-ketoglutarate dehydrogenase while decreasing glutamate decarboxylase activity under both normal and drought conditions. This results in higher levels of endogenous glutamate and GABA, bolstering plant defenses (Yong et al. 2017). But the saga of GABA does not end there. It acts as a downstream signaling molecule for stress-related transcription factors like WRKY, MYB, and bZIP, which enhance drought tolerance in plants. GABA also increases the activity of calcium-dependent protein kinase 26 (CDPK26) and mitogen-activated protein kinase 1 (MAPK1) by threefold and fivefold, respectively, during drought stress (Li et al. 2018). These proteins are integral to the response of plants to drought.

The story of GABA in drought-stressed plants is one of resilience and adaptation. From mitigating oxidative damage to acting as a crucial signaling molecule, GABA orchestrates a multifaceted defense strategy that enhances plant survival under adverse conditions. This fascinating interplay of biochemical, physiological, and molecular mechanisms showcases the remarkable ways in which plants respond to and overcome environmental challenges. Thus, GABA comprises a critical and essential component of the intricate defense network of the plants.

9.3.1 GABA and Osmotic Adjustment

In the challenging world of plant survival, maintaining a favorable water status is crucial for plants to withstand drought stress. This intricate balance not only maintains the integrity of cell membranes but also heralds a process known as osmotic adjustment via accumulation of compatible solutes such as sugars and proline, which help plants manage water deficit conditions (Nilsen and Orcutt 1996).

Table 9.1 Effects of GABA on plant responses under drought stress

S. No.	Plants	Mechanism of GABA	References
1.	*Lolium perenne* (Ryegrass)	Mitigation of drought stress damage by reducing lipid peroxidation and electrolyte leakage and improving relative water content (RWC) and antioxidant activity	Krishnan et al. (2013)
2.	Creeping bentgrass (*Agrostis stolonifera*)	Increased amino acid (GABA, glycine, valine, proline, 5-oxoproline, serine, threonine, aspartic acid, and glutamic acid) and organic acid (malic acid, lactic acid, gluconic acid, malonic acid, and ribonic acid)accumulation	Li et al. (2016a,b,c)
3.	Black pepper (*Piper nigrum L.*)	Reduced wilting percentageIncreased leaf RWC and antioxidant enzyme activity; more rapidly decreased cell osmotic potential; reduced lipid peroxidation rate; significantly decreased inhibition of photosynthetic and mitochondrial activity	Vijayakumari and Puthur (2016)
4.	*Trifolium repens* (White clover)	Improvement of PEG-induced drought tolerance	Yong et al. (2017)
5.	Creeping bentgrass (*Agrostis stolonifera*)	Increased turf quality, leaf water content, cell membrane permeability, photosynthetic pigments, and expression of CDPK26, MAPK1, ABF3, WRKY75, MYB13, HSP70, MT1, and 14–3-3	Li et al. (2018)
6.	*Nigella sativa L.* (Black cumin)	Improved growth and productivity under water deficit stress conditions through maintaining osmoregulation and antioxidant defense systems	Rezaei-Chiyaneh et al. (2018)
7.	*Daucus carota L.* (Carrot)	Decreased MDA levels and electrolyte leakage; increased plastid pigmentation, protein content, enzymatic antioxidants, and relative water content	Bashir et al. (2019)
8.	German chamomile (*Matricaria recutita L.*)	Positive proline response to severe and mild stress in the presence of GABA	Sadaghiani et al. (2019)
9.	Kidney bean (*Phaseolusvulgaris L.*)	Increased leaf area, fresh and dry shoot weight,and improved osmotic adjustment, membrane permeability, uptake of nutrients, and antioxidantdefense	Abd El-Gawad et al. (2020)
10.	Job's tears (*Coix lacryma-jobi L.*)	Preserved the electron transport chain and minimized oxidative damage caused by reactive oxygen species	Ferreira et al. (2020)
11.	Creeping bentgrass (*Agrostis stolonifera*)	Enhanced accumulation of amino acids and organic acids, including proline, valine, glycine, and GABA	Li et al. (2021)

(continued)

Table 9.1 (continued)

S. No.	Plants	Mechanism of GABA	References
12.	Sunflower (*Helianthus annuus L.*)	Increased plant height, fresh and dry weight of shoot and root;improved osmolyte metabolism, expression of genes, and antioxidant enzyme activity	Abdel Razik et al. (2021)
13.	*Arabidopsis thaliana*	Guard cell-derived GABA reduces stomatal opening and transpiration by negatively regulating a tonoplast-localized anion transporter, enhancing water use efficiency and drought tolerance	Xu et al. (2021)
14.	Apple (*Malus domestica*)	GABA enhanced drought resistance by promoting GABA shunt activity and secondary cell wall biosynthesis	Chen et al. (2022)
15.	Apple (*Malus domestica*)	Enhanced fruit quality by increasing sugar–acid ratio, firmness, pericarp malleability, and GABA content, while reducing acidity	Cheng et al. (2023)
16.	Pomegranate (*Punica granatum*)	Increases compatible osmolytes (carbohydrates, starch, glucose, fructose, sucrose) for cellular defense and enhances the absorption of key minerals (potassium, magnesium, phosphorus, manganese, zinc, iron)	Zarbakhsh and Shahsavar (2023)

In creeping bentgrass, the relative water content progressively and significantly declines as the water stress intensifies (Li et al. 2016a,b,c). In consonance, previous investigation also reports a drop in relative water content in mature leaves in creeping bentgrass under heat stress (Merewitz et al. 2011). The decline in relative water content indicates that creeping bentgrass suffers from a mild water deficit due to both drought and heat stress. Interestingly, the application of GABA to creeping bentgrass exposed to heat or drought stress produces promising results. GABA-treated plants obtain an increase in leaf relative water content along with positive osmotic adjustment, suggesting that GABA plays a significant role in maintaining higher water content under stress conditions (Li et al. 2016a,b,c). According to Chen and Murata (2002), compatible osmolytes such as water-soluble carbohydrates and proline are pivotal for osmotic adjustment. Furthermore, drought-preconditioned Kentucky bluegrass noticeably obtains heat tolerance attributed to higher levels of water-soluble carbohydrates and proline, which facilitate active osmotic adjustment during heat stress (Jiang and Huang 2001).

Further, the exogenous GABA application mitigates drought stress damage by maintaining cell turgor (Krishnan et al. 2013). Although GABA-treated plants gain significant levels of water-soluble carbohydrates and proline under heat stress, the exact mechanism other than osmotic adjustment eludes a comprehensive explanation for GABA action (Li et al. 2016a,b,c). Nevertheless, GABA-induced osmotic adjustment arising from elevated endogenous proline levels under heat stress gains currency for conferring improved tolerance in rice seedlings (Nayyar et al. 2014).

However, higher water-soluble carbohydrate content rather than proline for osmotic adjustment in GABA-treated plants gains recognition under drought stress. Overall, the foliar application of GABA efficaciously enhances drought tolerance in creeping bentgrass, which results from the maintenance of cell membrane stability, suppression of leaf senescence, and enhanced osmotic adjustment through the accumulation of water-soluble carbohydrates or proline. This intricate mechanism underscores the resilience of plants and their ability to adapt to and survive under adverse environmental conditions.

9.3.2 GABA and ROS Scavenging

Under drought stress, plants encounter various physiological and biochemical challenges that significantly impact their growth and productivity. The culprits behind much of this stress are ROS, which are generated in multiple plant cell organelles, including plasma membranes, peroxisomes, chloroplasts, and mitochondria. Both under normal and stressful conditions, ROS are produced, but their excessive accumulation under drought can lead to oxidative damage. According to Gill and Tuteja (2010), the primary sources of ROS under normal conditions are chloroplasts and peroxisomes. Excessive ROS production, influenced by factors such as genotype, growth phase, and environmental conditions like drought, wreaks havoc on plant cells. To counteract ROS, plants boost their antioxidant defenses, which include a variety of compounds and enzymes.

Plants recruit various protective compounds like carotenoids, proline, anthocyanins, ascorbate, tocopherols, flavonols, and amino acids to combat excessive ROS (Capell et al. 2004; Slabbert and Krüger 2014). In this pursuit, plants also engage a votary of powerful enzymes such as SOD, CAT, POD, APX, mon-dehydroascorbate reductase (MDHAR), and dehydroascorbate reductase (DHAR). Both work in tandem, protecting the cell membranes from damage caused by increased ROS levels during stress. Among these, ascorbate stands out because it is widely found in plants and plays a key role in neutralizing ROS and supporting essential processes in plants, both under stress and normal conditions.

Amid this biochemical battlefield, GABA modulates various physiological and biochemical processes to cope with drought-induced oxidative stress, maintaining cellular homeostasis. GABA-induced modulation integrates different key metabolic and signaling pathways that significantly bolster plant resilience to drought stress. For example, the exogenous GABA application improves the activities of enzymes such as CAT, POD, APX, MDHAR, and GR, while reducing H_2O_2 production that manifests as greater drought tolerance in plants (Razik et al. 2020). On the contrary, GABA application in perennial ryegrass provides protection against drought conditions without influencing SOD and CAT activities, apparently activating glyoxalase and antioxidant pathways (Krishnan et al. 2013). Presumably, GABA commands diverse signaling modulation strategies specific to plant species and stress conditions and illuminates the complex yet fascinating world of plant biochemistry under various stress scenarios.

9.3.3 GABA and Stomatal Regulation

GABA in plants induces stomatal closure and adjusts transpiration, thereby reducing water loss and improving water use efficiency during drought stress. Thus, stomatal regulation stands as a critical player in water conservation and ensures the survival and productivity of plants in water-deficient conditions. Besides, it also significantly impacts global carbon and water cycles, mitigating the adverse effects of climate change to a great extent (Keenan et al. 2013; Papanatsiou et al. 2019). Stomatal pores, bordered by pairs of guard cells, open and close in response to environmental cues such as light and darkness. This fine-tuned mechanism controls ion and water transport across guard cell membranes, facilitated by specialized transport proteins (Sussmilch et al. 2019; Hasan et al. 2021; Xu et al. 2021). Because of their key role in plant signaling and their ability to process multiple signals, guard cells are widely used as a model system to study how plants respond to both biotic and abiotic stresses (Murata et al. 2015; Yao et al. 2021).

GABA controls stomatal opening/closing across various plant species, including both dicot and monocot crops, and binds with the ALMT1 transporter that influences plant development (Renault et al. 2011; Ramesh et al. 2015). The interaction with ALMT family proteins helps alter stomatal movement. For example, ALMT6, ALMT9, and ALMT12 proteins, located in the vacuolar and plasma membranes, regulate malate influx/efflux that enables stomatal movement (Meyer et al. 2010; Mekonnen et al. 2016). Meyer et al. (2011) advocate that the action of the ALMT6 protein is controlled by the binding of a cofactor of either cytosolic or vacuolar origin because the removal of Ca^{2+} from the activated protein does not affect its activity. A recent investigation by Xu et al. (2021) reveals a new cytosolic GABA signaling route in plants generated by GAD2 (glutamate decarboxylase 2) that influences stomatal opening, water use efficiency (WUE), and drought tolerance. The transduction of GABA signals under both well-watered and drought conditions negatively regulates the activity of ALMT9.

Despite these advancements, the specific mechanisms by which GABA induces stomatal closure under drought stress remain elusive. Some studies suggest that GABA regulation of stomata might be triggered by light, darkness, and low signal intermediate concentrations, but more research is needed to fully elucidate these pathways. GABA-induced stomatal regulation under drought stress involves a complex interplay of biochemical, physiological, and molecular mechanisms that enhance plant resilience. By promoting stomatal closure, GABA modulates ion channels, interacts with hormonal pathways, and influences gene expression. These actions help plants conserve water and maintain cellular integrity under drought conditions. Understanding these processes better could help develop drought-resilient crops, which are crucial for sustainable farming as the climate changes. By studying how GABA works, we can find new ways to help plants survive and thrive under tough conditions.

9.3.4 GABA and Nutrient Uptake

Drought stress is an alarming challenge for plants, significantly impacting their ability to uptake and translocate essential nutrients. The intricate relationship between water use efficiency and nitrogen uptake becomes particularly evident under such stress conditions. Pioneering research by Hu and Schmidhalter (2005) points out how drought adversely affects nutrient translocation in plants. Both ammonium and nitrate forms of nitrogen uptake perform a critical role in maintaining high water use efficiency during drought (Ding et al. 2018; Plett et al. 2020). High-affinity nitrate transporters play a pivotal role here, regulating aquaporin activity and thereby modulating water transport within the plant. As a matter of fact, water stress condition decreases the activity of nitrate uptake proteins, NRT1 and NRT2 (Bista et al. 2018). Subsequently, snap bean plants experiencing water stress register a significant reduction in contents of essential nutrients such as N, P, K, Ca, Fe, and Zn in their leaves (Abd El-Gawad et al. 2021).

Interestingly, endogenous GABA levels can trigger nitrate uptake and enhance nitrogen metabolism in *Arabidopsis* roots (Barbosa et al. 2010) and facilitate nitrogen-fixing efficiency in legumes under nutrient stress (Sulieman 2011). The findings advocate for GABA to be a strategic key player for nutrient uptake in plants. Further, GABA application upregulates enzymes associated with nitrogen metabolism, such as NR, GS, and NADH-GOGAT (Li et al. 2016a,b,c), and modulates nitrogen uptake in snap bean leaves (Abd El-Gawad et al. 2021) as well.

This modulation is critical during drought stress, as N levels in leaves typically decrease significantly. GABA application helps partially recover these levels, likely through increased nitrate uptake via roots and the remobilization of nitrogen from other plant organs (Abd El-Gawad et al. 2021). Drought-induced decreases in phosphorus and potassium content are also notable, often attributed to reduced absorption by roots and impaired nutrient transport through the xylem (Ge et al. 2012a,b). Under drought stress, phosphorus tends to convert from an immobile to an insoluble form, reducing the effectiveness of P-uptake proteins like PHT1 (Bista et al. 2018). While this hypothesis is compelling, further investigations are necessary to fully understand the routes of nitrogen uptake and mobilization induced by GABA treatments.

GABA emerges as a multifaceted protector, enhancing nutrient uptake and translocation under drought conditions. Its ability to modulate key enzymes, improve water use efficiency, and stabilize essential nutrient levels under stress underscores its potential in agricultural practices aimed at mitigating the impacts of drought.

9.3.5 GABA Crosstalk with Signaling Molecules

Under drought conditions, plants resort to a series of complex physiological and biochemical responses orchestrated by a network of signaling molecules, with GABA playing a pivotal role. GABA interacts with other signaling molecules to help plants manage and adapt to drought stress effectively.

GABA critically interacts with Ca^{2+} under water deficit conditions, for calcium as an immobile element requires ample water for optimal uptake (Naeem et al. 2018). When water is scarce, plants must find alternative ways to maintain their calcium levels, for which GABA plays a significant role, facilitating Ca^{2+} uptake by roots and its subsequent redistribution within plant tissues. For instance, the application of GABA enhances the levels of Ca^{2+} and calmodulin (a protein that binds Ca^{2+} and is vital for cellular processes) in NaCl-stressed barley plants (Ma et al. 2019) and leads to a notable rise in leaf Ca^{2+} in water-stressed snap bean plants (Abd El-Gawad et al. 2021). The rise in endogenous Ca^{2+} levels possibly activates an enzyme, namely GAD, that boosts GABA accumulation, as previously reported in heat-stressed *Arabidopsis* plants (Locy et al. 2000). This interplay between GABA and Ca^{2+}, referred to as GABA-Ca^{2+} crosstalk, emerges as a key pathway for stress tolerance in plants.

Further research indicates that GABA signaling works through the negative control of ALMT activity, which in turn regulates anion flux within plant cells (Ramesh et al. 2015). The mechanism legitimizes the role of GABA as a signaling molecule that becomes crucial for the acquisition and uptake of essential nutrient elements—like nitrogen, calcium, phosphorus, potassium, iron, and zinc—particularly in water-stressed snap bean plants (Abd El-Gawad et al. 2021). GABA also influences the expression of the 14-3-3 gene family proteins in *Arabidopsis*, orchestrating the regulation of enzymes involved in carbon and nitrogen metabolism and thereby modulating stress responses of plants (Lancien and Roberts 2006). Presumably, GABA also interweaves with PAs to facilitate plant responses to various stresses. Specific decarboxylase enzymes facilitate PA synthesis through the decarboxylation of ornithine or arginine. PAs, such as Put, Spd, and Spm, are critical during stress conditions. Their biosynthesis and catabolism ramp up under such stresses, indicating their essential role in adaptive mechanisms of plants (Hu et al. 2015; Wang et al. 2014).

The relationship between GABA and PAs is particularly intriguing. PA catabolism, which involves the enzymes CuAO and PAO, generates GABA, creating a direct link between these molecules. For example, PA degradation under hypoxic stress significantly contributes to the GABA content in fava beans (Yang et al. 2013). This regulation suggests a feedback mechanism where elevated GABA levels curb the breakdown of PAs, thereby maintaining higher PA levels under stress (Wang et al. 2014). Further, drought conditions activate several enzymes involved in PA production and breakdown (Yong et al. 2017). Interestingly, GABA not only boosts PA synthesis but also inhibits its catabolism, leading to an overall increase in PA content. The accumulation of PAs, observed across various plants under environmental stresses, underscores their critical role in plant tolerance (Yang et al. 2021). It appears that GABA may serve as an additional modulator of plant stress tolerance by influencing PA metabolism.

While the exact mechanisms through which GABA regulates these processes remain unclear, the emerging picture highlights the multifaceted role of GABA in helping plants to cope with drought. By mediating nutrient uptake, regulating crucial metabolic pathways, and modulating PA levels, GABA acts as a key player in the plant's intricate response to drought stress. Through its interactions with Ca^{2+} and PAs, GABA helps maintain cellular integrity and function, facilitating osmotic adjustment and enhancing overall stress tolerance.

9.4 Conclusion

Hence, GABA occupies a center stage in the complex network of responses that plants employ to combat drought stress and prevails from osmoprotection to stomatal regulation and ROS scavenging, improving crop resilience in the face of water scarcity. As our understanding of the diverse physio-biochemical functions of GABA as a signaling molecule continues to grow, so too does its potential to revolutionize agriculture and mitigate the impact of drought on global food security.

References

Abd El-Gawad HG, Mukherjee S, Farag R, Abd Elbar OH, Hikal M, Abou El-Yazied A, Abd Elhady SA, Helal N, ElKelish A, El Nahhas N et al (2020) Exogenous γ-aminobutyric acid (GABA)-induced signaling events and field performance associated with mitigation of drought stress in Phaseolus vulgaris L. Plant Signal Behav 16:1853384

Abd El-Gawad HG, Mukherjee S, Farag R, Abd Elbar OH, Hikal M, Abou El-Yazied A, Abd Elhady SA, Helal N, ElKelish A, El Nahhas N, Azab E, Ismail IA, Mbarki S, Ibrahim MFM (2021) Exogenous γ-aminobutyric acid (GABA)-induced signaling events and field performance associated with mitigation of drought stress in Phaseolus vulgaris L. Plant Signal Behav. Feb 1;16(2):1853384. https://doi.org/10.1080/15592324.2020.1853384. Epub 2020 Dec 23. PMID: 33356834; PMCID: PMC7849733

Abdallah MB, Trupiano D, Polzella A, De Zio E, Sassi M, Scaloni A, Zarrouk M, Youssef NB, Scippa GS (2018) Unraveling physiological, biochemical and molecular mechanisms involved in olive (Olea europaea L. cv. Chétoui) tolerance to drought and salt stresses. J Plant Physiol 220:83–95

Abdel Razik ES, Alharbi BM, Pirzadah TB, Alnusairi GSH, Soliman MH, Hakeem KR (2021) γ-Aminobutyric acid (GABA) mitigates drought and heat stress in sunflower (*Helianthus annuus* L.) by regulating its physiological, biochemical and molecular pathways. Physiol Plant 172(2):505–527. https://doi.org/10.1111/ppl.13216

Abdullah W, K.I., Naeem, M. et al (2025) From neurotransmitter to plant protector: the intricate world of GABA signaling and its diverse functions in stress mitigation. J Plant Growth Regul 44:403–418. https://doi.org/10.1007/s00344-024-11470-0

Abid G, M'hamdi M, Mingeot D, Aouida M, Aroua I, Muhovski Y, Sassi K, Souissi F, Mannai K, Jebara M (2017) Effect of drought stress on chlorophyll fluorescence, antioxidant enzyme activities and gene expression patterns in faba bean (*Vicia faba* L.). Arch Agron Soil Sci 63:536–552

Akhtar I, Nazir N (2013) Effect of waterlogging and drought stress in plants. Int J Water Res Environ Eng 2:34–40

Anjum F, Yaseen M, Rasul E, Wahid A, Anjum S (2003) Water stress in barley (*Hordeum vulgare* L.). I. Effect on morphological characters. Pak J Agric Sci 40:43–44

Anjum SA, Xie XY, Wang LC, Saleem MF, Man C, Lei W (2011) Morphological, physiological and biochemical responses of plants to drought stress. Afr J Agric Res 6:2026–2032

Asadi S, Lebaschy MH, Khourgami A, Rad AHS (2012) Effect of drought stress on the morphology of three Salvia sclarea populations. Ann Biol Res 3:4503–4507

Bahreininejad B, Razmjou J, Mirza M (2013) Influence of water stress on morpho-physiological and phytochemical traits in Thymus daenensis. Int J Plant Prod 7:151–166

Bangar P, Chaudhury A, Tiwari B, Kumar S, Kumari R, Bhat KV (2019) Morphophysiological and biochemical response of mungbean [Vigna radiata (L.) Wilczek] varieties at different developmental stages under drought stress. Turk J Biol 43:58–69

Barbosa JM, Singh NK, Cherry JH, Locy RD (2010) Nitrate uptake and utilization is modulated by exogenous γ-aminobutyric acid in Arabidopsis thaliana seedlings. Plant Physiol Biochem 48:443–450

Bashir R, Riaz HN, Shafiq M, Parveen N, Alenazi MM, Anwar S, Alebidi AI (2019) Foliar application of γ-aminobutyric acid (GABA) improves vegetative growth, and the physiological and antioxidative potential of *Daucus Carota* L. under water deficit conditions. https://doi.org/10.20944/preprints201903.0227.v2

Bhargava S, Sawant K (2013) Drought stress adaptation: metabolic adjustment and regulation of gene expression. Plant Breed 132:21–32

Bhargavi B, Kalpana K, Reddy JK (2017) Influence of water stress on morphological and physiological changes in Andrographis paniculata. Int J Pure Appl Biosci 5:1550–1556

Bhatt RM, Rao NKS (2005) Influence of pod load on response of okra to water stress. Indian J Plant Physiol 10:54–59

Bista DR, Heckathorn SA, Jayawardena DM, Mishra S, Boldt JK (2018) Effects of drought on nutrient uptake and the levels of nutrient-uptake proteins in roots of drought-sensitive and -tolerant grasses. Plants 7:28

Bota J, Medrano H, Flexas J (2004) Is photosynthesis limited by decreased Rubisco activity and RuBP content under progressive water stress? New Phytol 162:671–681

Bouche N, Fromm H (2004) GABA in plants: Just a metabolite? Trends Plant Sci 9:110–115

Brito C, Dinis LT, Moutinho-Pereira J, Correia CM (2019) Drought stress effects and olive tree acclimation under a changing climate. Plants 8:232

Buragohain K, Tamuly D, Sonowal S et al (2024) Impact of drought stress on plant growth and its management using plant growth promoting rhizobacteria. Indian J Microbiol 64:287. https://doi.org/10.1007/s12088-024-01136-w

Cao Y, Luo Q, Tian Y, Meng F (2017) Physiological and proteomic analyses of the drought stress response in *Amygdalus Mira* (Koehne) Yü et Lu roots. BMC Plant Biol 17:53

Capell T, Bassie L, Christou P (2004) Modulation of the polyamine biosynthetic pathway in transgenic rice confers tolerance to drought stress. Proc Natl Acad Sci USA 101:9909–9914

Caverzan A, Casassola A, Brammer SP (2016) Reactive oxygen species and antioxidant enzymes involved in plant tolerance to stress. In: Shanker A, Shanker C (eds) Abiotic and biotic stress in plants—recent advances and future perspectives. InTech Open, London, pp 463–480

Chaves MM, Flexas J, Pinheiro C (2009) Photosynthesis under drought and salt stress: regulation mechanisms from whole plant to cell. Ann Bot 103:551–560

Chen THH, Murata N (2002) Enhancement of tolerance of abiotic stress by metabolic engineering of betaines and other compatible solutes. Curr Opin Plant Biol 5:250–257

Chen X, Li N, Liu C, Wang H, Li Y, Xie Y, Ma F, Liang J, Li C (2022) Exogenous GABA improves the resistance of apple seedlings to long-term drought stress by enhancing GABA shunt and secondary cell wall biosynthesis. Tree Physiol 42(12):2563–2577. https://doi.org/10.1093/treephys/tpac096

Cheng P, Yue Q, Zhang Y, Zhao S, Khan A, Yang X, He J, Wang S, Shen W, Qian Q, Du W, Ma F, Zhang D, Guan Q (2023) Application of γ-aminobutyric acid (GABA) improves fruit quality and rootstock drought tolerance in apple. J Plant Physiol 280:153890. https://doi.org/10.1016/j.jplph.2022.153890

Chowdhury J, Karim M, Khaliq Q, Ahmed A (2017) Effect of drought stress on biochemical change and cell membrane stability of soybean genotypes. Bangladesh J Agric Res 42:475–485

Cruz de Carvalho MH (2008) Drought stress and reactive oxygen species: production, scavenging and signaling. Plant Signal Behav 3:156–165

Cuéllar T, Pascaud F, Verdeil JL, Torregrosa L, Adam-Blondon AF, Thibaud JB, Sentenac H, Gaillard I (2010) A grapevine Shaker inward K^+ channel activated by the calcineurin B-like

calcium sensor 1-protein kinase CIPK23 network is expressed in grape berries under drought stress conditions. Plant J 61:58–69
Cuellar-Ortiz SM, De La Paz A-MM, Acosta-Gallegos J, Covarrubias AA (2008) Relationship between carbohydrate partitioning and drought resistance in common bean. Plant Cell Environ 31:1399–1409
Da Silva EC, Nogueira RJMC, da Silva MA, de Albuquerque MB (2011) Drought stress and plant nutrition. Plant Stress 5:32–41
Deepak SB, Thakur A, Singh S, Bakshi M, Bansal S (2019) Changes in crop physiology under drought stress: a review. J Pharmacogn Phytochem 8:1251–1253
Deka D, Singh AK, Singh AK (2018) Effect of drought stress on crop plants with special reference to drought avoidance and tolerance mechanisms: a review. Int J Curr Microbiol Appl Sci 7:2703–2721
Ding L, Lu Z, Gao L, Guo S, Shen SQ (2018) Is nitrogen a key determinant of water transport and photosynthesis in higher plants upon drought stress? Front Plant Sci 9:1143
Du Y, Zhao Q, Chen L, Yao X, Zhang W, Zhang B, Xie F (2020) Effect of drought stress on sugar metabolism in leaves and roots of soybean seedlings. Plant Physiol Biochem 146:1–12
Fahad S, Bajwa AA, Nazir U, Anjum SA, Farooq A, Zohaib A, Sadia S, Nasim W, Adkins S, Saud S, Ihsan MZ, Alharby H, Wu C, Wang Dand Huang J (2017) Crop Production under Drought and Heat Stress: Plant Responses and Management Options. Front Plant Sci 8:1147. https://doi.org/10.3389/fpls.2017.01147
Farooq M, Wahid A, Kobayashi N, Fujita D, Basra SMA (2009) Plant drought stress: effects, mechanisms and management. In: Lichtfouse E (ed) Sustainable agriculture. Springer, Dordrecht, pp 153–188
Ferreira RA, Borella J, Hüther CM, do Canto ACB, da Costa Correa NP, Correia DM, Borges RP, de Pinho CF, Machado TB, Pereira CR (2020) Drought induced stress in leaves of *Coix lacryma-jobi* L. under exogenous application of proline and GABA amino acids. Braz J Bot 43:513–521
Flexas J, Bota J, Galmes J, Medrano H, Ribas-Carbó M (2006) Keeping a positive carbon balance under adverse conditions: responses of photosynthesis and respiration to water stress. Physiol Planta 127:343–352
Flexas J, Gallé A, Galmés J, Ribas-Carbo M, Medrano H (2012) The response of photosynthesis to soil water stress. In: Aroca R (ed) Plant responses to drought stress. Springer, Berlin/Heidelberg, pp 129–144
Gao S, Wang Y, Yu S, Huang Y, Liu H, Chen W, He X (2020) Effects of drought stress on growth, physiology and secondary metabolites of two *Adonis* species in Northeast China. Sci Hortic 259:108795
Garg BK (2003) Nutrient uptake and management under drought: nutrient-moisture interaction. Curr Agric 27:1–8
Ge T, Sui F, Bai L, Tong C, Sun N (2012a) Effects of water stress on growth, biomass partitioning, and water-use efficiency in summer maize (*Zea mays* L.) throughout the growth cycle. Acta Physiol Plant 34:1043–1053
Ge T, Sun NB, Bai LP, Tong CL, Sui FG (2012b) Effects of drought stress on phosphorus and potassium uptake dynamics in summer maize (*Zea mays*) throughout the growth cycle. Acta Physiol Plant 34:2179–2186
Gheidary S, Akhzari D, Pessarakli M (2017) Effects of salinity, drought, and priming treatments on seed germination and growth parameters of *Lathyrus sativus* L. J Plant Nutr 40:1507–1514
Gill SS, Tuteja N (2010) Reactive oxygen species and antioxidant machinery in abiotic stress tolerance in crop plants. Plant Physiol Biochem 48:909–930
Gurumurthy S, Sarkar B, Vanaja M, Lakshmi J, Yadav S, Maheswari M (2019) Morpho-physiological and biochemical changes in black gram (*Vigna mungo* L. Hepper) genotypes under drought stress at flowering stage. Acta Physiol Plant 41:42
Hasan MM, Alharby HF, Uddin MN, Ali MA, Anwar Y, Fang XW, Hakeem KR, Alzahrani Y, Hajar AS (2020) Magnetized water confers drought stress tolerance in *Moringa* biotype via modulation of growth, gas exchange, and antioxidant activity. Pol J Environ Stud 29(2):1625–1636

Hasan MM, Hajar AS, Alharby HF, Hakeem KR (2018) Effects of magnetized water on phenolic compounds, lipid peroxidation and antioxidant activity of *Moringa* species under drought stress. J Anim Plant Sci 28:803–810

Hasan MM, Skalicky M, Jahan MS, Hossain MN, Anwar Z, Nie ZF, Alabdallah NM, Brestic M, Hejnak V, Fang XW (2021) Spermine: its emerging role in regulating drought stress responses in plants. Cells 10:261

Hossain Z, Nouri MZ, Komatsu S (2012) Plant cell organelle proteomics in response to abiotic stress. J Proteome Res 11:37–48

Hosseini MS, Samsampour D, Ebrahimi M, Abadía J, Khanahmadi ME (2018) Effect of drought stress on growth parameters, osmolyte contents, antioxidant enzymes and glycyrrhizin synthesis in licorice (*Glycyrrhiza glabra* L.) grown in the field. Phytochemistry 156:124–134

Hsiao TC (1973) Plant responses to water stress. Annu Rev Plant Physiol 24:519–570

Hu X, Xu Z, Xu W, Li J, Zhao N, Zhou Y (2015) Application of γ-aminobutyric acid demonstrates a protective role of polyamine and GABA metabolism in muskmelon seedlings under $Ca(NO_3)_2$ stress. Plant Physiol Biochem 92:1–10

Hu Y, Schmidhalter U (2005) Drought and salinity: a comparison of their effects on mineral nutrition of plants. J Plant Nutr Soil Sci 168:541–549

Impa SM, Nadaradjan S, Jagadish SVK (2012) Drought stress induced reactive oxygen species and antioxidants in plants. In: Abiotic stress responses in plants. Springer, New York, pp 131–147

Jiang YW, Huang BR (2001) Osmotic adjustment and root growth associated with drought preconditioning-enhanced heat tolerance in Kentucky bluegrass. Crop Sci 41:1168–1173

Karimpour M (2019) Effect of drought stress on RWC and chlorophyll content in wheat (*Triticum durum* L.) genotypes. World Ess J 7:52–56

Keenan TF, Hollinger DY, Bohrer G, Dragoni D, Munger JW, Schmid HP, Richardson AD (2013) Increase in forest water-use efficiency as atmospheric carbon dioxide concentrations rise. Nature 499:324–327

Khalid KA (2006) Influence of water stress on growth, essential oil, and chemical composition of herbs (*Ocimum* sp.). Int Agrophys 20:289–296

Khan A, Anwar Y, Hasan M, Iqbal A, Ali M, Alharby HF, Hakeem KR, Hasanuzzaman M (2017) Attenuation of drought stress in *Brassica* seedlings with exogenous application of Ca^{2+} and H_2O_2. Plants 6:20

Kramer PJ (1969) Plant and soil water relationship. Tata McGraw-Hill, Bombay/New Delhi, p 360

Krishnan S, Laskowski K, Shukla V, Merewitz EB (2013) Mitigation of drought stress damage by exogenous application of a non-protein amino acid γ-aminobutyric acid on perennial ryegrass. J Am Soc Hortic Sci 138:358–366

Kusaka M, Ohta M, Fujimura T (2005) Contribution of inorganic components to osmotic adjustment and leaf folding for drought tolerance in pearl millet. Physiol Plant 125:474–479

Lancien M, Roberts MR (2006) Regulation of *Arabidopsis thaliana* 14-3-3 gene expression by γ-aminobutyric acid. Plant Cell Environ 29:1430–1436

Li W, Liu J, Ashraf U, Li G, Li Y, Lu W, Gao L, Han F (2016a) Exogenous γ-aminobutyric acid (GABA) application improved early growth, net photosynthesis, and associated physiobiochemical events in maize. Front Plant Sci 7:919

Li Y, Sun C, Huang Z, Pan J, Wang L, Fan X (2009) Mechanisms of progressive water deficit tolerance and growth recovery of Chinese maize foundation genotypes Huangzao 4 and Chang 7-2, which are proposed on the basis of comparison of physiological and transcriptomic responses. Plant Cell Physiol 50:2092–2111

Li Z, Peng Y, Huang B (2016b) Physiological effects of γ-aminobutyric acid application on improving heat and drought tolerance in creeping bentgrass. J Am Soc Hortic Sci 141(1):76–84

Li Z, Peng Y, Huang B (2018) Alteration of transcripts of stress-protective genes and transcriptional factors by γ-aminobutyric acid (GABA) associated with improved heat and drought tolerance in creeping bentgrass (*Agrostis stolonifera*). Int J Mol Sci 19:1623

Li Z, Tang M, Cheng B, Han L (2021) Transcriptional regulation and stress- defensive key genes induced by γ-aminobutyric acid in association with tolerance to water stress in creeping bentgrass. Plant Signal Behav 16:1858247. https://doi.org/10.1080/15592324.2020.1858247

Li Z, Yu J, Peng Y, Huang B (2016c) Metabolic pathways regulated by γ-aminobutyric acid (GABA) contributing to heat tolerance in creeping bentgrass (*Agrostis stolonifera*). Sci Rep 6:30338
Locy R, Wu S, Bisnette J, Barger T, McNabb D, Zik M, Fromm H, Singh N, Cherry J (2000) The regulation of GABA accumulation by heat stress in *Arabidopsis*. In: Hohnston M (ed) Plant tolerance to abiotic stresses in agriculture: role of genetic engineering. Springer, Dordrecht, pp 39–52
Ma Y, Wang P, Gu Z, Tao Y, Shen C, Zhou Y, Han Y, Yang R (2019) Ca^{2+} involved in GABA signal transduction for phenolics accumulation in germinated hulless barley under NaCl stress. Food Chem X 2:100023
Mafakheri A, Siosemardeh A, Bahramnejad B, Struik P, Sohrabi Y (2010) Effect of drought stress on yield, proline and chlorophyll contents in three chickpea cultivars. Aust J Crop Sci 4:580
Maleki M, Ebrahimzade H, Gholami M, Niknam V (2011) The effect of drought stress and exogenous abscisic acid on growth, protein content and antioxidative enzyme activity in saffron (*Crocus sativus* L.). Afr J Biotechnol 10:9068–9075
Marcińska I, Czyczyło-Mysza I, Skrzypek E, Filek M, Grzesiak S, Grzesiak MT, Janowiak F, Hura T, Dziurka M, Dziurka K (2013) Impact of osmotic stress on physiological and biochemical characteristics in drought-susceptible and drought-resistant wheat genotypes. Acta Physiol Plant 35:451–461
Mekonnen DW, Flügge UI, Ludewig F (2016) Gamma-aminobutyric acid depletion affects stomata closure and drought tolerance of *Arabidopsis thaliana*. Plant Sci 245:25–34
Merewitz EB, Gianfagna T, Huang BR (2011) Photosynthesis, water use, and root viability under water stress as affected by expression of *SAG12-ipt* controlling cytokinin synthesis in *Agrostis stolonifera*. J Exp Bot 62:383–395
Meyer S, Mumm P, Imes D, Endler A, Weder B, Al-Rasheid KAS, Geiger D, Marten I, Martinoia E, Hedrich R (2010) AtALMT12 represents an R-type anion channel required for stomata movement in *Arabidopsis* guard cells. Plant J 63:1054–1062
Meyer S, Scholz-Starke J, De Angeli A, Kovermann P, Burla B, Gambale F, Martinoia E (2011) Malate transport by the vacuolar AtALMT6 channel in guard cells is subject to multiple regulation. Plant J 67:247–257
Møller IM, Jensen PE, Hansson A (2007) Oxidative modifications to cellular components in plants. Annu Rev Plant Biol 58:459–481
Moussa HR, Abdel-Aziz SM (2008) Comparative response of drought tolerant and drought sensitive maize genotypes to water stress. Aust J Crop Sci 1:31–36
Murata Y, Mori IC, Munemasa S (2015) Diverse stomatal signaling and the signal integration mechanism. Annu Rev Plant Biol 66:369–392
Naeem M, Naeem MS, Ahmad R, Ihsan MZ, Ashraf MY, Hussain Y, Fahad S (2018) Foliar calcium spray confers drought stress tolerance in maize via modulation of plant growth, water relations, proline content and hydrogen peroxide activity. Arch Agron Soil Sci 64(1):116–131
Najla S, Sanoubar R, Murshed R (2012) Morphological and biochemical changes in two parsley varieties upon water stress. Physiol Mol Biol Plants 18:133–139
Nayyar H, Kaur R, Kaur S, Singh R (2014) γ-aminobutyric acid (GABA) imparts partial protection from heat stress injury to rice seedling by improving leaf turgor and upregulating osmoprotectants and antioxidants. J Plant Growth Regul 33:408–419
Nezhadahmadi A, Prodhan ZH, Faruq G (2013) Drought tolerance in wheat. Sci World J 2013:610721
Nilsen ET, Orcutt DM (1996) The physiology of plants under stress. Wiley, Hoboken, NJ
Osakabe Y, Osakabe K, Shinozaki K, Tran L-SP (2014) Response of plants to water stress. Front Plant Sci 5:86
Papanatsiou M, Petersen J, Henderson L, Wang Y, Christie JM, Blatt MR (2019) Optogenetic manipulation of stomatal kinetics improves carbon assimilation, water use, and growth. Science 363:1456–1459
Patanè C, Saita A, Sortino O (2013) Comparative effects of salt and water stress on seed germination and early embryo growth in two cultivars of sweet sorghum. J Agron Crop Sci 199:30–37

Plett DC, Ranathunge K, Melino VJ, Kuya N, Uga Y, Kronzucker HJ (2020) The intersection of nitrogen nutrition and water use in plants: new paths toward improved crop productivity. J Exp Bot 71(15):4452–4468

Prathyusha IVSN, Chaitanya KV (2019) Effect of water stress on the physiological and biochemical responses of two different *Coleus* (Plectranthus) species. Biol Fut 70:312–322

Pulido P, Domínguez F, Cejudo FJ (2009) A hydrogen peroxide detoxification system in the nucleus of wheat seed cells: protection or signaling role? Plant Signal Behav 4:23–25

Qayyum A, Razzaq A, Ahmad M, Jenks MA (2011) Water stress causes differential effects on germination indices, total soluble sugar and proline content in wheat (*Triticum aestivum* L.) genotypes. Afr J Biotechnol 10:14038–14045

Qi J, Sun S, Yang L, Li M, Ma F, Zou Y (2019) Potassium uptake and transport in apple roots under drought stress. Hortic Plant J 5:10–16

Queiroz MS, Oliveira CES, Steiner F, Zuffo AM, Zoz T, Vendruscolo EP, Menis VS, Mello BFFR, Cabral RC, Menis TF (2019) Drought stresses on seed germination and early growth of maize and sorghum. J Agric Sci 11:310–318

Rahmati M, Mirás-Avalos JM, Valsesia P, Lescourret F, Génard M, Davarynejad GH, Bannayan M, Azizi M, Vercambre G (2018) Disentangling the effects of water stress on carbon acquisition, vegetative growth, and fruit quality of peach trees by means of the QualiTree model. Front Plant Sci 9:3

Ramesh SA, Tyerman SD, Xu B, Bose J, Kaur S, Conn V, Domingos P, Ullah S, Wege S, Shabala S et al (2015) GABA signaling modulates plant growth by directly regulating the activity of plant-specification transporters. Nat Commun 6:7879

Razik ESA, Alharbi BM, Pirzadah TB, Alnusairi GSH, Soliman MH, Hakeem KR (2020) γ-Aminobutyric acid (GABA) mitigates drought and heat stress in sunflower (*Helianthus annuus* L.) by regulating its physiological, biochemical and molecular pathways. Physiol Plant 172:505–527

Razmjoo K, Heydarizadeh P, Sabzalian MR (2008) Effect of salinity and drought stresses on growth parameters and essential oil content of *Matricaria chamomile*. Int J Agric Biol 10:451–454

Renault H, El Amrani A, Palanivelu R, Updegraff EP, Yu A, Renou JP, Preuss D, Bouchereau A, Deleu C (2011) GABA accumulation causes cell elongation defects and a decrease in expression of genes encoding secreted and cell wall-related proteins in *Arabidopsis thaliana*. Plant Cell Physiol 52:894–908

Rezaei-Chiyaneh E, Seyyedi SM, Ebrahimian E, Moghaddam SS, Damalas CA (2018) Exogenous application of gamma aminobutyric acid (GABA) alleviates the effect of water deficit stress in black cumin (*Nigella sativa* L.). Ind Crop Prod 112:741–748

Sadaghiani FM, Dehaghi MA, Pirzad A, Fotokian MH (2019) Variation in yield and biochemical factors of German chamomile (*Matricaria recutita* L.) under foliar application of osmolytes and drought stress conditions. J Herbmed Pharmacol 8:90–100

Salazar C, Hernández C, Pino MT (2015) Plant water stress: associations between ethylene and abscisic acid response. Chil J Agric Res 75:71–79

Samarah N, Mullen R, Cianzio S (2004) Size distribution and mineral nutrients of soybean seeds in response to drought stress. J Plant Nutr 27:815–835

Shao HB, Chu LY, Shao MA, Abdul Jaleel C, Hong-Mei M (2008) Higher plant antioxidants and redox signaling under environmental stresses. C R Biol 331:433–441

Sharma A, Zheng B (2019) Melatonin mediated regulation of drought stress: physiological and molecular aspects. Plants 8:190

Shi S, Fan M, Iwama K, Li F, Zhang Z, Jia L (2015) Physiological basis of drought tolerance in potato grown under long-term water deficiency. Int J Plant Prod 9:305–320

Sicher RC, Barnaby JY (2012) Impact of carbon dioxide enrichment on the responses of maize leaf transcripts and metabolites to water stress. Physiol Planta 144:238–253

Siddiqui MH, Al-Khaishany MY, Al-Qutami MA, Al-Whaibi MH, Grover A, Ali HM, Al-Wahibi MS, Bukhari NA (2015) Response of different genotypes of faba bean plant to drought stress. Int J Mol Sci 16:10214–10227

Singh J, Thakur JK (2018) Photosynthesis and abiotic stress in plants. In: Biotic and abiotic stress tolerance in plants. Springer, Singapore, pp 27–46

Skopelitis DS, Paranychianakis NV, Paschalidis KA, Pliakonis ED, Delis ID, Yakoumakis DI, Kouvarakis A, Papadakis AK, Stephanou EG, Roubelakis-Angelakis KA (2006) Abiotic stress generates ROS that signal expression of anionic glutamate dehydrogenases to form glutamate for proline synthesis in tobacco and grapevine. Plant Cell 18:2767–2781

Slabbert MM, Krüger GHJ (2014) Antioxidant enzyme activity, proline accumulation, leaf area and cell membrane stability in water stressed *Amaranthus* leaves. S Afr J Bot 95:123–128

Sourour A, Afef O, Mounir R, Mongi BY (2017) A review: morphological, physiological, biochemical and molecular plant responses to water deficit stress. Int J Eng Sci 6:1–4

Srivastava S, Srivastava M (2014) Morphological changes and antioxidant activity of *Stevia rebaudiana* under water stress. Am J Plant Sci 5:3417

Sulieman S (2011) Does GABA increase the efficiency of symbiotic N2 fixation in legumes? Plant Signal Behav 6:32–36

Sussmilch FC, Schultz J, Hedrich R, Roelfsema MRG (2019) Acquiring control: The evolution of stomatal signalling pathways. Trends Plant Sci 24:342–351

Suzuki N, Mittler R (2006) Reactive oxygen species and temperature stresses: a delicate balance between signaling and destruction. Physiol Plant 126:45–51

Tezara W, Mitchel V, Driscul SP, Lawlor DW (2002) Effects of water deficit and its interaction with CO_2 supply on the biochemistry and physiology of photosynthesis in sunflower. Exp Bot 53:1781–1791

Vijayakumari K, Puthur JT (2016) γ-Aminobutyric acid (GABA) priming enhances the osmotic stress tolerance in *Piper nigrum* Linn. plants subjected to PEG-induced stress. Plant Growth Regul 78:57–67. https://doi.org/10.1007/s10725-015-0074-6

Vurayai R, Emongor V, Moseki B (2011) Effect of water stress imposed at different growth and development stages on morphological traits and yield of bambara groundnuts (*Vigna subterranea* L. Verdc). Am J Plant Physiol 6:17–27

Wang C, Fan L, Gao H, Wu X, Li J, Lv G, Gong B (2014) Polyamine biosynthesis and degradation are modulated by exogenous gamma-aminobutyric acid in root-zone hypoxia-stressed melon roots. Plant Physiol Biochem 82:17–26

Wang Q, Zhao R, Chen Q, da Silva JAT, Chen L, Yu X (2019a) Physiological and biochemical responses of two herbaceous peony cultivars to drought stress. HortSci 54:492–498

Wang X, Liu H, Yu F, Hu B, Jia Y, Sha H, Zhao H (2019b) Differential activity of the antioxidant defence system and alterations in the accumulation of osmolyte and reactive oxygen species under drought stress and recovery in rice (*Oryza sativa* L.) tillering. Sci Rep 9:8543

Xiang DB, Peng LX, Zhao JL, Zou L, Zhao G, Song C (2013) Effect of drought stress on yield, chlorophyll contents and photosynthesis in tartary buckwheat (*Fagopyrum tataricum*). J Food Agric Environ 11:1358–1363

Xu B, Long Y, Feng X, Zhu X, Sai N, Chirkova L, Betts A, Herrmann J, Edwards EJ, Okamoto M et al (2021) GABA signalling modulates stomatal opening to enhance plant water use efficiency and drought resilience. Nat Commun 12:1952

Yang R, Guo Q, Gu Z (2013) GABA shunt and polyamine degradation pathway on γ-aminobutyric acid accumulation in germinating fava bean (*Vicia faba* L.) under hypoxia. Food Chem 136:152–159

Yang YJ, Bi MH, Nie ZF, Jiang H, Liu XD, Fang XW, Brodribb TJ (2021) Evolution of stomatal closure to optimise water use efficiency in response to dehydration in ferns and seed plants. New Phytol 230:2001–2010

Yao GQ, Nie ZF, Turner NC, Li FM, Gao TP, Fang XW, Scoffoni C (2021) Combined high leaf hydraulic safety and efficiency provides drought tolerance in *Caragana* species adapted to low mean annual precipitation. New Phytol 229:230–244

Yigit N, Sevik H, Cetin M, Kaya N (2016) Determination of the effect of drought stress on the seed germination in some plant species. In: Water stress in plants. IntechOpen, London, pp 43–62

Yong B, Xie H, Li Z, Li YP, Zhang Y, Nie G, Zhang XQ, Ma X, Huang LK, Yan YH et al (2017) Exogenous application of GABA improves PEG-induced drought tolerance positively

associated with GABA-shunt, polyamines, and proline metabolism in white clover. Front Physiol 8:1107
Zarbakhsh S, Shahsavar AR (2023) Exogenous γ-aminobutyric acid improves the photosynthesis efficiency, soluble sugar contents, and mineral nutrients in pomegranate plants exposed to drought, salinity, and drought-salinity stresses. BMC Plant Biol 23(1):543. https://doi.org/10.1186/s12870-023-04568-2
Zare M, Azizi MH, Bazrafshan F (2011) Effect of drought stress on some agronomic traits in ten barley (*Hordeum vulgare* L.) cultivars. Tech J Eng Appl Sci 1:57–62
Zhang L, Becker DF (2015) Connecting proline metabolism and signaling pathways in plant senescence. Front Plant Sci 6:552
Zsófi Z, Tóth E, Rusjan D, Bálo B (2011) Terroir aspects of grape quality in a cool climate wine region: relationship between water deficit, vegetative growth and berry sugar concentration. Sci Hortic 127:494–499

GABA as a Thermo-Protectant and Protectant Against Radiation

10

Abstract

Heat stress and radiation exposure are critical environmental stressors that adversely affect plant growth, development, and yield. Gamma-aminobutyric acid (GABA) contributes to plant resilience by modulating stress-responsive pathways and enhancing cellular protection mechanisms. We delve into the mechanisms by which GABA mitigates the impacts of heat stress, including its role in osmotic regulation, stabilization of heat shock proteins (HSPs), and interaction with oxidative stress pathways. Similarly, the chapter examines GABA's protective effects against radiation-induced damage, focusing on its involvement in scavenging reactive oxygen species (ROS) and supporting DNA repair processes. Through a review of experimental studies and case studies, the chapter highlights the potential of GABA-based strategies in developing crops with improved resistance to both thermal and radiation stresses. The discussion also focuses on the integrative role of GABA in stress crosstalk and the implications for future research and biotechnological applications. This comprehensive overview underscores the importance of GABA in enhancing plant resilience and offers insights into its practical applications in agriculture.

Keywords

Antioxidant defense · DNA repair · GABA · Radiation protection · Stress signaling · Thermotolerance

S. U. Jalil et al., *GABA Signaling System and Abiotic Stress Tolerance in Plants*,
https://doi.org/10.1007/978-981-95-0437-4_10

Highlights

- Gamma-aminobutyric acid (GABA) normalizes plant responses to heat stress by modulating osmotic balance, enhancing antioxidant defense, and stabilizing cellular structures, thereby improving thermotolerance.
- GABA-induced stimulation of antioxidant enzymes helps to mitigate oxidative stress triggered by heat and radiation, reducing cellular damage.
- GABA increases the synthesis of protective metabolites (e.g., flavonoids and phenolics), which absorb UV radiation and reduce radiation-induced DNA damage in plant cells.
- GABA signaling influences gene expression related to HSPs and DNA repair enzymes, contributing to cellular homeostasis under thermal and radiation stress.
- By preventing chlorophyll degradation and protecting photosystem integrity, GABA helps sustain photosynthetic activity under heat and radiation stress conditions.

10.1 Introduction

Plants, as sessile organisms, are constantly exposed to fluctuating environmental conditions and cannot relocate to avoid abiotic or biotic stresses. Consequently, their growth, development, and productivity are significantly impacted when subjected to extreme environmental factors (Lippmann et al. 2019; Jalil and Ansari 2020). Among these, heat and radiation are two major abiotic stress challenges that adversely affect plant physiology and survival. Rising global temperatures due to climate change have intensified heat stress, while increased UV radiation, exacerbated by stratospheric ozone depletion, poses a significant threat to plant health (Hedhly et al. 2009; Singh et al. 2020). Both stressors induce oxidative damage, disrupt metabolic pathways, and compromise plant productivity, ultimately affecting food security and ecosystem stability.

Heat stress occurs when temperatures exceed a plant's optimal physiological range, triggering a cascade of cellular and molecular disturbances. Elevated temperatures disrupt protein stability, membrane integrity, and enzymatic activities, leading to a decline in photosynthetic efficiency and metabolic homeostasis (Zhao et al. 2020). Key physiological processes, including respiration, transpiration, and nutrient uptake, are significantly altered under heat stress, ultimately affecting plant growth and development (Akter and Islam 2017; Liu et al. 2020). At the molecular level, plants respond to heat stress by upregulating heat shock proteins (HSPs) that function as molecular chaperones, stabilizing proteins and preventing their aggregation. The activation of HSFs, such as HsfA1 and HsfA2 in *Arabidopsis thaliana*, plays a central role in orchestrating these protective responses (Zhao et al. 2020). Furthermore, excessive heat leads to the accumulation of reactive oxygen species (ROS), which can cause lipid peroxidation, protein oxidation, and DNA damage. To mitigate this oxidative stress, plants enhance the activity of antioxidant defense systems, including enzymes such as SOD, CAT, and APX (Chaffai et al. 2024).

Radiation stress, particularly from UV-B radiation, poses a significant challenge to plant survival. UV-B exposure leads to direct DNA damage, the formation of cyclobutane pyrimidine dimers, and impairment of the photosynthetic machinery, resulting in reduced rubisco activity and overall carbon assimilation (Kataria et al. 2014). Plants have evolved several protective mechanisms to mitigate the harmful effects of UV-B radiation, including the synthesis of UV-absorbing secondary metabolites such as flavonoids and phenolic compounds, as well as the activation of DNA repair pathways (Shi and Liu 2021). Additionally, oxidative stress induced by UV radiation results in the excessive accumulation of ROS, necessitating the upregulation of antioxidant enzymes and protective metabolites to safeguard cellular components (Singh et al. 2020).

Gamma-aminobutyric acid (GABA) has emerged as a crucial molecule in plant stress responses (Khan et al. 2021), particularly under heat and radiation stress (Table 10.1). GABA levels increase significantly when plants are exposed to environmental stressors, indicating their pivotal role in mitigating damage and enhancing tolerance (Al-Quraan et al. 2012; Jalil et al. 2019). One of the primary ways GABA aids in stress tolerance is by modulating oxidative stress. Under heat and UV-B stress, ROS accumulation can cause severe cellular damage; however, GABA enhances the activity of antioxidant enzymes such as SOD and CAT, which scavenge ROS and maintain cellular redox homeostasis (Bouché et al. 2003; Fait et al. 2008). Additionally, GABA regulates ion transport by modulating ion channels, ensuring ionic balance and osmotic stability under stress conditions (Li et al. 2016; Jalil et al. 2017). GABA also plays a fundamental role in metabolic adaptations to stress. During heat stress, it stabilizes cellular membranes, enhances photosynthetic efficiency, and contributes to osmoprotection by promoting the accumulation of compatible solutes such as proline and trehalose (Cekic et al. 2018; Nayyar et al. 2014; Ansari et al. 2021). In *Vigna mungo*, GABA supplementation increases proline and trehalose levels, reduces oxidative damage, and maintains turgor pressure, thereby improving reproductive success and carbon fixation (Priya et al. 2019). Moreover, GABA functions as an alternative metabolic pathway via the GABA shunt, supplying intermediates to the TCA cycle and sustaining cellular energy production during prolonged heat stress (Cekic et al. 2018).

In the context of radiation stress, GABA mitigates UV-induced damage by enhancing antioxidant defenses and promoting DNA repair mechanisms. In cyanobacteria, exogenous GABA application significantly restored PSII activity and reduced oxidative damage under UV-B exposure, highlighting its protective role in maintaining photosynthetic efficiency under radiation stress (Pandey et al. 2022). Furthermore, studies on *Arabidopsis thaliana* have revealed that Ca^{2+}-dependent activation of the GABA shunt plays a crucial role in ROS scavenging under UV-light stress, further underscoring its importance in radiation tolerance (Al-Quraan 2015). This chapter provides an in-depth analysis of the molecular and physiological mechanisms by which GABA enhances thermo-protection and radiation stress tolerance, offering valuable insights for the development of climate-resilient crop varieties.

Table 10.1 Role of GABA on heat and radiation stress tolerance

Stress	Species	Mechanism	References
Heat stress	*Oryza sativa* (Rice)	Reduced heat stress injury by enhancing leaf turgor, osmoprotectants, and antioxidants	Nayyar et al. (2014)
	Agrostis stolonifera (Creeping bentgrass)	Enhance heat stress tolerance by boosting photosynthesis, the ascorbate-glutathione cycle, osmotic adjustment, and the GABA shunt	Li et al. (2016)
	Agrostis stolonifera (Creeping bentgrass)	Exogenous GABA upregulates heat stress-responsive genes, including ABF3, POD, APX, HSP90, DHN3, and MT1	Li et al. (2018)
	Vigna radiata (Mung beans)	GABA enhances reproductive function, aiding heat stress tolerance in plants	Priya et al. (2019)
	Agrostis stolonifera (Creeping bentgrass)	Exogenous GABA boosts endogenous levels, reducing oxidative damage and enhancing heat tolerance	Liu et al. (2019)
	Helianthus annuus (Sunflower)	Elevated antioxidants mitigate oxidative stress induced by heat stress	Abdel Razik et al. (2021)
	Arabidopsis thaliana	Facilitates metabolic homeostasis and autophagy, enhancing stress tolerance	Balfagón et al. (2022)
Radiation stress	*Arabidopsis thaliana*	Ca^{2+}-dependent GABA shunt activation enhances ROS scavenging, protecting cells and maintaining photosynthetic efficiency under UV stress	Al Quraan (2015)
	Cyanobacteria	GABA treatment enhances growth recovery and PSII activity while reducing oxidative damage under UV-B stress	Pandey et al. (2022)

10.2 Environmental Stressors: Heat and Radiation

Plants are continuously exposed to a range of environmental stressors, among which heat and radiation are particularly detrimental to their growth, development, and productivity. Heat stress, exacerbated by global climate change, disrupts cellular homeostasis, impairs photosynthesis, and accelerates oxidative damage, ultimately leading to reduced crop yields. Similarly, radiation stress, including UV and ionizing radiation, induces DNA damage, membrane instability, and oxidative stress, further challenging plant survival. To cope with these adverse conditions, plants have evolved intricate defense mechanisms involving heat shock proteins, antioxidant systems, and metabolic adjustments (Fig. 10.1). Understanding the molecular and physiological responses of plants to heat and radiation stress is crucial for developing strategies to enhance stress tolerance and sustain agricultural productivity under changing climatic conditions.

10.2.1 Heat Stress in Plants

Heat stress, a prevalent environmental challenge, significantly impacts plant physiology and growth. As temperatures rise beyond the optimal range, plants experience a range of physiological and biochemical disturbances. The primary mechanisms underlying heat stress involve the destabilization of cellular proteins and membranes, disruption of enzyme activities, and alteration of metabolic processes (Zhao et al. 2020).

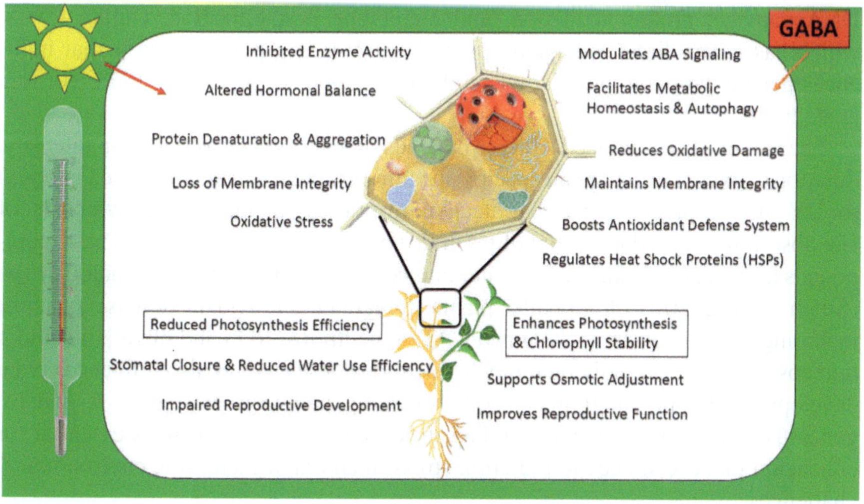

Fig. 10.1 GABA alleviation strategies counteract heat/radiation stress effects on plants. This schematic representation highlights the detrimental effects of heat stress on plants and the protective role of GABA in enhancing stress tolerance

At the molecular level, heat stress induces the denaturation of proteins, which impairs their function and can lead to the accumulation of misfolded proteins. This stress also affects membrane fluidity, resulting in increased permeability and leakage of essential cellular components (Kan et al. 2023). To counteract these effects, plants activate HSPs, which function as molecular chaperones to stabilize and refold damaged proteins, thereby mitigating heat-induced damage (Scharf et al. 2012; Ohama et al. 2017; Ren et al. 2019).

Additionally, heat stress disrupts photosynthesis by affecting the photosynthetic machinery, including the light-harvesting complexes and the Calvin cycle enzymes. This disruption leads to reduced photosynthetic efficiency and diminished energy production (Zahra et al. 2023). To manage this stress, plants may alter their photosynthetic pathways or activate stress-responsive signaling pathways to enhance their tolerance (Bita and Gerats 2013; Perdomo et al. 2017).

10.2.2 Radiation Stress in Plants

Radiation stress in plants encompasses various types, including ultraviolet and ionizing radiation, each exerting distinct impacts on plant cells.

10.2.2.1 UV Radiation Stress in Plants

UV radiation, particularly UV-B (280–315 nm), is a critical environmental factor affecting plant growth and metabolism. While the stratospheric ozone layer absorbs most UV-C (<280 nm) and a significant portion of UV-B, UV-A (315–400 nm) reaches the Earth's surface with minimal attenuation (Meyer et al. 2021). UV radiation interacts with cellular molecules, inducing photochemical reactions that lead to DNA damage, protein oxidation, and membrane destabilization. Unlike UV-A, which has lower direct mutagenic potential, UV-B is a potent stressor that influences gene expression and disrupts cellular function by generating ROS (Rastogi et al. 2010).

The extent of UV-induced damage depends on factors such as radiation intensity, duration of exposure, and plant species-specific defense mechanisms. When exposed to UV stress, plants activate protective responses, including the synthesis of UV-absorbing secondary metabolites like flavonoids and phenolics. These compounds act as natural sunscreens, shielding cellular components from photodamage (Singh et al. 2020). Additionally, plants upregulate antioxidant defense systems, including enzymes such as SOD and CAT, to neutralize ROS and maintain cellular homeostasis. Beyond individual biochemical responses, UV-B also influences plant development by modulating hormone signaling pathways and stress-responsive genes (Dai et al. 2024). Understanding these adaptive mechanisms is essential for enhancing plant resilience to radiation stress in an era of increasing environmental challenges.

10.2.2.2 Ionizing Radiation Stress in Plants

Ionizing radiation, including gamma rays and X-rays, possesses high energy that can penetrate plant tissues, leading to structural and functional disruptions at the cellular and molecular levels (Singh et al. 2020). Unlike nonionizing radiation, ionizing radiation generates charged particles (ions) that interact with biological molecules, causing oxidative stress and DNA damage. This form of radiation originates from unstable atoms undergoing radioactive decay, releasing energy in various forms, including alpha particles, beta particles, gamma rays, and neutron emissions. The extent of damage caused by ionizing radiation depends on factors such as radiation intensity, duration of exposure, and the plant's ability to activate protective responses (Tulchinsky et al. 2023).

Among the different types, alpha radiation consists of heavy particles (two protons and two neutrons) emitted during radioactive decay. While these particles cannot penetrate plant tissues deeply, their ingestion through soil-contaminated water can cause localized cellular damage (Singh et al. 2020). Beta radiation, which involves the emission of high-speed electrons, is more penetrating than alpha particles and can impact plant roots and shoots (Van Hoeck et al. 2017). Studies have shown that submerged roots of *Lemna minor* exhibit greater sensitivity to β-radiation compared to their floating fronds, highlighting the differential impact of radiation exposure on plant tissues (Van Hoeck et al. 2015, 2017).

Gamma radiation is one of the most penetrating forms, consisting of high-energy photons that can travel significant distances through air and biological tissues. Gamma rays interact with cellular components, leading to excessive ROS production, which in turn damages proteins, lipids, and nucleic acids (Kovacs and Keresztes 2002). Exposure to high doses of gamma radiation has been reported to cause chlorophyll degradation, impairing photosynthesis and reducing plant vitality (Salter and Hewitt 1992). Similarly, X-rays, although like gamma radiation in their ability to penetrate tissues, originate from energy transitions within electron clouds rather than nuclear decay. Their effects on plants include chromosomal aberrations, mutations, and disruption of metabolic pathways.

To counteract the detrimental effects of ionizing radiation, plants have evolved complex defense mechanisms, including DNA repair pathways, antioxidant enzyme activation, and stress-responsive signaling cascades. However, excessive radiation exposure can overwhelm these protective systems, leading to inhibited growth, reproductive failure, and long-term genetic instability. Understanding these responses is crucial for developing radiation-resistant plant varieties and mitigating the impact of environmental radiation stress on ecosystems.

10.3 GABA as a Thermo-Protectant in Plants

Plants frequently encounter heat stress, which significantly impacts their growth, development, and productivity. GABA has emerged as a crucial regulator in plant stress responses (Fig. 10.1). GABA modulates metabolic pathways, enhances stress tolerance, and mitigates heat-induced cellular damage by regulating redox balance and osmotic adjustment.

10.3.1 Mechanisms of Thermo-Protection

Plants deploy multiple mechanisms to counteract heat stress, and GABA plays a pivotal role in these adaptive responses. It enhances antioxidant defense, stabilizes membrane integrity, regulates gene expression, and promotes osmolyte accumulation to maintain cellular homeostasis. Exogenous GABA application has been shown to improve plant survival under high temperatures by modulating physiological and biochemical pathways. These mechanisms collectively contribute to the overall thermo-protective role of GABA in plants.

10.3.1.1 Antioxidant Defense Enhancement

GABA plays a critical role in mitigating oxidative stress induced by high temperatures by enhancing the activity of antioxidant enzymes. Exogenous application of GABA has been shown to significantly improve the activities of CAT, SOD, APX, and GR in rice leaves under short-term heat stress, thereby reducing heat-induced ROS accumulation (Nayyar et al. 2014). Similarly, GABA treatment in sunflowers results in elevated antioxidant levels, effectively alleviating oxidative stress caused by high temperatures (Abdel Razik et al. 2021). In creeping bentgrass, foliar application of GABA significantly enhances the activities of SOD, POD, APX, DHAR, and GR, contributing to decreased oxidative damage during heat stress (Li et al. 2016). Similarly, GABA application also increases endogenous GABA accumulation, which further confers heat tolerance by reducing oxidative damage in creeping bentgrass (Liu et al. 2019). Another study demonstrates that exogenous spermidine treatment improved heat tolerance in white clover (*Trifolium repens*) by modulating polyamine and GABA metabolism, enhancing antioxidant defense mechanisms, and regulating HSP expression (Luo et al. 2020). The combination of high light and heat stress also triggers GABA accumulation in *Arabidopsis thaliana*, which facilitates metabolic homeostasis and autophagy, thereby improving stress tolerance (Balfagón et al. 2022).

10.3.1.2 Osmolyte Accumulation

GABA contributes to osmotic regulation by promoting the accumulation of osmoprotectants such as proline and soluble sugars. These compounds help maintain cellular water balance and stability, improving heat stress tolerance. Exogenous GABA application effectively mitigates heat-induced damage in creeping bentgrass by increasing leaf relative water content and photosynthetic efficiency (Liu et al. 2019). Additionally, GABA-mediated osmolyte accumulation enhances cell turgor and protects plants from dehydration under high-temperature stress (Abdel Razik et al. 2021).

10.3.1.3 Gene Expression Regulation

Under heat stress, plants adaptively encourage GABA biosynthesis through calcium-induced GAD activation, as observed in *Arabidopsis* roots, where elevated cytosolic Ca^{2+} levels promote GAD activity via calmodulin interactions (Locy et al. 2000). A transient heat shock expressively increases endogenous GABA levels in cowpea and *Sesamum indicum* plants (Mayer et al. 1990; Bor et al. 2009). Transcriptomic analyses reveal that exogenous GABA application upregulates genes associated with heat stress response, including those encoding abscisic acid-responsive transcription factors (ABF3), antioxidant enzymes (POD, APX), HSP90, dehydrin (DHN3), and metallothioneins (MT1) in creeping bentgrass (Li et al. 2018). Heat stress also induces GABA accumulation in lichens, where reduced GAD and GDH activities prevent chlorophyll degradation and lipid peroxidation, thereby enhancing heat tolerance (Cekic et al. 2018). Further transcriptional studies indicate that exogenous GABA upregulates heat shock factor (HSFA-6a, HSFA-2c, HSFB-2b) and heat shock protein (HSP17.8, HSP26.7, HSP70, HSP90.1-b1) genes in creeping bentgrass, while inhibition of GABA biosynthesis suppresses these genes under heat stress conditions (Liu et al. 2019). This indicates that GABA is vital for controlling the expression of genes involved in stress responses, particularly those involved in protein protection and oxidative stress mitigation (Abdel Razik et al. 2021).

10.3.1.4 Membrane Integrity Maintenance

GABA helps plants tolerate heat stress by maintaining the stability of cellular membranes and mitochondrial function. When applied externally, GABA helps restore membrane integrity in plants stressed by heat, preventing chlorophyll degradation and sustaining photosynthesis. This protective effect is evident in creeping bentgrass, where endogenous GABA accumulation, induced by exogenous application, significantly reduces heat damage and improves membrane stability (Liu et al. 2019). Similar protective effects have been reported in rice, where GABA application enhanced cell membrane stability under heat stress (Nayyar et al. 2014).

10.3.1.5 Reproductive Function Protection

Heat stress negatively impacts plant reproductive processes, reducing pollen viability and seed set. However, exogenous GABA application has been shown to improve reproductive functions under high temperatures. In mung bean (*Vigna radiata*), GABA treatment increases pollen viability and germination, leading to improved pod and seed development under heat stress (Priya et al. 2019).

10.4 GABA as a Protectant Against Radiation

Plants are constantly exposed to various environmental stressors, including UV radiation, which induces oxidative stress by generating excessive ROS. To counteract this damage, plants have evolved sophisticated metabolic and signaling pathways that enhance stress tolerance. GABA regulates these metabolic and signaling pathways in such a fashion that helps mitigate oxidative stress under UV exposure.

A functional relationship between GABA metabolism and oxidative stress regulation has been demonstrated in *Arabidopsis thaliana*, particularly in the SSADH mutant. This mutant exhibits an impaired ability to regulate ROS levels, highlighting the necessity of an operational GABA shunt to limit oxidative intermediates during stress conditions such as heat and UV exposure (Bouche et al. 2003; Fait et al. 2008). The GABA shunt acts as a metabolic bypass for the TCA cycle, allowing plants to maintain energy homeostasis while avoiding oxidative damage to sensitive enzymes such as aconitase. By providing alternative carbon flux through NADH and succinate production, the GABA shunt ensures cellular redox balance under adverse conditions.

Further investigations suggest that the Ca^{2+}-dependent activation of the GABA shunt plays a pivotal role in ROS scavenging under UV-light stress in *A. thaliana* (Al Quraan 2015). This activation enhances the ability of plants to manage oxidative bursts, thereby preventing cellular damage and preserving photosynthetic efficiency. The link between Ca^{2+} signaling and GABA metabolism underscores the complexity of plant stress responses and suggests that GABA is not merely a metabolic intermediate but also an important signaling molecule in stress mitigation. Moreover, exogenous supplementation of GABA significantly improves plant resilience against UV-B radiation. In cyanobacteria, for instance, GABA treatment results in enhanced recovery of growth and PSII activity, accompanied by a substantial reduction in oxidative damage under UV-B stress (Pandey et al. 2022). This protective role of GABA is likely mediated by its ability to upregulate antioxidant enzyme activities, stabilize membrane integrity, and reinforce photoprotective mechanisms.

10.5 Conclusion

In conclusion, GABA emerges as a crucial biomolecule in enhancing plant resilience against both heat and radiation stress, offering promising biotechnological applications for crop improvement. Its role in regulating oxidative stress, stabilizing cellular structures, and modulating stress-responsive pathways positions it as a valuable tool for developing heat- and radiation-resistant crops. While genetic engineering and agronomic strategies incorporating GABA hold significant potential, challenges such as scalability, species-specific responses, and long-term field applicability must be addressed. Future research should focus on elucidating the precise molecular mechanisms of GABA-mediated stress tolerance, optimizing its application through advanced delivery systems, and conducting large-scale field trials to validate its efficacy. By integrating GABA-based strategies into agricultural practices, it is possible to enhance crop productivity and sustainability in the face of increasing environmental challenges.

References

Abdel Razik ES, Alharbi BM, Pirzadah TB, Alnusairi GSH, Soliman MH, Hakeem KR (2021) γ-Aminobutyric acid (GABA) mitigates drought and heat stress in sunflower (Helianthus annuus L.) by regulating its physiological, biochemical and molecular pathways. Physiol Plant 172(2):505–527. https://doi.org/10.1111/ppl.13216. Epub 2020 Oct 12

Akter N, Islam MR (2017) Heat stress effects and management in wheat. A review. Agron Sustain Dev 37:37. https://doi.org/10.1007/s13593-017-0443-9

Al-Quraan NA (2015) GABA shunt deficiencies and accumulation of reactive oxygen species under UV treatments: insight from Arabidopsis thaliana calmodulin mutants. Acta Physiol Plant 37:1–11

Al-Quraan NA, Locy RD, Singh NK (2012) Heat and cold stresses phenotypes of Arabidopsis thaliana calmodulin mutants: regulation of gamma-aminobutyric acid shunt pathway under temperature stress. Int J Plant Biol 3:9–17

Ansari MI, Jalil SU, Ansari SA et al (2021) GABA shunt: a key-player in mitigation of ROS during stress. Plant Growth Regul 94:131–149

Balfagón D, Gómez-Cadenas A, Rambla JL, Granell A, de Ollas C, Bassham DC, Mittler R, Sara IZ (2022) γ-Aminobutyric acid plays a key role in plant acclimation to a combination of high light and heat stress. Plant Physiol 188(4):2026–2038. https://doi.org/10.1093/plphys/kiac010

Bita CE, Gerats T (2013) Plant tolerance to high temperature in a changing environment: scientific fundamentals and production of heat stress-tolerant crops. Front Plant Sci 4:273. https://doi.org/10.3389/fpls.2013.00273

Bor M, Seckin B, Ozgur R et al (2009) Comparative effects of drought, salt, heavy metal and heat stresses on gamma-aminobutryric acid levels of sesame (*Sesamum indicum* L.). Acta Physiol Plant 31:655–659. https://doi.org/10.1007/s11738-008-0255-2

Bouché N, Fait A, Bouché D, Moller SG, Fromm H (2003) Mitochondrial succinate-semialdehyde dehydrogenase of the gamma-aminobutyrate shunt in required to restricts levels of reactive oxygen intermediates in plants. Proc Natl Acad Sci USA 100:6843–6848

Cekic FO, Goren-Saglam N, Torun H, Yigit E, Unal D (2018) Gamma-amino butyric acid metabolism under high temperature stress in two lichen species. Appl Ecol Environ Res 16(5):5529–5538. https://doi.org/10.15666/aeer/1605_55295538

Chaffai R, Ganesan M, Cherif A (2024) Mechanisms of plant response to heat stress: recent insights. In: Plant adaptation to abiotic stress: from signaling pathways and microbiomes to molecular mechanisms. Springer, Singapore. https://doi.org/10.1007/978-981-97-0672-3_5

Dai Y, Li M, Li H, Xia L, Zhang S (2024) Phytohormone-regulated UV-B photomorphogenesis and acclimation. Environ Exp Bot 224:105830. https://doi.org/10.1016/j.envexpbot.2024.105830

Fait A, Fromm H, Walter D, Galili G, Fernie AR (2008) Highway or byway: the metabolic role of the GABA shunt in plants. Trends Plant Sci 13:14–19

Hedhly A, Hormaza J, Herrero M (2009) Global warming and sexual plant reproduction. Trends Plant Sci 14:30–36. https://doi.org/10.1016/j.tplants.2008.11.001

Jalil SU, Ansari MI (2020) Stress implication and crop productivity. In: Plant ecophysiology and adaptation under climate change: mechanism and prespectives, 1st edn. Springer Nature Singapore, pp 73–86

Jalil SU, Ahmad I, Ansari MI (2017) Functional loss of GABA transaminase (GABA-T) expressed early leaf senescence under various stress conditions in *Arabidopsis thaliana*. Curr Plant Biol 9–10:11–22

Jalil SU, Khan MAA, Ansari MI (2019) Tremendous role of GABA shunt metabolic pathway in plant system. J Biol Chem Res 36:188–196

Kan Y, Mu X-R, Gao J, Lin H-X, Lin Y (2023) The molecular basis of heat stress responses in plants. Mol Plant 16(10):1612–1634. https://doi.org/10.1016/j.molp.2023.09.013

Kataria S, Jajoo A, Guruprasad KN (2014) Impact of increasing ultraviolet-B (UV-B) radiation on photosynthetic processes. J Photochem Photobiol B 137:55–66. https://doi.org/10.1016/j.jphotobiol.2014.02.004. Epub 2014 Feb 15

Khan MIR, Jalil SU, Chopra P, Chhillar H, Ferrante A, Khan NA, Ansari MI (2021) Role of GABA in plant growth, development and senescence, plant. Gene 26:100283

Kovacs E, Keresztes A (2002) Effect of gamma and UV-B/C radiations on plant cells. Micron 33(199):210

Li Z, Yu J, Peng Y, Huang B (2016) Metabolic pathways regulated by γ-aminobutyric acid (GABA) contributing to heat tolerance in creeping bentgrass (*Agrostis stolonifera*). Sci Rep 6:30338

Li Z, Peng Y, Huang B (2018) Alteration of transcripts of stress-protective genes and transcriptional factors by ?-aminobutyric acid (GABA) associated with improved heat and drought tolerance in creeping Bentgrass (*Agrostis stolonifera*). Int J Mol Sci 19(6):1623. https://doi.org/10.3390/ijms19061623

Lippmann R, Babben S, Menger A, Delker C, Quint M (2019) Development of wild and cultivated plants under global warming conditions. Curr Biol 29:R1326–R1338. https://doi.org/10.1016/j.cub.2019.10.016

Liu T, Liu Z, Li Z, Peng Y, Zhang X, Ma X, Huang L, Liu W, Nie G, He L (2019) Regulation of Heat Shock Factor Pathways by γ-aminobutyric Acid (GABA) Associated with Thermotolerance of Creeping Bentgrass. Int J Mol Sci 20(19):4713. https://doi.org/10.3390/ijms20194713. PMID: 31547604; PMCID: PMC6801925.

Liu J, Zhang R, Xu X, Fowler JC, Miller TEX, Dong T (2020) Effect of summer warming on growth, photosynthesis and water status in female and male Populus cathayana: implications for sex-specific drought and heat tolerances. Tree Physiol 40:1178–1191. https://doi.org/10.1093/treephys/tpaa069

Locy RD, Wu S, Bisnette J, Barger TW, McNabb DS, Zik M, Fromm H, Singh NK, Cherry JH (2000) The regulation of GABA accumulation by heat stress in arabidopsis. In: Cherry JH, Locy RD, Rychter A (eds) Plant tolerance to abiotic stresses in agriculture: role of genetic engineering. NATO science series, vol 83. Springer, Dordrecht. https://doi.org/10.1007/978-94-011-4323-3_3

Luo L, Li Z, Tang MY, Cheng BZ, Zeng WH, Peng Y, Nie G, Zhang XQ (2020) Metabolic regulation of polyamines and Γ-aminobutyric acid in relation to spermidine-induced heat tolerance in white clover. Plant Biol 22:794–804

Mayer RR, Cherry JL, Rhodes D (1990) Effects of heat shock onamino acid metabolism of cowpea cells. Phytochem 94:796–810

Meyer P, Van de Poel B, De Coninck B (2021) UV-B light and its application potential to reduce disease and pest incidence in crops. Hortic Res 8:194. https://doi.org/10.1038/s41438-021-00629-5

Nayyar H, Kaur R, Kaur S et al (2014) γ-Aminobutyric acid (GABA) imparts partial protection from heat stress injury to rice seedlings by improving leaf turgor and upregulating osmoprotectants and antioxidants. J Plant Growth Regul 33:408–419. https://doi.org/10.1007/s00344-013-9389-6

Ohama N, Sato H, Shinozaki K, Yamaguchi-Shinozaki K (2017) Transcriptional regulatory network of plant heat stress response. Trends Plant Sci 22:53–65. https://doi.org/10.1016/j.tplants.2016.08.015

Pandey A, Kumar S, Singh G, Prasad SM (2022) Regulatory role of gamma-aminobutyric acid in cyanobacteria challenged with UV-B: the implication of nitric oxide. Plant Stress 5:100094

Perdomo JA, Capó-Bauçà S, Carmo-Silva E, Galmés J (2017) Rubisco and rubisco activase play an important role in the biochemical limitations of photosynthesis in rice, wheat, and maize under high temperature and water deficit. Front Plant Sci 8:490. https://doi.org/10.3389/fpls.2017.00490

Priya M, Sharma L, Kaur R, Bindumadhava H, Nair RM, Siddique KHM, Nayyar H (2019) GABA (γ-aminobutyric acid), as a thermo-protectant, to improve the reproductive function of heat-stressed mungbean plants. Sci Rep 9(1):7788. https://doi.org/10.1038/s41598-019-44163-w

Rastogi RP, Richa KA, Tyagi MB, Sinha RP (2010) Molecular mechanisms of ultraviolet radiation-induced DNA damage and repair. J Nucleic Acids 2010:592980. https://doi.org/10.4061/2010/592980. PMID: 21209706; PMCID: PMC3010660

Ren S, Ma K, Lu Z, Chen G, Cui J, Tong P, Wang L, Teng N, Jin B (2019) Transcriptomic and-metabolomic analysis of the heat-stress response of Populus tomentosa Carr. Forests 10:383. https://doi.org/10.3390/f10050383

Salter L, Hewitt CN (1992) Ozone-hydrocarbon interactions in plants. Photochemistry 31(4):4045–4050

Scharf KD, Berberich T, Ebersberger I, Nover L (2012) The plant heat stress transcription factor (Hsf) family: structure, function and evolution. Biochim Biophys Acta 1819:104–119. https://doi.org/10.1016/j.bbagrm.2011.10.002

Shi C, Liu H (2021) How plants protect themselves from ultraviolet-B radiation stress. Plant Physiol 187(3):1096–1103. https://doi.org/10.1093/plphys/kiab245. PMID: 34734275; PMCID: PMC8566272

Singh S, Fatima A, Tiwari S, Prasad SM (2020) Plant responses to radiation stress and its adaptive mechanisms. In: Plant life under changing environment. Academic Press, pp 105–122. https://doi.org/10.1016/B978-0-12-818204-8.00006-0

Tulchinsky TH, Varavikova EA, Cohen MJ (2023) Environmental and occupational health. In: Tulchinsky TH, Varavikova EA, Cohen MJ (eds) The new public health, 4th edn. Academic Press, pp 681–750. https://doi.org/10.1016/B978-0-12-822957-6.00016-8

Van Hoeck A, Horemans N, Van Hees M, Nauts R, Knapen D, Vandenhove H, Blust R (2015) β-Radiation stress responses on growth and antioxidative defense system in plants: a study with Strontium-90 in Lemna minor. Int J Mol Sci 16(7):15309–15327. https://doi.org/10.3390/ijms160715309. PMID: 26198226; PMCID: PMC4519901

Van Hoeck A, Horemans N, Nauts R, Van Hees M, Vandenhove H, Blust R (2017) Lemna minor plants chronically exposed to ionising radiation: RNA-seq analysis indicates a dose rate dependent shift from acclimation to survival strategies. Plant Sci 257:84–95. https://doi.org/10.1016/j.plantsci.2017.01.010

Zahra N, Hafeez MB, Ghaffar A, Kausar A, Al Zeidi M, Siddique KHM, Farooq M (2023) Plant photosynthesis under heat stress: effects and management. Environ Exp Bot 206:105178. https://doi.org/10.1016/j.envexpbot.2022.105178

Zhao J, Lu Z, Wang L, Jin B (2020) Plant responses to heat stress: physiology, transcription, non coding RNAs, and epigenetics. Int J Mol Sci 22(1):117. https://doi.org/10.3390/ijms22010117. PMID: 33374376; PMCID: PMC7795586

GABA as Protectant Against Metal/ Metalloid Stress

11

Abstract

Metal and metalloid stress pose a significant threat to plant growth, development, and productivity. Among various plant defense strategies, gamma-aminobutyric acid (GABA) has gained attention for its crucial role in mitigating metal-induced toxicity. This chapter explores the protective functions of GABA under heavy metal and metalloid stress. GABA enhances stress tolerance by modulating key metabolic enzymes (GABA transaminase (GABA-T) and succinate semialdehyde dehydrogenase (SSADH)), strengthening antioxidant defense systems (superoxide dismutase (SOD), catalase (CAT), ascorbate peroxidase (APX), ascorbate, glutathione), and regulating gene expression within the GABA shunt. Furthermore, GABA acts as a signaling molecule, activating downstream pathways that alleviate oxidative damage, restrict metal uptake, and promote plant growth under stress conditions. It also contributes to cellular redox homeostasis and ion balance, reinforcing overall plant resilience. The chapter highlights recent research advances and discusses the potential of GABA-based strategies in phytoremediation and sustainable agriculture, offering valuable insights into improving plant performance in metal-stressed environments.

Keywords

Antioxidant defense · GABA shunt · Metal stress · Phytoremediation

Highlights

Gamma-aminobutyric acid (GABA) performs critical roles in:

- Mitigating metal and metalloid stress by modulating ion transport and maintaining ion homeostasis.

S. U. Jalil et al., *GABA Signaling System and Abiotic Stress Tolerance in Plants*,
https://doi.org/10.1007/978-981-95-0437-4_11

- Promoting the activity of key metabolic enzymes (GABA-T, SSADH) on their exogenous application, reducing the toxic impacts of heavy metals in various plant species.
- Enhancing antioxidant defense systems, including enzymatic (superoxide dismutase (SOD), CAT, APX) and nonenzymatic (ascorbate, glutathione) mechanisms, to scavenge reactive oxygen species (ROS) and prevent oxidative damage.
- Regulating gene expression in stress-responsive pathways, particularly the GABA shunt, to enhance stress tolerance by reducing toxic metal uptake and accumulation.
- Promoting nutrient uptake and root development, balancing the uptake of essential nutrients such as calcium, magnesium, and iron under metal stress.
- Acting as a signaling molecule that triggers downstream pathways, leading to reduced metal toxicity and improved plant growth and productivity under stress conditions.

11.1 Introduction

The environment in which plants grow is often far from ideal, especially when soils are contaminated with toxic metals and metalloids. The heavy metal stress has become one of the most alarming threats to plant health and agricultural productivity (Angon et al. 2024). Metals like Cd, Pb, Hg, and As, introduced through industrial waste, mining, and extensive use of metal-based fertilizers and pesticides, accumulate in the soil, wreaking havoc on plant physiology (Nagajyoti et al. 2010; Khanna et al. 2018). As a result, the disruption occurs in essential processes, such as nutrient uptake, and triggers the reactive oxygen species (ROS) overproduction, leading to oxidative damage, cellular dysfunction, and, in severe cases, plant death. Stunted growth, chlorosis, and necrosis become the unfortunate outcomes as plants fail to cope with the overwhelming toxicity, reducing not only their yield but also their capacity to survive in metal-polluted environments (Kumar and Aery 2016; Angon et al. 2024).

However, plants have developed a range of protective mechanisms to survive such hostile conditions, one of which is the activation of gamma-aminobutyric acid (GABA) signaling (Ansari et al. 2021; Khan et al. 2021). GABA has now gained a reputation as a vital player in plant stress physiology (Ansari et al. 2014). Under metal/metalloid stress, glutamate decarboxylase (GAD) catalyzes glutamate decarboxylation to produce GABA. As a result, GABA levels surge. This increase is more than just a stress response; it represents a sophisticated survival strategy. By modulating ion transport and enhancing the activity of antioxidant enzymes, GABA helps plants maintain cellular homeostasis and minimize oxidative damage. This protective mechanism is not just limited to reducing metal toxicity but also involves complex signaling pathways, where GABA interacts with plant hormones and stress-responsive genes, delaying leaf senescence and reinforcing the plant's overall resilience (Jalil et al. 2017, 2019a, b; Jalil and Ansari 2020).

Recent studies have provided compelling evidence of GABA's role in protecting plants from heavy metal stress. For instance, research on microalgae has shown that GABA can significantly reduce Cd absorption, promoting lipid biosynthesis and nutrient uptake under stress conditions (Zhao et al. 2020). In barley seedlings, GABA mitigates oxidative damage caused by proton and aluminum toxicity by activating key antioxidant defenses such as superoxide dismutase (SOD) and catalase (CAT) (Song et al. 2010). Similarly, GABA reduces Cr uptake in mustard plants exposed to chromium stress and boosts both enzymatic and nonenzymatic antioxidants, thereby minimizing oxidative damage (Mahmud et al. 2017).

Several plant species endorse the protective functions of GABA under stress conditions. Studies in rice have shown that GABA levels spike in response to chromium stress, while in *Nicotiana tabacum*, GABA accumulation takes place at intermediate Zn concentrations, suggesting a dose-dependent response (Daş et al. 2016). Under As stress, GABA not only activates the antioxidant enzyme system but also inhibits As accumulation, providing a robust defense mechanism against this toxic metalloid (Kumar et al. 2017). The upregulation of GABA shunt-related genes, such as GAD, further underscores the critical role of GABA in metal stress tolerance. Overexpression of these genes in Cd-sensitive tobacco plants enhances Cd tolerance, emphasizing the potential of GABA in developing stress-resilient crops through genetic engineering (Cheng et al. 2018). These findings highlight GABA's versatile role as both a protector and regulator of cellular defense mechanisms under metal/metalloid stress.

As the role of GABA in plant defense becomes clearer, it opens up new avenues for exploring how this molecule can be harnessed to mitigate heavy metal stress. The chapter explores these protective mechanisms, from ion homeostasis and ROS scavenging to gene regulation, and discusses the potential of biotechnological interventions to enhance GABA levels in plants, offering a promising solution to the growing challenge of heavy metal stress in agriculture.

11.2 Mechanisms of Metal/Metalloid Toxicity

The toxicity of metals and metalloids in plants, a multifaceted challenge, disrupts a wide range of physiological and biochemical processes, ultimately compromising growth, development, and productivity. These toxic elements, when present in excess, disturb the delicate balance within plant cells, affecting everything from nutrient uptake to energy production. The key mechanisms through which heavy metal toxicity exerts its harmful effects are intertwined with physiological, biochemical, and metabolic anomalies, leading to the imposition of severe plant stress and often irreversible damage.

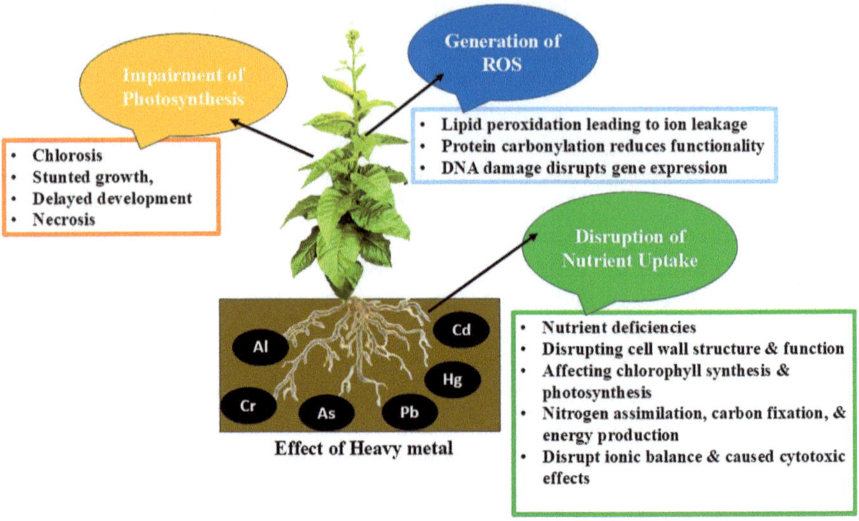

Fig. 11.1 The detrimental effects of metals/metalloids on plant physiology. Metals/metalloids enter the plant system through roots and disrupt essential cellular processes due to oxidative stress and damage to cellular components (lipids, proteins, nucleic acids) through induced ROS overproduction. An impairment in nutrient uptake also occurs, which disrupts ion homeostasis and inhibits photosynthesis by damaging chloroplasts. The resulting stress causes stunted growth, chlorosis, and impaired root and shoot development, ultimately reducing plant vitality and productivity

11.2.1 Functional and Metabolic Adjustments Under Metal/Metalloid Stress

Plants encountering metals/metalloids, including Cd, Pb, Cr, Al, Hg, and As, experience widespread physiological disruptions (Ghori et al. 2019). They accumulate within plant tissues, initiating a cascade of harmful reactions (Fig. 11.1). One of the first responses is the impairment of photosynthesis due to metal-induced chlorosis, where the synthesis of chlorophyll is inhibited, and the photosynthetic apparatus becomes dysfunctional (Pourrut et al. 2011). This is accompanied by stunted growth, delayed development, and, in some cases, necrosis (Singh et al. 2010; Kumar et al. 2019a, b). At the biochemical level, toxic metals interfere with essential metabolic pathways, particularly those involving enzyme activities, as many enzymes depend on metal cofactors for proper function (Kumar and Aery 2016). Metals can displace these essential cofactors, inactivating enzymes and disrupting critical metabolic reactions. Additionally, metals bind to proteins and other cellular macromolecules, altering their structure and function, leading to protein denaturation and impaired cellular metabolism (Nagajyoti et al. 2010; Kumar and Aery 2016).

11.2.2 Generation of ROS and Oxidative Damage

A hallmark of metal/metalloid stress is the overproduction of ROS, which are highly reactive molecules that can cause significant damage to plant cells (Shahid et al. 2014). Under normal conditions, plants maintain a delicate balance between the production and scavenging of ROS. However, the stress offsets this equilibrium, leading to an excessive accumulation of ROS, including $O_2^{\cdot-}$, H_2O_2, and $^{\cdot}OH$ (Kováčik et al. 2010; Pucciariello et al. 2012). The ROS attack vital cellular components, including lipids, proteins, and nucleic acids, causing oxidative damage (Martínez Domínguez et al. 2009; Cuypers et al. 2011; Carrasco-Gil et al. 2012; Chen et al. 2012). Lipid peroxidation, in particular, damages cellular membranes, increasing their permeability and leading to ion leakage, which further exacerbates the stress response (Pourrut et al. 2013). Proteins exposed to ROS undergo carbonylation, losing their functionality (Hasan et al. 2017), while DNA damage can result in mutations and impaired gene expression (Zhu et al. 2011; Shen et al. 2013). The list of affected species is extensive, with plants like *Trifolium repens*, *Allium cepa*, and *Solanum tuberosum* showing clear signs of metal-induced DNA damage (Aina et al. 2004; Barbosa et al. 2010; Gichner et al. 2006). An increase in ROS levels leads to heightened oxidative stress, which significantly damages plant cell structures and threatens their survival. Despite the plant's natural defense mechanisms, the overwhelming onslaught of ROS under metal stress proves to be a critical factor in reduced growth and productivity, leaving plants vulnerable to further environmental challenges.

11.2.3 Disruption of Essential Nutrient Uptake and Metabolic Processes

Stress caused by metals and metalloids greatly disrupts the absorption and movement of nutrients, which are crucial for supporting plant development. Harmful metals like Cd, Pb, and As compete with vital elements such as Ca, Mg, Fe, and Zn for entry into plant roots, leading to nutrient deficiencies (Kabata-Pendias and Mukherjee 2007; Das et al. 2019). For instance, Cd can inhibit the uptake of Ca by blocking Ca channels, thereby disrupting cell wall structure and function, which depend on Ca for stability (Ahmad et al. 2016; Huybrechts et al. 2021). Similarly, Pb interferes with magnesium uptake, affecting chlorophyll synthesis and photosynthesis (Shen et al. 2016; Dalyan et al. 2020). This competition for nutrient uptake impairs several metabolic processes, including nitrogen assimilation, carbon fixation, and energy production. Furthermore, metals interfere with ion transporters and channels, disrupting the ionic balance and causing cytotoxic effects (Guerinot 2000; Pittman 2005). This disruption of nutrient uptake not only leads to metabolic imbalances but also weakens the plant's overall stress tolerance mechanisms, making it more susceptible to oxidative damage and other stress-related injuries.

11.3 GABA's Role in Metal/Metalloid Stress Alleviation

GABA confers a vital mechanism in plant stress responses against metal and metalloid toxicity (Table 11.1). GABA acts both as a signaling molecule and a metabolic regulator. Under conditions of metal/metalloid stress, GABA accumulation occurs

Table 11.1 GABA and heavy metal stress

S. No.	Plants	Heavy metal	Mechanism of GABA	References
1.	*Hordeum vulgare* (Barley)	Aluminum/proton	Alleviates oxidative damage caused by aluminum/proton	Song et al. (2010)
2.	*Cucumis melo* (Muskmelon)	Calcium nitrate [Ca $(NO_3)^2$]	A protective role of polyamine and GABA metabolism	Hu et al. (2015)
3.	*Brassica juncea* (Brown mustard)	Chromium	Improve chromium stress tolerance by regulating the antioxidant defense and glyoxalase systems	Mahmud et al. (2017)
4.	*Oryza sativa* (Rice)	Arsenic	GABA reduces Lsi-1 and Lsi-2 gene expressions and modulates physiological responses in rice to provide tolerance against arsenic	Kumar et al. (2017)
5.	*Oryza sativa* (Rice)	Arsenic	GABA modulated fatty acids, stress-responsive amino acids, and polyamine biosynthesis	Kumar et al. (2019a)
6.	*Zea mays* (Maize)	Cadmium	Improved cadmium tolerance by regulation of polyamine metabolism and antioxidant defense systems	Seifikalhor et al. (2020)
7.	*Monoraphidium* (Green algae)	Cadmium	Regulates lipid production and cadmium uptake under cadmium stress	Zhao et al. (2020)
8.	*Nicotiana tabacum* (Tobacco)	Cadmium	Induces proline and glutathione synthesis in alleviating Cd-induced photosynthetic inhibition and oxidative damage	He et al. (2021)
9.	*Liriodendron chinense tulipifera* (Chinese Tuliptree)	Aluminum	GABA enhanced the activity of antioxidant enzymes	Wang et al. (2021)
10.	*Oryza sativa* (Rice)	Lead	Induced modulations in the performance of aromatic rice	Ashraf et al. (2022)
11.	*Malus pumila* (Apple)	Cadmium	GABA reduces the net Cd^{2+} fluxes, causing inhibition of cadmium uptake and accumulation in apples	Li et al. (2022)

(continued)

Table 11.1 (continued)

S. No.	Plants	Heavy metal	Mechanism of GABA	References
12.	*Lactuca sativa* (Lettuce)	Cadmium	Regulated physio-biochemical attributes	Waris et al. (2023)
13.	*Agrostis stolonifera* (Creeping bentgrass)	Acid-AluminumAluminum	Regulated Metabolic Homeostasis and alleviated acid-aluminum toxicity to roots associated with enhanced antioxidant metabolism as well as accumulation and transportation of citric and malic acid	Zhou et al. (2023, 2024)
14.	*Trigonella foenum-graecum* (Fenugreek)	Nickel	Reduced nickel uptake in roots and leaves, alongside stimulation of the plant's antioxidant system	Danish et al. (2024)
15.	*Oryza sativa* (Rice)	Copper	Regulates antioxidant defense mechanisms, gene regulation, mineral uptake, and copper homeostasis	Khan et al. (2024)
16.	*Chlorella sorokiniana* (Microalgae)	Chromium [Cr (III)]	Improved microalgal protein and lipid accumulation	Chakravorty et al. (2024)

significantly and provides a protective shield to mitigate the toxic effects of these elements on plant cells. The mechanisms by which GABA confers protection are diverse and multifaceted, involving ion homeostasis, ROS detoxification, and modulation of antioxidant defense systems (Fig. 11.2).

11.3.1 GABA's Role in Regulating Ion Transport and Homeostasis

GABA ensures that the toxic ions, such as Cd^{2+}, As^{3+}, and Pb^{2+}, do not overwhelm the plant's delicate cellular machinery. Metal stress often leads to the accretion of these toxic ions in plant cells, triggering ionic imbalances that can cripple key physiological processes. Yet, GABA steps in, playing a crucial role in regulating ion transporters and channels that control the influx and efflux of ions, effectively shielding plants from excessive metal accumulation. Zhao et al. (2020) have provided evidence that the increased cellular GABA levels in microalgae facilitate lipid synthesis while simultaneously downregulating Cd accumulation by modulating the expression of genes involved in lipogenesis. In angiosperms, the exogenous GABA significantly reduces the net Cd^{2+} flux in apple roots, preventing Cd from overloading the cellular systems (Li et al. 2022). Remarkably, the Cd content in these roots significantly declines under GABA treatment compared to that in untreated plants exposed to Cd stress, highlighting GABA's protective action.

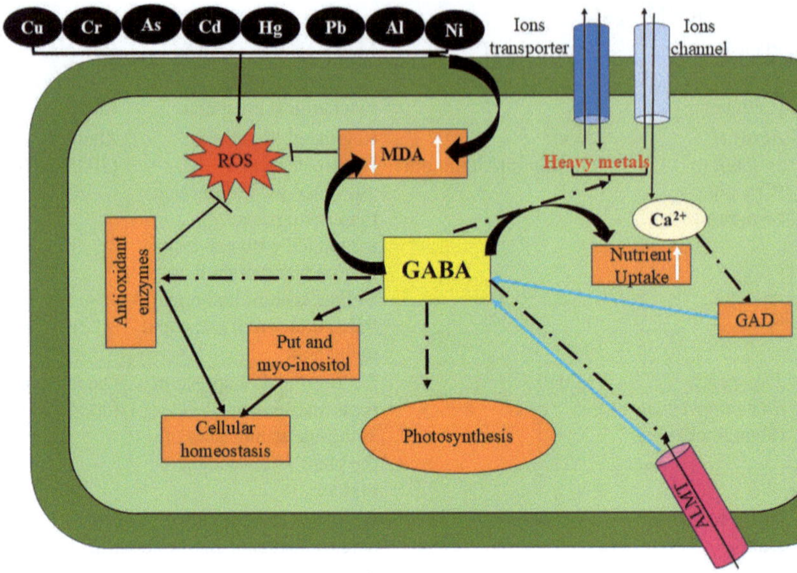

Fig. 11.2 Role of GABA in mitigating metal/metalloid stress in plants. GABA regulates ion transporters and channels to limit metal accumulation, maintains ionic balance by modulating ALMT transporters, and influences putrescine and myo-inositol levels. Under metal/metalloid stress, GABA enhances antioxidant defenses, reduces ROS and MDA levels, and supports nutrient uptake by controlling metal ion transport. Ca^{2+}-induced upregulation of GAD genes boosts GABA biosynthesis, activating pathways that strengthen stress tolerance. Regulation of processes is indicated by dotted lines, while GABA synthesis and accumulation are indicated by blue lines

The intriguing role of GABA extends even to more complex interactions, such as in creeping bentgrass under acid-Al stress, wherein GABA priming not only decreases the accumulation of malic and citric acids—organic acids critical in detoxifying Al—but also regulates the expression of key genes, such as STOP1-like and those encoding transporters like Aluminum-activated malate transporter 9 (ALMT9)-like for malic acid and MATE19-like for citric acid transport. These genes orchestrate a response that allows the plant to detoxify Al by forming Al-oxalate complexes, thus maintaining ionic homeostasis (Zhou et al. 2023). Moreover, GABA priming helps regulate key metabolites like putrescine and myo-inositol, critical for aboveground growth and stress defense, further reinforcing the plant's resilience against metal toxicity. Through such precise regulation of ion transport and homeostasis, GABA not only mitigates the harmful effects of metal stress but also safeguards the plant's overall metabolic balance, ensuring survival in harsh environments.

11.3.2 ROS Scavenging and Detoxification

One of the primary toxic effects of metal and metalloid stress is the ROS overproduction, which leads to oxidative damage in plant cells. In this stressful environment, GABA emerges as a critical protector, acting as a potent modulator of the plant's antioxidant defense system. Through boosting the function of major antioxidant enzymes like SOD, CAT, and GPX, GABA plays a pivotal role in neutralizing ROS and preventing them from damaging essential cellular components like lipids, proteins, and DNA (Wu et al. 2020). This upregulation of antioxidant defenses is crucial for maintaining redox homeostasis within plant cells, as it minimizes the harmful effects of ROS accumulation under metal/metalloid stress (Song et al. 2010; Kumar et al. 2019b). The effectiveness of GABA in mitigating oxidative stress is evident in various studies, where GABA treatment significantly reduces MDA levels—an indicator of lipid peroxidation and oxidative damage. Moreover, GABA's role extends beyond mere antioxidant modulation. The GABA shunt pathway, which supplies succinate and NADH during respiration, hints at GABA's involvement in directly managing ROS, particularly under oxidative stress (Shi et al. 2010). For instance, in rice plants subjected to copper stress, GABA-treated specimens exhibit a noticeable decrease in oxidative damage compared to untreated plants (Khan et al. 2024). GABA treatment in fenugreek leads to reduced nickel uptake in roots and aboveground parts of plants, alongside stimulation of the plant's antioxidant system (Danish et al. 2024). Overall, these findings highlight GABA's crucial role in removing ROS and protecting plants against the harmful effects of metal/metalloid-induced oxidative stress.

11.3.3 Modulation of Antioxidant Defenses Through GABA Signaling

Under metal/metalloid stress, GABA signaling modulates the plant's antioxidant defense system, enhancing antioxidants that mitigate oxidative damage. In plants subjected to metal toxicity, such as mustard under Cr stress, GABA stimulates the activation of various antioxidant enzymes, including APX, GR, and MDHAR (Mahmud et al. 2017). This upregulation facilitates efficient ROS scavenging, minimizing cellular damage and enhancing the plant's ability to tolerate stress. GABA also interacts with nonenzymatic antioxidants such as ascorbate and glutathione, which are key to detoxifying harmful ROS and maintaining cellular homeostasis under metal stress (He et al. 2021). Additionally, GABA treatment explicitly increases protein and GSH contents while reducing ROS and Cd^{2+} levels in microalgae, highlighting its role in antioxidant defense (Zhao et al. 2020). In rice exposed to Cu stress, GABA enhances SOD, GSH, APX, and CAT, effectively boosting the plant's resilience (Khan et al. 2024). Similarly, GABA alleviates Cd toxicity in maize by concertedly regulating antioxidant systems (Seifikalhor et al. 2020), while rice plants exhibit significant antioxidant enzyme adjustments with GABA application, further aiding ROS removal (Ashraf et al. 2022). In hybrid *Liriodendron* plants under Al stress, GABA activates antioxidant enzymes even under escalation of stress-induced ROS levels (Wang et al. 2021). A study on creeping bentgrass also demonstrates that GABA priming upregulates the activities of SOD, POD, CAT, and APX under prolonged acid-Al stress (Zhou et al. 2023). Through these mechanisms, GABA signaling strengthens plants' tolerance to metal/metalloid stress by maintaining redox homeostasis and diminishing oxidative damage.

11.3.4 GABA's Role in Improving Nutrient Uptake Under Metal/Metalloid Stress

GABA is crucial in improving nutrient absorption during metal and metalloid stress by regulating different physiological functions. It influences the function of metal ion transporters, reducing the influx of heavy metals like Cd and Pb, thereby minimizing their harmful impact on nutrient absorption. GABA not only mitigates metal toxicity but also promotes the absorption of vital nutrients like calcium, magnesium, and iron by regulating specific transport channels, thereby alleviating the competition posed by metal stress. Moreover, GABA promotes healthier root development, increasing the surface area available for nutrient absorption, and modulates key metabolic pathways to balance metabolites necessary for efficient nutrient utilization. For example, GABA improves mineral nutrient uptake in microalgae under Cd stress (Zhao et al. 2020). Through its interaction with stress signaling pathways, including those involving plant hormones like abscisic acid (ABA), GABA enhances the plant's capacity to manage nutrient acquisition and utilization effectively, which is crucial for maintaining overall plant health and productivity in metal-contaminated environments. Research on rice plants has shown that applying GABA can alleviate the harmful effects of copper stress on the uptake of vital elements such as calcium, magnesium, and potassium (Khan et al. 2024). A recent study suggests that GABA may enhance Ca uptake in lettuce under Cd stress (Waris et al. 2023). In various plants such as mustard (Mahmud et al. 2017), rice (Kumar et al. 2019a), tobacco (He et al. 2021), and maize (Seifikalhor et al. 2020), GABA influences the absorption and movement of critical minerals under metal stress, possibly reducing the harmful impacts of metal toxicity on plant growth.

11.3.5 Gene Expression and Signaling Pathways

GABA acts as an important signaling molecule that regulates the expression of genes related to stress responses, especially those associated with the GABA shunt pathway, a key metabolic pathway connecting carbon and nitrogen metabolism (Ansari et al. 2021).

Exogenous GABA application elevates GABA-T activity during $Ca(NO_3)_2$ stress in muskmelon (Hu et al. 2015) and reduces chromium absorption in the roots and shoots of mustard seedlings (Mahmud et al. 2017). Under metal and metalloid stress, the upregulation of GAD genes, responsible for GABA biosynthesis, often results in enhanced GABA accumulation, triggering downstream signaling pathways that bolster stress tolerance (Jalil et al. 2016, 2017). In rice exposed to As stress, GABA-induced gene expression has been found to inhibit As accumulation and activate antioxidant defense mechanisms, enhancing As tolerance (Kumar et al. 2017, 2019a). Additionally, GABA treatment notably boosts the expression of GABA T2 and succinate semialdehyde dehydrogenase (SSADH), aiding in the plant's ability to withstand As (III) toxicity (Kumar et al. 2017). Li et al. (2022) have further demonstrated that GABA treatment significantly reduces the expression of Cd-related uptake and transport genes in apples under Cd stress while increasing amino acids in roots. Overexpression of the MdGAD1 gene in transgenic apple roots reduces Cd accumulation and lowers Cd uptake due to suppressed expression of Cd transport-related genes. A feedback inhibition loop involving GABA and GAD expression exists that presumably suppresses GAD expression and confers resilience during Cu stress in rice (Khan et al. 2024). Nevertheless, exogenous GABA during Cu stress upregulates the expression of GABA-T and SSADH. The findings underscore GABA's role in regulating gene expression and signaling pathways, helping plants mitigate the detrimental effects of metal stress.

11.4 Conclusion

GABA mitigates metal and metalloid toxicity, engaging multifaceted mechanisms. By regulating ion transport, GABA limits the absorption and buildup of toxic metals while promoting the uptake of vital nutrients such as Ca, Mg, and Fe. Its ability to upregulate both antioxidant systems helps to neutralize ROS, thereby protecting cellular components from oxidative damage. Furthermore, GABA's involvement in regulating stress-responsive gene expression and metabolic pathways, particularly the GABA shunt, underscores its importance in improving plant resilience under toxic metal conditions. The implications of these protective roles are far-reaching, particularly in relation to crop resilience and sustainable agriculture. By leveraging GABA's ability to enhance stress tolerance, future strategies could focus on using GABA-based treatments or breeding GABA-responsive crops to improve growth and productivity in metal-contaminated environments. This could pave the way for more sustainable agricultural practices, ensuring food security in regions affected by soil contamination and reducing the dependency on chemical interventions for stress management.

References

Ahmad P, Abdel Latef AA, Abd-Allah EF, Hashem A, Sarwat M, Anjum NA, Gucel S (2016) Calcium and potassium supplementation enhanced growth, osmolyte secondary metabolite production, and enzymatic antioxidant machinery in cadmium-exposed chickpea (Cicer arietinum L.). Front Plant Sci 7:187060

Aina R, Sgorbati S, Santagostino A, Labra M, Ghiani A, Citterio S (2004) Specific hypomethylation of DNA is induced by heavy metals in white clover and industrial hemp. Physiol Plant 121:472–480

Angon PB, Islam MS, Kc S, Das A, Anjum N, Poudel A, Suchi SA (2024) Sources, effects and present perspectives of heavy metals contamination: soil, plants and human food chain. Heliyon 10(7):e28357

Ansari MI, Hasan S, Jalil SU (2014) Leaf senescence and GABA shunt. Bioinformation 10:730–732

Ansari MI, Jalil SU, Ansari SA et al (2021) GABA shunt: a key-player in mitigation of ROS during stress. Plant Growth Regul 94:131–149

Ashraf U, Mahmood S, Anjum SA, Abbas RN, Rasul F, Iqbal J, Mo Z, Tang X (2022) Exogenous gamma-aminobutyric acid application induced modulations in the performance of aromatic rice under lead toxicity. Front Plant Sci 13:933694

Barbosa JS, Cabral TM, Ferreira DN, Agnez-Lima LF, De Medeiros SRB (2010) Genotoxicity assessment in aquatic environment impacted by the presence of heavy metals. Ecotoxicol Environ Saf 73:320–325

Carrasco-Gil S, Estebaranz-Yubero M, Medel-Cuesta D, Millán R, Hernández LE (2012) Influence of nitrate fertilization on Hg uptake and oxidative stress parameters in alfalfa plants cultivated in a Hg-polluted soil. Environ Exp Bot 75:16–24

Chakravorty M, Jaiswal KK, Bhatnagar P, Parveen A, Upadhyay S, Vlaskin MS, Alajmi MF, Chauhan PK, Nanda M, Kumar V (2024) Exogenous GABA supplementation to facilitate Cr (III) tolerance and lipid biosynthesis in Chlorella sorokiniana. J Environ Manag 355:120441. https://doi.org/10.1016/j.jenvman.2024.120441. Epub 2024 Mar 1. PMID: 38430879

Chen F, Gao J, Zhou Q (2012) Toxicity assessment of simulated urban runoff containing polycyclic musks and cadmium in Carassius auratus using oxidative stress biomarkers. Environ Pollut 162:91–97

Cheng B, Li Z, Liang L et al (2018) The γ-aminobutyric acid (GABA) alleviates salt stress damage during seeds germination of white clover associated with Na + /K + transportation, dehydrins accumulation, and stressrelated genes expression in white clover. Int J Mol Sci 19:2520

Cuypers A, Smeets K, Ruytinx J, Opdenakker K, Keunen E, Remans T, Horemans N, Vanhoudt N, Van Sanden S, Van Belleghem F, Yvese G, Jana C, Jacoa V (2011) The cellular redox state as a modulator in cadmium and copper responses in Arabidopsis thaliana seedlings. J Plant Physiol 168:309–316

Dalyan E, Yüzbaşıoğlu E, Akpınar I (2020) Physiological and biochemical changes in plant growth and different plant enzymes in response to lead stress. In: Gupta D, Chatterjee S, Walther C (eds) Lead in plants and the environment. Radionuclides and heavy metals in the environment. Springer, Cham. https://doi.org/10.1007/978-3-030-21638-2_8

Danish S, Hussain GS, Hussain MB, Elgorban AM, Datta R (2024) Unveiling the potential of A. fabrum and γ-aminobutyric acid for mitigation of nickel toxicity in fenugreek. Sci Rep 14:11042

Daş ZA, Dimlioğlu G, Bor M, Özdemir F (2016) Zinc induced activation of GABA-shunt in tobacco (Nicotiana tabaccum L.). Environ Exp Bot 122:78–84. https://doi.org/10.1016/j.envexpbot.2015.09.006

Das P, Samantaray S, Rout GR (2019) Heavy metal-induced oxidative stress and protection by mycorrhizal fungi. In: Hossain MA, Hasanuzzaman MZ, Fujita S (eds) Oxidative damage to plants: antioxidant networks and signaling. Academic Press, pp 377–397

Ghori NH, Ghori T, Hayat MQ et al (2019) Heavy metal stress and responses in plants. Int J Environ Sci Technol 16:1807–1828

Gichner T, Patková Z, Száková J, Demnerová K (2006) Toxicity and DNA damage in tobacco and potato plants growing on soil polluted with heavy metals. Ecotoxicol Environ Saf 65:420–426

Guerinot ML (2000) The ZIP family of metal transporters. Biochim Biophys Acta 1465:190–198

Hasan MK, Cheng Y, Kanwar MK, Chu XY, Ahammed GJ, Qi ZY (2017) Responses of plant proteins to heavy metal stress – a review. Front Plant Sci 8:1492

He G, Zhang H, Liu S, Li H, Huo Y, Guo K, Xu Z, Zhang H (2021) Exogenous γ-glutamic acid (GABA) induces proline and glutathione synthesis in alleviating Cd-induced photosynthetic inhibition and oxidative damage in tobacco leaves. J Plant Interact 16:296–306

Hu X, Xu Z, Xu W, Li J, Zhao N, Zhou Y (2015) Application of γ-aminobutyric acid demonstrates a protective role of polyamine and GABA metabolism in muskmelon seedlings under Ca(NO3)2 stress. Plant Physiol Biochem 92:1–10

Huybrechts M, Hendrix S, Kyndt T, Demeestere K, Vandamme D, Cuypers A (2021) Short-term effects of cadmium on leaf growth and nutrient transport in rice plants. Plant Sci 313:111038

Jalil SU, Ahmad I, Ansari MI (2016) Phenotypic characterization of GABA-transaminase mutants of *Arabidopsis thaliana*. Adv Life Sci 5:10005–10008

Jalil SU, Ahmad I, Ansari MI (2017) Functional loss of GABA transaminase (GABA-T) expressed early leaf senescence under various stress conditions in *Arabidopsis thaliana*. Curr Plant Biol 9–10:11–22

Jalil SU, Ansari MI (2020) Physiological role of gamma aminobutyric acid in salt stress tolerance. In: Hasanuzzaman M, Tanveer M (eds) Salt and drought stress tolerance in plants, signaling and communication in plants. Springer, Switzerland, pp 337–350

Jalil SU, Khan MAA, Ansari MI (2019a) Tremendous role of GABA shunt metabolic pathway in plant system. J Biol Chem Res 36:188–196

Jalil SU, Khan MQR, Ansari MI (2019b) Role of GABA transaminase in the regulation of development and senescence in *Arabidopsis thaliana*. Curr Plant Biol 19:100119

Kabata-Pendias A, Mukherjee AB (2007) Trace elements from soil to human. Springer, Berlin

Khan MIR, Jalil SU, Chopra P, Chhillar H, Ferrante A, Khan NA, Ansari MI (2021) Role of GABA in plant growth, development and senescence. Plant Gene 26:100283

Khan Z, Jan R, Asif S, Farooq M, Kim KM (2024) Exogenous GABA enhances copper stress resilience in rice plants via antioxidant defense mechanisms, gene regulation, mineral uptake, and copper homeostasis. Antioxidants 13:700. https://doi.org/10.3390/antiox13060700

Khanna K, Kohli SK, Bali S, Kaur P, Saini P, Bakshi P, Bhardwaj R (2018) Role of micro-organisms in modulating antioxidant defence in plants exposed to metal toxicity. In: Hasanuzzaman M, Fujita M, Nahar K, Biswas JK (eds) Plants under metal and metalloid stress. Springer, Singapore, pp 303–335

Kováčik J, Klejdus B, Hedbavny J, Bačkor M (2010) Effect of copper and salicylic acid on phenolic metabolites and free amino acids in *Scenedesmus quadricauda* (Chlorophyceae). Plant Sci 178:307–311

Kumar A, Aery NC (2016) Impact, metabolism, and toxicity of heavy metals in plants. In: Hakeem KR, Akhtar MS, Abdullah SN (eds) Plant responses to xenobiotics. Springer, Singapore, pp 141–176

Kumar N, Dubey AK, Upadhyay AK, Gautam A, Ranjan R, Srikrishna S, Sahu N, Behera SK, Mallick S (2017) GABA accretion reduces Lsi-1 and Lsi-2 gene expressions and modulates physiological responses in *Oryza sativa* to provide tolerance towards arsenic. Sci Rep 7:8786

Kumar N, Gautam A, Dubey AK, Ranjan R, Pandey A, Kumari B, Singh G, Mandotra S, Chauhan PS, Srikrishna S (2019a) GABA mediated reduction of arsenite toxicity in rice seedling through modulation of fatty acids, stress responsive amino acids and polyamines biosynthesis. Ecotoxicol Environ Saf 173:15–27

Kumar V, Singh J, Kumar P (2019b) Heavy metals accumulation in crop plants: sources, response mechanisms, stress tolerance and their effects. In: Kumar V, Kumar R, Singh J, Kumar P (eds) Contaminants in agriculture and environment: health risks and remediation. Agro Environ Media, Haridwar, pp 38–57. https://doi.org/10.26832/AESA-2019-CAE-0161-04

Li Y, Li Y, Cui Y, Xie Y, Shi Y, Shang Y, Ma F, Zhang J, Li C (2022) GABA-mediated inhibition of cadmium uptake and accumulation in apples. Environ Pollut 300:118867. https://doi.org/10.1016/j.envpol.2022.118867

Mahmud JA, Hasanuzzaman M, Nahar K, Rahman A, Hossain MS, Fujita M (2017) γ-Aminobutyric acid (GABA) confers chromium stress tolerance in *Brassica juncea* L. by modulating the antioxidant defense and glyoxalase systems. Ecotoxicology 26:675–690

Martínez Domínguez D, Torronteras Santiago R, Córdoba García F (2009) Modulation of the antioxidative response of *Spartina densiflora* against iron exposure. Physiol Plant 136:169–179

Nagajyoti PC, Lee KD, Sreekanth TVM (2010) Heavy metals occurrence and toxicity for plants: a review. Environ Chem Lett 8:199–216

Pittman JK (2005) Managing the manganese: molecular mechanisms of manganese transport and homeostasis. New Phytol 167:733–742. https://doi.org/10.1111/j.1469-8137.2005.01453.x
Pourrut B, Jean S, Silvestre J, Pinelli E (2011) Lead-induced DNA damage in Vicia faba root cells: potential involvement of oxidative stress. Mutat Res 726:123–128
Pourrut B, Shahid M, Douay F, Dumat C, Pinelli E (2013) Molecular mechanisms involved in lead uptake, toxicity and detoxification in higher plants. In: Corpas FJ, Palma JM, Gupta DK (eds) Heavy metal stress in plants. Springer, Berlin, pp 121–147
Pucciariello C, Banti V, Perata P (2012) ROS signaling as common element in low oxygen and heat stresses. Plant Physiol Biochem 59:3–10
Seifikalhor M, Aliniaeifard S, Bernard F, Seif M, Latifi M, Hassani B, Li T (2020) γ-Aminobutyric acid confers cadmium tolerance in maize plants by concerted regulation of polyamine metabolism and antioxidant defense systems. Sci Rep 10:3356
Shahid M, Pourrut B, Dumat C, Nadeem M, Aslam M, Pinelli E (2014) Heavy-metal-induced reactive oxygen species: phytotoxicity and physicochemical changes in plants. Rev Environ Contam Toxicol 232:1–44. https://doi.org/10.1007/978-3-319-06746-9_1
Shen J, Song L, Müller K, Hu Y, Song Y, Yu W, Wang H, Wu J (2016) Magnesium alleviates adverse effects of lead on growth, photosynthesis, and ultrastructural alterations of Torreya grandis seedlings. Front Plant Sci 7:1819. https://doi.org/10.3389/fpls.2016.01819
Shen Y, Zhang Y, Chen J, Lin H, Zhao M, Peng H, Liu L, Yuan G, Zhang S, Zhang Z, Pan G (2013) Genome expression profile analysis reveals important transcripts in maize roots responding to the stress of heavy metal Pb. Physiol Plant 147:270–282. https://doi.org/10.1111/j.1399-3054.2012.01670.x
Shi SQ, Shi Z, Jiang ZP, Qi LW, Sun XM, Li CX, Liu JF, Xiao WF, Zhang SG (2010) Effects of exogenous GABA on gene expression of Caragana intermedia roots under NaCl stress: regulatory roles for H_2O_2 and ethylene production. Plant Cell Environ 33:149–162
Singh NK, Rai UN, Tewari A, Singh M (2010) Metal accumulation and growth response in Vigna radiata L. inoculated with chromate tolerant rhizobacteria and grown on tannery sludge amended soil. Bull Environ Contam Toxicol 84:118–124
Song H, Xu X, Wang H, Wang H, Tao Y (2010) Exogenous γ-aminobutyric acid alleviates oxidative damage caused by aluminium and proton stresses on barley seedlings. J Sci Food Agric 90:1410–1416
Wang P, Dong Y, Zhu L, Hao Z, Hu L, Hu X, Wang G, Cheng T, Shi J, Chen J (2021) The role of γ-aminobutyric acid in aluminum stress tolerance in a woody plant, Liriodendron chinense × tulipifera. Hortic Res 8:80
Waris Z, Noreen Z, Shah AA, Usman S, Shah AN, Rizwan M, Casini R, Elansary HO (2023) Efficacy of γ-aminobutyric acid (GABA) on physio-biochemical attributes of lettuce (Lactuca sativa L.) under cadmium toxicity. J Plant Growth Regul 42:5041–5057
Wu X, Jia Q, Ji S, Gong B, Li J, Lü G, Gao H (2020) Gamma-aminobutyric acid (GABA) alleviates salt damage in tomato by modulating Na^+ uptake, the GAD gene, amino acid synthesis and reactive oxygen species metabolism. BMC Plant Biol 20:465
Zhao Y, Song X, Zhong DB, Yu L, Yu X (2020) γ-Aminobutyric acid (GABA) regulates lipid production and cadmium uptake by Monoraphidium sp. QLY-1 under cadmium stress. Bioresour Technol 297:122500. https://doi.org/10.1016/j.biortech.2019.122500
Zhou M, Yuan Y, Lin J, Lin L, Zhou J, Li Z (2023) γ-Aminobutyric acid priming alleviates acid-aluminum toxicity to creeping bentgrass by regulating metabolic homeostasis. Int J Mol Sci 24:14309. https://doi.org/10.3390/ijms241814309
Zhou M, Huang C, Lin J, Yuan Y, Lin L, Zhou J, Li Z (2024) γ-Aminobutyric acid (GABA) priming alleviates acid-aluminum toxicity to roots of creeping bentgrass via enhancements in antioxidant defense and organic metabolites remodeling. Planta 260(1):33. https://doi.org/10.1007/s00425-024-04461-8. PMID: 38896325
Zhu C, Ding Y, Liu H (2011) MiR398 and plant stress responses. Physiol Plant 143:1–9

GABA as Protectant Against Wounding Stress

12

Abstract

Wounding stress in plants, caused by mechanical injury, herbivory, or environmental disturbances, triggers complex physiological and biochemical responses aimed at restoring cellular integrity and preventing pathogen invasion. Gamma-aminobutyric acid (GABA) mitigates the adverse effects of wounding stress because of its multifaceted role in plant defense, signaling cascades, ion flux regulation, reactive oxygen species (ROS) scavenging, and modulation of metabolic pathways. Upon wounding, GABA rapidly accumulates at the injury site, facilitating early signaling responses and enhancing tissue repair. GABA also influences the expression of wound stress-responsive genes and interacts with key phytohormones such as jasmonate (JA) and salicylic acid (SA) to orchestrate an integrated defense response. Overall, the protective role of GABA against wounding stress underscores its significance as a target for enhancing plant defense mechanisms in agricultural systems challenged by mechanical injuries and pest attacks, along with the restoration of crop growth and productivity.

Keywords

GABA · Plant defense · Phytohormones · ROS · Stress signaling · Wounding stress

Highlights

- Gamma-aminobutyric acid (GABA) regulates metabolic and signaling pathways to counteract wounding stress.
- Wounded tissues trigger rapid GABA biosynthesis through enzymes like GAD.
- GABA interacts with key signaling molecules, including calcium, reactive oxygen species (ROS), and phytohormones like JA and ABA, to enhance stress responses.

S. U. Jalil et al., *GABA Signaling System and Abiotic Stress Tolerance in Plants*,
https://doi.org/10.1007/978-981-95-0437-4_12

- It strengthens antioxidant defenses, helping to control ROS levels and prevent oxidative damage.
- Crosstalk between GABA and hormone pathways amplifies plant defense mechanisms, aiding in efficient wound healing.
- Enhancing GABA pathways may offer new strategies for improving crop resilience against physical damage in agricultural systems.

12.1 Introduction

Plants, like all living organisms, face numerous challenges in their environment, and wounding stress stands as one of the most pervasive threats (Savatin et al. 2014). Whether caused by herbivores, mechanical damage, or environmental factors such as storms or human activity, wounds disrupt the plant's physical integrity and expose internal tissues to pathogens and dehydration (Mostafa et al. 2022). The plant's response to such injury is swift and complex, involving a series of biochemical and physiological changes that aim to restore damaged tissues, prevent infection, and maintain overall plant homeostasis (Arimura et al. 2005; Spoel and Dong 2012; Belete 2018; Savatin et al. 2014; Mostafa et al. 2022). In essence, the ability to heal from wounds is crucial for plant survival, especially in the wild, where constant exposure to threats makes recovery an ongoing battle.

One of the fascinating aspects of plant biology is how efficiently plants have evolved mechanisms to respond to wounding. The wound-healing processes engage intricate signaling networks (Peña-Cortés et al. 1995), where hormones like jasmonate (JA) and ethylene play pivotal roles in activating defense pathways (Onkokesung et al. 2010). Recent research, however, suggests that gamma-aminobutyric acid (GABA) not only acts as a metabolite but also acts as a signaling molecule that mediates multiple aspects of wound repair (Bown et al. 2006; Scholz et al. 2015; Jalil et al. 2017).

GABA's involvement in wounding stress offers a unique dimension to plant defense. Upon wounding, GABA levels rapidly increase at the site of injury, influencing a variety of processes from oxidative stress management to cellular repair. GABA's ability to modulate calcium levels, interact with reactive oxygen species (ROS), and crosstalk with other signaling molecules makes it an indispensable part of the plant's defensive arsenal (Bown et al. 2006; Scholz et al. 2015). As scientists continue to unravel the role of GABA in plant biology, its potential in enhancing wound recovery has become a growing focus of interest.

Thus, exploring the protective role of GABA in mitigating the consequences of wounding stress in plants assumes paramount significance. By examining its role in wound signaling pathways, tissue repair, and interaction with other key molecules, we aim to provide a comprehensive understanding of how GABA contributes to plant resilience. Understanding GABA's role in wound healing not only deepens our knowledge of plant stress physiology but also opens up new avenues for improving crop resistance against physical damage in agriculture.

12.2 Mechanisms of Wounding Stress in Plants

Wounding stress in plants, whether caused by mechanical damage, herbivory, or environmental factors, triggers a complex and immediate set of defensive mechanisms. These responses are essential for the plant's survival, as they mitigate further injury, prevent pathogen invasion, and promote tissue repair. Upon wounding, plants activate a cascade of physiological and molecular processes, including Ca^{2+} release, ROS, and key stress hormones like JA and ethylene, all of which orchestrate the healing and defense pathways. Understanding these mechanisms is crucial for exploring how plants cope with and recover from physical damage (Fig. 12.1).

12.2.1 Types of Wounding Stress

Plants constantly experience wound stress due to physical abrasion and tissue damage. Wound stress restricts the growth and survival of plants. Wounds can arise from multiple sources, e.g., mechanical damage, such as that caused by human activity (e.g., pruning, cultivation) or natural events (e.g., wind, hailstorms), disrupting the structural integrity of plants. Another significant source of wounding stress is herbivory, where insects and animals feed on plant tissues, leaving behind open wounds

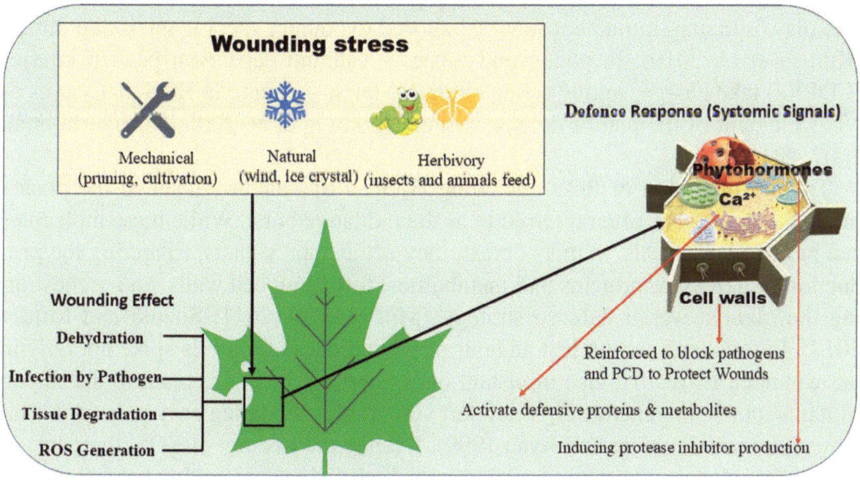

Fig. 12.1 Effect of wounding stress on plants and the synergistic defense response. Wounding stress triggers physiological and biochemical changes, including immediate activation of defense-related hormones in plants. Upon wounding, a rapid accumulation of secondary metabolites and ROS synergizes with the activation of Ca^{2+} signaling pathways to modulate the defense response by regulating ion fluxes, mitigating oxidative damage, and enhancing resistance against herbivores or pathogens

that serve as entry points for pathogens (Vega-Muñoz et al. 2020). Additionally, environmental factors such as abrasions from sand in high winds or ice crystal formation in frost conditions can cause wounding (Vega-Muñoz et al. 2020). Each of these types of damage leaves the plant vulnerable, initiating a cascade of complex responses aimed at mitigating further harm and promoting recovery.

12.2.2 Physiological and Molecular Responses to Wounding

When a plant experiences wounding, its response is immediate and multifaceted. On a physiological level, the plant acts quickly to prevent dehydration, infection, and further tissue degradation. Cell wall fortification prevents pathogen entry. The damaged cells often undergo programmed cell death to create a protective barrier around the wound (Bellincampi et al. 2014).

Concurrently, molecular signals spread from the wound site to activate systemic responses throughout the plant. When a plant is wounded, it triggers an immediate and intricate defense response. The first signal comes in the form of a rapid wave of Ca^{2+}, which spreads from the injury site, much like an alarm, activating early defense mechanisms across the plant (Kiep et al. 2015). Acting as molecular sentinels, calcium sensors such as CaM-binding proteins in *Arabidopsis* detect this influx and help accumulate SA, a critical player in enhancing the plant's resistance, particularly against pathogens like *Pseudomonas syringae* (Wang et al. 2009). In rice, calcineurin B-like protein (CBL)-interacting protein kinases (CIPKs) 14 and 15 come into play, initiating immune responses tailored to counter specific microbial threats (Kurusu et al. 2010). In potato and tobacco, calcium-dependent protein kinases (CDPKs) take charge, setting off an oxidative burst—a surge in ROS that serves as a key element of the plant's defense arsenal (Ludwig et al. 2005; Kobayashi et al. 2007, 2012).

The accumulation of ROS, including H_2O_2, forms another layer of the plant's immediate response, often referred to as the oxidative burst. While these molecules can be harmful to cells, in this context, they act as vital signals, triggering the production of defensive proteins and metabolites, fortifying cell walls, and orchestrating the plant's overall defense strategy (Mittler et al. 2011; Suzuki and Mittler 2012). In tomatoes, within just an hour of wounding, H_2O_2 levels spike not only in the wounded leaves but also in distant ones, alerting the entire plant to the threat. Mutants unable to generate H_2O_2 are left vulnerable, becoming easy prey for herbivores (Orozco-Cardenas and Ryan 1999). Despite the toxicity of ROS, their role as signaling molecules must be carefully controlled. Calcium signaling helps fine-tune this process, while enzymes such as RBOHs regulate the production and balance of ROS, ensuring they remain potent yet nonlethal to the plant's own cells (Minibayeva et al. 2014). Through this tightly orchestrated system, plants execute a swift and precise molecular defense, ensuring survival amidst continuous external threats.

12.2.3 Activation of Plant Defense Mechanisms by Phytohormones

When a plant is wounded, its defense machinery kicks into action almost instantly, orchestrating a complex hormonal symphony to protect itself. The key response conductors, namely, JA and ethylene, swiftly elevate their titers to trigger a powerful cascade of defenses (Fig. 12.2). JA immediately swings into action by prompting the production of protease inhibitors—specialized proteins that dissuade herbivores from further feeding by disrupting their digestion. But that is not all—JA also strengthens the plant's fortifications by reinforcing cell walls and accumulating secondary metabolites, making the plant less appealing and more resilient (Creelman and Mullet 1997; Wasternack 2007). In *Arabidopsis*, this response is incredibly fast: wounding a single leaf triggers a spike in JA levels in a connected leaf within just 90 seconds (Dengler and Kang 2001). The enzyme 13-LOX gets to work quickly, oxygenating fatty acids as the first step in JA production. LOX2 powers the processes near the wound, while LOX6 ensures that both the wounded leaf and distant tissues ramp up JA defenses (Glauser et al. 2009). Within mere seconds, JA is

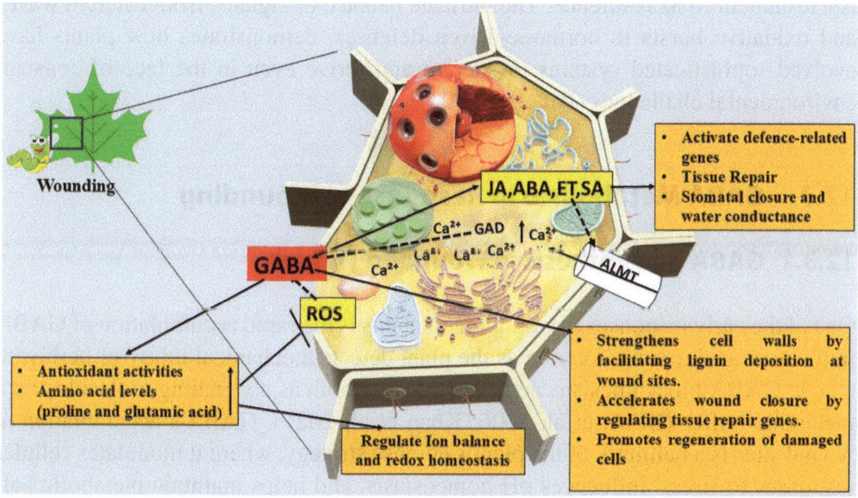

Fig. 12.2 Role of GABA in mitigating wounding stress in plants. GABA plays a central role in alleviating wounding stress, with its levels rising rapidly and spectacularly upon wounding. The elevated GABA levels interact with Ca^{2+} signaling to modulate ion fluxes, stabilize cellular homeostasis, and also regulate the expression of defense-related genes, promoting the accumulation of amino acids (proline and glutamic acid) and antioxidants. Consequently, ROS-induced damage gets mitigated. Doing so, GABA synergizes with phytohormones such as JA, ABA, ethylene, and SA to activate defense pathways, securing resilience against herbivores and pathogens in addition to facilitating tissue recovery and overall protection. This hormonal interaction also induces the expression of ALMT genes, which accumulate Ca^{2+}, stimulating the GAD enzyme, thereby further increasing GABA levels

converted to its active form, which then launches a full-scale systemic defense response (Chauvin et al. 2013).

While JA is busy coordinating the defense forces, ABA joins the battle. ABA, usually known for its role in drought tolerance, steps in to confine the damage. It triggers programmed cell death in the cells around the wound, preventing any infections or injuries from spreading further (Bostock and Stermer 1989). Meanwhile, ethylene orchestrates wound healing and makes sure the plant knows when to seal the site and when to let go of damaged tissue (O'Donnell et al. 1996). JA and ethylene do more than just protect the plant. They carefully balance the plant's need for defense with its drive for growth, ensuring that resources are allocated wisely, even in the face of danger (Onkokesung et al. 2010). The plant's defense response, once triggered, becomes a marvel of coordination and speed, mobilizing every available resource not just to survive but to thrive amidst adversity.

Together, these signals form a robust defense network that not only addresses the immediate threat posed by wounding but also primes the plant for future attacks. By coordinating physiological and molecular responses, the plant can effectively seal off wounds, limit the spread of damage, and initiate the repair processes needed for recovery. In essence, the plant's ability to mount an effective response to wounding is a testament to its resilience. The intricate balance of signals, from calcium waves and oxidative bursts to hormone-driven defenses, demonstrates how plants have evolved sophisticated systems to survive and thrive even in the face of constant environmental challenges (Pal et al. 2023).

12.3 GABA Metabolism in Response to Wounding

12.3.1 GABA Biosynthesis in Wounded Tissues

One of the early responses to wounding in plants is the rapid accumulation of GABA in the damaged tissues. As soon as the plant detects mechanical injury or herbivore attack, GABA biosynthesis is activated, serving both as a signaling molecule and a metabolic buffer (Bown et al. 2006; Khan et al. 2021). GABA's accumulation at wound sites is a hallmark of the plant's defense strategy, where it modulates cellular responses to stress, influences pH homeostasis, and helps maintain metabolic balance during recovery (Fig. 12.2).

12.3.2 Enzymes Involved in the GABA Shunt: GAD and GABA-T

In response to wounding stress, GAD initiates glutamate conversion into GABA, a reaction triggered by elevated cytosolic calcium levels under stress (Jalil et al. 2017). GABA transaminase (GABA-T) then catalyzes the conversion of GABA into succinic semialdehyde, allowing GABA to re-enter the TCA cycle and support cellular energy metabolism (Jalil et al. 2019a, b). This metabolic loop not only fuels the plant's energy needs but also enhances its defense mechanisms (Ansari et al.

2014). ROS act as second messengers to induce defense genes, as shown in wounded tomato plants (Orozco-Cardenas et al. 2001). Studies on *Arabidopsis* GABA-T mutants (*pop2–1* and *pop2–3*) reveal that these mutants exhibit reduced chlorophyll content, GABA levels, and GAD and GABA-T activity, alongside increased ion leakage and MDA levels, indicating heightened wounding stress sensitivity compared to wild-type plants (Jalil et al. 2017). The absence of GABA-T disrupts stress-induced metabolic adjustments, accelerating leaf senescence by impairing carbon fixation and intensifying oxidative damage. Wild-type plants, with functional GABA-T, display greater tolerance to stress, highlighting the enzyme's role in mitigating senescence under wounding conditions.

12.3.3 Regulation of GABA Levels Post-wounding: Interplay Between Biosynthesis and Degradation

A sharp rise in GABA accumulation following mechanical manipulation or injury to soybean leaf tissue is the first indication of plant response to the wound injury (Wallace et al. 1984). Regulation of GABA levels following wounding is tightly controlled by the balance between its biosynthesis and degradation. After its rapid production in response to injury, the plant must modulate GABA concentrations to ensure proper metabolic function and avoid excessive accumulation. The interplay between GABA biosynthesis, mediated by GAD, and its subsequent degradation via the GABA shunt ensures that GABA levels remain dynamically regulated to meet the metabolic demands of the wounded tissue (Jalil et al. 2017). This regulation allows GABA not only to function as an immediate stress signal but also to integrate into broader metabolic pathways that sustain the plant's recovery and growth after wounding (Abdullah et al. 2024).

12.4 GABA as a Signaling Molecule in Wounding Stress

When a plant is wounded, either through mechanical damage or herbivore attack, it triggers a rapid and coordinated defense system, in which GABA emerges as a central player, not only as a metabolite but also as a signaling molecule. Upon wounding, GABA levels at the site of injury surge almost immediately, activating a cascade of defense responses (Fig. 12.2). This surge is closely tied to the influx of Ca^{2+}, one of the earliest signals during wounding (Jiang and Ding 2023). GABA biosynthesis begins with the activation of GAD, which converts glutamate into GABA. GAD is uniquely designed to respond to changes in calcium levels due to its Ca^{2+}/CaM-binding domain (Baum et al. 1993), allowing it to amplify the wounding response by stimulating further GABA production and modulating ion fluxes (Snedden et al. 1995). The activation of GAD follows two pathways depending on the condition of the cytosol. Under normal, neutral pH conditions, GAD activation is calcium-dependent, requiring the binding of calmodulin to its Ca^{2+}/CaM site. However, when plant cells are wounded, the cytosolic pH drops, and GAD becomes active even without calcium due to the release of vacuolar contents into the cytosol (Carroll et al. 1994). In soybeans, this process was observed with the cytosol acidifying and leading to rapid GABA accumulation at the wound site (Shelp 1987; Ramputh and Bown 1996). In *Asparagus sprengeri*, GAD activation is Ca^{2+}-dependent, reinforcing the idea that calcium signaling is a key component of GABA biosynthesis during stress (Cholewa et al. 1997).

Another intriguing aspect of the signaling function of GABA is its interaction with ROS, molecules that rapidly produce following injury as part of the oxidative burst. ROS play a dual role in the plant's defense, acting both as signaling molecules and as direct protectants by reinforcing cell walls. However, unchecked ROS levels can cause significant cellular damage. GABA, with its regulatory properties, helps keep ROS levels in check, ensuring their function as effective signals without overwhelming the plant cells (Bown et al. 2002). In tomato plants, for instance, GABA regulates H_2O_2 levels, a key ROS involved in wound signaling, helping to manage the balance between defense and damage (Orozco-Cardenas and Ryan 1999). Herbivore attacks further highlight GABA's crucial role in plant defense. When larvae crawl across the leaves of tobacco (*Nicotiana tabacum*) or soybean plants, GABA levels surge dramatically within minutes. In tobacco, GABA increases four- to fivefold within 10 minutes of larval contact, while in soybean, the increase is even more pronounced, rising 10- to 12-fold (Bown et al. 2002). This rapid response is not a reaction to the larval feet but rather the result of tiny hook-like structures on the larvae's legs, which inflict wounds to the leaf surface (Hall et al. 2004). GABA accumulates at these wound sites, initiating a rapid defense response that helps the plants restrict further damage.

Beyond its interaction with ROS and calcium, GABA coordinates the plant's defense by fine-tuning the interaction with other signaling molecules and hormones. During wounding, GABA operates alongside phytohormones like JA, strengthening the plant's overall defense strategy. In *Arabidopsis*, both GABA and JA work in harmony, forming a highly responsive defense network that enables the plant to swiftly react to injury, regulate oxidative stress, and activate nearby tissues for potential threats (Scholz et al. 2015). Notably, GABA's influence is not limited to the wounded site—it can spread systemically throughout the plant, preparing undamaged tissues for possible future damage. For example, mechanical wounding in *Arabidopsis* triggers GABA accumulation in distant unwounded leaves, contributing to a whole-plant defense system (Scholz et al. 2015). This systemic GABA signaling mirrors the action of JA, a key phytohormone known to travel from the injured site to surrounding tissues, increasing the plant's readiness to confront further harm (Jimenez-Aleman et al. 2015). The synergistic interaction between GABA and JA adds layers of complexity to the plant's defense response, optimizing its capacity to deal with wounding.

Through its role in signaling, GABA creates an intricate defense network that connects calcium signaling, ROS production, and hormone pathways to ensure a rapid and effective response to wounding. This coordinated system allows plants not only to heal and manage oxidative stress but also to prepare for future challenges, making GABA a critical player in plant survival and resilience against injury.

12.5 Protective Role of GABA in Wound Healing

12.5.1 GABA's Role in Cell Wall Fortification and Tissue Repair

In response to wounding, plants must rapidly reinforce their cell walls to prevent further damage and pathogen invasion. GABA plays a crucial role in this process by contributing to the synthesis of cell wall components and regulating cellular turgor pressure. After an injury, GABA facilitates the deposition of lignin and other structural polymers, which strengthens the cell walls at the wound site, acting as a physical barrier against external threats. GABA also modulates the expression of genes associated with tissue repair processes, accelerating the closure of the wound and promoting regeneration of damaged cells (Abdullah et al. 2024; Islam et al. 2024).

12.5.2 Influence of GABA on Antioxidant Defense Systems and ROS Scavenging

Following wound infliction from a pest attack or physical damage, the plant reacts swiftly to defend itself. The initial response is a surge of ROS, which act as warning signals to help the plant recognize and address the injury. However, ROS levels rising beyond a threshold start damaging plant cells and enhancing further stress levels. In this sequence of events, GABA serves an essential function in plants by helping to regulate ROS levels, primarily through enhancing the activity of antioxidant enzymes SOD and CAT, preventing oxidative stress, and gearing up recovery of plants from stress. Hence, GABA ensures that ROS are used for signaling/alleviation of stress rather than accumulating to harmful levels.

Recently, GABA treatment to rice crops infested with *Sogatella furcifera* (white-backed planthopper) demonstrates a spectacular gain in antioxidant enzyme activity (APX, GPX, chloramphenicol acetyltransferase, POD, SOD, ABTS, and DPPH) that reduces ROS levels and pest attacks (Jan et al. 2024). Moreover, GABA activates a biochemical pathway known as the GABA shunt, which is particularly important during stress. This pathway helps the plant produce energy and manage ROS levels. For example, GABA treatment also elevates levels of NADPH and succinate, i.e., molecules essential for energy balance and ROS control, in the infested rice crop. Thus, GABA not only helps reduce ROS but also ensures requisite energy availability to plants to combat the stress. The investigation also corroborates an earlier observation in citrus plants, where GABA helps activate related stress-response genes (Hijaz and Killiny 2019). Earlier reports on GABA treatment also demonstrate a similar increase in antioxidant activity and lower oxidative stress in tomato and pear plants infected with fungi (Yu et al. 2014; Fu et al. 2017; Yang et al. 2017).

The protective role of GABA also extends to heavy metal stress, wherein GABA treatment significantly enhances the activity of CAT and SOD that reduces harmful compounds like H_2O_2 and $O_2^{\bullet -}$ (Seifikalhor et al. 2020). GABA acts as a powerful protector during wounding stress by boosting antioxidant defense, reducing oxidative damage, and ensuring that energy production remains intact. By activating key pathways like the GABA shunt, the plant manages ROS levels and recovers more effectively from injuries. This multilayered response ensures better plant survival in stressful environments.

12.5.3 GABA as Benevolent to Plant Chemical Defenses and Amino Acid Levels

GABA critically orchestrates a complex network of biochemical changes that regulates chemical defenses in response to wounding. When a plant is injured, GABA helps modulate the levels of key amino acids essential for stress response and recovery. These amino acids, such as proline, arginine, and glutamic acid, play crucial roles that enable plants to manage damage and stress.

GABA influences the accumulation of free amino acids during stress conditions. Research by Jan et al. (2024) demonstrates that stress from the *S. furcifera* initially reduces levels of essential amino acids like proline, arginine, aspartic acid, and glutamic acid in rice. However, GABA supplementation enhances titers of these amino acids. Particularly, proline is vital for protecting cells against oxidative stress and plays a role in maintaining ion balance and redox homeostasis (Signorelli et al. 2015; Che-Othman et al. 2020; Ansari et al. 2021). Proline interacts with the $NADP^+/NADPH$ ratio and scavenges ROS, helping to mitigate oxidative damage (Seifikalhor et al. 2020). The increased levels of proline, arginine, and glutamic acid with GABA treatment suggest a strengthened defense mechanism, allowing plants to better manage and recover from wounding stress. Thus, GABA enables plants to recover from wounding.

12.5.4 GABA Interaction with Phytohormones

In the complex response of plants to wounding stress, GABA and phytohormones interact and operate in tandem. Phytohormones, such as JA, ethylene, and ABA, are key regulators of plant stress responses, coordinating various physiological and biochemical processes to manage damage and promote recovery (Jalil and Ansari 2019; Pal et al. 2023). GABA intersects with these hormonal pathways to fine-tune the plant's response to injury by mounting effective defense mechanisms, tissue repair, and growth promotion. Understanding how GABA interacts with these phytohormones enhances our knowledge of plant stress biology and reveals potential strategies for improving plant resilience through targeted adjustment of these signaling networks.

12.5.4.1 Synergistic Effect of GABA and Phytohormones

Plants have an intricate system that responds to injuries through interactions of phytohormones, with JA playing a pivotal role. When a plant is wounded, JA levels increase dramatically, triggering a series of events that activate defense-related genes and enhance the production of GABA, a critical molecule involved in stress responses.

JA boosts GABA production by stimulating the enzyme GAD, which converts glutamate into GABA. This increase in GABA not only helps the plant manage the immediate stress of the injury but also primes it for future challenges by strengthening its metabolic and signaling pathways (Meyer et al. 2010). This interaction between JA and GABA forms a feedback loop: JA enhances GABA levels, and GABA, in turn, supports the plant's wound-healing processes.

However, GABA does not operate in isolation. It works together with other key phytohormones like ABA and ET to orchestrate an effective recovery. ABA is vital for stress tolerance and water conservation. It collaborates with GABA to regulate cellular responses, such as closing stomata to reduce water loss from wounded tissues (Che-Othman et al. 2020; Seifikalhor et al. 2020). Ethylene, known for regulating stress responses and senescence, also works in concert with GABA. Ethylene pathways enhance tissue repair and defense, with GABA modulating these responses to ensure a balanced and controlled healing process (Kathiresan et al. 1997).

Further studies have highlighted additional interactions between GABA and other phytohormones. For instance, auxins and ABA can induce ALMT genes, which are involved in GABA regulation under stress conditions (Podlesakova et al. 2019). In barley, overexpression of the *Arabidopsis* cytokinin dehydrogenase 1 gene (AtCKX1) facilitates expression of GABA-related genes such as GAD and ALMT (Pospíšilová et al. 2016). Moreover, GABA's role in regulating ethylene production through ACC synthase demonstrates a direct interaction between GABA and ethylene (Kathiresan et al. 1997).

GABA also influences ABA levels, which affect stomatal closure and water conductance. ABA typically induces stomatal closure to minimize water loss, a response that GABA appears to enhance (Mekonnen et al. 2016). Under drought stress, ABA accumulation promotes calcium flux and activates GAD to increase GABA synthesis (Shelp et al. 2012; Daszkowska-Golec and Szarejko 2013). The *Arabidopsis* GAD1 mutant exhibits higher transpiration even under drought, but ABA application can close stomata and reduce water loss, highlighting the significant role of GABA in this regulatory process (Meyer et al. 2010).

GABA is a crucial player in plant wound recovery. It synergizes with JA, ABA, and ethylene to create a finely tuned network that manages injury responses, conserves resources, and prepares the plant for future stress. Through its interactions with these hormones, GABA enhances the ability of plants to heal and adapt, underscoring its importance in plant defense and resilience.

12.5.4.2 Hormonal Regulation of GABA Synthesis

In response to wounding, the key growth-retardant phytohormones like JA, ABA, and ethylene regulate the synthesis of GABA in proportion to the severity through the activation of crucial enzymes such as GAD and GABA-T. As a result, a well-coordinated defense system gains activation that strikes a balance for resource allocation between defense and recovery due to wound stress. It prevents excess GABA accumulation while maintaining its vital role in signaling and metabolism.

Phytohormonally induced GABA synthesis does more than help plants heal from physical damage. It also serves as a metabolic regulator and signaling molecule, interacting with multiple defense pathways. For instance, the accumulated GABA in wounded plant tissues can inhibit bacterial infection by regulating bacterial quorum-sensing systems, such as the *Agrobacterium tumefaciens* attKLM operon (Chevrot et al. 2006).

S. furcifera infestation enhances GABA synthesis in rice and is related to the expression of GAD, GABA-T, and SSADH genes (Jan et al. 2024). Further, GABA metabolism supports a higher accumulation of key metabolites like NADPH and succinate. NADPH detoxifies ROS and maintains cellular redox balance, while succinate feeds into the TCA cycle, providing energy for recovery and repair (Fait et al. 2008).

In addition to metabolic regulation, GABA influences amino acid levels and stress-related gene expression. For example, the exogenous GABA application boosts the levels of defense-related amino acids such as proline and glutamic acid of plants to respond effectively in counteracting the stress situation in rice infested with *S. furcifera.* GABA treatment also significantly increases the expression of JA biosynthesis genes (LOX, AOS, AOC, and OPR), indicating that both SA and JA pathways contribute to GABA-mediated defense against *S. furcifera* (Jan et al. 2024). The complementary role between SA and JA in rice under pest attack suggests a more nuanced relationship, where GABA helps orchestrate their synergistic effects (Kanno et al. 2012).

The regulatory network does not stop at JA and SA. GABA treatment also regulates the expression of melatonin biosynthesis genes and melatonin level, a potent antioxidant that further enhances plant resilience (Jan et al. 2024). Melatonin and methyl jasmonate, a derivative of JA, interact in a positive feedback loop, where each compound amplifies the role of the other in stress mitigation (Liu et al. 2019; Shomali et al. 2021). Thus, GABA gains eminence as a master regulator to fine-tune hormonal signals for both immediate and long-term defense strategies to mitigate wound stress and recover healthy plants. In essence, the hormonal regulation of GABA synthesis during wounding ensures that the plant can efficiently respond to injuries and attacks. By modulating defense gene expression, metabolite production, and phytohormonal interactions, GABA emerges as a key player in plant survival, coordinating a sophisticated response to both internal damage and external threats.

12.6 Conclusion

GABA performs a function both as a metabolic regulator and a key signaling molecule to mitigate and recover from wound stress in diverse plants. Its rapid accumulation in response to wound injury triggers a cascade of protective mechanisms, including cell wall fortification, ROS scavenging, and the modulation of secondary metabolites, all of which contribute to tissue repair and defense. GABA operates in concert with phytohormones such as JA, ABA, and ethylene for its synthesis, accumulation, and development of the effective defense physio-molecular mechanisms to overcome wound stress in plants. The mechanisms encompass GABA integration into broader plant stress physiology, allowing for a well-coordinated response to physical damage. By mediating crosstalk between various hormonal pathways, GABA ensures a comprehensive defense strategy that not only addresses immediate damage but also primes plants for long-term resilience. Understanding the mechanisms by which GABA operates under wounding stress provides valuable insights into improving plant resilience, particularly in agricultural systems where physical damage from environmental factors and herbivory can significantly impact crop productivity. Enhancing GABA pathways through biotechnological or agronomic interventions could offer new approaches for developing crops with better stress tolerance and recovery, ultimately contributing to more sustainable and resilient agricultural practices.

References

Abdullah WKI, Naeem M et al (2024) From neurotransmitter to plant protector: the intricate world of GABA signaling and its diverse functions in stress mitigation. J Plant Growth Regul. https://doi.org/10.1007/s00344-024-11470-0

Ansari MI, Hasan S, Jalil SU (2014) Leaf senescence and GABA shunt. Bioinformation 10:730–732

Ansari MI, Jalil SU, Ansari SA et al (2021) GABA shunt: a key-player in mitigation of ROS during stress. Plant Growth Regul 94:131–149

Arimura G, Kost C, Boland W (2005) Herbivore-induced, indirect plant defences. Biochim Biophys Acta 1734:91–111. https://doi.org/10.1016/j.bbalip.2005.03.001

Baum G et al (1993) A plant glutamate decarboxylase containing a calmodulin binding domain. J Biol Chem 268:19610–19617

Belete T (2018) Defence mechanisms of plants to insect pests: from morphological to biochemical approach. Trends Tech Sci Res 2:555584. https://doi.org/10.19080/TTSR.2018.02.555584

Bellincampi D, Cervone F, Lionetti V (2014) Plant cell wall dynamics and wall-related susceptibility in plant–pathogen interactions. Front Plant Sci 5:228. https://doi.org/10.3389/fpls.2014.00228

Bostock RM, Stermer BA (1989) Perspectives on wound healing in resistance to pathogens. Annu Rev Phytopathol 27:343–371. https://doi.org/10.1146/annurev.py.27.090189.002015

Bown AW et al (2002) Insect footsteps on leaves stimulate the accumulation of 4-aminobutyrate and can be visualized through increased chlorophyll fluorescence and superoxide production. Plant Physiol 129:1430–1434

Bown AW, Macgregor KB, Shelp BJ (2006) Gamma-aminobutyrate: defence against invertebrate pests? Trends Plant Sci 11:424–427. https://doi.org/10.1016/j.tplants.2006.07.002

Carroll AD et al (1994) Ammonium assimilation and the role of g-aminobutyric acid in pH homeostasis in carrot cell suspensions. Plant Physiol 106:513–520

Chauvin A, Caldelari D, Wolfender JL, Farmer EE (2013) Four 13 lipoxygenases contribute to rapid jasmonate synthesis in wounded Arabidopsis thaliana leaves: a role for lipoxygenase 6 in responses to long-distance wound signals. New Phytol 197:566–575. https://doi.org/10.1111/nph.1202

Che-Othman MH, Jacoby RP, Millar AH, Taylor NL (2020) Wheat mitochondrial respiration shifts from the tricarboxylic acid cycle to the GABA shunt under salt stress. New Phytol 225:1166–1180. https://doi.org/10.1111/nph.15713

Chevrot R, Rosen R, Haudecoeur E et al (2006) GABA controls the level of quorum-sensing signal in Agrobacterium tumefaciens. Proc Natl Acad Sci USA 103:7460–7464. https://doi.org/10.1073/pnas.0602410103

Cholewa E, Bown AW, Cholewinski AJ, Shelp BJ, Snedden WA (1997) Cold-shock-stimulated γ-aminobutyric acid synthesis is mediated by an increase in cytosolic Ca2+, not by an increase in cytosolic H+. Can J Bot 75:375–382. https://doi.org/10.1139/b97-040

Creelman RA, Mullet JE (1997) Biosynthesis and action of jasmonates in plants. Annu Rev Plant Physiol Plant Mol Biol 48:355–381. https://doi.org/10.1146/annurev.arplant.48.1.355

Daszkowska-Golec A, Szarejko I (2013) Open or close the gate–stomata action under the control of phytohormones in drought stress conditions. Front Plant Sci 4:138. https://doi.org/10.3389/fpls.2013.00138

Dengler N, Kang J (2001) Vascular patterning and leaf shape. Curr Opin Plant Biol 4:50–56. https://doi.org/10.1016/S1369-5266(00)00135-7

Fait A, Fromm H, Walter D, Galili G, Fernie AR (2008) Highway or byway: the metabolic role of the GABA shunt in plants. Trends Plant Sci 13:14–19. https://doi.org/10.1016/j.tplants.2007.10.005

Fu D, Sun Y, Yu C et al (2017) Comparison of the effects of three types of aminobutyric acids on the control of Penicillium expansum infection in pear fruit. J Sci Food Agric 97:1497–1501. https://doi.org/10.1002/jsfa.7891

Glauser G, Dubugnon L, Mousavi SA, Rudaz S, Wolfender JL, Farmer EE (2009) Velocity estimates for signal propagation leading to systemic jasmonic acid accumulation in wounded Arabidopsis. J Biol Chem 284:34506–34513. https://doi.org/10.1074/jbc.M109.061432

Hall DE et al (2004) Footsteps from insect larvae damage leaf surfaces and initiate rapid responses. Eur J Plant Pathol 110:441–447

Hijaz F, Killiny N (2019) Exogenous GABA is quickly metabolized to succinic acid and fed into the plant TCA cycle. Plant Signal Behav 14:e1573096. https://doi.org/10.1080/15592324.2019.1573096

Islam SN, Kouser S, Hassan P et al (2024) Gamma-aminobutyric acid interactions with phytohormones and its role in modulating abiotic and biotic stress in plants. Stress Biol 4:36. https://doi.org/10.1007/s44154-024-00180-y

Jalil SU, Ahmad I, Ansari MI (2017) Functional loss of GABA transaminase (GABA-T) expressed early leaf senescence under various stress conditions in Arabidopsis thaliana. Curr Plant Biol 9–10:11–22

Jalil SU, Ansari MI (2019) Role of phytohormones in recuperating salt stress. In: Salt stress, microbes, and plant interaction: mechanism and molecular approaches. Springer, pp 91–104

Jalil SU, Khan MAA, Ansari MI (2019a) Tremendous role of GABA shunt metabolic pathway in plant system. J Biol Chem Res 36:188–196

Jalil SU, Khan MQR, Ansari MI (2019b) Role of GABA transaminase in the regulation of development and senescence in Arabidopsis thaliana. Curr Plant Biol 19:100119

Jan R, Asif S, Asaf S, Lubna KZ, Khan W, Kim KM (2024) Gamma-aminobutyric acid treatment promotes resistance against Sogatella furcifera in rice. Front Plant Sci 15:1419999. https://doi.org/10.3389/fpls.2024.1419999

Jiang Y, Ding P (2023) Calcium signaling in plant immunity: a spatiotemporally controlled symphony. Trends Plant Sci 28(1):74–89. https://doi.org/10.1016/j.tplants.2022.11.001

Jimenez-Aleman GH, Scholz SS, Heyer M, Reichelt M, Mithöfer A, Boland W (2015) Synthesis, metabolism and systemic transport of a fluorinated mimic of the endogenous jasmonate precursor OPC-8:0. Biochim Biophys Acta 12:1545–1553. https://doi.org/10.1016/j.bbalip.2015.09.002

Kanno H, Hasegawa M, Kodama O (2012) Accumulation of salicylic acid, jasmonic acid, and phytoalexins in rice, Oryza sativa, infested by the white-backed planthopper, Sogatella furcifera (Hemiptera: Delphacidae). Appl Entomol Zool 47:27–34. https://doi.org/10.1007/s13355-011-0085-3

Kathiresan A, Tung P, Chinnappa CC, Reid DM (1997) Gamma-aminobutyric acid stimulates ethylene biosynthesis in sunflower. Plant Physiol 115:129–135

Khan MIR, Jalil SU, Chopra P, Chhillar H, Ferrante A, Khan NA, Ansari MI (2021) Role of GABA in plant growth, development and senescence. Plant Gene 26:100283

Kiep V, Vadassery J, Lattke JP, Maaß JP, Boland W, Peiter E et al (2015) Systemic cytosolic Ca2+ elevation is activated upon wounding and herbivory in Arabidopsis. New Phytol 207:996–1004. https://doi.org/10.1111/nph.13493

Kobayashi M, Ohura I, Kawakita K, Yokota N, Fujiwara M, Shimamoto K et al (2007) Calcium-dependent protein kinases regulate the production of reactive oxygen species by potato NADPH oxidase. Plant Cell 19:1065–1080. https://doi.org/10.1105/tpc.106.048884

Kobayashi M, Yoshioka M, Asai S, Nomura H, Kuchimura K, Mori H et al (2012) StCDPK5 confers resistance to late blight pathogen but increases susceptibility to early blight pathogen in potato via reactive oxygen species burst. New Phytol 196:223–237. https://doi.org/10.1111/j.1469-8137.2012.04226.x

Kurusu T, Hamada J, Hamada H, Hanamata S, Kuchitsu K (2010) Roles of calcineurin B-like protein-interacting protein kinases in innate immunity in rice, Plant Signaling & Behavior 5:8, 1045–1047, https://doi.org/10.4161/psb.5.8.12407

Liu C, Chen L, Zhao R, Li R, Zhang S, Yu W et al (2019) Melatonin induces disease resistance to Botrytis cinerea in tomato fruit by activating jasmonic acid signaling pathway. J Agric Food Chem 67:6116–6124. https://doi.org/10.1021/acs.jafc.9b00058

Ludwig AA, Saitoh H, Felix G, Freymark G, Miersch O, Wasternack C et al (2005) Ethylene-mediated cross-talk between calcium-dependent protein kinase and MAPK signaling controls stress responses in plants. Proc Natl Acad Sci USA 102:10736–10741. https://doi.org/10.1073/pnas.0502954102

Mekonnen DW, Flügge U-I, Ludewig F (2016) Gamma-aminobutyric acid depletion affects stomata closure and drought tolerance of Arabidopsis thaliana. Plant Sci 245:25–34. https://doi.org/10.1016/j.plantsci.2016.01.005

Meyer S, Mumm P, Imes D, Endler A, Weder B, Al-Rasheid KA et al (2010) AtALMT12 represents an R-type anion channel required for stomatal movement in Arabidopsis guard cells. Plant J 63:1054–1062. https://doi.org/10.1111/tpj.2010.63.issue-6

Minibayeva F, Beckett RP, Kranner I (2014) Roles of apoplastic peroxidases in plant response to wounding. Phytochemistry 112:122. https://doi.org/10.1016/j.phytochem.2014.06.008

Mittler R, Vanderauwera S, Suzuki N, Miller G, Tognetti VB, Vandepoele K et al (2011) ROS signaling: the new wave? Trends Plant Sci 16:300–309. https://doi.org/10.1016/j.tplants.2011.03.007

Mostafa S, Wang Y, Zeng W, Jin B (2022) Plant responses to herbivory, wounding, and infection. Int J Mol Sci 23:7031. https://doi.org/10.3390/ijms23137031

O'Donnell PJ, Calvert C, Atzorn R, Wasternack C, Leyser HMO, Bowles DJ (1996) Ethylene as a signal mediating the wound response of tomato plants. Science 274:1914–1917. https://doi.org/10.1126/science.274.5294.1914

Onkokesung N, Baldwin IT, Galis I (2010) The role of jasmonic acid and ethylene crosstalk in direct defence of Nicotiana attenuata plants against chewing herbivores. Plant Signal Behav 5:1305–1307. https://doi.org/10.4161/psb.5.10.13124

Orozco-Cardenas M, Ryan CA (1999) Hydrogen peroxide is generated systemically in plant leaves by wounding and systemin via the octadecanoid pathway. Proc Natl Acad Sci USA 96:6553–6555. https://doi.org/10.1073/pnas.96.11.6553

Orozco-Cardenas ML, Narváez-Vásquez J, Ryan CA (2001) Hydrogen peroxide acts as a second messenger for the induction of defence genes in tomato plants in response to wounding, systemin, and methyl jasmonate. Plant Cell 13:179–191. https://doi.org/10.1105/tpc.13.1.179

Pal P, Ansari SA, Jalil SU, Ansari MI (2023) Regulatory role of phytohormones in plant growth and development. In: Plant hormones in crop improvement. https://doi.org/10.1016/B978-0-323-91886-2.00016-1

Peña-Cortés H, Fisahn J, Willmitzer L (1995) Signals involved in wound induced proteinase inhibitor II gene expression in tomato and potato plants. Proc Natl Acad Sci USA 92:4106–4113. https://doi.org/10.1073/pnas.92.10.4106

Podlesakova K, Ugena L, Spíchal L, Dolezal K, De Diego N (2019) Phytohormones and polyamines regulate plant stress responses by altering GABA pathway. New Biotechnol 48:53–65. https://doi.org/10.1016/j.nbt.2018.07.003

Pospíšilová H, Jiskrová E, Vojta P, Mrízová K, Kokáš F, Cudejková MM et al (2016) Transgenic barley overexpressing a cytokinin dehydrogenase gene shows greater tolerance to drought stress. New Biotechnol 33:692–705. https://doi.org/10.1016/j.nbt.2015.12.005

Ramputh AI, Bown AW (1996) Rapid γ-aminobutyric acid synthesis and the inhibition of the growth and development of oblique-banded leaf-roller larvae. Plant Physiol 111:1349–1352

Savatin DV, Gramegna G, Modesti V, Cervone F (2014) Wounding in the plant tissue: the defence of a dangerous passage. Front Plant Sci 5:470. https://doi.org/10.3389/fpls.2014.00470

Scholz SS, Reichelt M, Mekonnen DW, Ludewig F, Mithöfer A (2015) Insect herbivory-elicited GABA accumulation in plants is a wound-induced, direct, systemic, and jasmonate-independent defence response. Front Plant Sci 6:1128. https://doi.org/10.3389/fpls.2015.01128

Seifikalhor M, Aliniaeifard S, Bernard F, Seif M, Latifi M, Hassani B et al (2020) γ-Aminobutyric acid confers cadmium tolerance in maize plants by concerted regulation of polyamine metabolism and antioxidant defence systems. Sci Rep 10:59592. https://doi.org/10.1038/s41598-020-59592-1

Shelp BJ (1987) The composition of phloem exudate and xylem sap from broccoli (Brassica oleracea var. italica) supplied with NH_4^+, NO_3^- or NH_4NO_3. J Exp Bot 38:1619–1636

Shelp BJ, Bozzo GG, Trobacher CP, Chiu G, Bajwa VS (2012) Strategies and tools for studying the metabolism and function of γ-aminobutyrate in plants. I. Pathway structure. Botany 90:651–668. https://doi.org/10.1139/b2012-030

Shomali A, Aliniaeifard S, Didaran F, Lotfi M, Mohammadian M, Seif M et al (2021) Synergistic effects of melatonin and gamma-aminobutyric acid on protection of photosynthesis system in response to multiple abiotic stressors. Cells 10:1631. https://doi.org/10.3390/cells10071631

Signorelli S, Dans PD, Coitiño EL, Borsani O, Monza J (2015) Connecting proline and γ-aminobutyric acid in stressed plants through non-enzymatic reactions. PLoS One 10:e0115349. https://doi.org/10.1371/journal.pone.0115349

Snedden WA, Fromm H, Furuya T, Richardson A (1995) Calcium/calmodulin regulation of soybean glutamate decarboxylase. Plant Physiol 108:543–549

Spoel S, Dong X (2012) How do plants achieve immunity? Defence without specialized immune cells. Nat Rev Immunol 12:89–100. https://doi.org/10.1038/nri3141

Suzuki N, Mittler R (2012) Reactive oxygen species-dependent wound responses in animals and plants. Free Radic Biol Med 53:2269–2276. https://doi.org/10.1016/j.freeradbiomed.2012.10.538

Vega-Muñoz I, Duran-Flores D, Fernández-Fernández ÁD, Heyman J, Ritter A, Stael S (2020) Breaking bad news: dynamic molecular mechanisms of wound response in plants. Front Plant Sci 11:610445. https://doi.org/10.3389/fpls.2020.610445

Wallace W, Smith D, Brown JW (1984) Rapid accumulation of γ-aminobutyric acid and alanine in soybean leaves in response to an abrupt transfer to lower temperature, darkness, or mechanical manipulation. Plant Physiol 75:170–175

Wang L, Tsuda K, Sato M, Cohen JD, Katagiri F, Glazebrook J et al (2009) Arabidopsis CaM-binding protein CBP60g contributes to MAMP-induced SA accumulation and is involved in disease resistance. PLoS Pathog 5:e1000301. https://doi.org/10.1371/journal.ppat.1000301

Wasternack C (2007) Jasmonates: an update on biosynthesis, signal transduction and action in plant stress response, growth and development. Ann Bot 100:681–697. https://doi.org/10.1093/aob/mcm079

Yang J, Sun C, Zhang Y, Fu D, Zheng X, Yu T (2017) Induced resistance in tomato fruit by γ-aminobutyric acid for the control of Alternaria rot caused by *Alternaria alternata*. Food Chem 221:1014–1020. https://doi.org/10.1016/j.foodchem.2016.11.061

Yu C, Zeng L, Sheng K, Chen F, Zhou T, Zheng X et al (2014) γ-Aminobutyric acid induces resistance against *Penicillium expansum* by priming of defence responses in pear fruit. Food Chem 159:29–37. https://doi.org/10.1016/j.foodchem.2014.03.011

GABA and Physiological Response Against Hypoxia Stress

13

Abstract

Hypoxic stress, often resulting from waterlogging or submergence, severely limits oxygen availability and disrupts cellular respiration, leading to metabolic imbalances and impaired growth. Emerging evidence highlights gamma-aminobutyric acid (GABA)'s multifaceted role as a signaling molecule and metabolic intermediate that helps plants adapt to oxygen-deficient conditions. This chapter explores the biosynthesis and catabolism of GABA under hypoxic stress, emphasizing its involvement in energy homeostasis, redox balance, cytosolic pH regulation, and reactive oxygen species (ROS) scavenging. The interaction of GABA with hormonal signaling pathways, calcium signaling, and its contribution to stress memory and priming also elicits interest in hypoxia stress management. The GABA shunt pathway and its regulatory significance in enhancing plant resilience deserve special attention. Understanding the intricate GABA-mediated physiological and molecular mechanisms opens new avenues for developing hypoxia-tolerant crop varieties through genetic and metabolic engineering strategies.

Keywords

Cytosolic pH · Hypoxia · Homeostasis · GABA · ROS

Highlights

- Gamma-aminobutyric acid (GABA) is synthesized via the GABA shunt pathway and is crucial for stress physiology during hypoxia (low oxygen availability).
- Hypoxia leads to disruptions in cellular respiration, causing energy deficits and accumulation of harmful byproducts.

S. U. Jalil et al., *GABA Signaling System and Abiotic Stress Tolerance in Plants*,
https://doi.org/10.1007/978-981-95-0437-4_13

- GABA accumulation stabilizes cytosolic pH, modulates carbon and nitrogen metabolism, and maintains cellular redox balance under hypoxia.
- GABA acts as a signaling molecule, triggering adaptive responses to enhance plant survival in oxygen-deprived environments.

13.1 Introduction

Imagine a plant submerged in water during a flood, its roots gasping for oxygen that is no longer easily accessible. In such conditions, hypoxia, or low oxygen availability, becomes a serious threat to plant survival. Hypoxia interferes with the very foundation of a plant's physiology, disrupting cellular respiration and energy production (Puyang et al. 2015). Waterlogged soil generates a harsh environment where oxygen diffusion is drastically limited, resulting in an energy deficit that can diminish crop production by up to 80% in susceptible species (Setter and Waters 2003; Shabala 2011; Wu et al. 2021). The root zone becomes an arena where vital nutrients struggle to reach developing shoots, further compromising plant growth (Colmer and Greenway 2011; Wu et al. 2021). As the plant's mitochondria struggle to function under these oxygen-deprived conditions, a cascade of stress responses unfolds. Mitochondrial respiration slows, redox chains saturate, and reactive oxygen species (ROS) burst forth (Voesenek and Bailey-Serres 2015; Yamauchi et al. 2017). This oxidative assault disrupts cellular ion homeostasis, throwing metabolism into disarray. But plants are not helpless in the face of such adversity. One of their key allies in this struggle is gamma-aminobutyric acid (GABA), which surges within plant tissues when oxygen levels drop (Ham et al. 2012). GABA is not just a bystander in the plant's response to hypoxia—it is a vital player.

Under hypoxic conditions, GABA is synthesized at a rapid pace via the GABA shunt pathway, an alternative route that bypasses segments of the conventional Krebs cycle (Wu et al. 2021). This pathway is crucial, providing a metabolic lifeline that keeps energy production afloat when oxygen is scarce. But GABA does much more than maintain energy flow. It stabilizes cellular pH, balances ions, and regulates stress-responsive genes, orchestrating a sophisticated defense system that boosts the plant's resilience to hypoxia (Wu et al. 2021). GABA is also a powerful signaling molecule, triggering protective mechanisms that enhance survival during stress (Kinnersley and Turano 2000; Kreuzwieser et al. 2009). Recent research has revealed that GABA's influence extends even further, shaping critical physiological processes such as stomatal closure, root growth, and antioxidant defense activation (Jalil et al. 2017; Salah et al. 2019; Khan et al. 2021; Ansari et al. 2021). These responses are crucial for combating the oxidative stress that accompanies hypoxia. Moreover, GABA forms complex interactions with other signaling molecules and pathways (Jalil et al. 2019a; Jalil and Ansari 2020), creating a network of responses that fine-tune the plant's adaptation to low-oxygen environments (Wang et al. 2018).

A critical discussion deserves to address the intricate relationship between GABA metabolism and the plant's response to hypoxia. We will unravel how GABA helps plants navigate the challenges of oxygen deprivation and explore the potential biotechnological strategies to harness this pathway for crop improvement. By delving into recent scientific discoveries, this chapter seeks to illuminate the multifaceted role of GABA in enhancing plant resilience to hypoxic stress, offering insights that could pave the way for developing flood-tolerant crops.

13.2 Hypoxia Stress in Plants: A Silent Struggle for Oxygen

Hypoxia in plants refers to a condition where oxygen availability is insufficient to meet the metabolic needs of plant tissues, disrupting vital physiological processes. This shortage of oxygen is primarily caused by environmental factors that restrict oxygen diffusion, particularly in waterlogged soils or flooded environments. When soil becomes saturated with water, the air spaces in the soil are filled with water, limiting the movement of oxygen to the roots. As a result, the roots become oxygen-starved, unable to perform essential functions like nutrient uptake and respiration (Loreti and Perata 2020). Other factors, such as compacted soil, poor drainage, or excessive irrigation, can also contribute to hypoxic conditions. Plants growing in wetland areas may experience long-term hypoxia, whereas terrestrial plants typically face this stress during transient flooding or waterlogging events (van den Dries et al. 2013). Prolonged hypoxia impairs growth, energy production, and nutrient absorption, making it a significant challenge in both natural ecosystems and agricultural settings.

13.2.1 Physiological and Biochemical Effects of Hypoxia on Plant Growth and Development

When oxygen is scarce, a cascade of physiological and biochemical changes begins to unfold. At the cellular level, hypoxia disrupts mitochondrial respiration—the process plants rely on to generate energy (Fig. 13.1). Without enough oxygen, the efficiency of adenosine triphosphate (ATP) production declines, creating an energy crisis within the plant (Greenway and Armstrong 2018). The metabolic machinery begins to falter, forcing the plant to rely on anaerobic pathways like fermentation to produce energy, an inefficient alternative that generates toxic byproducts, such as ethanol, further impairing cell function (Voesenek and Bailey-Serres 2015). In response, plants activate survival mechanisms. One of the most immediate effects of hypoxia is the buildup of ROS, leading to oxidative stress that damages cellular structures. ROS accumulation can disturb ion homeostasis and lead to membrane dysfunction (Yamauchi et al. 2017). To mitigate this, plants initiate the production of antioxidants and activate protective pathways that attempt to balance the redox state. Furthermore, hypoxia slows down nutrient transport, especially nitrogen and phosphorus, which are essential for growth, leading to stunted development (Colmer and Greenway 2011). For instance, waterlogged soils can impede root function, decreasing the plant's ability to absorb nutrients, and this imbalance reverberates throughout the plant, from roots to shoots (Fig. 13.1).

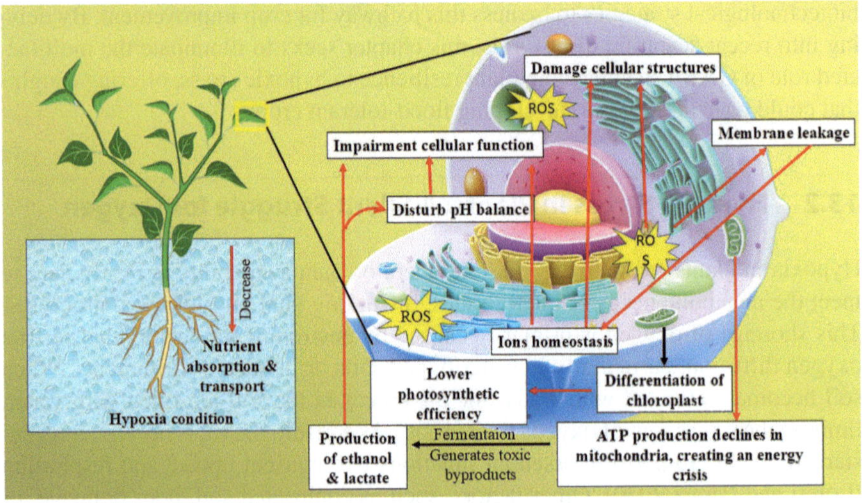

Fig. 13.1 The impact of hypoxia stress on plant physiology. The reduced oxygen availability in waterlogged or flooded soils induces the physiological effects of hypoxia stress in plants. Oxygen deficiency impairs mitochondrial respiration, leading to decreased ATP production and forcing plants to switch to anaerobic pathways, such as fermentation, which generate harmful byproducts like ethanol. The transition to anaerobic metabolism also causes intracellular pH to drop due to the accumulation of acidic metabolites like lactate. Additionally, hypoxia induces the accumulation of ROS, causing oxidative stress, damaging cellular structures, disrupting ion homeostasis, and impairing membrane integrity. These factors collectively compromise nutrient uptake, energy production, and overall plant growth, leading to significant physiological stress

Growth and development are particularly vulnerable in hypoxia. In crops such as rice and tomatoes, hypoxia triggers a slowdown in growth as a metabolic conservation strategy (Fukushima et al. 2020). However, some species, like rice, have evolved specific adaptations such as aerenchyma formation—air-filled cavities in roots that allow oxygen to flow through tissues (Mignolli et al. 2020). This structural change acts as an escape mechanism, helping the plant to survive in submerged environments. Yet many terrestrial plants, such as maize or tomato, lack such mechanisms and suffer more acutely under long-term hypoxia, showing symptoms such as chlorosis, reduced leaf size, and poor root health (Shabala et al. 2014).

Hypoxia not only affects individual plants but also holds profound implications for agriculture. With climate change and the increasing frequency of flooding events, understanding the physiological and biochemical impacts of hypoxia on plant systems is more critical than ever. The following sections explore the intricate adaptations and molecular pathways that plants activate to combat oxygen deprivation, shedding light on potential strategies to improve crop resilience.

13.2.2 Plant Adaptive Mechanisms to Hypoxia Stress

When oxygen levels plummet, plants are thrust into a survival mode, sparking a cascade of biochemical and physiological responses. This hypoxic condition, often induced by waterlogged soils or flooding, triggers a unique set of adaptive mechanisms that allow plants to balance energy demands and mitigate cellular damage. Like a well-rehearsed choreography, these responses unfold in a systematic and sophisticated manner, varying across species based on their natural environments and tolerance to oxygen deprivation.

The first line of defense during hypoxia is the rapid shift in metabolism. With mitochondria starved of oxygen, plants quickly reduce energy-consuming processes like protein synthesis and cell wall formation, reallocating resources toward glycolysis and fermentation, where ATP can be generated without oxygen. However, this switch comes at a cost: anaerobic metabolism is far less efficient, and the buildup of byproducts like ethanol and lactate can become toxic. Key enzymes such as pyruvate decarboxylase and alcohol dehydrogenase play central roles in these alternative pathways, helping sustain basic cellular activities (Ismond et al. 2003). In species like rice, which thrives in submerged environments, these metabolic adjustments are fine-tuned, allowing the plant to conserve carbohydrates and endure long periods of flooding (Bailey-Serres and Voesenek 2008).

One of the most remarkable structural adaptations is the formation of aerenchymas that allow oxygen to diffuse more effectively from aboveground tissues to roots. Plants like rice and wetland species rely heavily on this adaptation, creating internal air channels that act like natural snorkels, enabling the delivery of oxygen to parts of the plant that would otherwise suffocate (Fukao and Bailey-Serres 2004). This structural change is often accompanied by the development of adventitious roots, which form near the soil surface and capitalize on any available oxygen to improve root function during prolonged submergence.

At the molecular level, a finely tuned response is activated as plants sense oxygen depletion. The key regulatory system that governs this response is the hypoxia-induced signaling pathway, which includes a group of transcription factors known as ethylene response factors. The response factors are responsible for turning on genes that encode proteins involved in anaerobic metabolism, stress protection, and cell survival (Licausi et al. 2011). Ethylene, a gaseous plant hormone, plays a vital role in the process, triggering rapid responses to low oxygen and preparing the plant for the harsh conditions ahead.

Another fascinating element of the plant's hypoxia response is the role of ROS and reactive nitrogen species. Under hypoxia, a paradox occurs: despite the lack of oxygen, plants experience oxidative stress as small amounts of oxygen are used inefficiently, leading to the formation of ROS. These reactive molecules can cause cellular damage, but they also serve as important signaling agents, activating antioxidant defenses and stress-responsive genes (Chang et al. 2011). Plants have developed a suite of antioxidant enzymes, such as superoxide dismutase (SOD) and catalase (CAT), to neutralize ROS and prevent cellular damage.

Plants also rely on hormonal crosstalk to manage the energy crisis induced by hypoxia. Abscisic acid (ABA) levels rise during flooding, reducing stomatal conductance to conserve water and carbohydrates. Simultaneously, the accumulation of GABA plays a protective role by stabilizing ion homeostasis, contributing to a fine-tuned metabolic balance that helps plants navigate the precarious path between survival and energy depletion (Wu et al. 2021). This multifaceted hormonal and metabolic regulation allows plants to efficiently balance growth and stress responses under oxygen-limited conditions.

The diversity of adaptive strategies is a testament to the complexity of plant survival under hypoxia. While some plants rely heavily on morphological changes, others depend on metabolic flexibility or intricate hormonal signaling networks. These adaptive mechanisms not only enable survival during hypoxia but also position plants to recover swiftly when oxygen levels are restored, ensuring resilience in a constantly changing environment. This ability to adapt and evolve has allowed plants to conquer diverse habitats, from wetlands to deserts, and continue their growth under some of the most extreme environmental conditions.

13.3 GABA in Hypoxia-Induced Physiological Responses

13.3.1 GABA and pH Regulation and Cellular Homeostasis

Under hypoxic conditions, plants encounter a significant challenge—a drop in intracellular pH caused by the shift from aerobic to anaerobic metabolism. As oxygen availability diminishes, the accumulation of lactate and other acidic metabolites disrupts the pH balance, threatening the stability of vital cellular functions. This shift could severely impair enzyme activities, metabolic pathways, and overall cellular homeostasis if left unchecked. Nevertheless, plants have developed advanced strategies to mitigate such disruptions, with GABA emerging as a central component in these defense mechanisms.

GABA's role in buffering pH changes during hypoxia is particularly crucial. When oxygen levels drop, GABA accumulates rapidly, helping to neutralize the acidic environment and stabilize intracellular pH (Fig. 13.2). This buffering action allows enzymatic reactions to proceed despite the anaerobic conditions, ensuring that metabolic processes remain functional. For instance, Michaeli and Fromm (2015) and Shelp and Zarei (2017) have observed that GABA's surge during hypoxia serves as a critical factor in maintaining cellular equilibrium, acting as a biochemical safeguard against the damaging effects of pH imbalances.

The GABA shunt, an alternative metabolic pathway, also plays a significant role in this process. The GABA shunt produces succinate, which can be further metabolized even under oxygen-limited conditions (Fig. 13.2). This not only helps sustain energy production but also supports the maintenance of cellular homeostasis. Breitkreuz et al. (2003) have highlighted that, during hypoxia, the GABA shunt becomes an essential component of the plant's adaptive response, providing a continuous flow of metabolites that stabilize both pH and energy levels.

Through these actions, GABA not only prevents the harmful accumulation of acidic compounds but also ensures the continuity of metabolic functions, playing a central role in the plant's ability to withstand and recover from hypoxic stress. Its dual function as a pH buffer and metabolic regulator underscores its importance in maintaining cellular homeostasis during such environmental challenges.

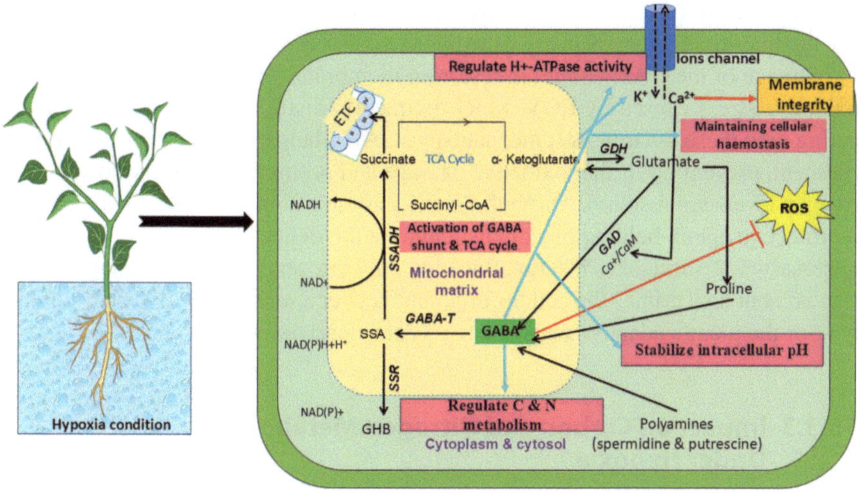

Fig. 13.2 Role of GABA in hypoxia stress tolerance. Under oxygen-deficient conditions, GABA accumulates rapidly, neutralizing the acidic environment and stabilizing intracellular pH. The GABA shunt provides a crucial alternative metabolic pathway, bypassing two steps of the tricarboxylic acid (TCA) cycle to produce succinate, which sustains energy production even under limited oxygen availability. GABA also regulates ion homeostasis, preventing excessive K^+ efflux through guard cell outward-rectifying K^+ (GORK) channels and modulating Ca^{2+} signaling. Its interaction with the ROS-Ca^{2+} hub helps prevent reactive oxygen species (ROS) accumulation, reducing oxidative damage. Additionally, GABA influences H^+-ATPase activity to restore membrane potential, preserving cellular integrity. By controlling ion transport and maintaining energy production, GABA plays a critical role in mitigating hypoxia-induced stress and supporting plant survival. Exogenous GABA also replenishes TCA cycle intermediates and maintains metabolic balance under hypoxic conditions

13.3.2 GABA-Mediated Modulation of Ion Transport and Balance

Under hypoxic conditions, ion transport plays a pivotal role in determining whether a plant cell can survive or succumb to stress-induced damage. The reduced oxygen availability disrupts normal cellular processes, including the regulation of essential ions like Ca^{2+} and K^+. These imbalances can severely impact membrane integrity, leading to cellular damage or even death. However, GABA, acting as a modulator of ion transport, preserves ion balance and mitigates hypoxia-induced stress (Fig. 13.2).

GABA's interaction with ion channels, particularly those involved in regulating Ca^{2+} and K^+ transport, is a key factor in maintaining cellular ion homeostasis during hypoxia. Shabala et al. (2014) have observed that GABA-pretreated barley roots obtain tolerance to hypoxic stress. This underscores GABA's ability to reduce ROS-induced disruptions in Ca^{2+} signaling and to mitigate K^+ leakage, which is commonly observed under stress conditions. The controlled extrusion of Ca^{2+} in GABA-treated roots prevents cytotoxic imbalances, while the reduction in K^+ leakage helps sustain membrane potential.

This balance of Ca^{2+} and K^+ is crucial not only for immediate survival but also for facilitating cellular recovery post-hypoxia. Wu et al. (2021) also highlight the importance of the GABA shunt in restoring membrane potential and alleviating the oxidative stress caused by ROS, which frequently disrupts ion channels. By interacting with these ion transport mechanisms, GABA helps stabilize the cellular environment, preventing further oxidative damage and promoting recovery once normal oxygen levels are restored. Through its modulation of ion transport, GABA acts as a protective agent, helping plant cells maintain the delicate balance of Ca^{2+} and K^+ necessary for survival. This function not only prevents cellular damage during hypoxia but also primes the cell for efficient recovery, making GABA a key component of the plant's adaptive response to oxygen deprivation.

13.3.3 Impact on Carbon and Nitrogen Metabolism Under Hypoxia

During hypoxia, the disruption of both carbon and nitrogen metabolism poses a significant challenge to plant survival. The decreased availability of oxygen impairs the TCA cycle. However, GABA plays a crucial compensatory role in mitigating these disruptions, ensuring that metabolic processes can adapt to the oxygen-limited conditions.

GABA's integration into the GABA shunt offers a metabolic alternative when the TCA cycle is compromised. As hypoxia restricts the full oxidation of carbon in the TCA cycle, the GABA shunt steps in by bypassing two stages of the cycle, providing an alternative route for carbon flow. This allows for the continued production of succinate, which can enter the TCA cycle downstream and contribute to the generation of NADH and energy, even under reduced oxygen levels (Salvatierra et al. 2016). Applying GABA externally helps melons absorb nitrate more effectively, which in turn helps them cope with hypoxia stress (Song et al. 2012). Thus, the GABA shunt helps maintain energy production and supports overall carbon metabolism during hypoxia.

In addition to its role in carbon metabolism, GABA is intricately linked to nitrogen metabolism. Its precursor, glutamate, is a key player in nitrogen assimilation, and the accumulation of GABA during hypoxia reflects its role in balancing nitrogen metabolism when normal processes are disrupted. The GABA shunt aids in stabilizing nitrogen flow, ensuring that nitrogenous compounds are effectively cycled and utilized during stress. This dual function of GABA, mediating both carbon and nitrogen metabolism, provides plants with a robust mechanism to adapt their metabolic processes to the adverse conditions imposed by hypoxia.

Furthermore, studies by Luu et al. (2019) highlight how the GABA shunt's bypassing of the TCA cycle helps maintain the balance between carbon and nitrogen under these conditions. By ensuring that both carbon and nitrogen metabolism continue to function, albeit through alternative pathways, GABA enables plants to sustain vital metabolic processes, promoting survival and recovery from oxygen deprivation.

13.4 GABA as a Signaling Molecule in Hypoxia Response

13.4.1 GABA in Signal Transduction Pathways

Beyond its well-documented metabolic functions, GABA emerges as a pivotal signaling molecule during hypoxic stress, playing a key role in regulating signal transduction pathways. One of the primary ways GABA exerts this influence is through its interaction with Ca^{2+} signaling, a critical pathway that orchestrates plant responses to various environmental stresses, including hypoxia. Hypoxic conditions trigger a rapid influx of calcium ions, which serve as second messengers, activating downstream stress response mechanisms aimed at mitigating damage and restoring cellular function.

GABA modulates these calcium fluxes, fine-tuning the cellular signaling to ensure that the plant mounts an appropriate stress response. Miyashita and Good (2008) have demonstrated GABA's involvement in regulating calcium signaling during hypoxia, where it helps coordinate the plant's physiological adjustments to oxygen deficiency. By ensuring that calcium fluxes remain controlled, GABA aids in preventing the overactivation of stress pathways, which could otherwise exacerbate cellular damage.

The GABA shunt has been implicated in this regulatory process. As an alternative metabolic route during hypoxia, the GABA shunt not only helps sustain energy production but also plays a role in the control of stress signaling pathways. By modulating calcium signaling, the GABA shunt contributes to the restoration of cellular homeostasis and the alleviation of oxidative damage during and after hypoxia. This dual role of GABA—balancing metabolic demands while also serving as a key regulator of signal transduction—underscores its importance in the plant's overall response to hypoxic stress. Through its interactions with calcium signaling and other pathways, GABA helps plants navigate the complex molecular landscape of hypoxia, ensuring both survival and recovery.

13.4.2 Interaction Between GABA and Other Signaling Molecules

The intricate relationship between GABA and other signaling molecules, particularly ROS and Ca^{2+}, becomes especially significant under hypoxic conditions. As oxygen levels drop, the production of ROS, including superoxide and hydrogen peroxide, escalates. Although ROS are notorious for causing oxidative damage, they also play a dual role as signaling molecules that help plants adapt to stress. In this balancing act, GABA emerges as a key player. By modulating ROS accumulation, GABA not only reduces oxidative damage but also fine-tunes ROS-related signaling pathways, allowing plants to mount a controlled response to hypoxia.

In hypoxic environments, the interaction between GABA and calcium signaling becomes even more crucial. Calcium ions serve as ubiquitous secondary messengers that regulate numerous physiological processes, including stress responses. GABA's influence on Ca^{2+} dynamics, particularly in the context of ROS, helps stabilize cytosolic homeostasis. Wu et al. (2021) have conducted a study on several *Arabidopsis* mutants impaired in the GABA shunt pathway (*pop2*, *gad1*, and *gad2*), subjecting them to hypoxic conditions. Their findings reveal that an elevated GABA content and enhanced GABA shunt activity are essential for restoring membrane potential, which ROS have disrupted. Moreover, GABA helps alleviate ROS-induced disturbances in cytosolic potassium (K^+) homeostasis and plays a role in stabilizing Ca^{2+} signaling, highlighting its protective function in mitigating both oxidative and ionic imbalances.

In this intricate crosstalk between GABA, ROS, and Ca^{2+}, GABA serves as both a protector and a regulator. It ensures that ROS levels remain within a manageable range, preventing excessive oxidative damage while preserving their role as stress signals. At the same time, GABA's interaction with Ca^{2+} ensures that ion homeostasis is maintained, a critical factor for cellular function and survival under hypoxic stress. This dynamic interplay underscores GABA's multifaceted role in the plant's defense strategy, enabling it to navigate the challenges of oxygen deprivation with a finely tuned response.

13.4.3 Influence of GABA on Gene Expression Related to Hypoxia Tolerance

GABA's role during hypoxia extends beyond its traditional metabolic functions and delves into gene regulation. When plants experience oxygen deprivation, a rapid increase in GABA levels acts as a molecular signal, triggering the expression of specific genes linked to stress tolerance, ion transport, and metabolic adjustments. One of the most critical genes involved in this process is glutamate decarboxylase (GAD), an enzyme responsible for GABA synthesis. During hypoxia, the expression of GAD genes is modulated in a way that allows plants to adapt and survive in low-oxygen conditions. Additionally, malate dehydrogenase (MDH) activity declines, disrupting the TCA cycle and energy production during hypoxia stress. Exogenous GABA application further reduces MDH levels, acting as an anaplerotic molecule to replenish Krebs cycle intermediates. This helps maintain metabolic balance and supports plant survival under oxygen-deficient conditions (Fan et al. 2015).

Miyashita and Good (2008) provide an in-depth look at how different GAD genes respond to hypoxia in *Arabidopsis*. They emphasize that under prolonged oxygen deprivation, GAD2 expression decreases, while GAD1 remains constant, and GAD4 is upregulated as much as 20 times the normal level, specifically in the roots. This differential expression suggests a finely tuned genetic response to hypoxia, where certain GAD isoforms become more important than others, depending on the duration and intensity of stress. The GABA shunt, bypassing the traditional TCA cycle, helps regulate these responses by limiting the accumulation of alanine, a hypoxia-induced metabolite, which would otherwise disrupt metabolic balance. In another study by Salvatierra et al. (2016), the application of exogenous

GABA to *Prunus* rootstocks demonstrates a fascinating genetic interplay. While the treatment improves root hypoxia tolerance in a sensitive ecotype, it has little effect on a naturally tolerant one. The researchers claim that one of the early genes responding to GABA application is highly homologous to the *Arabidopsis* GAD1, emphasizing GABA's influence on specific genetic pathways under stress. Further insights reveal that GABA application to muskmelon roots under hypoxia not only alleviates stress but also engages mitochondrial MDH, an enzyme critical for maintaining TCA cycle intermediates (Luu et al. 2019). This finding highlights how GABA not only influences gene expression but also connects it to broader metabolic processes, ensuring energy flow under oxygen-deficient conditions.

GABA-accumulating mutants, such as the *pop2–5* line, reveal a particularly striking observation of reduced H_2O_2 levels and improved hypoxia tolerance compared to GABA-deficient mutants like *gad1/2*, which are more sensitive to hypoxic conditions (Wu et al. 2021). This difference in stress tolerance underscores GABA's genetic and physiological significance during hypoxia. The story of GABA's influence on gene expression does not end here. Under hypoxic conditions, another metabolite, e.g., gamma-hydroxybutyrate (GHB), accumulates in the leaves of several plant species, including *Arabidopsis*, tobacco, green tea, and soybean (Breitkreuz et al. 2003; Allan et al. 2003, 2008, 2012). The observation by Breitkreuz et al. (2003) highlights an increase in GABA, alanine, and GHB content but a decrease in the expression of GABA-T and GHBDH just after 4 hours of dark flooding in *Arabidopsis* leaves. Since GABA-T activity declines during hypoxia, the normal conversion of succinate semialdehyde (SSA) to succinate is interrupted, leading to the conversion of SSA into GHB instead. This redirection in metabolic flow suggests that GABA and its byproducts play a broader role in maintaining redox balance and influencing gene activity under stress.

GABA's role in hypoxia-induced gene expression reveals a complex network of molecular signaling and metabolic adjustments, where this small molecule acts as both a buffer and a trigger for genetic responses, ensuring plants can endure and recover from oxygen deprivation. Through these intricate genetic and metabolic pathways, GABA stands as a vital component of the plant's survival strategy under hypoxia.

13.5 Mechanistic Overview of the GABA Shunt Pathway During Hypoxia

Plants face the relentless challenge of hypoxia that disrupts metabolic balance. In this scenario, the GABA shunt pathway emerges as a vital player in stress adaptation. This metabolic pathway, often overshadowed by more prominent stress responses, reveals its crucial role in helping plants navigate the complexities of low-oxygen environments. Under normal conditions, the GABA shunt pathway functions as a secondary route in the TCA cycle (Ansari et al. 2014; Jalil et al. 2019a) and facilitates the conversion of glutamate to GABA via glutamate GAD, followed by the conversion of GABA to SSA and then to succinate. The GABA shunt effectively bypasses two key steps in the TCA cycle (Jalil et al. 2016; Jalil et al. 2019b), allowing continuous energy production when oxygen levels are insufficient.

Hypoxia triggers a profound and distinct alteration in GABA metabolism. Unlike other stressors, which elevate GABA levels to a certain extent, hypoxia induces a dramatic surge in GABA concentration. For instance, *Arabidopsis* subjected to salt stress for 4 days shows a modest two-fold increase in GABA (Zarei et al. 2016). In stark contrast, hypoxic conditions can propel GABA levels up to six-fold (Breitkreuz et al. 2003) to 12-fold (Allan et al. 2012), underscoring the exceptional magnitude of GABA accumulation under oxygen deprivation.

This heightened response highlights the GABA shunt pathway's critical role in adapting to hypoxia. Yet, this is not without its complications. In the context of submerged and recovered *pop2–8 Arabidopsis* leaves, the GABA concentration often exceeds the optimal functional range. This excess might explain why *pop2–8* demonstrates better resilience to salt stress compared to submergence. Similarly, the *pop2–1* variant, originating from the Landsberg erecta (Ler) ecotype, exhibits heightened sensitivity to both salinity and GABA treatment relative to the Columbia ecotype and *pop2–8* (Renault et al. 2010, 2011; Jalil et al. 2017). This heightened sensitivity is linked to GABA-induced cell elongation defects and repression of genes related to cell wall proteins (Renault et al. 2011).

The hypoxia-driven accumulation of alanine, which is partially derived from the GABA shunt pathway, further complicates the picture. Studies reveal that under hypoxic conditions, plants, such as *Arabidopsis*, not only accumulate GABA but also convert it to GHB instead of succinate (Breitkreuz et al. 2003; Allan et al. 2008). This conversion reflects the metabolic shift during hypoxia, where succinate's production is hampered. *Arabidopsis* mutants deficient in SSADH, an enzyme crucial for GHB degradation, exhibit increased levels of GHB and ROS, alongside stunted growth (Bouché et al. 2003; Ludewig et al. 2008). Interestingly, the double mutant (*ssadh pop2*) shows less toxic SSA accumulation, illustrating a potential compensatory mechanism.

Moreover, various factors influence GABA accumulation. For instance, in fava beans and tea leaves, the hypoxia- and anoxia-induced GABA levels are significantly reduced by aminoguanidine, an inhibitor of diamine oxidase that blocks putrescine production, a precursor for GABA biosynthesis (Yang et al. 2013; Liao et al. 2017). This reduction underscores the complex interplay between GABA biosynthesis and stress responses, revealing how the GABA shunt pathway adapts uniquely to hypoxic conditions, balancing metabolic needs with stress tolerance.

13.6 Conclusion

GABA plays a pivotal role in plant adaptation to hypoxic stress, acting through multiple mechanisms to enhance survival and recovery. Beyond its metabolic functions, GABA modulates key signaling pathways involved in stress responses. It interacts with ROS and calcium signaling to balance oxidative stress and ion transport, thereby maintaining cellular homeostasis. GABA also influences the expression of stress-responsive genes, orchestrating the plant's response to hypoxia. However, while GABA's accumulation is beneficial, excessive levels can be detrimental, underscoring the need for precise regulation. The GABA shunt pathway, which bypasses key steps in the TCA cycle, ensures continued energy production and stabilizes cellular pH, crucial for maintaining enzymatic activity under low oxygen conditions. This pathway supports metabolic adjustments and helps prevent cellular damage. Understanding GABA's multifaceted roles offers valuable insights into enhancing plant resilience to hypoxia. By harnessing this knowledge, we can develop crops with improved tolerance to low oxygen conditions, advancing agricultural productivity and sustainability in challenging environments.

References

Allan W, Peiris C, Bown A, Shelp B (2003) Gamma-hydroxybutyrate accumulates in green tea and soybean sprouts in response to oxygen deficiency. Can J Plant Sci 83:951–953

Allan WL, Breitkreuz KE, Waller JC, Simpson JP, Hoover GJ, Rochon AD et al (2012) Detoxification of succinate semialdehyde in Arabidopsis glyoxylate reductase and NAD kinase mutants subjected to submergence stress. Botany 90:51–61

Allan WL, Simpson JP, Clark SM, Shelp BJ (2008) γ-hydroxybutyrate accumulation in Arabidopsis and tobacco plants is a general response to abiotic stress: putative regulation by redox balance and glyoxylate reductase isoforms. J Exp Bot 59:2555–2564

Ansari MI, Hasan S, Jalil SU (2014) Leaf senescence and GABA shunt. Bioinformation 10:730–732

Ansari MI, Jalil SU, Ansari SA et al (2021) GABA shunt: a key-player in mitigation of ROS during stress. Plant Growth Regul 94:131–149

Bailey-Serres J, Voesenek L (2008) Flooding stress: acclimations and genetic diversity. Annu Rev Plant Biol 59:313–339

Bouché N, Fait A, Bouchez D, Møller SG, Fromm H (2003) Mitochondrial succinic semialdehyde dehydrogenase of the γ-aminobutyrate shunt is required to restrict levels of reactive oxygen intermediates in plants. Proc Natl Acad Sci USA 100:6843–6848

Breitkreuz KE, Allan WL, Van Cauwenberghe OR, Jakobs C, Talibi D, André B, Shelp BJ (2003) A novel γ-hydroxybutyrate dehydrogenase: identification and expression of an Arabidopsis cDNA and potential role under oxygen deficiency. J Biol Chem 278:41552–41556

Chang R, Jang CJH, Branco-Price C, Nghiem P, Bailey-Serres J (2011) Transient MPK6 activation in response to oxygen deprivation and reoxygenation is mediated by mitochondria and aids seedling survival in Arabidopsis. Plant Mol Biol 78:109–122

Colmer TD, Greenway H (2011) Ion transport in seminal and adventitious roots of cereals during O2 deficiency. J Exp Bot 62:39–57

Fan L, Wu X, Tian Z, Jia K, Pan Y, Li J, Gao H (2015) Comparative proteomic analysis of gamma-aminobutyric acid responses in hypoxia-treated and untreated melon roots. Phytochemistry 116:28–37

Fukao T, Bailey-Serres J (2004) Plant responses to hypoxia—is survival a balancing act? Trends Plant Sci 9:449–456

Fukushima A, Kuroha T, Nagai K et al (2020) Metabolite and phytohormone profiling illustrates metabolic reprogramming as an escape strategy of deepwater rice during partially submerged stress. Meta 10:68

Greenway H, Armstrong W (2018) Energy-crises in well-aerated and anoxic tissue: does tolerance require the same specific proteins and energy-efficient transport? Funct Plant Biol 45:877–894

Ham TH, Chu SH, Han SJ, Ryu SN (2012) γ-aminobutyric acid metabolism in plants under environmental stresses. Korean J Crop Sci 57:144–150

Ismond KP, Dolferus R, Pauw MD, Dennis ES, Good AG (2003) Enhanced low oxygen survival in Arabidopsis through increased metabolic flux in the fermentative pathway. Plant Physiol 132:1292–1302

Jalil SU, Ahmad I, Ansari MI (2016) Phenotypic characterization of GABA-transaminase mutants of Arabidopsis thaliana. Adv Life Sci 5:10005–10008

Jalil SU, Ahmad I, Ansari MI (2017) Functional loss of GABA transaminase (GABA-T) expressed early leaf senescence under various stress conditions in Arabidopsis thaliana. Curr Plant Biol 9–10:11–22

Jalil SU, Ansari MI (2020) Physiological role of gamma aminobutyric acid in salt stress tolerance. In: Hasanuzzaman M, Tanveer M (eds) Salt and drought stress tolerance in plants, signaling and communication in plants. Springer, pp 337–350

Jalil SU, Khan MAA, Ansari MI (2019a) Tremendous role of GABA shunt metabolic pathway in plant system. J Biol Chem Res 36:188–196

Jalil SU, Khan MQR, Ansari MI (2019b) Role of GABA transaminase in the regulation of development and senescence in Arabidopsis thaliana. Curr Plant Biol 19:100119

Khan MIR, Jalil SU, Chopra P, Chhillar H, Ferrante A, Khan NA, Ansari MI (2021) Role of GABA in plant growth, development and senescence. Plant Gene 26:100283

Kinnersley AM, Turano FJ (2000) Gamma aminobutyric acid (GABA) and plant responses to stress. Crit Rev Plant Sci 19:479–509

Kreuzwieser J, Hauberg J, Howell KA, Carroll A, Rennenberg H, Millar AH, Whelan J (2009) Differential response of grey poplar leaves and roots underpins stress adaptation during hypoxia. Plant Physiol 149:461–473

Liao J, Wu X, Xing Z, Li Q, Duan Y, Fang W, Zhu X (2017) γ-Aminobutyric acid (GABA) accumulation in tea (Camellia sinensis L.) through the GABA shunt and polyamine degradation pathways under anoxia. J Agric Food Chem 65:3013–3018

Licausi F, Kosmacz M, Weits DA, Giuntoli B, Giorgi FM, Voesenek LACJ, Perata P, van Dongen JT (2011) Oxygen sensing in plants is mediated by an N-end rule pathway for protein destabilization. Nature 479:419–422

Loreti E, Perata P (2020) The many facets of hypoxia in plants. Plants 9:745

Ludewig F, Hüser A, Fromm H, Beauclair L, Bouché N (2008) Mutants of GABA transaminase (POP2) suppress the severe phenotype of succinic semialdehyde dehydrogenase (ssadh) mutants in Arabidopsis. PLoS One 3:e3383

Luu K, Rajagopalan N, Ching JC, Loewen MC, Loewen ME (2019) The malate activated ALMT12 anion channel in the grass Brachypodium distachyon is co-activated by Ca2+/calmodulin. J Biol Chem 294:6142–6156

Michaeli S, Fromm H (2015) Closing the loop on the GABA shunt in plants: are GABA metabolism and signaling entwined? Front Plant Sci 6:419

Mignolli F, Todaro JS, Vidoz ML (2020) Internal aeration and respiration of submerged tomato hypocotyls are enhanced by ethylene-mediated aerenchyma formation and hypertrophy. Physiol Plant 169:49–63

Miyashita Y, Good AG (2008) Contribution of the GABA shunt to hypoxia induced alanine accumulation in roots of Arabidopsis thaliana. Plant Cell Physiol 49:92–102

Puyang X, An M, Xu L, Han L, Zhang X (2015) Antioxidant responses to waterlogging stress and subsequent recovery in two Kentucky bluegrass (Poa pratensis L.) cultivars. Acta Physiol Plant 37:1–12

Renault H, El Amrani A, Palanivelu R, Updegraff EP, Yu A, Renou JP, Preuss D, Bouchereau A, Deleu C (2011) GABA accumulation causes cell elongation defects and a decrease in expression of genes encoding secreted and cell wall-related proteins in Arabidopsis thaliana. Plant Cell Physiol 52:894–908

Renault H, Roussel V, El Amrani A, Arzel M, Renault D, Bouchereau A, Deleu C, Morin H, Lemonnier P, Bonnin E, El Amrani A (2010) The Arabidopsis pop2-1 mutant reveals the involvement of GABA transaminase in salt stress tolerance. BMC Plant Biol 10:20

Salah A, Li J, Ge J, Cao C, Li H, Wang Y, Liu Z, Zhan M, Zhao M (2019) Morphological and physiological responses of maize seedlings under drought and waterlogging. J Agric Sci Tech 21:1199–1214

Salvatierra A, Pimentel P, Almada R, Hinrichsen P (2016) Exogenous GABA application transiently improves the tolerance to root hypoxia on a sensitive genotype of Prunus rootstock. Environ Exp Bot 125:52–66

Setter TL, Waters I (2003) Review of prospects for germplasm improvement for waterlogging tolerance in wheat, barley and oats. Plant Soil 253:1–34

Shabala S (2011) Physiological and cellular aspects of phytotoxicity tolerance in plants: the role of membrane transporters and implications for crop breeding for waterlogging tolerance. New Phytol 190:289–298

Shabala S, Shabala L, Barcelo J, Poschenrieder C (2014) Membrane transporters mediating root signalling and adaptive responses to oxygen deprivation and soil flooding. Plant Cell Environ 37:2216–2233

Shelp BJ, Zarei A (2017) Subcellular compartmentation of 4-aminobutyrate (GABA) metabolism in Arabidopsis: an update. Plant Signal Behav 12:e1322244

Song S, Li J, Gao H, Li QY, Yang LW, Gong RJ (2012) Effect of exogenous γ-aminobutyric acid on inorganic nitrogen metabolism and mineral elements contents of melon seedling under hypoxia stress. Acta Hort Sin 39:695–704

van den Dries N, Giannì S, Czerednik A, Krens FA, de Klerk GJ (2013) Flooding of the apoplast is a key factor in the development of hyperhydricity. J Exp Bot 64:5221–5230

Voesenek LACJ, Bailey-Serres J (2015) Flood adaptive traits and processes: an overview. New Phytol 206:57–73

Wang P, Liu K, Gu Z, Yang R (2018) Enhanced γ-aminobutyric acid accumulation, alleviated componential deterioration and technofunctionality loss of germinated wheat by hypoxia stress. Food Chem 269:473–479

Wu Q, Su N, Huang X, Cui J, Shabala L, Zhou M, Yu M, Shabala S (2021) Hypoxia induced increase in GABA content is essential for restoration of membrane potential and preventing ROS-induced disturbance to ion homeostasis. Plant Commun 2:100188

Yamauchi T, Yoshioka M, Fukazawa A, Mori H, Nishizawa NK, Tsutsumi N, Yoshioka H, Nakazono M (2017) An NADPH oxidase RBOH functions in rice roots during lysigenous aerenchyma formation under oxygen-deficient conditions. Plant Cell 29:775–790

Yang R, Guo Q, Gu Z (2013) GABA shunt and polyamine degradation pathway on γ-aminobutyric acid accumulation in germinating fava bean (Vicia faba L.) under hypoxia. Food Chem 136:152–159

Zarei A, Trobacher CP, Shelp BJ (2016) Arabidopsis aldehyde dehydrogenase 10 family members confer salt tolerance through putrescine-derived 4-aminobutyrate (GABA) production. Sci Rep 6:35115

GABA and Leaf Senescence

14

Abstract

As the concluding stage in leaf development, senescence plays a fundamental role in balancing plant health and yield. Gamma-aminobutyric acid (GABA) serves as a key regulator in this process, exerting diverse functions. It interacts with plant hormones like ethylene, abscisic acid, and salicylic acid, contributing to intricate signalling networks that control both stress responses and senescence progression. Environmental conditions such as light intensity, temperature fluctuations, and nutrient status significantly affect GABA metabolism, thereby altering its role in regulating senescence. This overview also highlights recent findings on GABA-responsive receptors and transcription factors involved in signalling cascades, offering promising insights for enhancing crop performance and resilience.

Keywords

C/N metabolism · GABA · GABA shunt · Leaf senescence · ROS

Highlights

- GABA plays a pivotal role in leaf senescence, functioning as a signalling molecule that modulates various physiological and metabolic activities.
- It participates in crosstalk with hormones, thereby influencing pathways related to senescence.
- GABA enhances plant stress tolerance during leaf senescence by mitigating oxidative stress and maintaining cellular homeostasis.
- Environmental factors like light, temperature, and nutrient availability significantly influence GABA metabolism and its regulatory role in senescence.

S. U. Jalil et al., *GABA Signaling System and Abiotic Stress Tolerance in Plants*,
https://doi.org/10.1007/978-981-95-0437-4_14

14.1 Introduction

Leaf senescence represents a critical developmental phase that is precisely regulated to ensure optimal nutrient allocation and plant survival, which is governed by a complex interplay of molecular, physiological, and environmental signals (Guo et al. 2021). Among the numerous compounds involved, gamma-aminobutyric acid (GABA) has emerged as a significant metabolite contributing to this process (Ansari et al. 2014). Its multifaceted roles in plant physiology, including responses to stress, regulation of growth, and coordination of carbon-nitrogen metabolism and nutrient uptake, have drawn considerable attention from plant scientists (Li et al. 2021). Deciphering how GABA regulates leaf senescence is essential not only for advancing plant biology but also for its potential implications in agriculture and ecosystem management.

As leaves progress toward the end of their lifespan, a sophisticated sequence of molecular events unfolds within the plant (Lim et al. 2007). These events orchestrate the systematic dismantling of cellular components and metabolic systems (Jing et al. 2005; Lim et al. 2007). Programmed and finely tuned, senescence plays a vital role in reallocating nutrients, optimizing resource use, and maintaining overall plant vigour (Buchanan-Wollaston 1997; Ansari et al. 2014). Emerging research has spotlighted the compelling role of GABA in this context, revealing its involvement in regulatory pathways that influence both natural and stress-induced senescence.

This chapter provides an in-depth exploration of GABA's functions during leaf senescence, covering its biosynthesis, signalling roles, and interactions with key regulatory networks. Special focus is given to how GABA interfaces with hormonal signals, environmental stimuli, and cellular responses during the ageing of leaves. Additionally, we evaluate how altering GABA metabolism or signalling pathways could potentially delay or modulate senescence, offering promising strategies for crop enhancement and sustainability.

By consolidating recent findings and highlighting current advancements, this chapter seeks to present a holistic understanding of GABA's involvement in leaf senescence. It also aims to identify unresolved questions and propose directions for future studies to clarify the comprehensive role of GABA in this crucial physiological process. Ultimately, the insights gained here may support the development of innovative strategies to boost crop yield and resilience in the face of environmental challenges.

14.2 A Comprehensive Overview of Leaf Senescence: Process and Significance

In plants, there are two types of ageing processes: one that happens when cells are still dividing (called mitotic senescence) and another that occurs after cells stop dividing (called post-mitotic senescence) (Gan and Amasino 1997). The first type occurs in the growing tips of plants, where new cells are made, and it's similar to the ageing process in animal and yeast cells (Guo and Gan 2005). The second type of ageing happens in parts of the plant, like leaves and flowers. Leaf ageing is a complex process that is influenced by both the plant's internal signals and changes in its environment (Jing et al. 2003; Lim et al. 2007; Nooden 1988). It involves the breakdown of chloroplasts, disruption of photosynthesis, degradation of cellular structures, and increased membrane permeability. Eventually, there is a systematic breakdown of macromolecules and organelles, such as mitochondria and nuclei, which also begin to deteriorate. Internal factors like developmental progression, phytohormones, and age, along with adverse environmental conditions (Fig. 14.1).

Leaf senescence is vital for plant survival, primarily facilitating nutrient recycling and mobilization (Guo and Gan 2014). Leaf growth accounts for the synthesis and accumulation of essential biological components, including metabolites and nutrients. During senescence, leaves recycle their stored resources, redirecting nutrients and metabolites to sustain the growth and development of other parts of the plant (Smart 1994; Nam 1997; Himelblau and Amasino 2001; Lim et al. 2003; Ansari et al. 2014). Leaf abscission, the final phase of senescence, occurs after nutrient remobilization. Research has shown leaf senescence is reversible; treatments can revert fully yellow leaves, which have already recycled their nutrients, to green (Thomas and Donnison 2000). Chloroplasts can restructure, synthesize new proteins, and restart photosynthesis. This highlights the regulation of senescence

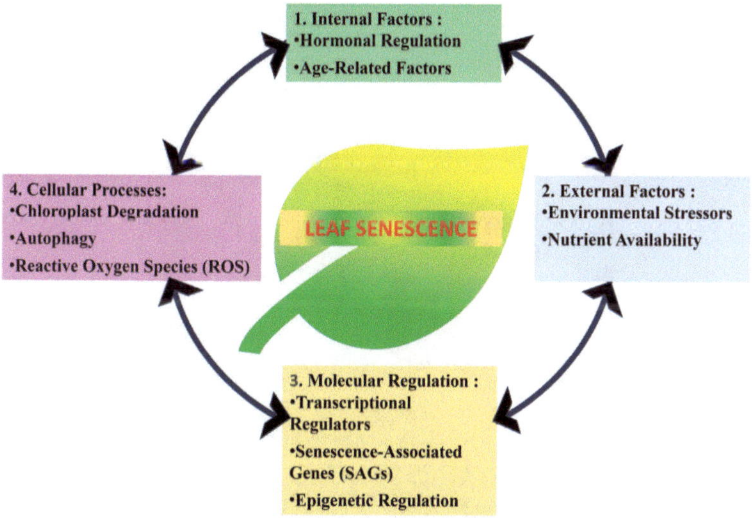

Fig. 14.1 Overview of major internal and external factors contributing to the initiation and progression of leaf senescence. A leaf transitioning from green to yellow symbolizes the senescence process. The four quadrants surrounding the leaf highlight major contributing factors. Internal factors include hormonal regulation and age-related mechanisms, while external factors encompass environmental stressors and nutrient availability. Molecular regulation involves transcriptional regulators, senescence-associated genes, and epigenetic modifications. Cellular processes represent chloroplast degradation, autophagy, and the complex interplay regulating leaf senescence and nutrient recycling. The structured representation emphasizes the multifaceted control of this process in plants

from initiation to completion. Environmental stresses can induce premature senescence, enabling plants to allocate nutrients to other parts for survival. For example, induced senescence in infected leaves allows nutrient transport away from pathogens, removing contamination sources (Buchanan-Wollaston et al. 2003).

14.3 Senescence Stages and Mechanisms

Leaf senescence refers to the natural and programmed process of ageing and deterioration that occurs in leaves as they reach the end of their lifespan. This progression is marked by a cascade of physiological, biochemical, and molecular alterations that eventually result in leaf death and abscission (Guo and Gan 2005; Lim et al. 2007). Senescence plays a vital role in plant development by facilitating nutrient redistribution and enhancing the plant's ability to adapt to changing environmental conditions. Singh et al. (2022) broadly identify three main stages of leaf senescence (Fig. 14.2):

- *Initiation*: This phase marks the beginning of the senescence process prompted by numerous internal and external influences, together with developmental cues, hormonal signals, and environmental stresses. Initiation involves the activation

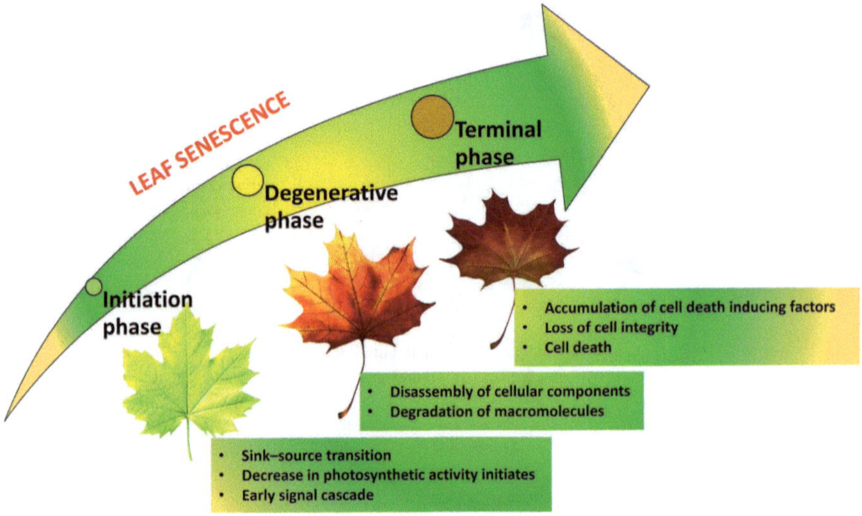

Fig. 14.2 Leaf senescence passes through three sequential stages, as shown on a gradient arrow. The leaf senescence process culminates with colour change progression from green to yellow, orange, and ultimately brown. Processes such as sink-source transition, decreased photosynthetic activity, and early signalling cascades mark the initiation stage. As senescence progresses, cellular components are dismantled, and macromolecules degrade. During the terminal phase, the gathering of apoptosis-inducing elements leads to cellular disintegration and controlled cell death. The structured representation highlights the dynamic nature of senescence, emphasizing its regulatory complexity and physiological significance

of specific genes and signalling pathways that initiate the breakdown of cellular components and the reprogramming of metabolic processes.
- *Progression*: During this phase, senescence progresses rapidly, accompanied by visible changes in leaf morphology, physiology, and metabolism. These changes include yellowing of the leaf due to chlorophyll degradation, loss of turgor pressure, breakdown of cellular membranes, and remobilization of nutrients from senescing tissues to other parts of the plant. Metabolic activities such as photosynthesis, protein synthesis, and nucleic acid synthesis decline, while catabolic processes, including protein degradation and autophagy, are upregulated.
- *Final Phase*: Also known as the terminal phase, this is the last stage of leaf senescence where the leaf undergoes complete breakdown and nutrient remobilization. Nutrients such as nitrogen and phosphorus are retrieved from senescing tissues and transported to other parts of the plant, such as young leaves or storage organs, to support new growth or reproductive processes. Eventually, the leaf completely withers and may fall off the plant.

14.4 Senescence and Physiological and Molecular Changes

Leaf senescence is executed through a well-organized sequence of physiological and molecular events, where cellular components are gradually dismantled in a controlled manner. One of the earliest and most significant markers of this process is the breakdown of chlorophyll. Notably, several senescence-related activities occur prior to the visible yellowing of leaves (Buchanan-Wollaston et al. 2003). This phase includes the degradation of proteins and other macromolecules, leading to the efficient redistribution of essential nutrients and minerals from ageing leaves to actively growing tissues. The eventual shutdown of photosynthetic machinery also marks this transition (Ansari et al. 2005; Lim et al. 2007; Ansari and Chen 2009; Ansari et al. 2014).

14.4.1 Chlorophyll Breakdown

Researchers have extensively investigated the mechanisms responsible for the disintegration of chlorophyll and proteins during leaf senescence. In recent years, attention has focused on the genetic regulation of this degradation process, resulting in the identification and cloning of several associated genes. However, some of these genes are not actively expressed during leaf senescence itself (Takamiya et al. 2000).

Instead, the senescence process relies on the enzyme pheophorbide a oxygenase, which plays a key role by cleaving the tetrapyrrole ring structure of chlorophyll to produce red chlorophyll catabolite (RCC) (Hörtensteiner et al. 1998). Experiments with *Festuca pratensis* mutants that lack this enzyme have revealed a "stay-green" phenotype, accompanied by the accumulation of pheophorbide a, confirming the enzyme's regulatory significance (Vicentini et al. 1995). Increased activity of pheophorbide a oxygenase during senescence further supports its central role in chlorophyll catabolism, and its detailed characterization is expected to enhance our understanding of this pathway.

The catabolic end-products of chlorophyll, primarily non-fluorescent chlorophyll catabolites, are sequestered in vacuoles. These compounds, while not contributing to nitrogen recycling, serve an essential detoxification function by neutralizing potentially harmful pigment-protein complexes, thus preserving cellular integrity during senescence.

Support for the importance of this degradation pathway comes from studies on the Acd2 gene, which encodes an enzyme involved in chlorophyll breakdown. Mutations affecting this enzyme's function result in the accumulation of toxic intermediates like RCC, triggering accelerated cell death (Hörtensteiner et al. 1998; Buchanan-Wollaston et al. 2003).

14.4.2 Chloroplast Protein Degradation

The breakdown of proteins is a fundamental catabolic event during leaf senescence and plays a pivotal role in nutrient remobilization, especially in reclaiming nitrogen (Wada and Ishida 2009; Diaz-Mendoza et al. 2016; Buet et al. 2019; Domínguez and Cejudo 2021). Throughout this process, the expression of various proteases is dynamically regulated within cellular compartments such as the nucleus, chloroplasts, and mitochondria, promoting the conversion of proteins into amino acids and ammonium (Diaz-Mendoza et al. 2014; Roberts et al. 2012).

This protein degradation depends on a sophisticated trafficking system that coordinates the movement and breakdown of proteins across different organelles (Roberts et al. 2012; Carrion et al. 2013; Avila-Ospina et al. 2014; Diaz-Mendoza et al. 2014). Senescence, often accelerated by environmental stresses, involves the dismantling of organelles, with chloroplasts being the first and most significantly affected (Diaz-Mendoza et al. 2016). As the central sites for photosynthesis, chloroplasts experience extensive structural and functional deterioration during senescence, particularly involving the degradation of their protein constituents (Roberts et al. 2012). Several distinct proteolytic pathways are responsible for chloroplast protein catabolism. Within plastids, key enzymes, including FtsH metalloproteases, Clp serine proteases, and DegP serine proteases, become upregulated as senescence progresses. These proteases are instrumental in degrading essential chloroplast proteins, such as components of the photosystem II complex (Roberts et al. 2012).

CND41 aspartic-protease breakdowns proteins like Rubisco to essentially release nitrogen for its translocation as efficient nutrient remobilization from senescing leaves (Kato et al. 2004, Ishida et al. 2008). Although senescence-related protease genes are expressed, their connection with chloroplast proteins occurs after membrane leakage due to senescence, suggesting a complex regulatory network (Ishida et al. 2002). Increased reactive oxygen species (ROS) production during senescence contributes to protein degradation that occurs following initiation of lipid degradation (Buchanan-Wollaston et al. 2003). However, the direct impact of increased ROS on Rubisco degradation remains uncertain.

14.4.3 Lipid Degradation and Nutrient Remobilization

During senescence, various cellular components, including lipids, undergo degradation to provide energy for new growth or storage (Thompson et al. 1998). Lipids, as crucial components of cell membranes and nutrient storage molecules, play significant roles in the senescence process (Cohen et al. 2022). Senescing leaves undergo extensive membrane remodelling, leading to the degradation of membrane lipids, including phospholipids and galactolipids. Enzymes like phosphatidic acid phosphatase, lipoxygenase, phospholipase D, and lytic acyl hydrolase are actively involved in degrading these membrane lipids, releasing fatty acids in the process. Many of the genes encoding these enzymes are notably upregulated during senescence (He and Gan 2002).

The elevated expression of senescence-related genes highlights the vital role of β-oxidation in breaking down lipids and redistributing their components. While β-oxidation operates throughout a plant's lifecycle, its function becomes especially critical during senescence (Graham and Eastmond 2002). This pathway mainly takes place in the peroxisomes of leaf tissue and in germinating seeds (Graham 2008). Triacylglycerols stored in lipid droplets serve as energy-rich reserves during senescence. These are hydrolysed by lipases into glycerol and free fatty acids, which are then utilized either as direct energy sources or as precursors for synthesizing other essential biomolecules (Xu et al. 2020; Domínguez and Cejudo 2021). Through β-oxidation, the liberated fatty acids are converted into acetyl-CoA, a metabolite that feeds into the tricarboxylic acid (TCA) cycle to generate energy or is used in the synthesis of amino acids and secondary metabolites.

Acetyl-CoA is central to maintaining aerobic respiration during senescence, while the activity of enzymes like pyruvate orthophosphate dikinase—enhanced during senescence—enables the conversion of pyruvate to sugars via gluconeogenesis, further supporting metabolic needs in ageing leaves (Hong et al. 2000; He and Gan 2002). Genetic modifications in lipid-degrading enzymes can influence the timing of leaf senescence. For example, *A. thaliana* lines with reduced expression of senescence-associated lipases exhibit delayed senescence, while overexpression of specific genes such as senescence-associated gene (SAG) 101 leads to accelerated ageing (Hong et al. 2000; He and Gan 2002).

During leaf senescence, increased lineolate:oxygen oxidoreductase activity results in the accumulation of lipid hydroperoxides, which can serve as signalling molecules or substrates for further lipid degradation pathways (Berger et al. 2001). ROS generated during senescence can oxidatively modify lipids, leading to lipid peroxidation. This process generates reactive aldehydes such as MDA, which can further damage cellular components if not detoxified (Ayala et al. 2014).

Analyses of *A. thaliana* reveal that senescent leaves show significant reductions in elemental nutrients—up to 50% in elements such as Cr, Mo, S, Fe, Zn, and

Cu—and even greater losses, up to 80%, in macronutrients including potassium (K), nitrogen (N), and phosphorus (P) compared to healthy green leaves (Himmelblau and Amasino 2001). Although the specific genes responsible for these nutrient remobilization processes remain unidentified, studies in *Brassica napus* show increased glutamine levels during late senescence, suggesting a shift in nitrogen transport dynamics (Clément et al. 2018). This is supported by the observation of elevated cytosolic glutamine synthetase (GS) activity during senescence, indicating its potential involvement in amino acid transformation and enhanced nitrogen translocation (Fortunato et al. 2023).

14.4.4 Nitrogen Transport

Nitrogen is a fundamental nutrient essential for vital plant functions such as protein formation and enzymatic reactions. In the context of leaf senescence—the terminal phase of leaf development—nitrogen is systematically reallocated from ageing leaves to support the development of younger tissues and reproductive organs (Masclaux-Daubresse and Chardon 2011). This internal recycling mechanism prevents the loss of nitrogen during senescence, ensuring it remains available for continued plant growth (Havé et al. 2017).

Within plants, nitrogen is transported in various molecular forms, including amino acids, proteins, amides, urea, and nitrate. The predominant forms vary depending on species and environmental context (Masclaux-Daubresse et al. 2010). To facilitate this movement, plants utilize specialized transport proteins such as amino acid transporters and nitrate transporters, which play a crucial role in the intracellular and intercellular movement of nitrogen (Fan et al. 2009; Lezhneva et al. 2014; Wu et al. 2014). These proteins are typically embedded in the plasma membrane, enabling the uptake of nitrogen into cells or its export to adjacent cells or tissues (Hajibarat and Saidi 2022).

Chloroplasts are the central site of protein biosynthesis in leaves (Peoples and Dalling 1988). As senescence progresses, the focus of nitrogen recycling shifts toward chloroplast proteins—particularly Rubisco—due to their abundance and high nitrogen content. The dismantling of chloroplasts is among the earliest visual signs of senescence, typically observed as leaf yellowing caused by chlorophyll degradation. Nonetheless, other cellular proteins and even nucleic acids also serve as potential nitrogen sources during this phase. Gene expression studies during senescence have revealed several candidates involved in proteolysis and nitrogen metabolism, emphasizing the regulatory complexity of this process (Roberts et al. 2012).

Although inorganic nitrogen contributes to the redistribution of nitrogen resources, its role appears limited due to its relatively low presence compared to organic nitrogen stored in proteins. Thus, the major processes contributing to nitrogen remobilization include proteolysis, amino acid conversion, and their subsequent transport via the phloem. The vascular system—especially the phloem—is central to this redistribution, where nitrogen compounds are loaded into sieve tubes and delivered to actively growing sink tissues (Masclaux et al. 2000; Masclaux-Daubresse et al. 2006). During senescence, the dynamic between source and sink tissues shifts: ageing leaves function as nitrogen sources, while developing tissues and seeds act as sinks, ensuring optimal nitrogen use across the plant.

14.4.5 Mitochondrial Dynamics and Cellular Integrity

Mitochondria play a crucial role as organelles involved in energy production through oxidative phosphorylation and various metabolic pathways (Buchanan-Wollaston 1997). Throughout leaf senescence, mitochondria undergo dynamic structural changes characterized by alterations in size, shape, and intracellular distribution, including fragmentation, elongation, and modifications in cristae morphology (Keech et al. 2007). Although protein loss is extensive during senescence, mitochondrial activity remains relatively high, likely to maintain energy generation and facilitate nutrient redistribution (Thompson et al. 1998; Keech et al. 2007, 2010). Additionally, despite interruptions in microtubule networks, mitochondria retain some degree of mobility (Keech et al. 2010), indicating the persistence of cytoplasmic streaming at least partially due to the preservation of actin filaments (Keech et al. 2010). As senescence progresses toward its terminal phase, the rupture of the tonoplast triggers the release of vacuolar hydrolytic enzymes into the cytoplasm, leading to swift degradation of cellular components and culminating in cell death (Hörtensteiner and Feller 2002; van Doorn and Woltering 2004).

14.5 Leaf Senescence Modulating Factors and Dynamics

Factors modulating leaf senescence dynamics encompass a broad range of influences that regulate the timing, intensity, and progression of this vital process in plants. Internal and external aspects intricately regulate senescence, orchestrating cellular and molecular changes and eliciting tissue degradation and nutrient remobilization. Understanding these factors is essential for elucidating the mechanisms governing leaf senescence and their impact on plant growth and adaptation.

14.5.1 The Natural Process of Age-Dependent Senescence

The developmental cycle of a leaf is divided into three distinct phases: expansion, maturation, and senescence (Guiboileau et al. 2010). During the maturation phase, the leaf remains responsive to both internal physiological cues and external environmental stimuli, often initiating reversible physiological changes. However, as the leaf progresses further, it transitions into a phase of irreversible decline, marking the onset of senescence and ultimately leading to cell death (Lim et al. 2007a). This senescence phase is characterized by the progressive breakdown of cellular components, particularly those involved in photosynthesis, and is driven by complex degradation and nutrient translocation mechanisms (Gan and Amasino 1997; Guo and Gan 2005; Hörtensteiner and Feller 2002; Buchanan-Wollaston et al. 2003; Lim et al. 2007a, b; Guiboileau et al. 2010; Fischer 2012).

In monocarpic annual species, the initiation of senescence in individual leaves—or in the entire plant—often coincides with seed development, marking the completion of one generation and preparing for the next (Noodén 1988a, b). In crops like soybean, specific leaves begin to senesce in synchrony with the development of reproductive organs, highlighting a coordinated developmental strategy governed by interactions between vegetative and reproductive tissues (Noodén 1988a). This type of reproductive control over senescence is a common feature in many annual crop species. However, in *A. thaliana*, the ageing of individual leaves appears to occur independently of reproductive organ formation, suggesting that this trait may be common among members of the Brassicaceae family (Noodén and Penney 2001).

Although senescence is largely pre-programmed within the plant's developmental framework, environmental stresses can significantly influence its onset and progression by interacting with stress-responsive signalling pathways (Jansson and Thomas 2008). Gene expression analyses have shown that many regulatory elements and catabolic processes associated with leaf senescence are conserved across monocot and dicot species (Buchanan-Wollaston et al. 2005; Jukanti et al. 2008; Gregersen et al. 2008; Breeze et al. 2011). Additionally, specific transcription factor families such as WRKY, NAC, C2H2 zinc-finger, AP2/EREBP, MYB, and bZIP have been identified as key regulators in controlling the expression of senescence-related genes (Buchanan-Wollaston et al. 2005; Balazadeh et al. 2008; Guo and Gan 2006; Breeze et al. 2011).

14.5.2 Senescence Triggered by Environmental Stresses

Stress-induced leaf senescence refers to the premature ageing of plant tissues in response to unfavourable environmental conditions (Bray et al. 2000; Gan 2003; Guo and Gan 2012). Exposure to challenges such as drought, salinity, extreme temperatures, oxidative stress, or pathogen attacks often activates signalling pathways that initiate senescence as a survival mechanism (Robert-Seilaniantz et al. 2011; Peleg and Blumwald 2011; Tan et al. 2023). This strategy enables plants to conserve energy and redirect nutrients toward essential growth and reproductive processes, enhancing their adaptability to stress.

A wide range of abiotic stressors can accelerate senescence (Sade et al. 2018; Tan et al. 2023). For example, drought limits water availability and reduces photosynthetic efficiency, prompting leaf senescence to minimize water loss and sustain newly forming tissues (Munné-Bosch and Alegre 2004; Hajibarat and Saidi 2022). High salinity disrupts ionic and osmotic balance, which promotes oxidative stress and accelerates cellular ageing (Balazadeh et al. 2010; Sakuraba et al. 2018; Park et al. 2022). Similarly, heat stress compromises membrane stability and photosynthetic machinery, resulting in increased ROS production and early senescence (Jespersen et al. 2015, 2016; Kim et al. 2020). Cold temperatures are known to lower cytokinin levels while increasing abscisic acid concentrations in leaf tissues—hormonal shifts that trigger senescence (Zhang et al. 2020; Wang et al. 2022).

Additional stressors such as UV radiation and heavy metal contamination also play a role in hastening senescence. These factors disrupt photosynthesis and nutrient uptake, often through the overproduction of ROS, leading to cellular damage and accelerated leaf decline (John et al. 2001; Sztatelman et al. 2015; Piacentini et al. 2020; Zhang et al. 2022; Fei et al. 2022).

In response to biotic stress, such as pathogen infection, plants may trigger localized or systemic senescence. This defensive strategy restricts pathogen spread while reallocating nutrients to unaffected areas. However, prolonged immune responses or disease symptoms can also lead to early senescence, negatively impacting crop yields (Noodén 1988a). Phytohormone signalling—particularly involving salicylic acid, jasmonic acid, and ethylene—links pathogen defence with the regulation of senescence (Robert-Seilaniantz et al. 2011).

A common feature across many stress responses is the overaccumulation of ROS due to oxidative stress (Jajic et al. 2015). These molecules—such as H_2O_2, $O_2\bullet^-$, and OH^-, are byproducts of cellular metabolism that become toxic when present in excess (Kinnersley and Turano 2000). If left uncontrolled, ROS can severely damage cellular lipids, proteins, and DNA, accelerating tissue senescence and eventual cell death (Sharma et al. 2012; Jajic et al. 2015).

To adapt, plants undergo metabolic reprogramming during stress-induced senescence. This includes the breakdown of large molecules such as proteins, fats, and nucleic acids to release energy and raw materials necessary for survival. Nutrients are also redirected from ageing tissues to developing or more critical parts of the plant. While this process is crucial for stress adaptation, prolonged or intense stress can reduce plant growth, limit reproductive success, and negatively impact yield and productivity.

14.5.3 Sugar Dichotomy and Senescence

Sugars serve dual roles in plant physiology: they are both vital energy sources and act as signalling molecules that influence leaf senescence. On one hand, they are essential for maintaining metabolic processes and supporting growth. On the other hand, when accumulated in excess, sugars can signal and even accelerate the onset of senescence.

High sugar concentrations are known to suppress photosynthesis and initiate early ageing in leaves. For example, mutations that disrupt the function of hexokinase 1 (HXK1) lead to delayed senescence, whereas plants with elevated HXK1 expression show faster ageing, indicating HXK1's involvement in sugar-dependent signalling pathways that regulate senescence (Moore et al. 2003; Dai et al. 1999; Wingler et al. 2006). Experimental results also show that when glucose is present alongside low nitrogen availability, it triggers visible leaf yellowing and stimulates the expression of senescence-associated genes like SAG12 and certain MYB transcription factors (Pourtau et al. 2006).

As senescence progresses, sugars like glucose and fructose tend to accumulate in the leaves, while sucrose levels remain comparatively stable (Wingler et al. 2006). Typically, sugar concentrations gradually increase, peaking during the late growth phases or just before leaf ageing becomes evident. However, the exact mechanisms by which plants regulate and manage sugar and starch reserves during senescence remain unclear (Kim 2019).

The influence of sugars on leaf ageing is highly context-dependent. Under some conditions, sugars can suppress senescence, while under others, they promote it (van Doorn and Woltering 2008). For example, in detached leaves undergoing senescence, sugar application has been shown to reduce SAG12 expression (Noh and Amasino 1999). Conversely, sugar build-up resulting from disrupted phloem transport, such as that caused by steam girdling, has been associated with enhanced senescence (Parrott et al. 2005).

Environmental conditions such as high light intensity and low nitrogen levels also contribute to accelerated leaf ageing. Interestingly, when nitrogen is supplied along with sugars, the progression of senescence slows down (Ono et al. 2001; Pourtau et al. 2004; Ono and Watanabe 1997). In whole-plant studies, increases in sugar content are often accompanied by a decline in chlorophyll and reductions in amino acid and protein levels, suggesting a close link between sugar-driven senescence and limited nitrogen availability (Masclaux et al. 2000).

14.6 GABA and Leaf Senescence

The involvement of GABA in leaf senescence has become an increasingly important focus in plant physiological research. Senescence is a precisely regulated form of programmed cell death. This process plays a critical role in reallocating nutrients, especially nitrogen from ageing tissues to actively growing parts of the plant (Sakamoto and Takami 2014; Ansari et al. 2014).

Among the various physiological processes influenced by GABA, its contribution to nitrogen metabolism during senescence is particularly significant (Ansari et al. 2005, 2014). The GABA shunt facilitates the remobilization of nitrogen by incorporating nitrogenous compounds into central metabolic pathways (Barbosa et al. 2010; Ansari et al. 2014). During senescence, nitrogen from macromolecules is broken down and transformed into transportable forms, mainly the amino acids glutamine and asparagine, which are delivered through the phloem to growing organs like seeds and fruits (Feller and Fischer 1994). These amino acids are further metabolized into glutamate and α-keto acids by the action of transaminases (Ansari et al. 2014). A portion of glutamate is converted to GABA via the enzyme GAD, feeding into the GABA shunt.

This pathway involves GABA-T and links to the TCA cycle, allowing for integration of carbon and nitrogen metabolism. The transformation of glutamate into GABA and its recycling through these pathways significantly affects the C/N balance within developing seeds (Fig. 14.3). This, in turn, influences fatty acid biosynthesis and the accumulation of storage compounds, making GABA a key player in

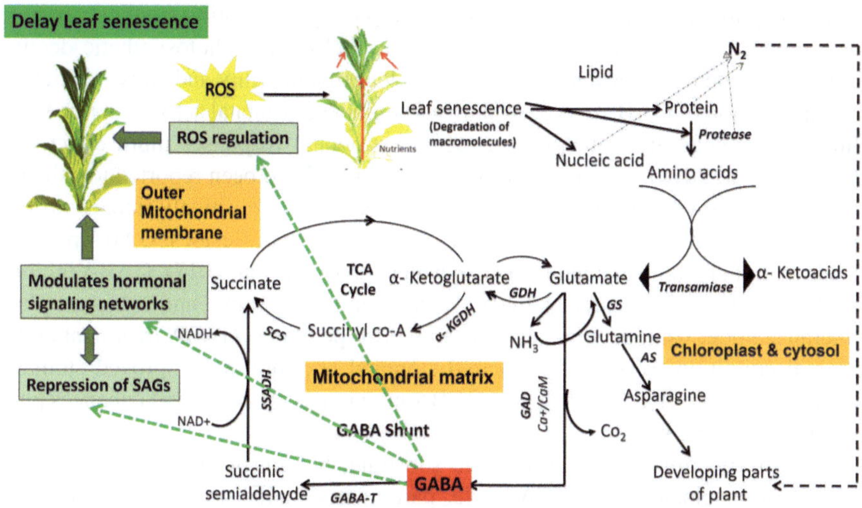

Fig. 14.3 GABA and metabolite recycling routes in leaf senescence. GABA accords leaf senescence through multiple biochemical and physiological pathways. Interaction with the TCA cycle impacts α-ketoglutarate and glutamate metabolism, energy production, and nitrogen assimilation. GABA modulates hormonal signalling networks, which regulate the balance between senescence acceleration and delay. GABA influences gene expression patterns, including the activation and repression of SAGs, which govern chlorophyll breakdown, nutrient recycling, and cellular longevity. Through its diverse interactions with metabolic, signalling, and gene regulatory networks, GABA emerges as a crucial determinant of the overall progression of senescence and leaf longevity

regulating metabolic flux during leaf senescence and seed development (Fait et al. 2008).

14.6.1 GABA and Photosynthetic System During Senescence

Leaf senescence, a developmental phase marked by chlorophyll degradation, leads to the yellowing or reddening of leaves. Chlorophyll breakdown is a significant event in this process; however, GABA can modulate the process, potentially delaying senescence by preserving chlorophyll content (Jalil et al. 2017, 2022).

Photosynthetic efficiency is a vital indicator of a plant's health and its ability to adapt to changing environmental conditions. GABA has been shown to contribute significantly to preserving photosynthetic activity during senescence. One way it does so is by influencing stomatal behaviour, enhancing stomatal opening to facilitate greater CO_2 intake, which supports the light-dependent reactions of photosynthesis (Xu et al. 2021a, b). Experimental evidence from Mekonnen et al. (2016), using gad1/2 mutants of *A. thaliana* (which are deficient in GABA synthesis), highlights this role. These mutants exhibit increased stomatal opening and are more susceptible to wilting under drought conditions, suggesting impaired stomatal closure. This observation underscores the importance of GABA in regulating stomatal function, particularly under environmental stress. By modulating stomatal aperture and supporting CO_2 assimilation, GABA may help delay photosynthetic decline during senescence, contributing to extended leaf functionality and stress resilience.

Application of GABA has been shown to influence stomatal activity in rice seedlings under heat stress, helping to regulate stomatal responses more effectively (Nayyar et al. 2014). In maize, GABA treatment has also been reported to enhance early plant growth, boost photosynthetic efficiency, and positively impact various physiological and biochemical processes (Li et al. 2016a). Besides, GABA priming leads to higher stomatal conductance in watermelon (Li et al. 2017). GABA inhibits ALMT9, a key ion transfer pathway during stomatal opening, and enhances plant tolerance under osmotic stress as well. In wild-type plants, GABA treatment under normal conditions results in increased stomatal aperture and improved stomatal conductance (Xu et al. 2019). Thus, GABA maintains cell turgor and enhances photosynthesis under stress.

GABA functions as an anaplerotic substrate, helping to ensure a continuous source of succinate to the TCA cycle, particularly under stress conditions (Guo et al. 2023). The enzyme GABA-T plays a pivotal role in this process by controlling both the levels of GABA and its metabolic conversion during stress. Studies involving *A. thaliana* mutants lacking GABA-T activity (*pop2-1* and *pop2-3*) have demonstrated reduced chlorophyll and GABA accumulation under stress, along with lower activities of both GABA-T and GAD. These stress-sensitive mutants also exhibit higher membrane ion leakage and elevated MDA content, indicative of oxidative damage (Khan et al. 2021). These observations highlight the importance of GABA-T in reducing reactive oxygen species (ROS) accumulation in chloroplasts, supporting both carbon fixation and electron transport during photosynthesis, thereby contributing to delayed senescence.

14.6.2 GABA and Hormonal Signalling

Intricate hormonal signalling pathways drive leaf senescence, with GABA emerging as a critical modulator of these pathways. GABA integrates internal metabolic changes and environmental cues to influence senescence dynamics. By networking with key plant hormones, including ethylene, ABA, SA, and cytokinin, GABA supports fine-tuning the balance between promoting and delaying senescence under normal and stress conditions.

Ethylene is a key factor in speeding up leaf ageing by controlling genes that break down chlorophyll, move nutrients around, and trigger cell death. This plant hormone has diverse roles, affecting both the development and ageing of plants, which vary based on its levels and the type of plant. Ethylene can either accelerate or inhibit these processes based on environmental and developmental cues (Gilliham and Tyerman 2016; Islam et al. 2024). Ethylene is produced from S-adenosyl methionine (SAM), an activated derivative of methionine. The key regulatory step in this biosynthetic pathway involves the transformation of SAM into

1-aminocyclopropane-1-carboxylic acid (ACC), catalysed by the enzyme ACC synthase. Interestingly, SAM also serves as a precursor for polyamines, highlighting a connection between ethylene biosynthesis and GABA metabolism (Islam et al. 2024). GABA influences ethylene production by modulating the activities of key enzymes such as ACC synthase and ACC oxidase. Under stress conditions, high levels of ethylene can accelerate senescence; however, GABA acts as a regulatory buffer, reducing excessive ethylene production, particularly during oxidative stress. This buffering effect delays premature senescence, maintains photosynthetic efficiency, and improves stress tolerance (Shi et al. 2010). Stress conditions often trigger a sequential increase in Ca^{2+}, GABA, and ethylene levels, indicating a tightly interconnected signalling mechanism (Kinnersley and Turano 2000). This interaction emphasizes GABA's role in maintaining a balance between growth and senescence by mitigating the negative impacts of elevated ethylene under stress.

14.6.3 Antioxidant Activity Regulation

ROS are metabolic byproducts generated in plant cells under stress conditions and during leaf senescence. These molecules exist either as free radicals (superoxide anion, hydroperoxyl radical, alkoxy radical, hydroxyl radical) or non-radical forms (hydrogen peroxide, singlet oxygen) (Khan and Khan 2017; Sachdev et al. 2021). While low ROS concentrations serve as signalling molecules that trigger defence responses, excessive accumulation causes oxidative damage to cellular components, including proteins, nucleic acids, lipids, and membranes, ultimately disrupting metabolic processes (Juan et al. 2021). Under abiotic stress conditions, the equilibrium between ROS generation and antioxidant defences becomes disrupted, leading to oxidative stress.

Plants have developed sophisticated antioxidant systems to counteract ROS toxicity, comprising both enzymatic and non-enzymatic components. The enzymatic defence system includes SOD, CAT, APX, POD, dehydroascorbate reductase (DHAR), and monodehydroascorbate reductase (MDHAR). Non-enzymatic antioxidants involve carotenoids, flavonoids, phenolic compounds, and ascorbic acid (Hasanuzzaman et al. 2020; Wongshaya et al. 2020; Kumar et al. 2017).

Emerging evidence indicates that GABA contributes significantly to antioxidant defence mechanisms during leaf senescence. As senescence progresses and ROS levels rise, GABA functions as an effective ROS scavenger, protecting cellular integrity and potentially delaying senescence (Jalil et al. 2019a,b). The accumulation of GABA appears to alleviate oxidative damage, thereby enhancing plant tolerance to oxidative stress (AL-Quraan 2015). The metabolic fate of GABA is particularly noteworthy—its breakdown products enter the tricarboxylic acid (TCA) cycle, and inhibition of GABA catabolism disrupts mitochondrial respiration while promoting ROS accumulation (Seifi et al. 2018). Studies have demonstrated GABA's ability to activate antioxidant defences under salt stress conditions, reducing ROS-mediated damage (Seifi et al. 2018).

The Ca^{2+}/CaM signalling pathway plays a crucial role in this process. Under UV-B radiation, Ca^{2+}/CaM pathway activation stimulates GABA biosynthesis. *Arabidopsis thaliana* mutants with impaired Ca^{2+}/CaM signalling exhibit reduced ROS scavenging capacity (Jalil et al. 2019a,b) and decreased UV-B resistance, suggesting GABA's involvement in regulating antioxidant pathways (AL-Quraan 2015).

Experimental evidence consistently demonstrates that exogenous GABA application helps maintain redox homeostasis by modulating antioxidant enzyme activities. GABA pretreatment has been shown to reduce ROS and malondialdehyde (MDA) levels, regulate photosynthetic processes, and enhance plant resilience to various stresses, including chilling injury, hypoxia, salinity, and heat stress (Li et al. 2016a, b). In muskmelon, GABA treatment regulates membrane lipid peroxidation and ion homeostasis while stimulating the antioxidant system to promote growth (Xu et al. 2021c). Similar protective effects have been observed in studies examining chilling injury and heavy metal stress, where GABA supplementation enhanced antioxidant enzyme activities and reduced ROS accumulation in leaves (Aghdam et al. 2016; Yang et al. 2011; Song et al. 2010).

Comparative studies in tobacco under water stress revealed that both GABA and proline accumulate significantly, with both compounds demonstrating ROS scavenging activity. However, GABA exhibited superior efficacy in eliminating superoxide anions and hydrogen peroxide (Liu et al. 2011). These findings collectively suggest that endogenous GABA accumulation represents an important component of plants' antioxidant defence mechanisms under stressful conditions.

14.6.4 Environmental Stressors and Leaf Senescence

Adverse environmental conditions negatively affect plant growth, development, and productivity by disrupting physiological, biochemical, and metabolic processes, as well as disturbing osmotic and oxidative balance (Choudhary et al. 2020). These stressors cause an imbalance between the production and scavenging of ROS, leading to oxidative damage (Choudhary et al. 2020). Chloroplasts are the primary sites of ROS generation under severe environmental stress (Partelli and Batista 2001), though other cellular compartments, including the cell wall, mitochondria, peroxisomes, endoplasmic reticulum, and apoplast, also contribute to ROS production (Pottosin et al. 2014; Navrot et al. 2007; Noctor et al. 2002; Mittler 2002; Hu et al. 2006). Accelerated leaf senescence is a key adaptive strategy plants use to counteract oxidative stress caused by excessive ROS accumulation (Kumar et al. 2019).

GABA plays a crucial role in plant responses to environmental stresses such as drought, salinity, and pathogen attack, which can accelerate leaf ageing. By influencing stress responses, GABA indirectly regulates the initiation and progression of senescence under unfavourable conditions. Disruptions in GABA shunt components can lead to ROS-induced cell death under heat and light stress. For example, *Arabidopsis* mutants lacking *SSADH* exhibit elevated reactive oxygen intermediates (ROIs), along with phenotypic defects such as stunted growth, smaller leaves, necrotic lesions, reduced chlorophyll content, leaf bleaching, and fewer flowers compared to wild-type plants (Fait et al. 2005). Treating these mutants with a GABA-T inhibitor reduces ROI accumulation, prevents cell death, and alleviates the biochemical and morphological defects caused by *SSADH* deficiency, ultimately improving plant growth (Fait et al. 2005). This demonstrates that the GABA shunt helps mitigate oxidative stress by scavenging excess ROIs.

CaM also contributes to oxidative stress tolerance, particularly under UV exposure (UVA and UVB), influencing seed germination, growth, and ROS levels. In *Arabidopsis*, *CaM* mutants with T-DNA insertions show increased GABA accumulation and altered GABA shunt metabolites in response to UV treatment, along with enhanced antioxidant defence mechanisms to reduce ROS levels (Al-Quraan 2015).

Exogenous GABA application has been shown to alleviate oxidative stress in plants. In pepper (*Capsicum annuum* L.), GABA supplementation under low-light conditions enhances stress tolerance by boosting antioxidant defences and improving photochemical efficiency. It reduces oxidative stress markers like MDA while increasing the activity of antioxidant enzymes such as SOD and CAT (Li et al. 2017). Additionally, GABA treatment promotes chlorophyll content, stomatal conductance, photochemical quenching, photosynthesis, and PSII activity. The study suggests that endogenous GABA may be metabolized into succinate, helping regulate the carbon-to-nitrogen balance in pepper plants (Li et al. 2017).

Impairments in the GABA shunt led to increased ROS levels under stress (Bouché et al. 2003). Exogenous GABA application has been found to reduce chilling injury in peach fruits by enhancing antioxidant enzyme activity and maintaining cellular energy levels, thereby improving stress tolerance and delaying senescence (Yang et al. 2011; Bouché et al. 2003). Similarly, GABA treatment enhances alkaline stress tolerance by regulating antioxidant systems, improving photosynthetic efficiency, and modulating TCA cycle enzyme activity (Li et al. 2021). These changes in TCA cycle enzymes help balance carbon and nitrogen pools, further linking GABA to senescence regulation under abiotic stress.

In tobacco leaves exposed to NaCl stress, growth retardation is accompanied by significant shifts in GABA levels and alterations in metabolic pathways, including the TCA cycle and glycolysis (Zhang et al. 2011). The antioxidant defence system plays a crucial role in counteracting stress-induced oxidative damage, thereby delaying senescence (Iqbal et al. 2012).

14.6.5 GABA and Cytosolic pH Regulation

Plant leaf senescence is a controlled, genetically programmed form of cell death that is influenced by various signalling compounds. Among these, GABA plays a significant role in managing different aspects of senescence, including the regulation of cytosolic pH, a crucial factor in cellular stability (Ramesh et al. 2015).

As leaves age, their cellular stability becomes disrupted. This includes changes to ion flow across membranes, which in turn disturb the internal pH environment of cells (Del Duca et al. 2014). Such pH imbalances can interfere with regular metabolic activity and plant development. However, GABA acts to help maintain pH balance within the cytosol, especially under stress.

When plants are exposed to low-oxygen conditions, like those caused by flooding, their cytosolic pH drops, triggering a rise in GABA production (Roberts et al. 1984; Kinnersley and Turano 2000). This process is often worsened by physical stress and the release of organic acids into the cytoplasm due to vacuolar rupture, which further lowers pH and intensifies cytosolic acidity.

Hypoxic conditions are known to reduce the cytosolic pH from normal physiological levels. This change activates L-GAD, an enzyme responsible for converting glutamate into GABA, thereby increasing GABA accumulation (Crawford et al. 1994). In one experiment, exposing plant cells to a weak acid led to a quick drop in cytosolic pH by about 0.6 units and a rapid surge in GABA production, with a response time of just two seconds (Crawford et al. 1994). Similarly, after a few minutes of hypoxia-induced pH decline, GABA synthesis was found to accelerate (Roberts et al. 1992).

Alongside pH changes, stress conditions also raise cytosolic Ca^{2+} levels (Gehring et al. 1990). This increase in Ca^{2+} may enhance the activity of L-GAD, further promoting GABA formation (Crawford et al. 1994). In carrot cells, for instance, glutamate decarboxylation is intensified when cytosolic pH drops (Carroll et al. 1994). This not only supports GABA biosynthesis but also helps remove excess protons from the cytosol, playing a critical role in maintaining intracellular pH balance during stress (Carroll et al. 1994).

14.6.6 GABA and C-N Pool Maintenance

Maintaining a balance between C and N is crucial for plant growth, development, and metabolic functions. An optimal C/N ratio ensures proper regulation of key physiological processes, including leaf senescence (Ansari et al. 2014). GABA plays a vital role in maintaining this balance by influencing both carbon and nitrogen metabolism. Through interactions with major metabolic pathways and signalling networks, GABA helps sustain plant growth, development, and metabolic functions, particularly during senescence (Coruzzi and Zhou 2001). Additionally, an optimal C/N ratio is essential for normal plant growth and development (Bao et al. 2015), as it regulates various metabolic processes, including the progression of leaf senescence (Aoyama et al. 2014). Under nitrogen-deficient conditions, plants accumulate excess sugars, leading to increased carbon levels (Ono et al. 2001). The interplay between carbon and nitrogen is further regulated by a complex signalling network involving nitrate, ammonium, and nitrogen-containing metabolites such as glutamate, glutamine, and aspartate (Miller et al. 2008). To adapt to fluctuating C and N availability, plants strategically allocate these resources and adjust their metabolic activities (Aoyama et al. 2014; Sato et al. 2011; Sulpice et al. 2013). Furthermore, carbon metabolism signals, including CO_2 and TCA cycle intermediates, play a significant role in nitrogen assimilation (Naliwajski and Skłodowska 2018).

GABA is a key regulator of the C/N balance (Fig. 14.3). The GABA shunt bypasses certain steps of the TCA cycle, connecting it to essential metabolic processes (Studart-Guimarães et al. 2007). For instance, in transgenic tomato plants with suppressed succinyl-CoA ligase (SCoAL) activity, GABA metabolism compensates for reduced TCA cycle function, highlighting its role in core metabolic pathways. Similarly, Araújo et al. (2008) demonstrated that inhibiting α-ketoglutarate dehydrogenase in potato tubers links GABA metabolism to the TCA cycle. Research by Fait et al. (2008) further supports this, showing that GABA interacts with primary carbon metabolism at both metabolomic and transcriptomic levels. In cocklebur (*Xanthium strumarium*) leaves, the GABA shunt regulates both carbon and nitrogen metabolism (Tcherkez et al. 2009). Enhanced GABA shunt activity compensates for deficiencies in α-ketoglutarate dehydrogenase, reinforcing GABA's role in central metabolic regulation (Michaeli and Fromm 2015). Additionally, GABA-related genes, such as *GAD2* and *SSADH*, exhibit coordinated expression with genes involved in carbon metabolism.

Exogenous GABA application (50 mM) in *Arabidopsis* seedlings promotes root growth, enhances nitrate uptake, and increases nitrogen accumulation while boosting the activity of key nitrogen metabolism enzymes, including nitrate reductase (NR), NADH-dependent glutamate synthase (NADH-GOGAT), and glutamine synthetase (GS) (Barbosa et al. 2010). These findings suggest that GABA modulates nitrogen metabolism enzymes to maintain cellular C/N balance. GABA also acts as a regulator of C/N balance by influencing organic acid breakdown and amino acid synthesis, as observed in post-harvest citrus fruits (Katz et al. 2011; Sun et al. 2013; Ansari et al. 2021). GABA transporters, such as AtGABP in *Arabidopsis*, are essential for integrating GABA into the TCA cycle. Studies by Michaeli et al. (2011) revealed that *Arabidopsis* mutants lacking AtGABP exhibit stunted growth, whereas wild-type plants with functional AtGABP thrive under carbon-limited conditions. Moreover, GABA transaminase (GABA-T), identified as a senescence-associated gene (SAG), contributes to leaf senescence (Ansari et al. 2005; Jalil et al. 2019a,b). These findings underscore GABA's importance in sustaining carbon levels and promoting plant growth.

Additionally, 14-3-3 proteins, which regulate C/N metabolism, interact with enzymes such as NR and GS, with GABA and calcium modulating these interactions (Lancien and Roberts 2006; Carillo 2018). This intricate regulatory network highlights GABA's critical role in maintaining C/N balance and optimizing plant growth and senescence. Several studies have demonstrated GABA's involvement in regulating cytosolic pH, abiotic stress responses, and primary C/N metabolism, key processes linked to senescence. Thus, GABA has a strong connection with plant senescence. Further research on GABA kinetics, dose-dependent effects, and GABA-induced changes in senescence-related gene expression could uncover additional links between GABA and senescence. Manipulating GABA-mediated regulatory mechanisms may offer new strategies to modify plant growth and productivity, enhancing agricultural sustainability.

14.7 Conclusion

GABA is a versatile signalling molecule that regulates leaf senescence by modulating ROS scavenging, maintaining cellular homeostasis, and influencing transcriptional networks. It delays premature senescence under stress conditions, while facilitating programmed senescence during natural ageing. GABA's interactions with phytohormones, ion fluxes, and carbon-nitrogen metabolism underline its significance in improving stress resilience and enhancing crop productivity. Recent research highlights its potential in sustainable agriculture through bio-regulation and stress adaptation. Future studies should explore its molecular mechanisms, crosstalk with hormonal pathways, and potential applications in innovative agricultural technologies.

References

Aghdam MS, Naderi R, Jannatizadeh A, Sarcheshmeh MAA, Babalar M (2016) Enhancement of postharvest chilling tolerance of anthurium cut flowers by γ-aminobutyric acid (GABA) treatments. Sci Hortic 198:52–60. https://doi.org/10.1016/j.scienta.2015.11.019

Al-Quraan NA (2015) GABA shunt deficiencies and accumulation of reactive oxygen species under UV treatments: insight from Arabidopsis thaliana calmodulin mutants. Acta Physiol Plant 37:86. https://doi.org/10.1007/s11738-015-1836-5

Ansari MI, Chen SCG (2009) Biochemical characterization of gamma-aminobutyric acid (GABA): pyruvate transaminase during rice leaf senescence. Int J Integr Biol 16:27–32

Ansari MI, Lee RH, Chen SCG (2005) A novel senescence-associated gene encoding gamma aminobutyric acid (GABA): pyruvate transaminase is upregulated during rice leaf senescence. Physiol Plant 123:1–8

Ansari MI, Hasan S, Jalil SU (2014) Leaf senescence and GABA shunt. Bioinformation 10:734–736

Ansari MI, Jalil SU, Ansari SA et al (2021) GABA shunt: a key-player in mitigation of ROS during stress. Plant Growth Regul 94:131–149

Aoyama S, Huarancca Reyes T, Guglielminetti L, Lu Y, Morita Y, Sato T, Yamaguchi J (2014) Ubiquitin ligase ATL31 functions in leaf senescence in response to the balance between atmospheric CO 2. Plant Cell Physiol 55:293–305

Araújo WL, Nunes-Nesi A, Trenkamp S, Bunik VI, Fernie AR (2008) Inhibition of 2-oxoglutarate dehydrogenase in potato tuber suggests the enzyme is limiting for respiration and confirms its importance in nitrogen assimilation. Plant Physiol 148:1782–1796

Avila-Ospina L, Moison M, Yoshimoto K, Masclaux-Daubresse C (2014) Autophagy, plant senescence, and nutrient recycling. J Exp Bot 65:3799–3811

Ayala A, Muñoz MF, Argüelles S (2014) Lipid peroxidation: production, metabolism, and signaling mechanisms of malondialdehyde and 4-hydroxy-2-nonenal. Oxidative Med Cell Longev 2014:360438. https://doi.org/10.1155/2014/360438

Balzadeh S, Riano-Pachon DM, Mueller-Roeber B (2008) Transcription factors regulating leaf senescence in Arabidopsis thaliana. Plant Biol 10:63–75

Balazadeh S, Siddiqui H, Allu AD, Matallana-Ramirez LP, Caldana C, Mehrnia M, Zanor MI, Kohler B, Mueller-Roeber B (2010) A gene regulatory network controlled by the NAC transcription factor ANAC092/AtNAC2/ORE1 during salt-promoted senescence. Plant J 62:250–264

Bao A, Zhao Z, Ding G, Shi L, Xu F, Cai H (2015) The stable level of glutamine synthetase 2 plays an important role in rice growth and in carbon-nitrogen metabolic balance. Int J Mol Sci 16:12713–12736

Barbosa JM, Singh NK, Cherry JH, Locy RD (2010) Nitrate uptake and utilization is modulated by exogenous gamma-aminobutyric acid in Arabidopsis thaliana seedlings. Plant Physiol Biochem 48(6):443–450

Berger S, Weichert H, Porzel A, Wasternack C, Kühn H, Feussner I (2001) Enzymatic and non-enzymatic lipid peroxidation in leaf development. Biochim Biophys Acta 1533(266):276. https://doi.org/10.1016/S1388-1981(01)00161-5

Bouche N, Fait A, Bouchez D, Møller SG, Fromm H (2003) Mitochondrial succinic- semialdehyde dehydrogenase of the γ-aminobutyrate shunt is required to restrict levels of reactive oxygen intermediates in plants. PNAS 100:6843–6848

Bray AB, Bailey-Serres J, Weretilnyk E (2000) Responses to abiotic stress. In: Buchanan BB, Gruissem W, Jones RL (eds) Biochemistry & molecular biology of plants. American Society of Plant Physiol, Rockville, p 1158–1203

Breeze E, Harrison E, McHattie S, Hughes L, Hickman R, Hill C (2011) High resolution temporal profiling of transcripts during Arabidopsis leaf senescence reveals a distinct chronology of processes and regulation. Plant Cell 23:873–894

Buchanan-Wollaston V (1997) The molecular biology of leaf senescence. J Exp Bot 48(181):199. https://doi.org/10.1093/jxb/48.2.181

Buchanan-Wollaston V, Earl S, Harrison E, Mathas E, Navabpour S, Page T et al (2003) The molecular analysis of leaf senescence—a genomics approach. Plant Biotechnol J 1:3–22. https://doi.org/10.1046/j.1467-7652.2003.00004.x

Buchanan-Wollaston V, Page T, Harrison E, Breeze E, Lim PO, Nam HG, Lin JF, Wu SH, Swidzinski J, Ishizaki K, Leaver CJ (2005) Comparative transcriptome analysis reveals significant differences in gene expression and signalling pathways between developmental and dark/starvation-induced senescence in Arabidopsis. Plant J 42:567–585

Buet A, Costa ML, Martínez DE, Guiamet JJ (2019) Chloroplast protein degradation in senescing leaves: proteases and lytic compartments. Front Plant Sci 10:747. https://doi.org/10.3389/fpls.2019.00747

Carillo P (2018) GABA shunt in durum wheat. Front Plant Sci 9:100

Carrion CA, Costa ML, Martinez DE, Mohr C, Humbeck K, Guiamet JJ (2013) In vivo inhibition of cysteine proteases provides evidence for the involvement of 'senescence associated vacuoles' in chloroplast protein degradation during dark-induced senescence of tobacco leaves. J Exp Bot 64:4967–4980

Carroll AD, Fox GG, Laurie S, Phillips R, Ratcliffe RG, Stewart GR (1994) Ammonium assimilation and the role of γ-aminobutyric acid in pH homeostasis in carrot cell suspensions. Plant Physiol 106:513–520

Choudhary A, Kumar A, Kaur N (2020) ROS and oxidative burst: roots in plant development. Plant Divers 42(1):33–43. https://doi.org/10.1016/j.pld.2019.10.002

Clément G, Moison M, Soulay F, Reisdorf-Cren M, Masclaux-Daubresse C (2018) Metabolomics of laminae and midvein during leaf senescence and source-sink metabolite management in Brassica napus L. leaves. J Exp Bot 69:891–903. https://doi.org/10.1093/jxb/erx253

Cohen M, Hertweck K, Itkin M, Malitsky S, Dassa B, Fischer AM, Fluhr R (2022) Enhanced proteostasis, lipid remodeling, and nitrogen remobilization define barley flag leaf senescence. J Exp Bot 73:6816–6837. https://doi.org/10.1093/jxb/erac329

Coruzzi GM, Zhou L (2001) Carbon and nitrogen sensing and signaling in plants: emerging 'matrix effects'. Curr Opin Plant Biol 4:247–253

Crawford LA, Bown AW, Breitkreuz KE, Cuine FC (1994) The synthesis of γ- aminobutyric acid in response to treatments reducing cytosolic pH. Plant Physiol 104:865–871

Dai N, Schaffer A, Petreikov M, Shahak Y, Giller Y, Ratner K, Levine A, Granot D (1999) Overexpression of Arabidopsis hexokinase in tomato plants inhibits growth, reduces photosynthesis, and induces rapid senescence. Plant Cell 11:1253–1266

Del Duca S, Serafini-Fracassini D, Cai G (2014) Senescence and programmed cell death in plants: polyamine action mediated by transglutaminase. Front Plant Sci 5:120

Diaz-Mendoza M, Arroyo-Velasco B, Gonzalez-Melendi P, Martinez M, Diaz I (2014) C1A cysteine protease-cystatin interactions in leaf senescence. J Exp Bot 65:3825–3833

Diaz-Mendoza M, Velasco-Arroyo B, Santamaria ME, González-Melendi P, Martinez M, Diaz I (2016) Plant senescence and proteolysis: two processes with one destiny. Genet Mol Biol 39:329–338. https://doi.org/10.1590/1678-4685-GMB-2016-0015

Domínguez F, Cejudo FJ (2021) Chloroplast dismantling in leaf senescence. J Exp Bot 72:5905–5918

Fait A, Yellin A, Fromm H (2005) GABA shunt deficiencies and accumulation of reactive oxygen intermediates: insight from Arabidopsis mutants. FEBS Lett 579:415–420

Fait H, Fromm D, Walter G, Galili AR (2008) Fernie, Highway or byway: the metabolic role of the GABA shunt in plants. Trends Plant Sci 13:14–19

Fan SC, Lin CS, Hsu PK, Lin SH, Tsay YF (2009) The Arabidopsis nitrate transporter NRT1.7, expressed in phloem, is responsible for source to-sink remobilization of nitrate. Plant Cell 21:2750–2761

Feller U, Fischer A (1994) Nitrogen metabolism in senescing leaves. Crit Rev Plant Sci 13:241–273

Fei L, Zuo S, Zhang J, Wang Z (2022) Phytoextraction by harvesting dead leaves: cadmium accumulation associated with the leaf senescence in Festuca arundinacea Schreb. Environ Sci Pollut Res Int 29:79214–79223

Fischer AM (2012) The complex regulation of senescence. Crit Rev Plant Sci 31:124–147

Fortunato S, Nigro D, Lasorella C, Marcotuli I, Gadaleta A, de Pinto MC (2023) The role of glutamine synthetase (GS) and glutamate synthase (GOGAT) in the improvement of nitrogen use efficiency in cereals. Biomolecules 13:1771. https://doi.org/10.3390/biom13121771

Gan S (2003) Mitotic and postmitotic senescence in plants. Sci SAGE KE 2003:RE7

Gan S, Amasino RM (1997) Making sense of senescence: molecular genetic regulation and manipulation of leaf senescence. Plant Physiol 113:313–319

Gehring AC, Irving HR, Parish RW (1990) Effects of auxin and abscisic acid on cytosolic calcium and pH in plants cells. Proc Natl Acad Sci USA 87:9645–9649

Gilliham M, Tyerman SD (2016) Linking metabolism to membrane signaling: the GABA malate connection. Trends Plant Sci 21:295–301. https://doi.org/10.1016/j.tplants.2015.11.011

Graham IA (2008) Seed storage oil mobilization. Annu Rev Plant Biol 59:115–142

Graham IA, Eastmond PJ (2002) Pathways of straight and branched chain fatty acid catabolism in higher plants. Prog Lipid Res 41(2):156–181. https://doi.org/10.1016/s0163-7827(01)00022-4. PMID: 11755682

Gregersen PL, Holm PB, Krupinska K (2008) Leaf senescence and nutrient remobilisation in barley and wheat. Plant Biol 10:37–49

Guiboileau A, Sormani R, Meyer C, Masclaux-Daubresse C (2010) Senescence and death of plant organs: nutrient recycling and developmental regulation. C R Biol 333:382–391

Guo Y, Gan S (2005) Leaf senescence: signals, execution, and regulation. Curr Top Dev Biol 71:83–112

Guo Y, Gan S (2006) AtNAP, a NAC family transcription factor, has an important role in leaf senescence. Plant J 46:601–612

Guo Y, Gan SS (2014) Translational researches on leaf senescence for enhancing plant productivity and quality. J Exp Bot 65:3901–3913

Guo Y, Ren G, Zhang K et al (2021) Leaf senescence: progression, regulation, and application. Mol Hortic 1:5. https://doi.org/10.1186/s43897-021-00006-9

Guo Z, Gong J, Luo S, Zuo Y, Shen Y (2023) Role of gamma-aminobutyric acid in plant defense response. Meta 13:741. https://doi.org/10.3390/metabo13060741

Hajibarat Z, Saidi A (2022) Senescence-associated proteins and nitrogen remobilization in grain filling under drought stress condition. J Genet Eng Biotechnol 20:101. https://doi.org/10.1186/s43141-022-00378-5

Hasanuzzaman M, Bhuyan MHMA, Zulfiqar F, Raza A, Mohsin SM, Mahmud JA, Fujita M, Fotopoulos V (2020) Reactive oxygen species and antioxidant defense in plants under abiotic stress: revisiting the crucial role of a universal defense regulator. Antioxidants 9:681

Havé M, Marmagne A, Chardon F, Masclaux-Daubresse C (2017) Nitrogen remobilization during leaf senescence: lessons from Arabidopsis to crops. J Exp Bot 68:2513–2529. https://doi.org/10.1093/jxb/erw365

He Y, Gan SA (2002) Gene encoding an acyl hydrolase is involved in leaf senescence in Arabidopsis. Plant Cell 14:805–815

Himelblau E, Amasino RM (2001) Nutrients mobilized from leaves of Arabidopsis thaliana during leaf senescence. J Plant Physiol 158:1317–1323

Hong Y, Wang T-W, Hudak KA, Schade F, Froese CD, Thompson JE (2000) An ethylene-induced cDNA encoding a lipase expressed at the onset of senescence. Proc Natl Acad Sci USA 97:8717–8722

Hortensteiner S (1998) NCC malonyltransferase catalyses the final step in chlorophyll breakdown in rape (Brassica napus). Phytochemistry 49:953–956

Hörtensteiner S, Feller U (2002) Nitrogen metabolism and remobilization during senescence. J Exp Bot 53(370):927–937. https://doi.org/10.1093/jexbot/53.370.927. PMID: 11912235

Hu X, Zhang A, Zhang J, Jiang M (2006) Abscisic acid is a key inducer of hydrogen peroxide production in leaves of maize plants exposed to water stress. Plant Cell Physiol 47:1484–1495

Iqbal N, Khan NA, Nazar R, Teixeira da Silva JA (2012) Ethylene-stimulated photosynthesis results from increased nitrogen and sulfur. Environ Exp Bot 78:84–90

Ishida H, Anzawa D, Kokubun N, Makino A, Mae T (2002) Direct evidence for non-enzymatic fragmentation of chloroplastic glutamine synthetase by a reactive oxygen species. Plant Cell Environ 25:625–631

Ishida H, Yoshimoto K, Izumi M, Reisen D, Yano Y, Makino A, Ohsumi Y, Hanson MR, Mae T (2008) Mobilization of Rubisco and stroma-localized fluorescent proteins of chloroplast to the vacuole by ATG gene-dependent autophagic process. Plant Physiol 148:142–155

Islam SN, Kouser S, Hassan P et al (2024) Gamma-aminobutyric acid interactions with phytohormones and its role in modulating abiotic and biotic stress in plants. Stress Biol 4:36. https://doi.org/10.1007/s44154-024-00180-y

Jajic I, Sarna T, Strzalka K (2015) Senescence, stress, and reactive oxygen species. Plants 4:393–411

Jalil SU, Ahmad I, Ansari MI (2017) Functional loss of GABA transaminase (GABA-T) expressed early leaf senescence under various stress conditions in Arabidopsis thaliana. Curr Plant Biol 9–10:11–22

Jalil SU, Khan MAA, Ansari MI (2019a) Tremendous role of GABA shunt metabolic pathway in plant system. J Biol Chem Res 36:188–196

Jalil SU, Khan MQR, Ansari MI (2019b) Role of GABA transaminase in the regulation of development and senescence in Arabidopsis thaliana. Curr Plant Biol 19:100119. https://doi.org/10.1016/j.cpb.2019.100119

Jalil SU, Ansari S, Ansari MI (2022) Role of environment stress, leaf senescence and crop productivity. In: Ansari MI (ed) Augmenting crop productivity in stress environment. Springer, Singapore. https://doi.org/10.1007/978-981-16-6361-1_2

Jansson S, Thomas H (2008) Senescence: developmental program or timetable? New Phytol 179:575–579

Jespersen D, Yu J, Huang B (2015) Metabolite responses to exogenous application of nitrogen, cytokinin, and ethylene inhibitors in relation to heat-induced senescence in creeping bentgrass. PLoS One 10:e0123744

Jespersen D, Zhang J, Huang B (2016) Chlorophyll loss associated with heat-induced senescence in bentgrass. Plant Sci 249:1–12

Jing HC, Hille J, Dijkwel RR (2003) Ageing in plants: conserved strategies and novel pathways. Plant Biol 5:455–464

Jing HC, Schippers JH, Hille J, Dijkwel PP (2005) Ethylene-induced leaf senescence depends on age-related changes and OLD genes in Arabidopsis. J Exp Bot 56:2915–2923. https://doi.org/10.1093/jxb/eri287

John CF, Morris K, Jordan BR, Thomas B, Mackerness SA-H (2001) Ultraviolet-B exposure leads to up-regulation of senescence-associated genes in Arabidopsis thaliana. J Exp Bot 52:1367–1373

Juan CA, Manuel J, Plou FJ (2021) The chemistry of reactive oxygen species (ROS) revisited: outlining their role in biological macromolecules (DNA, lipids and proteins) and induced pathologies. Int J Mol Sci 22:4642

Jukanti AK, Heidlebaugh NM, Parrott DL, Fischer IA, McInnerney K, Fischer AM (2008) Comparative transcriptome profiling of nearisogenic barley (Hordeum vulgare) lines differing in the allelic state of a major grain protein content locus identifies genes with possible roles in leaf senescence and nitrogen reallocation. New Phytol 177:333–349

Kato Y, Murakami S, Yamamoto Y, Chantani H, Kondo Y, Nakano T, Yokota A, Sato F (2004) The DNA-binding protease, CND41, and the degradation of ribulose-1,5-bisphosphate carboxylase/oxygenase in senescent leaves of tobacco. Planta 220:97–104

Katz E, Boo KH, Kim HY, Eigenheer RA, Phinney BS, Shulaev V, Negre- Zakharov F, Sadka BE (2011) Label-free shotgun proteomics and metabolite analysis reveal a significant metabolic shift during citrus fruit development. J Exp Bot 62:5367–5368

Keech O, Pesquet E, Ahad A, Askne A, Nordvall DAG, Vodnala SM et al (2007) The different fates of mitochondria and chloroplasts during dark-induced senescence in Arabidopsis leaves. Plant Cell Environ 30:1523–1534

Keech O, Pesquet E, Gutierrez L, Ahad A, Bellini C, Smith SM et al (2010) Leaf senescence is accompanied by an early disruption of the microtubule network in Arabidopsis thaliana. Plant Physiol 110:163402

Khan MIR, Khan NA (2017) An introduction to antioxidants and their roles in plant stress tolerance. In: Khan MIR, Khan NA (eds) Reactive oxygen species and antioxidant systems in plants: role and regulation under abiotic stress. Springer, Singapore, pp 1–23

Khan MIR, Jalil SU, Chopra P, Chhillar H, Ferrante A, Khan NA, Ansari MI (2021) Role of GABA in plant growth, development and senescence. Plant Gene 26:100283

Kim J (2019) Sugar metabolism as input signals and fuel for leaf senescence. Genes Genom 41:737–746

Kim C, Kim SJ, Jeong J, Park E, Oh E, Park YI, Lim PO, Choi G (2020) High ambient temperature accelerates leaf senescence via phytochrome-interacting factor 4 and 5 in Arabidopsis. Mol Cells 43:645–661

Kinnersley AM, Turano FJ (2000) Gamma aminobutyric acid (GABA) and plant responses to stress. Crit Rev Plant Sci 19:479–509. https://doi.org/10.1080/07352680091139277

Kumar V, Khare T, Sharma M, Wani SH (2017) ROS-induced signaling and gene expression in crops under salinity stress. In: Khan MIR, Khan NA (eds) Reactive oxygen species and antioxidant systems in plants: role and regulation under abiotic stress. Springer, Singapore, pp 159–184

Kumar V, Khare T, Srivastav A, Surekha C, Shriram V, Wani SH (2019) Oxidative stress and leaf senescence: important insights. In: Senescence signalling and control in plants. Academic Press, pp 139–163

Lancien M, Roberts MR (2006) Regulation of Arabidopsis thaliana 14-3-3 gene expression by gammaaminobutyric acid. Plant Cell Environ 29:1430–1436

Lezhneva L, Kiba T, Feria-Bourrellier AB, Lafouge F, Boutet-Mercey S, Zoufan P, Sakakibara H, Daniel-Vedele F, Krapp A (2014) The Arabidopsis nitrate transporter NRT2.5 plays a role in nitrate acquisition and remobilization in nitrogen-starved plants. Plant J 80:230–241

Li W, Liu J, Ashraf U, Li G, Li Y, Lu W, Gao L, Han F, Hu J (2016a) Exogenous γ aminobutyric acid (GABA) application improved early growth, net photosynthesis, and associated physio-biochemical events in maize. Front Plant Sci 7:919. https://doi.org/10.3389/fpls.2016.00919

Li Z, Yu J, Peng Y, Huang B (2016b) Metabolic pathways regulated by γ-aminobutyric acid (GABA) contributing to heat tolerance in creeping bentgrass (Agrostis stolonifera). Sci Rep 6:30338. https://doi.org/10.1038/srep30338

Li H, Chang J, Chen H, Wang Z, Gu X, Wei C, Zhang Y, Ma J, Yang J, Zhang X (2017) Exogenous melatonin confers salt stress tolerance to watermelon by improving photosynthesis and redox homeostasis. Front Plant Sci 8:295. https://doi.org/10.3389/fpls.2017.00295

Li L, Dou N, Zhang H, Wu C (2021) The versatile GABA in plants. Plant Signal Behav 16(3):1862565. https://doi.org/10.1080/15592324.2020.1862565

Lim PO, Woo HR, Nam HG (2003) Molecular genetics of leaf senescence in Arabidopsis. Trends Plant Sci 8:272–278

Lim PO, Kim HJ, Nam HG (2007) Leaf senescence. Annu Rev Plant Biol 58:115–136. https://doi.org/10.1146/annurev.arplant.57.032905.105316

Lim PO, Kim HJ, Nam HG (2007a) Leaf senescence. Annu Rev Plant Biol 58:115

Liu C, Zhao L, Yu G (2011) The dominant glutamic acid metabolic flux to produce gamma amino butyric acid over proline in Nicotiana tabacum leaves under water stress relates to its significant role in antioxidant activity. J Integr Plant Biol 53:608–618. https://doi.org/10.1111/j.1744-7909.2011.01049.x

Masclaux-Daubresse C, Chardon F (2011) Exploring nitrogen remobilization for seed filling using natural variation in Arabidopsis thaliana. J Exp Bot 62:2131–2142

Masclaux C, Valadier M-H, Brugière N, Morot-Gaudry JF, Hirel B (2000) Characterization of the sink/source transition in tobacco (Nicotiana tabacum L.) shoots in relation to nitrogen management and leaf senescence. Planta 211:510–518

Masclaux-Daubresse C, Reisdorf-Cren M, Pageau K, Lelandais M, Grandjean O, Kronenberger J, Valadier MH, Feraud M, Jouglet T, Suzuki A (2006) Glutamine synthetase/glutamate synthase pathway and glutamate dehydrogenase play distinct roles for sink/source nitrogen cycle in tobacco (Nicotiana tabacum L.). Plant Physiol 140:444–456

Masclaux-Daubresse C, Daniel-Vedele F, Dechorgnat J, Chardon F, Gaufichon L, Suzuki A (2010) Nitrogen uptake, assimilation and remobilization in plants: challenges for sustainable and productive agriculture. Ann Bot 105(7):1141–1157. https://doi.org/10.1093/aob/mcq028

Mekonnen DW, Flügge UI, Ludewig F (2016) Gamma-aminobutyric acid depletion affects stomata closure and drought tolerance of Arabidopsis thaliana. Plant Sci 245:25–34. https://doi.org/10.1016/j.plantsci.2016.01.005

Michaeli S, Fromm H (2015) Closing the loop on the GABA shunt in plants: are GABA metabolism and signalling entwined? Front Plant Sci 9:419

Michaeli S, Fait A, Lagor K, Nunes-Nesi A, Grillich N, Yellin A, Bar D, Khan M, Frank ARF, Turano H, Fromm J (2011) A mitochondrial GABA permease connects the GABA shunt and the TCA cycle, and is essential for normal carbon metabolism. Plant J 67:485–498

Miller AJ, Fan X, Shen Q, Smith SJ (2008) Amino acids and nitrate as signals for the regulation of nitrogen acquisition. J Exp Bot 59:111–119

Mittler R (2002) Oxidative stress, antioxidants and stress tolerance. Trends Plant Sci 7:405–410

Moore B, Zhou L, Rolland F, Hall Q, Cheng WH, Liu YX, Hwang I, Jones T, Sheen J (2003) Role of the Arabidopsis glucose sensor HXK1 in nutrient, light, and hormonal signaling. Science 300:332–336

Munné-Bosch S, Alegre L (2004) Die and let live: leaf senescence contributes to plant survival under drought stress. Funct Plant Biol 31(3):203–216. https://doi.org/10.1071/FP03236

Naliwajski MR, Skłodowska M (2018) The relationship between carbon and nitrogen metabolism in cucumber leaves acclimated to salt stress. PeerJ 6:e6043

Nam HG (1997) The molecular genetic analysis of leaf senescence. Curr Opin Biotechnol 8(8):200–207

Navrot N, Rouhier N, Gelhaye E, Jacquot JP (2007) Reactive oxygen species generation and antioxidant systems in plant mitochondria. Physiol Plant 129:185–195

Nayyar H, Kaur R, Kaur S, Singh R (2014) γ-Aminobutyric acid (GABA) imparts partial protection from heat stress injury to rice seedlings by improving leaf turgor and upregulating osmoprotectants and antioxidants. J Plant Growth Regul 33:408–419. https://doi.org/10.1007/s00344-013-9389-6

Noctor G, Veljovic-Jovanovic S, Driscoll S, Novitskaya L, Foyer CH (2002) Drought and oxidative load in the leaves of C3 plants: a predominant role for photorespiration? Ann Bot 89:841–850

Noh Y-S, Amasino R (1999) Identification of a promoter region responsible for the senescence-specific expression of SAG12. Plant Mol Biol 41:181–194

Nooden LD (1988) Whole plant senescence. In: Nooden LD, Leopold SC (eds) Senescence and aging in plants. Academic Press, San Diego, pp 391–439

Noodén LD, Penney JP (2001) Correlative controls of senescence and plant death in Arabidopsis thaliana (Brassicaceae). J Exp Bot 52:2151–2159
Ono K, Watanabe A (1997) Levels of endogenous sugars, transcripts of rbcS and rbcL, and of RuBisCO protein in senescing sunflower leaves. Plant Cell Physiol 38:1032–1038
Ono K, Nishi Y, Watanabe A, Terashima I (2001) Possible mechanisms of adaptive leaf senescence. Plant Biol 3:234–243
Park SJ, Park S, Kim Y, Hyeon DY, Park H, Jeong J, Jeong U, Yoon YS, You D, Kwak J et al (2022) Ethylene responsive factor34 mediates stress-induced leaf senescence by regulating salt stress-responsive genes. Plant Cell Environ 45:1719–1733
Parrott D, Yang L, Shama L, Fischer AM (2005) Senescence is accelerated, and several proteases are induced by carbon 'feast' conditions in barley (Hordeum vulgare L.) leaves. Planta 222:989–1000
Partelli FL, Batista S (2001) Characterization of the main lipid components of chloroplast membranes and cold induced changes in Coffea spp. Environ Exp Bot 74:194–204
Peleg Z, Reguera M, Tumimbang E, Walia H, Blumwald E (2011) Cytokinin-mediated source/sink modifications improve drought tolerance and increase grain yield in rice under water-stress. Plant Biotechnol J 9:747–758
Peoples MB, Dalling MJ (1988) The interplay between proteolysis and amino acid metabolism during senescence and nitrogen reallocation. In: Noodén LD, Leopold AC (eds) Senescence and aging in plants. Academic Press, San Diego, pp 181–217
Piacentini D, Corpas FJ, D'Angeli S, Altamura MM, Falasca G (2020) Cadmium and arsenic induced-stress differentially modulates Arabidopsis root architecture, peroxisome distribution, enzymatic activities and their nitric oxide content. Plant Physiol Biochem 148:312–323
Podlešáková K, Ugena L, Spíchal L, Doležal K, De Diego N (2019) Phytohormones and polyamines regulate plant stress responses by altering GABA pathway. New Biotechnol 48:53–65. https://doi.org/10.1016/j.nbt.2018.07.003
Pottosin I, Velarde-Buendía AM, Bose J, Zepeda-Jazo I, Shabala S, Dobrovinskaya O (2014) Cross-talk between reactive oxygen species and polyamines in regulation of ion transport across the plasma membrane: implications for plant adaptive responses. J Exp Bot 65:1271–1283
Pourtau N, Marès M, Purdy S, Quentin N, Ruël A, Wingler A (2004) Interactions of abscisic acid and sugar signalling in the regulation of leaf senescence. Planta 219(765):772
Pourtau N, Jennings R, Pelzer E, Pallas J, Wingler A (2006) Effect of sugar-induced senescence on gene expression and implications for the regulation of senescence in Arabidopsis. Planta 224:556–568
Ramesh SA, Tyerman SD, Xu B, Bose J, Kaur S, Conn V, Domingos P, Ullah S, Wege S, Shabala S et al (2015) GABA signalling modulates plant growth by directly regulating the of plant-specific anion transporters. Nat Commun 6:7879. https://doi.org/10.1038/ncomms8879
Roberts JKM, Callis J, Wemmer D, Walbot V, Jardetzky O (1984) Mechanism of cytoplasmic pH regulation in hypoxic maize root tips and its role in survival under hypoxia. Proc Natl Acad Sci USA 81:3379–3383
Roberts JKM, Hooks MA, Miaullis AP, Edwards S, Webster C (1992) Contribution of malate and amino acid metabolism to cytoplasmic pH regulation in hypoxic maize root tips studied using nuclear magnetic resonance spectroscopy. Plant Physiol 98:480–487
Robert-Seilaniantz A, Grant M, Jones JDG (2011) Hormone crosstalk in plant disease and defense: more than just Jasmonate– salicylate antagonism. In: VanAlfen NK, Bruening G, Leach JE (eds) Ann Rev Phytopatho, vol 49, pp 317–343
Roberts IN, Caputo C, Criado MV, Funk C (2012) Senescence-associated proteases in plants. Physiol Plant 145:130–139
Sachdev S, Ansari SA, Ansari MI, Fujita M, Hasanuzzaman M (2021) Abiotic stress and reactive oxygen species: generation, signaling, and defense mechanisms. Antioxidants 10:277
Sade N, Rubio-Wilhelmi MDM, Umnajkitikorn K, Blumwald E (2018) Stress-induced senescence and plant tolerance to abiotic stress. J Exp Bot 69:845–853
Sakamoto W, Takami T (2014) Nucleases in higher plants and their possible involvement in DNA degradation during leaf senescence. J Exp Bot 65:3835

Sakuraba Y, Kim D, Paek NC (2018) Salt treatments and induction of senescence. Methods Mol Biol 1744:141–149

Sato T, Maekawa S, Yasuda S, Yamaguchi J (2011) Carbon and nitrogen metabolism regulated by the ubiquitin–proteasome system. Plant Signal Behav 6:1465–1468

Schmulling T (2002) New insights into the functions of cytokinins in plant development. J Plant Growth Regul 21:40–49. https://doi.org/10.1007/s003440010046

Seifi KM, Aliniaeifard S, Mehdi S, Javadi AE, Françoise B, Batool H (2018) Enhanced salt tolerance and photosynthetic performance: implication of γ-aminobutyric acid application in salt-exposed lettuce (*Lactuca sativa* L.) plants. Plant Physiol Biochem 130:157

Sharma P, Jha AB, Dubey RS, Pessarakli M (2012) Reactive oxygen species, oxidative damage, and antioxidative defense mechanism in plants under stressful conditions. J Bot 2012:1–26

Shi SQ, Zheng S, Jiang ZP, Liu Y, Ding X, Fan YL, Wang SH (2010) Effects of exogenous GABA on gene expression of Caragana intermedia roots under NaCl stress: regulatory roles for H_2O_2 and ethylene production. Plant Cell Environ 33:149–162. https://doi.org/10.1111/j.1365-3040.2009.02065.x

Singh S, Singh M, Bisht S, Sharma JG (2022) Leaf senescence and its regulation with phytohormones and essential elements: an overview. J Appl Biol Biotechnol 10:185–197

Smart CM (1994) Gene expression during leaf senescence. New Phytol 126:419–448

Song H, Xu X, Wang H, Wang H, Tao Y (2010) Exogenous gamma-aminobutyric acid alleviates oxidative damage caused by aluminium and proton stresses on barley seedlings. J Sci Food Agric 90:1410–1416. https://doi.org/10.1002/jsfa.3951

Studart-Guimaraes C, Fait A, Nunes-Nesi A, Carrari F, Usadel B, Fernie AR (2007) Reduced expression of succinyl-coenzyme a ligase can be compensated for by up- regulation of the gamma-aminobutyrate shunt in illuminated tomato leaves. Plant Physiol 145:626–639

Sulpice R, Nikoloski Z, Tschoep H, Antonio C, Kleessen S, Larhlimi A (2013) Impact of the carbon and nitrogen supply on relationships and connectivity between metabolism and biomass in a broad panel of Arabidopsis accessions. Plant Physiol 162:347–363

Sun X, Zhu A, Liu S, Sheng L, Ma Q, Zhang L, Nishawy EM, Zeng Y, Xu J, Ma Z, Cheng Y, Deng X (2013) Integration of metabolomics and subcellular organelle expression microarray to increase understanding the organic acid changes in post-harvest citrus fruit. J Integr Plant Biol 55:1038–1053

Sztatelman O, Grzyb J, Gabryś H, Romanowska E (2015) The effect of UV-B on Arabidopsis leaves depends on light conditions after treatment. BMC Plant Biol 15:281. https://doi.org/10.1186/s12870-015-0667-2

Takamiya KI, Tsuchiya T, Ohta H (2000) Degradation pathway(s) of chlorophyll: what has gene cloning revealed? Trends Plant Sci 5:426–431

Tcherkez G, Mahe A, Gauthier P, Mauve C, Gout E, Bligny R (2009) In folio respiratory fluxomics revealed by 13C isotopic labeling and H/D isotope effects highlight the noncyclic nature of the tricarboxylic acid "cycle" in illuminated leaves. Plant Physiol 151:620–630

Tan S, Sha Y, Sun L, Li Z (2023) Abiotic stress-induced leaf senescence: regulatory mechanisms and application. Int J Mol Sci 24:11996. https://doi.org/10.3390/ijms241511996

Thomas H, Donnison I (2000) Back from the brink: plant senescence and its reversibility. In: Bryant JA, Hughes SG, Garland JM (eds) Programmed cell death in animals and plants. Bios, Oxford, pp 149–162

Thompson JE, Froese CD, Madey E, Smith MD, Hong Y (1998) Lipid metabolism during plant senescence. Prog Lipid Res 37:119–141. https://doi.org/10.1016/S0163-7827(98)00006

van Doorn WG, Woltering EJ (2004) Senescence and programmed cell death: substance or semantics? J Exp Bot 55(406):2147–2153. https://doi.org/10.1093/jxb/erh264. Epub 2004 Sep 10. PMID: 15361529

van Doorn WG, Woltering EJ (2008) Physiology and molecular biology of petal senescence. J Exp Bot 59(3):453–480. https://doi.org/10.1093/jxb/erm356. Epub 2008 Feb 28. PMID: 18310084

Vicentini F, Hörtensteiner S, Schellenberg M, Thomas H, Matile P (1995) Chlorophyll breakdown in senescent leaves: identification of the biochemical lesion in a stay-green genotype of

Festuca pratensis Huds. New Phytol 129:247–252. https://doi.org/10.1111/j.1469-8137.1995.tb04294.x
Wada S, Ishida H (2009) Chloroplast autophagy during senescence of individually darkened leaves. Plant Signal Behav 4:565–567. https://doi.org/10.1104/pp.108.130013
Wang H, Gao C, Ge Q (2022) Low temperature and short daylength interact to affect the leaf senescence of two temperate tree species. Tree Physiol 42:2252–2265
Wingler A, Purdy S, MacLean JA, Pourtau N (2006) The role of sugars in integrating environmental signals during the regulation of leaf senescence. J Exp Bot 57:391–399
Wongshaya P, Chayjarung P, Tothong C, Pilaisangsuree V, Somboon T, Kongbangkerd A, Limmongkon A (2020) Effect of light and mechanical stress in combination with chemical elicitors on the production of stilbene compounds and defensive responses in peanut hairy root culture. Plant Physiol Biochem 157:93–104
Wu T, Qin Z, Fan L, Xue C, Zhou X, Xin M, Du Y (2014) Involvement of CsNRT1.7 in nitrate recycling during senescence in cucumber. J Plant Nutr Soil Sci 177:714–721
Xu Y, Charles MT, Luo Z, Mimee B, Tong Z, Véronneau PY, Rolland D (2019) Ultraviolet-C priming of strawberry leaves against subsequent mycosphaerella fragariae infection involves the action of reactive oxygen species, plant hormones, and terpenes. Plant Cell Environ 42(3):815–831
Xu C, Fan J, Shanklin J (2020) Metabolic and functional connections between cytoplasmic and chloroplast triacylglycerol storage. Prog Lipid Res 80:101069
Xu B, Long Y, Feng X, Zhu X, Sai N, Chirkova L, Betts A, Herrmann J, Edwards EJ, Okamoto M et al (2021a) GABA signalling modulates stomatal opening to enhance plant water use efficiency and drought resilience. Nat Commun 12:1952. https://doi.org/10.1038/s41467-021-21694-3
Xu B, Sai N, Gilliham M (2021b) The emerging role of GABA as a transport regulator and physiological signal. Plant Physiol 187:2005–2016. https://doi.org/10.1093/plphys/kiab347
Xu J, Liu T, Qu F, Jin X, Huang N, Wang J, Hu X (2021c) Nitric oxide mediates γ-aminobutyric acid-enhanced muskmelon tolerance to salinity–alkalinity stress conditions. Sci Hortic 286:110229. https://doi.org/10.1016/j.scienta.2021.110229
Yang A, Cao S, Yang Z, Cai Y, Zheng Y (2011) γ-Aminobutyric acid treatment reduces chilling injury and activates the defence response of peach fruit. Food Chem 129:1619–1622
Zhang J, Zhang Y, Du Y, Chen S, Tang H (2011) Dynamic metabonomic responses of tobacco (Nicotiana tabacum) plants to salt stress. J Proteome Res 10:1904–1914
Zhang S, Dai J, Ge Q (2020) Responses of autumn phenology to climate change and the correlations of plant hormone regulation. Sci Rep 10:9039
Zhang J, Fei L, Dong Q, Zuo S, Li Y, Wang Z (2022) Cadmium binding during leaf senescence in Festuca arundinacea: promotion phytoextraction efficiency by harvesting dead leaves. Chemosphere 289:133253

GABA and Postharvest Management 15

Abstract

In the postharvest phase, maintaining the quality and extending the shelf life of agricultural products is paramount. Gamma-aminobutyric acid (GABA) plays a crucial role in postharvest management, exploring its influence on stress alleviation, enhancement of antioxidant defense mechanisms, and modulation of key metabolic pathways. By examining the biochemical and physiological underpinnings of GABA's action, a comprehensive understanding of how GABA can be harnessed to improve postharvest quality and reduce losses in various horticultural crops is warranted. Through this exploration, the potential for GABA-based strategies to revolutionize postharvest management practices will be highlighted, offering new avenues for research and application in the agricultural sector.

Keywords

Antioxidant defense · GABA · Horticultural crops · Postharvest stress · ROS · Shelf life extension

Highlights

- Gamma-aminobutyric acid (GABA) enhances antioxidant defense mechanisms and offsets Reactive oxygen species (ROS) accumulation during storage.
- GABA acts as a signaling molecule, regulating stress-responsive genes and enhancing physiological resilience during storage.
- GABA alleviates postharvest stress, maintaining cellular integrity and reducing quality deterioration.
- GABA modulates key metabolic pathways, influencing respiration rate, ethylene production, and delaying senescence.

S. U. Jalil et al., *GABA Signaling System and Abiotic Stress Tolerance in Plants*,
https://doi.org/10.1007/978-981-95-0437-4_15

- GABA improves shelf life and quality, contributing to better texture, color, and nutritional value of stored produce.
- GABA offers a promising postharvest strategy, with potential applications in reducing losses and improving storage outcomes in horticultural crops.

15.1 Introduction

In agriculture, managing crops after harvest is crucial for maintaining their quality, shelf life, and market value. This process becomes challenging due to the complex interactions of physiological, biochemical, and environmental factors. Effective postharvest practices are essential as they help prevent quality deterioration and economic losses in the horticultural industry. While techniques such as low temperatures and controlled atmospheres (low oxygen and high carbon dioxide) are commonly used to extend shelf life, they can also lead to issues like chilling injury, excessive water loss, and fungal decay (Lum et al. 2016; Tarkowski et al. 2020; Ziv and Fallik 2021). Gamma-aminobutyric acid (GABA) has recently gained attention for its potential benefits in postharvest management (Khan et al. 2021).

GABA's role, traditionally linked with stress mitigation in living plants, has increasingly been recognized for its significant impact during the postharvest phase. It plays a crucial role in regulating oxidative stress and ethylene production and maintaining cellular homeostasis (Li et al. 2021; Khan et al. 2021). As postharvest management strategies evolve, incorporating GABA-based approaches presents a promising opportunity to improve storage quality and reduce postharvest losses in fruits, vegetables, and other horticultural products. Recently, GABA has emerged as a safe and effective compound for preserving the quality of fresh produce during storage. GABA treatments substantially decrease decay rates, regulate carbon and nitrogen metabolism, and mitigate oxidative damage, thereby enhancing the storage performance of fruits (Aghdam et al. 2022; Zhang et al. 2022; Zhang et al. 2024).

The chapter explores the biochemical and physiological roles of GABA during postharvest storage, highlighting its impact on stress alleviation, antioxidant defenses, and metabolic modulation. By delving into the mechanisms by which GABA can be leveraged to improve postharvest quality, we aim to provide a comprehensive overview that bridges fundamental research with practical applications, ultimately contributing to more efficient and sustainable postharvest management practices.

15.2 Biochemistry of GABA in Plants

GABA is a nonprotein amino acid that plays a pivotal role in plant metabolism, particularly during stress responses. The synthesis of GABA in plants primarily occurs through the decarboxylation of glutamate, a reaction catalyzed by the enzyme glutamate decarboxylase (GAD) (Khan et al. 2021). This process is typically activated under conditions of stress, such as low pH, hypoxia, or mechanical damage, leading to a rapid accumulation of GABA in plant tissues. Once synthesized, GABA can be metabolized through the GABA shunt, a bypass of the tricarboxylic acid (TCA) cycle (Ansari et al. 2014, 2021). In this shunt, GABA transaminase (GABA-T) first converts GABA into succinic semialdehyde that subsequently acts as a substrate to succinic semialdehyde dehydrogenase (SSADH) for the synthesis of succinate (Jalil et al. 2016). Succinate enters the TCA cycle, linking GABA metabolism to the central carbon metabolism of the plant, which is crucial for energy production and biosynthetic processes (Ansari et al. 2021).

The regulation of GABA levels is critical during the postharvest period, as it influences the quality and shelf life of harvested crops (Aghdam et al. 2022). Postharvest conditions, such as changes in temperature, oxygen levels, and exposure to stressors, can lead to fluctuations in GABA concentration. This regulation is closely tied to the plant's response to oxidative stress and other postharvest challenges (Lum et al. 2016). For example, GABA accumulation modulates the antioxidative defense system, alleviating oxidative damage during postharvest storage (Zhang et al. 2024). Additionally, GABA acts as a signaling molecule, influencing processes like fruit ripening and stress response pathways (Han et al. 2018; Li et al. 2021). The fine-tuning of GABA synthesis and catabolism during postharvest is therefore essential for maintaining cellular homeostasis and preventing deterioration of the stored produce.

GABA does not act in isolation; it interacts with various metabolic pathways that are integral to plant growth and stress responses (Jalil et al. 2019a; Jalil and Ansari 2020). One of the key interactions is with the carbon and nitrogen metabolism pathways. GABA can influence the balance between carbon and nitrogen in the plant by interacting with the TCA cycle and amino acid metabolism (Jalil et al. 2017; Jalil et al. 2019b). This interaction is particularly important during postharvest, when energy and nitrogen requirements shift as the plant tissues experience the transition to enter the storage state. Moreover, GABA interacts with the polyamine biosynthesis pathway, where it can modulate levels of polyamines like putrescine, which are also involved in stress responses and postharvest quality (Ansari et al. 2021). Additionally, GABA regulates ion homeostasis, particularly in relation to calcium and potassium ions, and emphatically underscores its importance in maintaining cellular integrity during the postharvest period. Through these interactions, GABA contributes to a network of metabolic processes that collectively enhance the resilience of plants during storage, thereby improving postharvest outcomes.

15.2.1 Role of GABA in Stress Alleviation During Postharvest

GABA plays a pivotal role in alleviating postharvest stress by managing oxidative stress, modulating ethylene production, and maintaining cellular homeostasis (Fig. 15.1). It enhances the activity of antioxidant enzymes, reducing oxidative damage and prolonging the shelf life of the produce (Zhang et al. 2024). GABA also regulates ethylene levels, slowing down fruit ripening and preventing premature spoilage (Aghdam et al. 2019). By stabilizing pH and ion fluxes, GABA helps

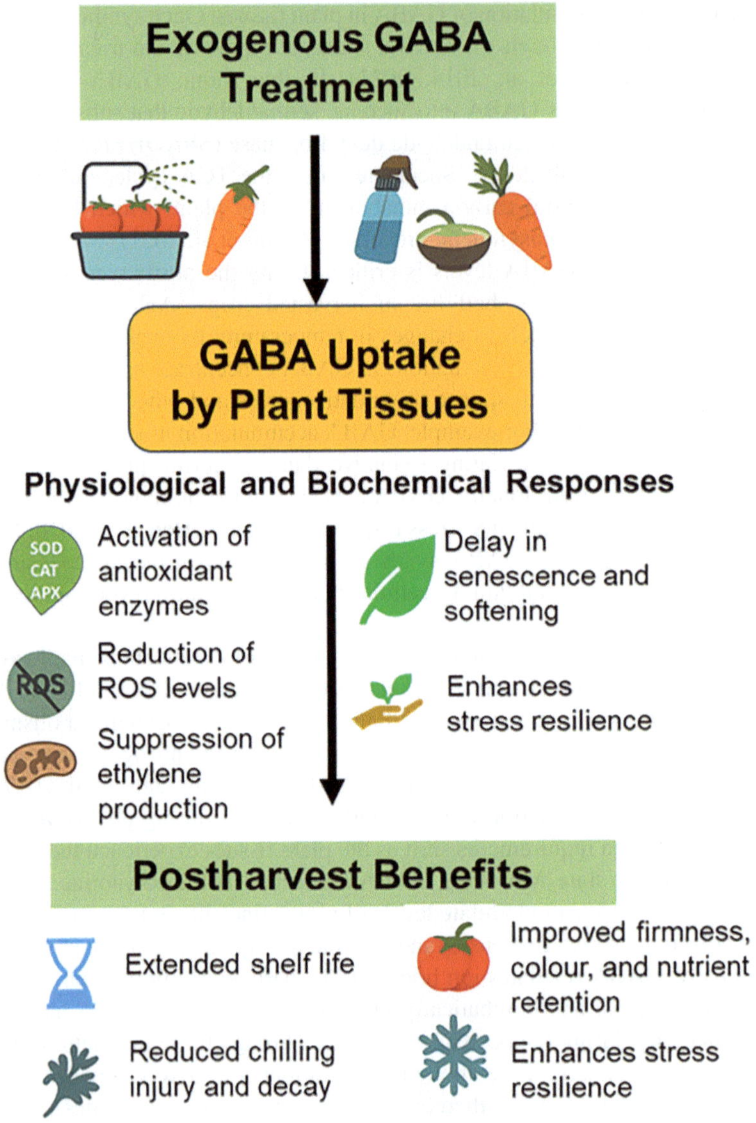

Fig. 15.1 GABA application and postharvest management of horticultural crops. Upon application through dipping, spraying, or vapor treatment, plant tissues take up GABA, which triggers various physiological and biochemical responses. These include activation of antioxidant enzymes (Superoxide dismutase (SOD), Catalase (CAT), and Peroxidase (POD)), ROS reduction, suppression of ethylene production, delayed senescence and softening, and enhanced stress resilience. Collectively, these effects contribute to improved postharvest quality, including extended shelf life, reduced chilling injury and decay, and better retention of firmness, color, and nutritional value

preserve cellular integrity, ensuring that stored produce remains of high quality during postharvest storage.

15.2.2 Involvement of GABA in Oxidative Stress Management

During the postharvest period, agricultural produce is particularly vulnerable to oxidative stress, which can lead to cellular damage and accelerated deterioration (Fig. 15.1). GABA plays a crucial role in managing oxidative stress, for it enhances the activity of key antioxidant enzymes, such as SOD, CAT, and POD, which help to ROS. The accumulation of GABA in response to postharvest stress serves as a protective measure, reducing the oxidative burden on cells and preventing lipid peroxidation, protein degradation, and nucleic acid damage. By mitigating oxidative stress, GABA contributes significantly to prolonging the shelf life and preserving the quality of stored produce.

15.2.3 Impact of GABA on Ethylene Production and Fruit Ripening

Ethylene is a critical hormone that regulates fruit ripening, and its production must be carefully controlled during postharvest storage to prevent premature spoilage. GABA influences ethylene biosynthesis and signaling pathways, thereby modulating the ripening process (Li et al. 2021). Studies have shown that GABA can delay ethylene production, slowing down the ripening of fruits and extending their shelf life (Han et al. 2018). This effect is particularly beneficial for climacteric fruits, where a rapid increase in ethylene production can lead to over-ripening and increased susceptibility to decay. By fine-tuning ethylene levels, GABA helps to maintain the desired ripeness of fruits during storage, ensuring they reach the market in optimal condition. For example, GABA treatments decrease ethylene production and delay loss of titratable acidity from 'Cripps Pink' apple fruit (Han et al. 2018). Exogenous GABA application preserves apple fruit quality by modulating ethylene synthesis and polyamine metabolism (Li et al. 2021).

15.2.4 Role of GABA in Maintaining Cellular Homeostasis

Maintaining cellular homeostasis is vital during the postharvest period to prevent physiological disorders and ensure the quality of produce. GABA plays a central role in this process by acting as a signaling molecule that regulates various physiological and metabolic pathways (Liu et al. 2023b). One of its key functions is in the stabilization of pH levels within cells, which is critical for enzyme activity and metabolic balance. GABA also helps in the regulation of ion fluxes, particularly calcium and potassium ions, which are essential for maintaining cell turgor and membrane integrity (Zhang et al. 2024). Furthermore, GABA interacts with metabolic pathways, such as the carbon and nitrogen cycles, and sustains the energy needs of the cell and metabolic processes as well during storage (Liu et al. 2023a). Through these mechanisms, GABA ensures that cells remain functional and resilient under postharvest conditions, thereby reducing the incidence of physiological disorders and enhancing the overall quality of the stored produce.

15.2.5 Regulating Stomatal Function and Enhancing Postharvest Quality with GABA

In the intricate dance of plant physiology, stomatal function is a critical factor in maintaining freshness and quality in cut flowers, leafy vegetables, and immature fruits. Excessive water loss through stomata can lead to wilting, reduced freshness, and diminished quality (Gahir et al. 2021). Here, GABA emerges as a key player, akin to a vigilant gatekeeper managing the delicate balance of water retention (Xu et al. 2021). By externally applying GABA, researchers have discovered that plants can better control stomatal openings, reducing water loss and preserving turgor. This not only extends the freshness of the produce but also enhances its resilience against environmental stress (van Meeteren and Aliniaeifard 2016). GABA's influence extends further, serving as a protective barrier against bacterial and fungal pathogens by preventing their entry through stomata. This dual role of GABA—preserving quality and enhancing safety—highlights its potential as a valuable tool for improving postharvest management.

15.2.6 Impact of GABA on Vacuolar Function and Flavor Profile in Postharvest Produce

In the quest for optimal flavor and quality, the vacuole plays a pivotal role in regulating acidity and taste in fruits (Shelp et al. 2021). In ripe apples, malate, an organic acid, accumulates in the vacuole, shaping the fruit's flavor profile. The transport of malate is managed by the protein ALMT9 (MdMa1), which operates like a meticulous transporter, ensuring the right balance of malate (Li et al. 2020). Applying GABA during cold storage reduces malate levels, resulting in lower acidity and a sweeter taste. This interaction underscores GABA's role in influencing fruit flavor and quality during storage. In tomatoes, the protein SlCAT9 functions as an exchanger for GABA and other compounds, with its activity increasing during ripening to elevate GABA levels. This contributes to the development of the umami taste. Conversely, higher GABA levels in immature fruit can deter pests, while lower levels in ripe fruit may attract animals for seed dispersal (Takayama and Ezura 2015). These findings reveal GABA's complex role in flavor development and postharvest resistance, offering exciting possibilities for enhancing fruit quality.

15.2.7 GABA and Antioxidant Defense Mechanisms

GABA upregulates SOD, CAT, and POD that detoxify ROS and prevent cellular damage during stress or postharvest conditions (Makino et al. 2008). Consequently, GABA orchestrates a robust antioxidant defense, which is essential for prolonging the shelf life and preserving the quality of postharvest produce.

During postharvest storage, the accumulation of ROS accelerates the deterioration of fruits and vegetables. GABA contributes not only to reducing ROS levels but also to directly interacting with ROS scavenging pathways (Zhang et al. 2024). By lowering ROS accumulation, GABA mitigates oxidative damage to cellular structures, such as lipids, proteins, and nucleic acids, thus preserving the integrity and functionality of cells during storage. This reduction in oxidative stress is key to maintaining the freshness and nutritional value of stored produce (Aghdam et al. 2022).

Several studies have highlighted the effectiveness of GABA in improving the postharvest quality of various crops through its antioxidant properties. For instance, in vine-ripe tomatoes, GABA treatment increases the activity of antioxidant enzymes, resulting in lower ROS levels and delayed ripening (Makino et al. 2008). GABA application has been associated with reduced oxidative browning and increased resistance to food-borne pathogens in fresh-cut apples and fresh-cut potatoes (Gao et al. 2018a; Gao et al. 2018b). Similarly, GABA enhances the antioxidant capacity, reducing decay and extending shelf life in strawberries (Wang et al. 2021a, b). GABA treatments facilitate antioxidant capacity by increasing the activities of SOD and CAT and DPPH radical scavenging activity in postharvest strawberry fruit (Zhang et al. 2024). Postharvest application of GABA elevates DPPH scavenging capacity and activities of CAT and POD and prolongs the shelf life during mango fruit storage (Rastegar et al., 2019). In pears, GABA-induced resistance against *Penicillium expansum* in harvested fruit arises from the enhancement of activities or gene expression of various defense-related enzymes (Yu et al. 2014). GABA treatment significantly protects against fruit chilling injury by increasing the antioxidant system and promoting proline accumulation in banana (Wang et al. 2014), peach (Shang et al. 2011; Yang et al. 2011), cucumber (Malekzadeh et al. 2017), and zucchini (Palma et al. 2019). These findings demonstrate that GABA confers a remarkable ROS scavenging ability through improving the antioxidant system in postharvest fruit.

15.2.8 Influence of GABA on Postharvest Quality and Shelf Life

GABA helps preserve the quality and extend the shelf life of fruits and vegetables after harvest. By maintaining fruit firmness, color retention, and texture, GABA helps keep produce fresh and visually appealing for longer periods. It also reduces postharvest losses by mitigating oxidative stress, slowing down ethylene production, and preventing spoilage caused by microorganisms. As a result, GABA-treated produce resists decay and maintains its marketability. Hence, GABA comprises a valuable tool in postharvest management.

15.2.9 Effects of GABA on Fruit Firmness, Color Retention, and Texture

GABA keeps fruits and vegetables firm, colorful, and fresh after harvest. When applied to produce, GABA helps maintain the firmness of fruits, which is important for their structure and resistance to damage. It also prevents discoloration by stabilizing natural pigments like chlorophyll and anthocyanins, which are responsible for the vibrant colors of fruits and vegetables. Additionally, GABA slows down the softening in produce, helping to maintain their texture and overall quality during storage. This means that GABA-treated fruits and vegetables stay visually appealing and retain their desirable taste and feel for longer periods.

A wealth of literature exists that associates GABA with delay in aging and the quality preservation of blueberries, cherries, mushrooms, and apples stored at low temperatures or room temperature (Ge et al. 2018; Han et al. 2018; Aghdam et al. 2019; Rabiei et al. 2019; Al Shoffe et al. 2021; Shekari et al. 2021). It also helps pears and mangoes to resist browning and maintain quality during low-temperature storage (Li et al. 2019; Rastegar et al. 2019). GABA also retains the acidity of apples by increasing malate levels and reducing ethylene production, which slows down the ripening process (Han et al. 2018).

15.2.10 Role of GABA in Reducing Postharvest Losses and Spoilage

Postharvest spoilage and decay cause significant losses in the horticultural industry. Interestingly, GABA comes to the rescue from adverse situations arising from postharvest stress and storage. GABA strengthens the natural defenses, slows down the rate of decay, and extends the shelf life of fruits and vegetables. It achieves this by reducing oxidative stress, lowering ethylene production, and inhibiting the growth of microorganisms that cause spoilage. By managing these factors, GABA helps protect produce from fungal infections and other disorders that can occur during storage, thereby reducing postharvest losses.

GABA extends its ability to enhance resistance against fungal infections in pears, oranges, strawberries, and tomatoes stored at low or room temperatures (Yu et al. 2014; Fu et al. 2017; Sheng et al. 2017; Yang et al. 2017). Furthermore, walnuts treated with GABA show better resistance against pathogens and browning during storage at room temperature (Ebrahimzadeh et al. 2019). The effectiveness of GABA in promoting pathogen resistance is attributable to its role in boosting salicylate signaling, disease resistance proteins, and maintaining the strength of the cell walls (Yu et al. 2014; Fu et al. 2017; Yang et al. 2017; Gao et al. 2018a; Zhao et al. 2021). For instance, the process of cutting fresh produce, such as carrots, does not affect their quality when stored at low temperatures, and in fact, seems to increase GABA production, which improves storage outcomes (Hou et al. 2022). This finding is consistent with previous studies showing that mechanical damage or wounding can increase GABA levels in plants (Shelp et al. 2012). Moreover, GABA strengthens resistance to browning and bacterial pathogens in fresh-cut pears, apples, and potatoes during extended storage under low temperatures, especially when combined with controlled atmosphere storage (Gao et al. 2018a, b; Wang et al. 2021a; Zhao et al. 2021).

15.2.11 Mechanisms of GABA-Mediated Metabolic Modulation

Research on how fruits and vegetables respond to low-temperature (LT) and controlled atmosphere (CA) storage has provided valuable insights into ways to extend their shelf life and keep them marketable. These storage methods often boost the activity of the GABA pathway, which can delay aging, preserve quality, and reduce damage from chilling. Several key genes and proteins, such as MdGAD1, MdGABA-T1, and others, play crucial roles in this process, especially in apples, regardless of the storage method used (Han et al. 2021). Studies suggest that under LT and CA conditions, specific factors like protein breakdown and calcium activation are particularly important in driving this GABA pathway (Seifikalhor et al. 2019; Suhel et al. 2022). Presumably, the GABA produced is mainly derived from the amino acid glutamate and is converted into substances that feed into the energy-producing pathways, increasing the levels of essential energy molecules like ATP and NADPH (Aghdam et al. 2018, 2020; Shelp et al. 2021). This energy is vital for boosting the plant's defense mechanisms, stabilizing cell membranes, and preventing damage from oxidative stress. As a result, fruits and vegetables are better protected from chilling injuries and decay, maintaining their quality for longer periods. Moreover, the presence of the tonoplast ALMT transporter offers a tantalizing opportunity for further manipulation (Takayama and Ezura 2015). Exogenous GABA application regulates stomatal opening/closing and reduces water loss, thus preserving the freshness of cut flowers and immature green fruit (Xu et al. 2021; Gahir et al. 2021; Aghdam et al. 2016). This transporter, along with the tonoplast CAT exchanger, also opens up possibilities for influencing fruit flavor, adding yet another dimension to the potential benefits of GABA application in postharvest storage (Aghdam et al. 2022). As research continues, the potential to improve the shelf life, quality, and marketability of fruits and vegetables only grows, offering a glimpse into the future of sustainable agriculture and food security.

15.3 Conclusion

To sum up, GABA emerges as a transformative organic compound in postharvest management, offering significant benefits for enhancing the quality and extending the shelf life of fruits and vegetables. GABA exhibits the ability to regulate stomatal function that helps in reducing water loss, maintaining firmness, and preserving freshness, while its role in modulating oxidative stress and pathogen resistance further safeguards produce during storage. GABA also has an impact on vacuolar functions and flavor profiles, revealing its potential to fine-tune the acidity and taste of stored fruits, contributing to both sensory appeal and market value. As we advance our understanding of the multifaceted roles of GABA, it becomes evident that integrating GABA-based strategies could revolutionize postharvest practices, reducing losses and improving product quality. Future research should focus on optimizing GABA application techniques and exploring its interactions with other physiological pathways to fully harness its potential. The promising outcomes of GABA application highlight its potential to be a key tool in sustainable agriculture, paving the way for enhanced postharvest management and quality preservation.

References

Aghdam MS, Naderi R, Jannatizadeh A, Sarcheshmeh MAA, Babalar M (2016) Enhancement of postharvest chilling tolerance of anthurium cut flowers by γ-aminobutyric acid (GABA) treatments. Sci Hortic 198:52–60. https://doi.org/10.1016/j.scienta.2015.11.019

Aghdam MS, Jannatizadeh A, Luo Z, Paliyath G (2018) Ensuring sufficient intracellular ATP supplying and friendly extracellular ATP signaling attenuates stresses, delays senescence and maintains quality in horticultural crops during postharvest life. Trends Food Sci Technol 76:67–81. https://doi.org/10.1016/j.tifs.2018.04.003

Aghdam MS, Kakavand F, Rabiei V, Zaare-Nahandi F, Razavi F (2019) γ-Aminobutyric acid and nitric oxide treatments preserve sensory and nutritional quality of cornelian cherry fruits during postharvest cold storage by delaying softening and enhancing phenols accumulation. Sci Hortic 246:812–817. https://doi.org/10.1016/j.scienta.2018.11.064

Aghdam MS, Palma JM, Corpas FJ (2020) NADPH as a quality footprinting in horticultural crops marketability. Trends Food Sci Technol 103:152–161. https://doi.org/10.1016/j.tifs.2020.07.002

Aghdam MS, Flaherty EJ, Shelp BJ (2022) γ-Aminobutyrate improves the postharvest marketability of horticultural commodities: advances and prospects. Front Plant Sci 13:884572. https://doi.org/10.3389/fpls.2022.884572

Al Shoffe Y, Nock JF, Zhang Y, Watkins CB (2021) Pre- and post-harvest γ-aminobutyric acid application in relation to fruit quality and physiological disorder development in 'Honeycrisp' apples. Sci Horti 289:110431. https://doi.org/10.1016/j.scienta.2021.110431

Ansari MI, Hasan S, Jalil SU (2014) Leaf senescence and GABA shunt. Bioinformation 10:730–732

Ansari MI, Jalil SU, Ansari SA et al (2021) GABA shunt: a key-player in mitigation of ROS during stress. Plant Growth Regul 94:131–149

Ebrahimzadeh A, Pirzad F, Tahanian H, Aghdam MS (2019) Influence of gum arabic enriched with GABA coating on oxidative damage of walnut kernels. Food Technol Biotechnol 57:554–560. https://doi.org/10.17113/ftb.57.04.19.6380

Fu D, Sun Y, Yu C, Zheng X, Yu T, Lu H (2017) Comparison of the effects of three types of aminobutyric acids on the control of *Penicillium expansum* infection in pear fruit. J Sci Food Agric 97:1497–1501. https://doi.org/10.1002/jsfa.7891

Gahir S, Bharath P, Raghavendra AS (2021) Stomatal closure sets in motion long-term strategies of plant defense against microbial pathogens. Front Plant Sci 12:761952. https://doi.org/10.3389/fpls.2021.761952

Gao H, Wu S, Zeng Q, Li P, Guan W (2018a) Effects of exogenous γ-aminobutyric acid treatment on browning and food-borne pathogens in fresh-cut apples. Postharvest Biol Technol 146:1–8. https://doi.org/10.1016/j.postharvbio.2018.08.007

Gao H, Zeng Q, Ren Z, Li P, Xu X (2018b) Effect of exogenous γ-aminobutyric acid treatment on the enzymatic browning of fresh-cut potato during storage. J Food Sci Technol 55:5035–5044. https://doi.org/10.1007/s13197-018-3442-x

Ge Y, Duan B, Li C, Tang Q, Li X, Wei M et al (2018) γ-Aminobutyric acid delays senescence of blueberry fruit by regulation of reactive oxygen species metabolism and phenylpropanoid pathway. Sci Horti 240:303–309. https://doi.org/10.1016/j.scienta.2018.06.044

Han S, Nan Y, Qu W, He Y, Ban Q, Lv Y et al (2018) Exogenous γ-aminobutyric acid treatment contributes to regulation of malate metabolism and ethylene synthesis in apple fruit during storage. J Agric Food Chem 66:13473–13482. https://doi.org/10.1021/acs.jafc.8b04674

Han S, Liu H, Han Y, He Y, Nan Y, Qu W, Rao J (2021) Effects of calcium treatment on malate metabolism and γ-aminobutyric acid (GABA) pathway in postharvest apple fruit. Food Chem 334:127479. https://doi.org/10.1016/j.foodchem.2020.127479. Epub 2020 Jul 5. PMID: 32688181

Hou Y, Ren H, Wang K, Cao S, Zheng Y, Wei Y et al (2022) Influence of fresh-cut process on γ-aminobutyric acid (GABA) metabolism and sensory properties in carrot. J Food Sci Technol 59:552–561. https://doi.org/10.1007/s13197-021-05039-y

Jalil SU, Ansari MI (2020) Physiological role of gamma aminobutyric acid in salt stress tolerance. In: Hasanuzzaman M, Tanveer M (eds) Salt and drought stress tolerance in plants, signaling and communication in plants. Springer, Cham, pp 337–350

Jalil SU, Ahmad I, Ansari MI (2016) Phenotypic characterization of GABA-transaminase mutants of *Arabidopsis thaliana*. Adv Life Sci 5:10005–10008

Jalil SU, Ahmad I, Ansari MI (2017) Functional loss of GABA transaminase (GABA-T) expressed early leaf senescence under various stress conditions in *Arabidopsis thaliana*. Curr Plant Biol 9–10:11–22

Jalil SU, Khan MAA, Ansari MI (2019a) Tremendous role of GABA shunt metabolic pathway in plant system. J Biol Chem Res 36:188–196

Jalil SU, Khan MQR, Ansari MI (2019b) Role of GABA transaminase in the regulation of development and senescence in *Arabidopsis thaliana*. Curr Plant Biol 19:100119

Khan M, Jalil SU, Chopra P, Chhillar H, Ansari MI (2021) Role of GABA in plant growth, development and senescence. Plant Gene 26:100283

Li J, Zhou X, Wei B, Cheng S, Zhou Q, Ji S (2019) GABA application improves the mitochondrial antioxidant system and reduces peel browning in 'Nanguo' pears after removal from cold storage. Food Chem 297:124903. https://doi.org/10.1016/j.foodchem.2019.05.177

Li C, Dougherty L, Coluccio AE, Meng D, El-Sharkawy I, Borejsza-Wysocka E et al (2020) Apple ALMT9 requires a conserved C-terminal domain. Plant Physiol 182:992–1006. https://doi.org/10.1104/pp.19.01300

Li C, Zhu J, Sun L, Cheng Y, Hou J, Fan Y et al (2021) Exogenous γ-aminobutyric acid maintains fruit quality of apples through regulation of ethylene anabolism and polyamine metabolism. Plant Physiol Biochem 169:92–101. https://doi.org/10.1016/j.plaphy.2021.11.008

Liu Q, Li X, Jin S, Dong W, Zhang Y, Chen W et al (2023a) γ-Aminobutyric acid treatment induced chilling tolerance in postharvest kiwifruit (Actinidia chinensis cv. hongyang) via regulating ascorbic acid metabolism. Food Chem 404:134661. https://doi.org/10.1016/j.foodchem.2022.134661

Liu X, Ma H, Liu J, Liu D, Wang C (2023b) The γ-aminobutyric acid (GABA) synthesis gene regulates the resistance to water core-induced hypoxia stress for pear fruits. Agronomy 13:1062. https://doi.org/10.3390/agronomy13041062

Lum GB, Shelp BJ, DeEll JR, Bozzo GG (2016) Oxidative metabolism is associated with physiological disorders in fruits stored under multiple environmental stresses. Plant Sci 245:143–152. https://doi.org/10.1016/j.plantsci.2016.02.005

Makino Y, Soga N, Oshita S, Kawagoe Y, Tanaka A (2008) Stimulation of γ-aminobutyric acid production in vine-ripe tomato (Lycopersicon esculentum Mill.) fruits under modified atmospheres. J Agric Food Chem 56:7189–7193. https://doi.org/10.1021/jf801516e

Malekzadeh P, Khosravi-Nejad F, Hatamnia AA, Sheikhakbari Mehr R (2017) Impact of postharvest exogenous γ-aminobutyric acid treatment on cucumber fruit in response to chilling tolerance. Physiol Mol Biol Plants 23:827–836. https://doi.org/10.1007/s12298-017-0475-2

Palma F, Carvajal F, Jiménez-Muñoz R, Pulido A, Jamilena M, Garrido D (2019) Exogenous γ-aminobutyric acid treatment improves the cold tolerance of zucchini fruit during postharvest storage. Plant Physiol Biochem 136:188–195. https://doi.org/10.1016/j.plaphy.2019.01.023

Rabiei V, Kakavand F, Zaare-Nahandi F, Razavi F, Aghdam M (2019) Nitric oxide and γ-aminobutyric acid treatments delay senescence of cornelian cherry fruits during postharvest cold storage by enhancing antioxidant system activity. Sci Hortic 243:268–273. https://doi.org/10.1016/j.scienta.2018.08.034

Rastegar S, Khankahdani HH, Rahimzadeh M (2019) Effect of γ-aminobutyric acid on the antioxidant system and biochemical changes of mango fruit during storage. J Food Meas Charact 14:778–789. https://doi.org/10.1007/s11694-019-00326-x

Seifikalhor M, Aliniaeifad S, Hassani B, Nikman V, Lastochkina O (2019) Diverse role of γ-aminobutyric acid in dynamic plant responses. Plant Cell Rep 38:847–867. https://doi.org/10.1007/s00299-019-02396-z

Shang H, Cao S, Yang Z, Cai Y, Zheng Y (2011) Effect of exogenous gamma-aminobutyric acid treatment on proline accumulation and chilling injury in peach fruit after long-term cold storage. J Agric Food Chem 59:1264–1268. https://doi.org/10.1021/jf104424z

Shekari A, Hassani RN, Aghdam MS (2021) Exogenous application of GABA retards cap browning in *Agaricus bisporus* and its possible mechanism. Postharvest Biol Technol 174:111434. https://doi.org/10.1016/j.postharvbio.2020.111434

Shelp BJ, Bozzo GG, Trobacher CP, Chiu G, Bajwa VS (2012) Strategies and tools for studying the metabolism and function of γ-aminobutyric acid in plants. I. Pathway structure. Botany 90:651–668. https://doi.org/10.1139/B2012-030

Shelp BJ, Aghdam MS, Flaherty EJ (2021) γ-Aminobutyrate (GABA) regulated plant defense: mechanisms and opportunities. Plants 10:1939. https://doi.org/10.3390/plants10091939

Sheng L, Shen D, Luo Y, Sun X, Wang J, Luo T et al (2017) Exogenous γ-aminobutyric acid treatment affects citrate and amino acid accumulation to improve fruit quality and storage performance of postharvest citrus fruit. Food Chem 216:138–145. https://doi.org/10.1016/j.foodchem.2016.08.024

Suhel M, Husain T, Pandey A, Singh S, Dubey NK, Prasad SM et al (2022) An appraisal of ancient molecule GABA in abiotic stress tolerance in plants, and its crosstalk with other signaling molecules. J Plant Growth Regul 42:614. https://doi.org/10.1007/s00344-022-10610-8

Takayama M, Ezura H (2015) How and why does tomato accumulate a large amount of GABA in the fruit? Front Plant Sci 6:612. https://doi.org/10.3389/fpls.2015.00612

Tarkowski KP, Signorelli S, Höfte M (2020) γ-Aminobutyric acid and related amino acids in plant immune responses: emerging mechanisms of action. Plant Cell Environ 43:1103–1116. https://doi.org/10.1111/pce.13734

van Meeteren U, Aliniaeifard S (2016) Stomata and postharvest physiology. In: Pareek S (ed) Postharvest ripening physiology of crops. CRC Press, Boca Raton, pp 157–191

Wang Y, Luo Z, Huang X, Yang K, Gao S, Du R (2014) Effect of exogenous γ-aminobutyric acid (GABA) treatment on chilling injury and antioxidant capacity in banana peel. Sci Hortic 168:132–137. https://doi.org/10.1016/j.scienta.2014.01.022

Wang D, Li D, Xu Y, Li L, Belwal T, Zhang X et al (2021a) Elevated CO_2 alleviates browning development by modulating metabolisms of membrane lipids, proline, and GABA in fresh-cut Asian pear fruit. Sci Hortic 281:109932. https://doi.org/10.1016/j.scienta.2021.109932

Wang F, Xiao J, Zhang Y, Li R, Liu L, Deng J (2021b) Biocontrol ability and action mechanism of Bacillus halotolerans against Botrytis cinerea causing grey mould in postharvest strawberry fruit. Postharvest Biol Technol 174:111456. https://doi.org/10.1016/j.postharvbio.2020.111456

Xu B, Long Y, Feng X, Zhu X, Sai N, Chirkova L et al (2021) GABA signalling modulates stomatal opening to enhance plant water use efficiency and drought resilience. Nat Commun 12:1952. https://doi.org/10.1038/s41467-021-21694-3

Yang A, Cao S, Yang Z, Cai Y, Zheng Y (2011) γ-Aminobutyric acid treatment reduces chilling injury and activates the defence response of peach fruit. Food Chem 129:1619–1622. https://doi.org/10.1016/j.foodchem.2011.06.018

Yang J, Sun C, Zhang Y, Fu D, Zheng X, Yu T (2017) Induced resistance in tomato fruit by γ-aminobutyric acid for the control of alternaria rot caused by *Alternaria alternata*. Food Chem 221:1014–1020. https://doi.org/10.1016/j.foodchem.2016.11.061

Yu C, Zeng L, Sheng K, Chen F, Zhou T, Zheng X et al (2014) γ-Aminobutyric acid induces resistance against *Penicillium expansum* by priming of defence responses in pear fruit. Food Chem 159:29–37. https://doi.org/10.1016/j.foodchem.2014.03.011

Zhang Y, Li S, Deng M, Gui R, Liu Y, Chen X et al (2022) Blue light combined with salicylic acid treatment maintained the postharvest quality of strawberry fruit during refrigerated storage. Food Chem X 15:100384. https://doi.org/10.1016/j.fochx.2022.100384

Zhang Y, Lin B, Tang G, Chen Y, Deng M, Lin Y, Li M et al (2024) Application of γ-aminobutyric acid improves the postharvest marketability of strawberry by maintaining fruit quality and enhancing antioxidant system. Food Chem X 21:101252. https://doi.org/10.1016/j.fochx.2024.101252

Zhao P, Li W, Zhen C, Wang K, Qin Z, Gao H (2021) Transcriptomic analysis of the effects of γ-aminobutyric acid treatment on browning and induced disease resistance in fresh-cut apples. Postharvest Biol Technol 181:111686. https://doi.org/10.1016/j.postharvbio.2021.111686

Ziv C, Fallik E (2021) Postharvest storage techniques and quality evaluation of fruits and vegetables for reducing food loss. Agronomy 11:133. https://doi.org/10.3390/agronomy11061133

Morphological Responses of Plants to Exogenous GABA

16

Abstract

Gamma-aminobutyric acid (GABA) plays a decisive role in plant growth regulation and stress adaptation, encouraging its exogenous administration for augmented crop performance and productivity. Exogenous GABA application proficiently modulates plant morphology under various environmental conditions. Its application influences root development, shoot architecture, leaf morphology, and overall biomass accumulation. GABA promotes root elongation, lateral root formation, and root hair development, facilitating enhanced water and nutrient uptake. In shoots, it regulates stem elongation, leaf expansion, and chlorophyll biosynthesis, leading to improved photosynthetic efficiency and biomass production. GABA application secures stress-induced morphological changes, mitigating the adverse effects of drought, salinity, and temperature extremes. By modulating cellular signalling pathways and interacting with plant hormones, GABA enhances plant structural resilience and adaptability. Understanding these morphological effects highlights GABA's potential as a tool for crop improvement, particularly in the face of global climate challenges.

Keywords

Biomass accumulation · Exogenous GABA · Flowering regulation · Leaf expansion · Leaf senescence · Root elongation · Shoot development

Highlights

- Exogenous gamma-aminobutyric acid (GABA) application promotes root elongation, lateral root formation, and shoot development under both normal and stress conditions.

S. U. Jalil et al., *GABA Signaling System and Abiotic Stress Tolerance in Plants*,
https://doi.org/10.1007/978-981-95-0437-4_16

- GABA influences leaf expansion by regulating chlorophyll content and stomatal regulation, contributing to enhanced photosynthetic efficiency.
- GABA treatment helps in maintaining leaf greenness and delaying senescence, particularly under stress conditions like drought and salinity.
- GABA modulates flowering time, enhances pollen viability, and improves fruit/seed set in various plant species.
- GABA affects cell division and expansion, contributing to overall plant architecture and biomass accumulation.

16.1 Introduction

Plants continuously adapt their growth and development to changing environmental conditions through intricate signalling networks. One such signalling molecule that has gained significant attention in recent years is gamma-aminobutyric acid (GABA), which engages in diverse modulation mechanisms, benefitting physiological and morphological processes in plants. As a result, GABA-treated plants obtain superior root and shoot growth, leaf morphology, and flowering, ushering in better stress adaptation (Khan et al. 2021). The exogenous application of GABA regulates cell expansion/ division, facilitates hormonal interactions, manifesting as an efficient plant architecture commensurate with plant growth and stress tolerance (Jin et al. 2023).

External environmental factors and internal hormonal regulation coordinate in shaping plant morphology. GABA interacts with key phytohormones such as auxins, abscisic acid, gibberellins, and ethylene to modulate morphological traits (Gray 2004). Under normal conditions, exogenous GABA promotes root elongation, shoot biomass accumulation, and improved photosynthetic efficiency by influencing stomatal behaviour and chlorophyll retention (Li et al. 2016a, b). In contrast, under abiotic stress conditions such as drought, salinity, and heavy metal toxicity, GABA helps plants adapt by enhancing root architecture, delaying leaf senescence, and strengthening reproductive structures (Renault et al. 2011; Biancucci et al. 2015; Jalil et al. 2017; Khan et al. 2021; Ullah et al. 2023).

Understanding the morphological responses of plants to exogenous GABA is essential for improving crop resilience and productivity. This chapter explores the effects of GABA on root and shoot development, leaf morphology, flowering, and stress-induced structural adaptations. The integration of GABA-mediated responses in plant growth regulation could provide novel strategies for sustainable agriculture and stress management in crops (Fig. 16.1).

16.2 Seed Germination and Seedling Growth

GABA has garnered significant attention for its role in enhancing seed germination and seedling growth across various plant species, especially under stress conditions. Matsuyama et al. (2009) have observed a notable increase in GABA levels during

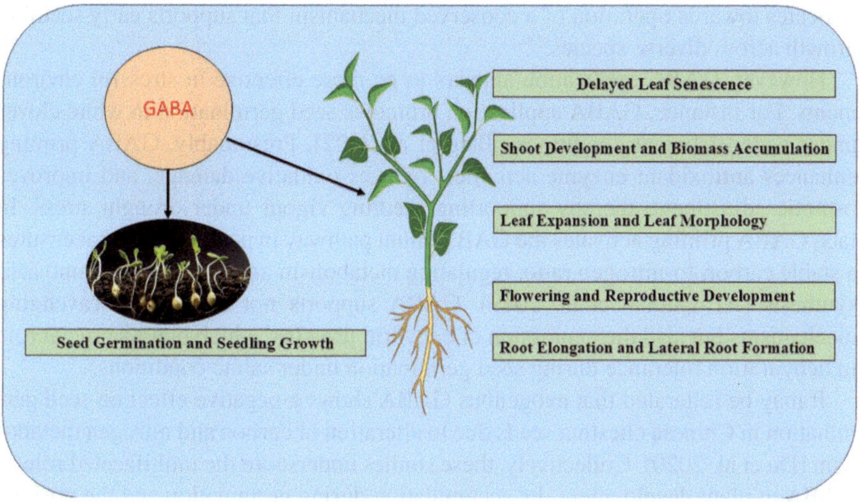

Fig. 16.1 Morphological responses of plants to exogenous GABA. GABA promotes seed germination and seedling growth, leading to enhanced root elongation and lateral root formation. It contributes to improved shoot development, biomass accumulation, and leaf expansion while influencing leaf morphology. Additionally, GABA plays a role in flowering, reproductive development, and delaying leaf senescence, collectively supporting overall plant growth and development

the germination of soybean seeds, suggesting its involvement in early seedling development. This foundational discovery obtains further strength from the investigation by Xu et al. (2010) that reports oat seeds enhancing GABA accumulation during their germination, implying a role in essential metabolic processes. The subsequent years witnessed a surge in research focusing on GABA's impact under stress conditions. Sharma et al. (2016) have highlighted the elevated GABA levels in barnyard millet during germination. Its elevated levels also promote vital enzymatic activities commensurate with stress tolerance. Further investigation demonstrates a parity between GABA accumulation and α-amylase activity germination of adlay seeds, highlighting efficient starch breakdown and requisite energy provision for seedling emergence (Xu et al. 2017).

Zhao et al. (2017) have expanded this understanding to rice, where germinating seeds gain GABA content, correlating with improved stress responses and metabolic adjustments in early growth stages. Around the same time, an investigation further reveals that the α-amylase gene expression corresponds with GABA accumulation in germinating barley seeds, promoting starch degradation and supplying energy for seedling development (Sheng et al. 2018). GABA-treated wheat and maize seeds exhibit accelerated germination rate and overcome salt-induced injuries during germination (Cheng et al. 2018). Chinese wild rice seeds, while germinating, also gain significantly in GABA levels (Chu et al. 2019). Efficiency of GABA in seed germination appears to enhance osmotic balance, boost antioxidant activity, and upregulate stress-responsive genes. Thus, GABA accumulation indicates towards operation of a conserved mechanism that supports early seedling growth across diverse species.

However, GABA application appears to be more effective in stressful environments. For instance, GABA application promotes seed germination in white clover under water-deficient conditions (Tang et al. 2022). Presumably, GABA priming enhances antioxidant enzyme activities, reduces oxidative damage, and improves osmotic adjustment, thereby supporting seedling vigour under drought stress. In fact, GABA priming activates the GABA shunt pathway in pepper seeds that ensures a stable carbon-to-nitrogen ratio, regulating metabolism and optimizing amino acid synthesis (AL-Quraan et al. 2019). GABA supports not only ROS scavenging mechanisms but also the expression of dehydrin proteins, which plays a crucial role in dehydration tolerance during seed germination under saline conditions.

It may be reiterated that exogenous GABA shows a negative effect on seed germination in Chinese chestnut seeds due to alteration of carbon and nitrogen metabolism (Du et al. 2020). Collectively, these studies underscore the multifaceted role of GABA in plant development. Its accumulation during germination and the positive effects of its exogenous application highlight GABA's potential as a vital regulator in seed germination and seedling growth.

16.3 Root Morphogenesis and Architecture

The plant root system provides anchorage to the soil and absorbs water and nutrients from the soil. A well-developed root system ensures plant survival and vigour that culminates in productivity. Hence, root development is of paramount importance for agricultural sustainability. GABA is essential for various aspects of root development, while exogenous application of GABA inhibits certain root growth processes, and its effects on root hair development remain to be elucidated. The influence of exogenous GABA on root architecture is complex and may depend on factors such as concentration, plant species, and developmental stage.

16.3.1 Primary Root Growth

Exogenous GABA inhibits primary root growth in *Arabidopsis thaliana* mutants deficient in GABA transaminase, suggesting that elevated GABA levels can negatively impact root elongation (Renault et al. 2011). GABA-treated maize seedlings improve root fresh weight (Li et al. 2016a, b). In mung bean, GABA application significantly improves root length and root fresh weight under salinity stress (Ullah et al. 2023).

16.3.2 Adventitious Root Formation

Exogenous GABA enhances root length and lateral root formation, improving water and nutrient uptake, especially under sub-optimal conditions (Michaeli et al. 2011).

However, excessive GABA levels inhibit *Brassica napu* adventitious root growth, where vigabatrin-induced GABA-T inhibition reduces primary root growth (Deleu et al. 2013). Similarly, exogenous GABA application and GAD2 overexpression in poplar downregulate ethylene-dependent pathways at the early stages and inhibit adventitious root (AR) formation with altered hormone-related gene expression and hormone levels (Xie et al. 2020). In Chinese chestnut seeds, exogenous GABA alters carbon and nitrogen metabolism, preventing AR growth and maintaining seed dormancy (Du et al. 2020). In *Malus xiaojinensis* and tobacco plants, exogenous GABA application delays root emergence, reduces root number and root length arising from polar auxin transport inhibition (Li et al. 2021). On the contrary, exogenous application of GABA influences the morphological growth of *Liriodendron* hybrid plantlets derived from somatic embryos. Specifically, GABA treatment has a pronounced luxuriant impact on lateral root numbers, root length, and stem diameter (Chen et al. 2022). However, GABA-associated metabolic factors appear to be critically, albeit inconclusively, involved in adventitious root formation (Pei et al. 2022). The current inconclusive situation arises due to the paucity of in-depth investigative scientific literature that may elucidate trigger points at which GABA acts as a metabolite or signal molecule controlling gene expression and the cellular environment responsible for adventitious root formation.

16.3.3 Root Hair Development

Root hairs are extensively responsible for maintaining the dynamics of water and nutrient absorption. Their development is highly responsive to exogenous GABA. The role of exogenous GABA in root hair development is complex and not yet fully understood. Specific effects of exogenous GABA application on root hair density and length require further investigation.

16.4 Shoot Development and Biomass Accumulation

Shoot development and biomass accumulation are critical determinants of plant productivity, influenced by various environmental and physiological factors. Exogenous GABA enhances shoot growth by modulating cell division, expansion, and hormonal signalling. It promotes stem elongation, leaf expansion, and overall biomass accumulation by improving nutrient uptake, water retention, and photosynthetic efficiency. These combined effects make GABA a crucial regulator of the shoot architecture in diverse plant species.

16.4.1 Stem Elongation

GABA in *Stellaria longipes* exhibits a concentration-dependent effect on stem elongation, mediated through its interaction with ethylene biosynthesis pathways. At lower concentrations, GABA promotes stem elongation at low doses and inhibits at high doses. Elevated GABA levels increase the expression of ACC synthase mRNA and promote ethylene production. The accumulation of ethylene subsequently suppresses stem elongation, suggesting that GABA operates its inhibitory effect on shoot elongation through modulation of ethylene biosynthesis (Kathiresan et al. 1998). Kinnersley and Lin (2000) reported in Lemna species that application of GABA improved mineral uptake, favorably influenced ion transport, and resulted in a two- to threefold increase in plant growth. When GABA antagonists such as bicuculline and picrotoxin, receptor modifiers known from animal systems were applied, the growth-promoting effects of GABA were negated. In contrast, baclofen, a GABA agonist in animals, significantly enhanced GABA-mediated growth in Lemna. Collectively, these findings indicate that GABA may act as a signaling molecule in plants, interacting with specific receptors to regulate ion transport and growth processes (Kinnersley and Lin 2000). GABA application, by inciting cell elongation and division and/or maintaining metabolic balance within plant tissues, promotes the growth of maize seedlings (Li et al. 2016a, b). Exogenous GABA improves the shoot length of mung bean plants under salinity stress (Ullah et al. 2023). It appears that low GABA levels synergize with signal transduction pathways associated with growth-promoting processes, whereas high GABA level senses adverse environmental conditions, triggering stress resilience processes and ensuring survival of plants.

16.4.2 Biomass Accumulation and Yield

The role of exogenous GABA in enhancing plant biomass and yield has been widely studied, with evidence accumulating over the years across various species. GABA influences key physiological processes, such as ion balance, photosynthesis, and nitrogen metabolism, ultimately contributing to improved growth and stress tolerance. Early research on the effects of exogenous GABA in cereals demonstrates its role in promoting biomass accumulation. In *Zea mays* (maize), GABA application activates key enzymes of nitrogen metabolism, confers photosynthetic efficiency, and facilitates better carbon and nitrogen assimilation that translates into augmented seedling biomass (Li et al. 2016a, b). This chapter provides a foundational understanding of GABA's influence on primary metabolism in cereal crops. Further studies expand this understanding to legumes. Abd El-Gawad et al. (2020) have investigated the role of exogenous GABA in *Phaseolus vulgaris* (snap bean) under drought stress. GABA treatment increases shoot biomass, leaf area, and relative water content by enhancing osmolyte accumulation and antioxidant enzyme activity, which improves stress resilience. Recent studies have confirmed these beneficial effects in additional crop species. In *Vigna radiata* (mung bean), GABA mitigates salinity stress by reducing sodium accumulation, enhancing photosynthetic pigment levels, and optimizing nitrogen metabolism, ultimately leading to greater biomass production (Ullah et al. 2023). GABA application in *Eriobotrya japonica* (loquat) stimulates cell division and elongation, leading to increased plant height, biomass accumulation, and a well-developed root system (Yang et al. 2023).

16.4.3 Leaf Expansion and Leaf Morphology

Leaf morphology and chlorophyll content are fundamental determinants of photosynthetic efficiency and overall plant performance. Exogenous GABA application has been shown to positively regulate leaf expansion, structure, and water retention, thereby contributing to improved plant growth and stress resilience. Under drought stress, GABA treatment significantly boosts leaf area and relative water content in snap bean (*Phaseolus vulgaris*), suggesting its role in leaf expansion and water retention under limited water availability (Abd El-Gawad et al. 2020). Similarly, in response to cold stress, exogenous GABA prevents excessive leaf thickening, maintaining optimal leaf structure and facilitating efficient CO_2 and H_2O exchange, which is essential for sustaining photosynthetic activity and leaf functionality (Li et al. 2022). In addition to its protective effects under abiotic stress, GABA has been linked to increased biomass accumulation in leaves. Studies on mung bean (*Vigna radiata*) demonstrate an improvement in both fresh and dry leaf weight as a result of GABA application under saline conditions, reinforcing its role in mitigating osmotic stress and supporting leaf growth (Ullah et al. 2023). Moreover, GABA-induced modifications in leaf architecture correlate with enhanced photosynthetic capacity and greater resilience to environmental stressors (Geng et al. 2025).

16.5 GABA-Induced Delayed Leaf Senescence

Leaf senescence is a complex and highly regulated process that is influenced by various environmental and endogenous factors, including oxidative stress and hormonal signalling. GABA has emerged as a crucial signalling molecule that modulates numerous metabolic and physiological pathways, including those associated with leaf senescence (Ansari et al. 2014). The role of exogenous GABA in delaying both natural and stress-induced senescence has widely been studied, with evidence pointing towards its ability to enhance chlorophyll retention, sustain photosynthetic efficiency, and regulate key hormonal and molecular networks.

16.5.1 Senescence Under Normal and Stress Conditions

GABA's role in delaying leaf senescence comes from its function as an antioxidant. Senescing leaves experience a rise in ROS, leading to cellular damage and accelerated ageing. GABA acts as a potent ROS scavenger, protecting plant cells from oxidative stress and delaying the onset of senescence (Jalil et al. 2017, 2019). Exogenous GABA application protects creeping bentgrass (*Agrostis stolonifera*) from heat and drought stress, where its foliar application significantly delays chlorophyll degradation and enhances turf quality, suggesting a direct role of GABA in preserving photosynthetic pigments (Li et al. 2016a, b). Subsequent studies reinforce these findings. For example, GABA contributes to chlorophyll retention under different environmental stress conditions. In mung bean (*Vigna radiata*), GABA application prevents premature leaf yellowing by mitigating oxidative stress and improving nitrogen assimilation (Priya et al. 2019). Similarly, GABA treatment in tomato (*Solanum lycopersicum*) significantly delays degradation of chlorophyll "a" content under salt stress and thereby maintains higher photosynthetic efficiency (Wu et al. 2020). These findings highlight the protective role of GABA against environmental stressors that induce premature leaf senescence.

16.5.2 Hormonal and Molecular Pathways

The regulation of leaf senescence is tightly controlled by phytohormones, particularly abscisic acid and ethylene. GABA has now gained acceptance as an important modulator for hormonal pathways. Ji et al. (2018) have demonstrated that exogenous GABA influences the expression of genes involved in ABA and ethylene biosynthesis, thereby delaying the hormonal signals that trigger senescence. Furthermore, GABA maintains cellular homeostasis and ion balance to mitigate oxidative stress. At the transcriptional level, GABA's impact on senescence-related gene expression is noteworthy in multiple plant species. Xu et al. (2020) report that GABA suppresses the expression of *CDF4*, a transcription factor known to promote leaf senescence through ABA and ROS signalling. This finding suggests that GABA may act upstream of ABA-mediated senescence regulation, delaying leaf ageing by modulating transcriptional networks. GABA-T, a crucial enzyme in the GABA metabolic pathway, also plays a pivotal role in regulating leaf longevity. In Arabidopsis (*Arabidopsis thaliana*), knockout mutants of *pop2-1* and *pop2-3* (which lack GABA-T activity) exhibit a significant reduction in stress-induced chlorophyll retention, along with increased ion leakage and lipid peroxidation, indicating enhanced oxidative stress (Khan et al. 2021). Therefore, GABA metabolism is closely linked to stress tolerance and delayed leaf senescence. More recently, Kabała and Janicka (2024) highlight the broader role of GABA as a signalling molecule in plant defence against unfavourable environmental factors. Their findings suggest that GABA interacts with other signalling molecules beyond ABA and ethylene, integrating various stress response pathways to enhance plant resilience and delay senescence.

16.6 Flowering and Reproductive Development

GABA plays a role in modulating flowering time through its interaction with hormonal and metabolic pathways. While the exact mechanisms remain under investigation, it is suggested that GABA influences ethylene production, a hormone known to affect flowering processes. For instance, GABA supplementation has been associated with developmental and functional aspects of flowers, indicating a potential role in reproductive functions (Priya et al. 2019). GABA contributes to reproductive success by enhancing processes such as pollination and fruit development. Notably, GABA gradients are crucial in guiding pollen tube growth towards the ovule, facilitating successful fertilization. Studies have shown that exogenous GABA affects pollen tube growth by modulating calcium-permeable membrane channels, which are essential for proper pollen tube navigation and fertilization (Biancucci et al. 2015).

16.7 Conclusion

In conclusion, exogenous GABA plays a crucial role in shaping plant morphology by modulating key developmental processes, including root and shoot growth, leaf expansion, flowering, and stress-induced structural adaptations. Its interaction with phytohormones such as auxins, ABA, gibberellins, and ethylene highlights its significance in regulating cell expansion, division, and overall plant architecture. Under normal conditions, GABA enhances biomass accumulation and photosynthetic efficiency, while under stress conditions, it aids in root restructuring, leaf senescence delay, and improved reproductive success. The growing body of research on GABA-mediated morphological responses provides valuable insights into its potential applications in agriculture, particularly in enhancing crop resilience and productivity under adverse environmental conditions. Understanding and harnessing the regulatory effects of GABA could pave the way for innovative strategies in sustainable crop management and stress mitigation.

References

Abd El-Gawad HG, Mukherjee S, Farag R, Abd Elbar OH, Hikal M, Abou El-Yazied A et al (2020) Exogenous γ-aminobutyric acid (GABA)-induced signaling events and field performance associated with mitigation of drought stress in *Phaseolus vulgaris* L. Plant Signal Behav 16(2). https://doi.org/10.1080/15592324.2020.1853384

AL-Quraan NA, AL-Ajlouni ZI, Obedat DI (2019) The GABA shunt pathway in germinating seeds of wheat (Triticum aestivum L.) and barley (Hordeum vulgare L.) under salt stress. Seed Sci Res 29:1–11. https://doi.org/10.1017/S0960258519000230

Ansari MI, Hasan S, Jalil SU (2014) Leaf senescence and GABA shunt. Bioinformation 10:734–736

Biancucci M, Mattioli R, Forlani G, Funck D, Costantino P, Trovato M (2015) Role of proline and GABA in sexual reproduction of angiosperms. Front Plant Sci 6:680. https://doi.org/10.3389/fpls.2015.00680

Chen T, Yang D, Fan R et al (2022) γ-Aminobutyric acid a novel candidate for rapid induction in somatic embryogenesis of *Liriodendron* hybrid. Plant Growth Regul 96:293–302. https://doi.org/10.1007/s10725-021-00776-8

Cheng BZ, Li Z, Liang LL, Cao Y, Zeng WH, Zhang XQ, Ma X, Huang LK, Nie G, Liu W (2018) The γ-aminobutyric acid (GABA) alleviates salt stress damage during seeds germination of white clover associated with Na+/K+ transportation, dehydrins accumulation, and stress-related genes expression in white clover. Int J Mol Sci 19:2520

Chu C, Yan N, Du Y, Liu X, Chu M, Shi J, Zhang H, Liu Y, Zhang Z (2019) iTRAQ-based proteomic analysis reveals the accumulation of bioactive compounds in Chinese wild rice (Zizania latifolia) during germination. Food Chem 289:635–644. https://doi.org/10.1016/j.foodchem.2019.03.092

Deleu C, Faes P, Niogret MF, Bouchereau A (2013) Effects of the inhibitor of the γ-aminobutyrate-transaminase, vinyl-γ-aminobutyrate, on development and nitrogen metabolism in Brassica napus seedlings. Plant Physiol Biochem 64(5):60–69. https://doi.org/10.1016/j.plaphy.2012.12.007

Du C, Chen W, Wu Y, Wang G, Zhao J, Sun J, Ji J, Yan D, Jiang Z, Shi S (2020) Effects of GABA and vigabatrin on the germination of Chinese chestnut recalcitrant seeds and its implications for seed dormancy and storage. Plants (Basel) 9:4

Geng W, Zhang Y, Li C, Song G, Shi S (2025) Effect of exogenous γ-aminobutyric acid (GABA) on the growth, photosynthetic pigment, antioxidant and GABA metabolism of *Festuca arundinacea* (Tall Fescues) under cadmium stress. Plants 14(3):383. https://doi.org/10.3390/plants14030383

Gray WM (2004) Hormonal regulation of plant growth and development. PLoS Biol 2(9):E311. https://doi.org/10.1371/journal.pbio.0020311. Epub 2004 Sep 14. PMID: 15367944; PMCID: PMC516799

Jalil SU, Ahmad I, Ansari MI (2017) Functional loss of GABA transaminase (GABA-T) expressed early leaf senescence under various stress conditions in *Arabidopsis thaliana*. Curr Plant Biol 9–10:11–22

Jalil SU, Khan MQR, Ansari MI (2019) Role of GABA transaminase in the regulation of development and senescence in *Arabidopsis thaliana*. Curr Plant Biol 19:100119

Jin Y, Zhi L, Tang X, Chen Y, Hancock JT, Hu X (2023) The Function of GABA in Plant Cell Growth, Development and Stress Response, Phyton-International Journal of Experimental Botany, 92(8):2211–2225, https://doi.org/10.32604/phyton.2023.026595

Ji J, Yue J, Xie T et al (2018) Roles of γ-aminobutyric acid on salinity-responsive genes at transcriptomic level in poplar: involving in abscisic acid and ethylene-signalling pathways. Planta 248:675–690. https://doi.org/10.1007/s00425-018-2915-9

Kabała K, Janicka M (2024) Relationship between the GABA pathway and signaling of other regulatory molecules. Int J Mol Sci 25(19):10749. https://doi.org/10.3390/ijms251910749. PMID: 39409078; PMCID: PMC11476557

Kathiresan A, Miranda J, Chinnappa CC, Reid DM (1998) γ–aminobutyric acid promotes stem elongation in Stellarialongipes: the role of ethylene. Plant Growth Regul 26:131–137. https://doi.org/10.1023/A:1006107815064

Khan MIR, Jalil SU, Chopra P, Chhillar H, Ferrante A, Khan NA, Ansari MI (2021) Role of GABA in plant growth, development and senescence. Plant Gene 26:100283

Kinnersley AM, Lin F (2000) Receptor modifiers indicate that 4-aminobutyric acid (GABA) is a potential modulator of ion transport in plants. Plant Growth Regul 32:65–76. https://doi.org/10.1023/A:1006305120202

Li W, Liu J, Ashraf U, Li G, Li Y, Lu W, Gao L, Han F, Hu J (2016a) Exogenous γ-aminobutyric acid (GABA) application improved early growth, net photosynthesis, and associated physio-biochemical events in maize. Front Plant Sci 7:919. https://doi.org/10.3389/fpls.2016.00919

Li Z, Peng Y, Huang B (2016b) Physiological effects of g-Aminobutyric acid application on improving heat and drought tolerance in creeping Bentgrass. J Amer Soc Hort Sci 141(1):76–84

Li X, Xu X, Shen F et al (2021) γ-Aminobutyric acid participates in the adult-phase adventitious rooting recalcitrance. J Plant Growth Regul 40:1981–1991. https://doi.org/10.1007/s00344-020-10251-9

Li Y, Yu X, Ma K (2022) Physiological effects of g-aminobutyric acid application on cold tolerance in Medicago ruthenica. Front Plant Sci 13:958029. https://doi.org/10.3389/fpls.2022.958029

Matsuyama A, Yoshimura K, Shimizu C, Murano Y, Takeuchi H, Ishimoto M (2009) Characterization of glutamate decarboxylase mediating γ-amino butyric acid increase in the early germination stage of soybean (Glycine max [L.] Merr). J Biosci Bioeng 107:538–543. https://doi.org/10.1016/j.jbiosc.2009.01.012

Michaeli S, Fait A, Lagor K, Nunes-Nesi A, Grillich N et al (2011) A mitochondrial GABA permease connects the GABA shunt and the TCA cycle, and is essential for normal carbon metabolism. Plant J 67(3):485–498. https://doi.org/10.1111/j.1365-313X.2011.04612.x

Pei L, Zhao Y, Shi X, Chen R, Yan J et al (2022) The role of γ-aminobutyric acid (GABA) in the occurrence of adventitious roots and somatic embryos in woody plants. Plants 11(24):3512. https://doi.org/10.3390/plants11243512

Priya M, Sharma L, Kaur R et al (2019) GABA (γ-aminobutyric acid), as a thermo-protectant, to improve the reproductive function of heat-stressed mungbean plants. Sci Rep 9:7788. https://doi.org/10.1038/s41598-019-44163-w

Renault H, El Amrani A, Palanivelu R, Updegraff EP, Yu A et al (2011) GABA accumulation causes cell elongation defects and a decrease in expression of genes encoding secreted and cell wall-related proteins in Arabidopsis thaliana. Plant Cell Physiol 52(5):894–908. https://doi.org/10.1093/pcp/pcr041

Sharma S, Saxena DC, Riar CS (2016) Analysing the effect of germination on phenolics, dietary fibres, minerals and γ-aminobutyric acid contents of barnyard millet (Echinochloa frumentaceae). Food Biosci 13:60–68. https://doi.org/10.1016/j.fbio.2015.12.007

Sheng Y, Xiao H, Guo C, Wu H, Wang X (2018) Effects of exogenous gamma-aminobutyric acid on α-amylase activity in the aleurone of barley seeds. Plant Physiol Biochem 127:39–46

Tang J, Li ML, Mao PS, Jiang YW (2022) Effects of gamma-aminobutyric acid on seed germination, ion balance, and metabolic activity in perennial ryegrass under salinity stress. J Plant Growth Regul 41:1835–1844

Ullah A, Ali I, Noor J, Zeng F, Bawazeer S, Eldin SM, Asghar MA, Javed HH, Saleem K, Ullah S, Ali H (2023) Exogenous g-aminobutyric acid (GABA) mitigated salinity-induced impairments in mungbean plants by regulating their nitrogen metabolism and antioxidant potential. Front Plant Sci 13:1081188. https://doi.org/10.3389/fpls.2022.1081188

Wu X, Jia Q, Ji S et al (2020) Gamma-aminobutyric acid (GABA) alleviates salt damage in tomato by modulating Na^+ uptake, the *GAD* gene, amino acid synthesis and reactive oxygen species metabolism. BMC Plant Biol 20:465. https://doi.org/10.1186/s12870-020-02669-w

Xie T, Ji J, Chen W, Yue J, Du C, Sun J, Shi S (2020) GABA negatively regulates adventitious root development. J Exp Bot 71:1459–1474

Xu JG, Hu QP, Duan JL, Tian CR (2010) Dynamic changes in gamma-aminobutyric acid and glutamate decarboxylase activity in oats (Avena nuda L.) during steeping and germination. J Agric Food Chem 58(17):9759–9763. https://doi.org/10.1021/jf101268a

Xu L, Chen L, Ali B, Yang N, Chen Y, Wu F, Jin Z, Xu X (2017) Impact of germination on nutritional and physicochemical properties of adlay seed (Coixlachryma-jobi L.). Food Chem 229:312–318. https://doi.org/10.1016/j.foodchem.2017.02.096. Epub 2017 Feb 21. PMID: 28372179

Xu P, Chen H, Cai W (2020) Transcription factor CDF4 promotes leaf senescence and floral organ abscission by regulating abscisic acid and reactive oxygen species pathways in Arabidopsis. EMBO Rep 21:e48967. https://doi.org/10.15252/embr.201948967

Yang Z, Xu Y, Song P, Li X, Zhou J, Lin L, Xia H, Liang D, Luo X, Zhang H et al (2023) Effects of gamma amino butyric acid (GABA) on nutrient uptake of Loquat [*Eriobotrya japonica* (Thunb.) Lindl.] seedlings. Horticulturae 9:196. https://doi.org/10.3390/horticulturae9020196

Zhao GC, Xie MX, Wang YC, Li JY (2017) Molecular mechanisms under lying γ-aminobutyric acid (GABA) accumulation in giant embryo rice seeds. J Agric Food Chem 65:4883–4889. https://doi.org/10.1021/acs.jafc.7b00013

Physiological Responses of Plants to Exogenous GABA

17

Abstract

Gamma-aminobutyric acid (GABA) modulates plant physiological responses under both optimal and stress conditions. Applying GABA externally has shown great potential in improving plant resistance to a range of abiotic stresses. GABA administration stimulates key processes of plant physiology such as nitrogen metabolism, amino acid synthesis, water relations, stomatal conductance, photosynthesis, and chlorophyll metabolism. GABA supports stress adaptation by influencing the expression of genes that regulate nutrient uptake, enhancing the activity of key metabolic enzymes, stabilizing cellular structures, and regulating hormone signalling pathways. It also promotes osmotic adjustment, maintains ion homeostasis, and protects the photosynthetic apparatus under adverse environmental conditions. GABA administration also activates antioxidant defence mechanisms, reduces oxidative stress, and supports cellular energy balance through the GABA shunt. The synergy among diverse physiological responses due to GABA application facilitates stress resilience and prompts growth/productivity augmentation. The mechanisms underlying GABA-mediated stress mitigation offer their potential applications in sustainable agriculture and the improvement of crop performance under challenging environmental conditions.

Keywords

Antioxidant defence · Exogenous GABA · Hormonal regulation · Ion homeostasis · Photosynthesis · Nitrogen metabolism · Stomatal regulation

S. U. Jalil et al., *GABA Signaling System and Abiotic Stress Tolerance in Plants*,
https://doi.org/10.1007/978-981-95-0437-4_17

Highlights

- External application of gamma-aminobutyric acid (GABA) strengthens plant resilience against different abiotic stresses by adjusting key physiological and metabolic processes.
- It also supports better nitrogen uptake, boosts amino acid production, and helps maintain ion balance during stressful conditions.
- GABA regulates stomatal conductance, enhances water use efficiency, and supports osmotic adjustment during drought.
- It preserves chlorophyll content and photosynthetic efficiency by protecting PS II and stabilizing cellular structures.
- GABA activates antioxidant defences and signalling networks, contributing to improved plant growth and stress resilience.

17.1 Introduction

Plants are constantly exposed to a wide array of environmental stresses, which severely affect growth, development, and productivity. To mitigate these adverse effects, plants have evolved complex physiological and biochemical mechanisms to maintain cellular homeostasis and adapt to stress conditions. Among these mechanisms, the role of signalling molecules in modulating plant responses has garnered significant attention.

Gamma-aminobutyric acid (GABA) regulates plant physiological processes, particularly under conditions of abiotic stress (Jalil and Ansari 2020). It is endogenously synthesized in plants via the GABA shunt pathway, which bypasses two steps of the TCA cycle and is tightly linked to central carbon metabolism (Ansari et al. 2014, 2021). Through this pathway, GABA serves as an important metabolic intermediate that facilitates the integration of carbon and nitrogen metabolism, especially under stress conditions where energy and resource allocation are critical (Jalil et al. 2017, 2019). Beyond its metabolic function, GABA also acts as a signalling molecule involved in modulating cytosolic pH, ion homeostasis, and membrane potential, all of which are vital for maintaining cellular stability during environmental perturbations (Khan et al. 2021; Islam et al. 2024; Abdullah et al. 2025).

Additionally, GABA contributes to redox homeostasis by regulating the balance between ROS production and scavenging, thereby protecting cells from oxidative damage (Ullah et al. 2023). Recent studies have shown that the exogenous application of GABA, either as a foliar spray or soil amendment, can significantly enhance plant tolerance to various abiotic stresses (Khan et al. 2021; Ahmad and Fariduddin 2024). This enhancement is achieved through modulation of physiological processes such as nitrogen assimilation, enhanced amino acid synthesis, maintenance of water status through uptake and stomatal regulation, photosynthetic efficiency, and antioxidant defence mechanisms (Li et al. 2017; Liu et al. 2021; Zarbakhsh and Shahsavar 2023; Khan et al. 2024; Jurkonienė et al. 2025; Wang et al. 2025; Zhang et al. 2025a). As such, GABA is increasingly recognized as a key player in plant

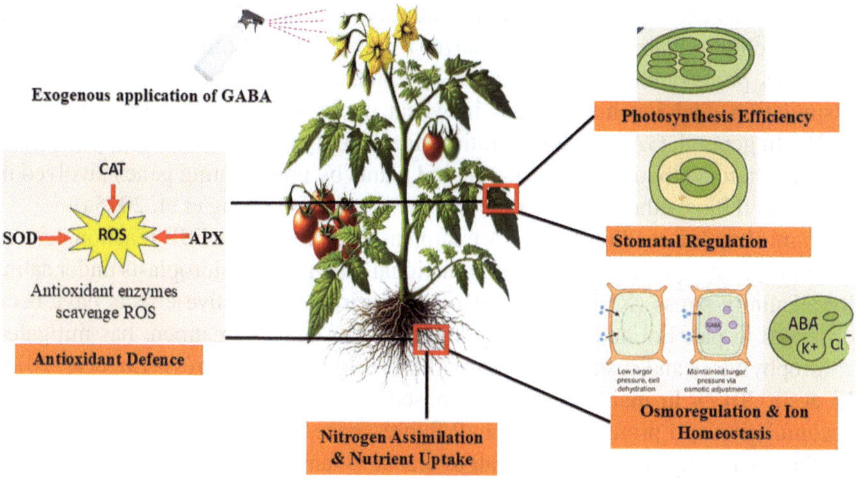

Fig. 17.1 Effect of exogenous GABA application on plant physiological responses. Exogenous GABA enhances plant tolerance to abiotic stress by boosting antioxidant defence, photosynthesis, stomatal regulation, osmoregulation, ion homeostasis, and nutrient uptake. Red boxes indicate plant parts where each physiological process primarily occurs

adaptive responses and is being explored for its potential use in developing stress-resilient crop varieties for sustainable agriculture.

This chapter provides a comprehensive overview of the effects of exogenous GABA on key physiological responses in plants (Fig. 17.1). Emphasis is placed on its roles in nitrogen metabolism, water relations, photosynthesis, and stress signalling, highlighting the potential of GABA as a promising bio-regulator for sustainable agriculture.

17.2 Exogenous GABA and Plant Stress Mitigation

Exogenously applied GABA has emerged as a potent modulator of various physiological processes, aiding plants in coping with environmental challenges. It influences key functions such as stomatal conductance, photosynthetic efficiency, nutrient uptake, and antioxidant defence. GABA enhances water use efficiency and maintains ion homeostasis, contributing to cellular stability and metabolic balance (Fig. 17.1). By modulating these physiological pathways, exogenous GABA significantly improves plant resilience to abiotic stresses.

17.2.1 Influence on Photosynthesis and Chlorophyll Metabolism

Exogenous GABA plays a vital role in enhancing photosynthesis and chlorophyll metabolism, particularly under various abiotic stress conditions. GABA application promotes gas exchange and net photosynthesis by maintaining cell turgor pressure and reducing oxidative damage in maize (Li et al. 2016). Similarly, its application improves photosynthetic performance by enhancing nitrogen metabolism, sulfur assimilation, and ion homeostasis in wheat under salinity stress (Khanna et al. 2021). In general, GABA administration significantly increases chlorophyll content and net photosynthetic rate in salt-stressed plants by upregulating genes involved in chlorophyll biosynthesis and photosynthetic processes (Zhang et al. 2025a).

In muskmelon seedlings, GABA treatment helps repair the PSII reaction centres and thus preserves the structural and functional integrity of chloroplasts under salinity-alkalinity stress (Xiang et al. 2016). Comparable protective effects have been observed in pakchoi and pepper seedlings, where GABA treatment has mitigated chlorophyll loss and sustained photosynthetic activity during stress (Liu et al. 2016; Li et al. 2017). In maize seedlings exposed to salinity, GABA reduces toxic ion accumulation and protects chlorophyll fluorescence parameters, preserving photosynthetic capacity and enhancing salt tolerance (Wang et al. 2017). Under waterlogging stress, GABA effectively reduces chlorophyll degradation, preserves pigment stability, and improves photosynthetic efficiency in maize leaves (Salah et al. 2019). Thus, exogenous GABA protects the photosynthetic apparatus, particularly PSII.

GABA also strengthens antioxidant defence systems, stabilizes cellular membranes, and helps maintain photosynthetic functionality under drought, as observed in chickpea plants (Seifikalhor et al. 2022). In pomegranate cultivars, GABA application sustains transpiration rates, net photosynthesis, intercellular CO_2 concentration, and stomatal conductance, thereby alleviating stress-induced declines in photosynthetic efficiency (Zarbakhsh and Shahsavar 2023).

GABA treatment in mungbean plants subjected to heat stress alleviates membrane damage and preserves photosynthetic components such as chlorophyll content, chlorophyll fluorescence, and rubisco activity. It also supports carbon assimilation by maintaining sucrose synthesis and utilization, ensuring sustained growth under heat stress (Priya et al. 2019). In Arabidopsis, GABA administration regulates the expression of genes related to light-harvesting complexes, optimizing energy absorption and minimizing photodamage under high light and temperature stress (Balfagón et al. 2022).

17.2.2 Water Relations and Stomatal Conductance

Exogenous GABA maintains plant water relations and stomatal conductance, which are critical components of drought tolerance and water use efficiency. Under water-deficit conditions, endogenous GABA levels often rise, particularly in guard cells, where it directly influences stomatal behaviour. GABA modulates stomatal aperture by regulating ion flux through channels such as aluminium-activated malate transporter 9 (ALMT9), resulting in partial stomatal closure and reduced transpiration rates (Xu et al. 2021a). This mechanism not only minimizes water loss but also maintains adequate internal CO_2 concentrations for photosynthesis under stress.

Moreover, GABA acts synergistically with ABA, a key drought-responsive phytohormone, by enhancing ABA signal transduction pathways involved in stomatal regulation. Studies have shown that GABA treatment increases the sensitivity of stomata to ABA, thereby amplifying the stomatal closure response and further conserving water under drought conditions (Liu et al. 2021). This GABA-ABA interaction is believed to involve both transcriptional regulation of ABA-responsive genes and post-translational modification of ion channels responsible for guard cell turgor changes.

In addition to its role in regulating stomatal behaviour, exogenous GABA plays a crucial part in maintaining plant water status through effective osmotic adjustment mechanisms. Under drought or water-deficit conditions, plants often experience cellular dehydration due to reduced water availability. GABA helps counteract this by promoting the synthesis and accumulation of compatible solutes, such as proline, soluble sugars, and polyols, which act as osmo-protectants. These compounds stabilize cellular structures, maintain osmotic balance, and preserve cell turgor pressure, thereby enabling plants to retain water and maintain physiological functions even under stress (Jurkonienė et al. 2025).

Moreover, GABA application has been shown to enhance root hydraulic conductivity, which facilitates the exploration and absorption of water from the soil. This is complemented by its positive effects on membrane integrity, helping to prevent lipid peroxidation and leakage of cellular contents that are the common outcomes of water stress. By preserving membrane stability and enhancing water transport efficiency through roots, GABA aids in sustaining the water balance within plant tissues and minimizing the extent of cellular damage in heat-stressed mungbean plants (Priya et al. 2019). Collectively, these mechanisms make GABA a key regulator in improving water retention, osmotic stability, and stress tolerance in plants.

17.2.3 Nitrogen Assimilation and Amino Acid Synthesis

Treatment of plants with GABA significantly improves nitrogen metabolism and amino acid synthesis, particularly in a stressful environment. GABA functions through the GABA shunt pathway, linking carbon and nitrogen metabolism by providing carbon skeletons and energy necessary for biosynthetic processes (Al-Khayri et al. 2024). This integration supports the synthesis of essential amino acids like glutamate, thereby maintaining metabolic balance during stress (Yu et al. 2024).

Beyond its role in stress signalling, exogenous GABA contributes to the optimization of nitrogen utilization and amino acid biosynthesis in plants. Its application influences key enzymatic and genetic components of nitrogen metabolism, particularly under challenging environmental conditions. The enzymatic activation leads to improved nitrogen assimilation efficiency, which translates into enhanced biomass accumulation and better overall plant performance (Li et al. 2016). Additionally, in muskmelon seedlings exposed to calcium nitrate-induced stress, exogenous GABA maintains nitrogen homeostasis by regulating transaminase activities, which are crucial for the interconversion of amino acids. This helps sustain amino acid pool balance, reduce oxidative damage, and improve resistance to ionic and osmotic stress (Zhen et al. 2018).

In addition, the application of GABA is crucial for mitigating nitrogen deficiency by regulating the expression of genes involved in nitrogen uptake and assimilation (Islam et al. 2024), thereby enhancing nitrogen use efficiency. In *Andrographis paniculata*, GABA acts as a signalling molecule that upregulates NRT genes via its specific receptor. This upregulation enhances NO_3^- uptake from the soil, which is subsequently assimilated more efficiently through increased enzymatic activity in the nitrogen metabolism pathway. As a result, GABA treatment significantly mitigates the effects of nitrogen deficiency by boosting both nitrate acquisition and assimilation (Huang et al. 2023). Similarly, exogenous application of GABA to sugar beet elevates the activity of key nitrogen metabolism enzymes, such as nitrate reductase and glutamine synthetase. This biochemical enhancement facilitates amino acid biosynthesis, supporting overall plant growth and stress tolerance under nutrient-limited conditions (Yu et al. 2024).

Under elevated rhizosphere CO_2 conditions, exogenous GABA application in melon seedlings significantly augments nitrogen assimilation by upregulating key genes involved in nitrogen uptake and transport, such as NRT and AMT genes (Wang et al. 2025). This regulatory effect helps mitigate the adverse impact of CO_2-induced stress on plant growth and nutrient balance. Similarly, in maize, GABA treatment stimulates the activities of nitrate reductase and glutamine synthetase, two central enzymes in nitrogen metabolism that facilitate the conversion of nitrate into ammonium and its subsequent incorporation into amino acids.

17.2.4 Regulation of Hormonal Crosstalk

Exogenous application of GABA also acts as a significant modulator of hormonal crosstalk in plants, influencing various physiological processes and enhancing stress tolerance. GABA interacts with multiple phytohormones, including ABA, ethylene, jasmonic acid, SA, and auxins, thereby integrating metabolic and hormonal signalling pathways to fine-tune plant responses under both normal and stress conditions. Furthermore, GABA supports cytokinin-regulated nutrient remobilization and stress adaptation, as seen in transgenic barley with elevated GABA-related gene expression (Pospíšilová et al. 2016), highlighting its integrative role in hormonal crosstalk during stress.

Under abiotic stress conditions, such as drought and salinity, GABA levels increase, which in turn modulates ABA signalling pathways. This modulation enhances stomatal closure, reducing water loss and improving drought tolerance. Additionally, GABA influences ethylene biosynthesis and signalling, which play a crucial role in plant development and stress responses (Islam et al. 2024). Applying GABA has been found to boost ethylene production, which in turn influences things like fruit ripening and leaf ageing. It modulates ethylene biosynthesis by influencing ACC synthase and oxidase, reducing excessive ethylene accumulation, and delaying senescence (Kinnersley and Turano 2000; Shi et al. 2010; Gilliham and Tyerman

2016). GABA also interrelates with ABA signalling in guard cells, maintaining stomatal function and redox balance during drought stress (Mekonnen et al. 2016; Ji et al. 2018; Xu et al. 2021a, b). Through its crosstalk with SA, GABA delays stress-induced senescence by enhancing antioxidant defences and maintaining cellular integrity (Babarabie et al. 2019).

GABA interacts with JA and SA pathways, which are integral to plant defence mechanisms against biotic stresses. By modulating these hormonal pathways, GABA enhances the plant's ability to respond to pathogen attacks and insect herbivory. Applying GABA to sweet orange plants increases levels of these hormones (Hijazbet al. 2018). In lettuce, GABA accumulation after *B. cinerea* infection, primed by specific compounds, depends on ET signalling (Tarkowski et al. 2019). This suggests that ET may enhance GABA's positive effects on plant defence by influencing the methionine cycle, which is crucial for ET production (Wang et al. 2019). ET also regulates calcium influx in cells, affecting GAD activity during symbiotic interactions (Oldroyd et al. 2001). Furthermore, GABA's interaction with auxin signalling influences root development and architecture, contributing to improved nutrient and water uptake under stress conditions. GABA exerts profound effects on auxin homeostasis and transport, influencing processes such as root architecture and AR formation. GABA treatment has been shown to suppress auxin-dependent upregulation, delaying AR development by reducing polar auxin transport (Li et al. 2021). This suggests that GABA-mediated modulation of auxin distribution is crucial in early rooting processes (Pei et al. 2022).

The crosstalk between GABA and phytohormones is complex and involves the regulation of gene expression related to hormone biosynthesis, signalling, and response. This intricate network allows plants to integrate various environmental signals and adjust their growth and development accordingly. For example, GABA's influence on ABA and ethylene pathways can lead to coordinated responses that balance growth and stress adaptation. Moreover, in sweet orange (*Citrus sinensis*), GABA application significantly increases auxin levels (IAA and IPA), along with SA, JA, and ABA, through the upregulation of biosynthetic genes, including CitTAA, CitYUC2 (auxin pathway), CitPAL, CitCS, CitICS (salicylate pathway), and CitLOX, CitAOS, CitAOC (jasmonate pathway), reinforcing its role in stress mitigation (Hijaz et al. 2018). Similarly, GABA administration in apple seedlings amplifies ABA signalling along with stomatal closure, reducing transpiration and improving oxidative stress tolerance (Liu et al. 2021). In Arabidopsis, BRL3 overexpression elevates GABA, proline, and tyrosine levels under oxidative stress, suggesting GABA's link to brassinosteroid signalling (Fàbregas et al. 2018). GABA priming in *Cassia italica* balances hormone levels by increasing auxins and gibberellins while decreasing ABA and jasmonic acid, thereby enhancing salt stress tolerance (Alqarawi et al. 2016). In *Caragana intermedia* and poplar, GABA modulates ABA receptor and ethylene biosynthesis genes, underscoring its hormonal regulatory role during salinity stress (Ji et al. 2018; Shi et al. 2010). Epibrassinolide treatment boosts GABA accumulation and antioxidant defences in *Cucurbita pepo* (Nejad-Alimoradi et al. 2019), while combined application of GABA, brassinolide, and SA improves grapevine performance under salinity (Zhao et al. 2023). In

Origanum vulgare, GABA confers superior salt stress protection, whereas SA appears more effective against heat stress (Garoosi et al. 2023). In *Capsicum annuum*, SA facilitates GAD activity under osmotic stress but not GABA levels directly, while in chilling-sensitive fruits, elevated GABA correlates with reduced seed browning, pointing to its role in auxin-mediated cold stress responses (Çekiç 2017).

17.2.5 Oxidative Stress Tolerance and Antioxidant Defence

Exogenous application of GABA has been widely recognized for its significant role in enhancing oxidative stress tolerance in plants by regulating the antioxidant defence system. Under abiotic stresses such as drought, salinity, and heavy metal exposure, plants often accumulate excessive ROS, leading to cellular and metabolic damage. Among cellular compartments, chloroplasts are the primary hotspots for ROS production under environmental stress (Partelli and Batista 2001), although ROS generation also occurs in apoplast (Hu et al. 2006), the cell wall (Pottosin et al. 2014), endoplasmic reticulum (Mittler 2002), mitochondria (Navrot et al. 2007), and peroxisomes (Noctor et al. 2002). GABA mitigates these harmful effects by enhancing the activities of major antioxidant enzymes such as CAT, POD, and SOD, which collectively scavenge ROS and safeguard cellular structures (Kalhor et al. 2018; Aljuaid and Ashour 2022; Ullah et al. 2023). In pepper plants exposed to low light stress, exogenous GABA significantly reduces oxidative damage by stimulating antioxidant defences and enhancing photochemical efficiency (Li et al. 2017).

Beyond ROS scavenging, GABA contributes to the stabilization of cellular membranes and the regulation of intracellular pH. These protective effects are particularly beneficial under stress conditions where ion imbalance exacerbates oxidative stress. In rice under copper toxicity, exogenous GABA alleviates oxidative damage by enhancing antioxidant enzyme activities and reducing ROS levels (Khan et al. 2024). GABA also stimulates the enzymes of the AsA–GSH cycle, such as APX, DHAR, and MDHAR. In peaches, under chilling stress, GABA treatment upregulates APX, DHAR, MDHAR, along with glutathione peroxidase, glutathione-S-transferase, CAT, and SOD (Yang et al. 2011). Likewise, in sunflowers exposed to combined drought and heat stress, exogenous GABA improves osmolyte metabolism and triggers expression of stress-responsive genes such as heat shock proteins, and elevated activities of antioxidants like SOD, POX, GR, APX, and MDHAR, thereby ameliorating the adverse effects of stress (Abdel Razik et al. 2021).

At the metabolic level, GABA enters the GABA shunt pathway, where it is converted to succinic acid and integrated into the TCA cycle. This provides an alternative route for energy generation and supports redox balance within the cell (Ansari et al. 2021). The GABA shunt also contributes to the production of NADH and NADPH, which are essential cofactors for antioxidant enzymes, such as glutathione reductase, thereby linking stress responses with energy metabolism (Gill and Tuteja 2010; Mittler et al. 2011). Additionally, GABA modulates ROS signalling in stress-responsive cascades without causing cellular damage by regulating ROS levels to maintain their role as signalling molecules e.g. H_2O_2.

17.2.6 Cellular Ion Homeostasis

Exogenous application of GABA plays a significant role in maintaining cellular ion homeostasis in plants, particularly under abiotic stress conditions such as salinity, drought, and hypoxia. GABA functions as a signalling molecule that modulates ion transport mechanisms, thereby enhancing plant resilience to environmental stresses.

One of the primary ways GABA contributes to ion homeostasis is by regulating the activity of specific ion channels. GABA interacts with aluminium-activated malate transporters (ALMTs) and gated outwardly rectifying K^+ (GORK) channels, which are crucial for maintaining K^+ balance within plant cells (Ramesh et al. 2018; Xu et al. 2021a, b). By modulating these channels, GABA helps prevent excessive K^+ efflux, a common issue under stress conditions that can lead to cellular dysfunction (Abdullah et al. 2025).

In addition to its role in potassium regulation, GABA significantly influences Ca^{2+} signalling pathways, which are central to numerous physiological and stress-responsive processes in plants. Calcium ions act as crucial secondary messengers in the perception and transmission of various environmental signals. Exogenous GABA treatment has been reported to enhance Ca^{2+} influx into plant cells (Kinnersley and Lin 2000; Kabała and Janicka 2024), thereby contributing to the rapid activation of downstream signalling cascades. This GABA-induced modulation of calcium signalling plays a vital role in stress signal transduction, facilitating the plant's ability to perceive and respond to abiotic stresses such as salinity, drought, and temperature fluctuations. Additionally, increased Ca^{2+} influx triggered by GABA has been linked to the activation of key defence-related pathways, enhancing the expression of stress-responsive genes and strengthening the plant's overall resistance to environmental challenges (Ma et al. 2019). Through this mechanism, GABA not only supports ionic equilibrium but also integrates into the broader network of plant stress adaptation.

Acting as a signalling molecule, GABA modulates ion transport and helps plants adapt to ionic imbalances triggered by environmental stress. Under salt stress, exogenous GABA application reduces Na^+ accumulation in plant tissues (Shi et al. 2024). By mitigating ROS levels and limiting Na^+ uptake, GABA contributes to maintaining cellular ion homeostasis. Moreover, it significantly upregulates the expression of key Na^+ and K^+ transport-related genes, including *HKT1*, *AKT1*, *NHX1*, *SOS1*, *SOS2*, and *SOS3*, particularly in apple plants (Shi et al. 2024). This coordinated regulation supports a favourable Na^+/K^+ ratio, which is vital for sustaining cellular function and overall plant health under salinity stress.

GABA plays a crucial role in stabilizing cellular membranes and maintaining intracellular pH under stress conditions (Khan et al. 2021; Islam et al. 2024). Abiotic stresses like flooding and hypoxia reduce cytosolic pH, triggering rapid GABA accumulation as a defence mechanism (Roberts et al. 1984; Kinnersley and Turano 2000). This decrease in pH, often resulting from the rupture of vacuolar membranes and the leakage of organic acids into the cytosol, activates GAD, which in turn promotes the synthesis of GABA (Roberts et al. 1992; Crawford et al. 1994). Additionally, a concurrent rise in cytosolic Ca^{2+} under low pH conditions further enhances GABA production through GAD activation (Gehring et al. 1990; Carroll et al. 1994). In *Daucus carota*, this glutamate decarboxylation, a proton-consuming reaction, aids in proton buffering and pH homeostasis (Carroll et al. 1994). Alongside, GABA enhances antioxidant enzyme activities, mitigating oxidative stress commonly associated with ionic imbalances during environmental stress (Yuan et al. 2023).

17.2.7 Role in Energy Metabolism and TCA Cycle Modulation

GABA serves as a metabolic intermediate in the TCA cycle, contributing to energy production and ATP synthesis (Ansari et al. 2014). It supports cellular energy demands under stress by maintaining carbon flux and metabolic balance. Upon application, GABA is rapidly absorbed by plant tissues and metabolized through the GABA shunt pathway. This pathway converts GABA into succinic acid, which then enters the TCA cycle, enhancing the plant's energy production capabilities (Hijaz and Killiny 2019).

The GABA shunt serves as an alternative route to the traditional TCA cycle, bypassing certain steps and directly supplying intermediates like succinate (Srivastava et al. 2021). This not only aids in maintaining the carbon-nitrogen balance but also ensures a continuous supply of energy, especially under stress conditions. In rice seedlings, exogenous GABA application has been observed to increase the levels of TCA cycle intermediates, thereby promoting growth and enhancing stress tolerance (Lee et al. 2021).

Moreover, GABA's influence extends to the regulation of key enzymes associated with the TCA cycle. By modulating the activities of enzymes like SSADH, GABA ensures efficient conversion of its metabolites into energy-yielding compounds (Lee et al. 2021). This regulation is crucial for maintaining cellular respiration and energy homeostasis, particularly during environmental stresses such as drought, salinity, and hypoxia (Wu et al. 2021).

In addition to its metabolic roles, exogenous GABA has been linked to the enhancement of antioxidant defence mechanisms. By boosting the activities of enzymes like CAT and SOD, GABA helps in mitigating oxidative stress, thereby protecting cellular components, and ensuring optimal metabolic function (Wu et al. 2021). The exogenous application of GABA significantly influences plant energy metabolism by integrating into the TCA cycle through the GABA shunt, regulating key metabolic enzymes, and bolstering antioxidant defences (Abdullah et al. 2025). These multifaceted roles underscore GABA's importance in enhancing plant resilience and productivity under various environmental conditions.

17.2.8 PCD and Senescence Regulation

The external application of GABA has been shown to significantly influence PCD and delay senescence in plants. Research indicates that GABA treatment helps reduce oxidative damage and inhibit PCD in plant tissues (Zhang et al. 2025b).

Liang et al. (2022) have reported that, in fresh-cut pumpkins, exogenous GABA reduces ROS accumulation, decreases electrolyte leakage, and inhibits DNA degradation and cytochrome c release, thereby mitigating PCD.

Similarly, in *Brassica* species subjected to cadmium stress, GABA application alleviates PCD by reducing cadmium accumulation, inhibiting ROS production, and activating the antioxidant defence system. These findings suggest that GABA's protective effects against PCD are associated with its ability to enhance antioxidant enzyme activities, such as SOD, CAT, and POD, which collectively reduce oxidative stress (Zhang et al. 2025b).

In terms of senescence regulation, exogenous GABA has been reported to delay leaf senescence by maintaining chlorophyll content and photosynthetic efficiency. For example, in gerbera cut flowers, pre-harvest GABA treatment extends vase life and reduces chilling injury by enhancing antioxidant enzyme activities and decreasing MDA content, a marker of lipid peroxidation (Mohammadi et al. 2021). Additionally, GABA's interaction with phytohormones, such as ABA and SA, may contribute to its role in delaying senescence (Hijaz et al. 2018).

17.3 Conclusion

Exogenous application of GABA has become a promising approach to enhance plant resilience against a broad spectrum of abiotic stresses. As both a metabolic intermediate and a signalling molecule, GABA plays a vital role in coordinating physiological responses that underpin stress tolerance. It contributes significantly to nitrogen metabolism, amino acid synthesis, photosynthetic efficiency, osmotic adjustment, and ion homeostasis. Moreover, GABA modulates stomatal behaviour, regulates water use efficiency, and reinforces antioxidant defences, thereby mitigating oxidative damage and maintaining cellular integrity during adverse environmental conditions. Evidence from various crop species underscores the potential of exogenous GABA in improving plant growth, yield stability, and metabolic balance under stress. Its ability to influence gene expression and enzyme activity further highlights its regulatory importance in stress-responsive pathways. Moving forward, integrating GABA-based treatments into crop management practices could offer an eco-friendly and effective tool for sustainable agriculture. However, further research is warranted to optimize application methods, dosages, and timing across diverse plant species and stress environments for maximum agronomic benefit.

References

Abdel Razik E, Alharbi B, Pirzadah T, Alnusairi G, Soliman M, Hakeem K (2021) γ-Aminobutyric acid (GABA) mitigates drought and heat stress in sunflower (Helianthus annuus L.) by regulating its physiological, biochemical and molecular pathways. Physiol Plant 172(2):505–527

Abdullah W, Wani KI, Naeem M et al (2025) From neurotransmitter to plant protector: the intricate world of GABA signaling and its diverse functions in stress mitigation. J Plant Growth Regul 44:403–418. https://doi.org/10.1007/s00344-024-11470-0

Ahmad S, Fariduddin Q (2024) Deciphering the enigmatic role of gamma-aminobutyric acid (GABA) in plants: synthesis, transport, regulation, signaling, and biological roles in interaction with growth regulators and abiotic stresses. Plant Physiol Biochem 208:108502. https://doi.org/10.1016/j.plaphy.2024.108502

Aljuaid BS, Ashour H (2022) Exogenous γ-aminobutyric acid (GABA) application mitigates salinity stress in maize plants. Life 12:1860

Al-Khayri JM, Abdel-Haleem M, Khedr EH (2024) Harnessing GABA pathways to improve plant resilience against salt stress. Horticulturae 10:1296. https://doi.org/10.3390/horticulturae10121296

Alqarawi AA, Hashem A, Abd-Allah EF, Al-Huqail AA, Alshahrani TS, Alshalawi SR, Egamberdieva D (2016) Protective role of gamma-aminobutyric acid on Cassia italica Mill under salt stress. Legum Res Int J 39:396–404

Ansari MI, Hasan S, Jalil SU (2014) Leaf senescence and GABA shunt. Bioinformation 10:730–732

Ansari MI, Jalil SU, Ansari SA et al (2021) GABA shunt: a key-player in mitigation of ROS during stress. Plant Growth Regul 94:131–149. https://doi.org/10.1007/s10725-021-00710-y

Babarabie M, Zarei H, Eskandari A (2019) The impact of pre-harvest treatment with gamma-aminobutyric acid (GABA) and salicylic acid on vase life and post-harvest traits of tuberose cut flowers. Acta Sci Pol Hortorum Cultus 18:83–92. https://doi.org/10.24326/asphc.2019.4.8

Balfagón D, Gómez-Cadenas A, Rambla JL, Granell A, de Ollas C, Bassham DC, Mittler R, Zandalinas SI (2022) γ-Aminobutyric acid plays a key role in plant acclimation to a combination of high light and heat stress. Plant Physiol 188:2026–2038. https://doi.org/10.1093/plphys/kiac010

Carroll AD, Fox GG, Laurie S, Phillips R, Ratcliffe RG, Stewart GR (1994) Ammonium assimilation and the role of γ-aminobutyric acid in pH homeostasis in carrot cell suspensions. Plant Physiol 106:513–520

Çekiç FÖ (2017) Effects of salicylic acid on GABA metabolism in pepper plants under stress conditions. Hacet J Biol Chem 3:359–364

Crawford LA, Bown AW, Breitkreuz KE, Cuine FC (1994) The synthesis of γ-aminobutyric acid in response to treatments reducing cytosolic pH. Plant Physiol 104:865–871

Fàbregas N, Lozano-Elena F, Blasco-Escámez D, Tohge T, Martínez-Andújar C, Albacete A, Osorio S, Bustamante M, Riechmann JL, Nomura T et al (2018) Overexpression of the vascular brassinosteroid receptor BRL3 confers drought resistance without penalizing plant growth. Nat Commun 9:4680

Garoosi MK, Sanjarian F, Chaichi M (2023) The role of γ-aminobutyric acid and salicylic acid in heat stress tolerance under salinity conditions in *Origanum vulgare* L. PLoS One 18:e0288169

Gehring AC, Irving HR, Parish RW (1990) Effects of auxin and abscisic acid on cytosolic calcium and pH in plant cells. Proc Natl Acad Sci USA 87:9645–9649

Gill SS, Tuteja N (2010) Polyamines and abiotic stress tolerance in plants. Plant Signal Behav 5:26–33

Gilliham M, Tyerman SD (2016) Linking metabolism to membrane signaling: the GABA–malate connection. Trends Plant Sci 21:295–301. https://doi.org/10.1016/j.tplants.2015.11.011

Hijaz F, Killiny N (2019) Exogenous GABA is quickly metabolized to succinic acid and fed into the plant TCA cycle. Plant Signal Behav 14:e1573096. https://doi.org/10.1080/15592324.2019.1573096

Hijaz F, Nehela Y, Killiny N (2018) Application of gamma-aminobutyric acid increased the level of phytohormones in *Citrus sinensis*. Planta 248:909–918. https://doi.org/10.1007/s00425-018-2947-1

Hu X, Zhang A, Zhang J, Jiang M (2006) Abscisic acid is a key inducer of hydrogen peroxide production in leaves of maize plants exposed to water stress. Plant Cell Physiol 47:1484–1495

Huang XJ, Jian SF, Wan S, Miao JH, Zhong C (2023) Exogenous γ-aminobutyric acid (GABA) alleviates nitrogen deficiency by mediating nitrate uptake and assimilation in *Andrographis paniculata* seedlings. Plant Physiol Biochem 198:107700. https://doi.org/10.1016/j.plaphy.2023.107700

Islam SNU, Kouser S, Hassan P, Asgher M, Shah AA, Khan NA (2024) Gamma-aminobutyric acid interactions with phytohormones and its role in modulating abiotic and biotic stress in plants. Stress Biol 4:36. https://doi.org/10.1007/s44154-024-00180-y

Jalil SU, Ansari MI (2020) Physiological role of gamma aminobutyric acid in salt stress tolerance. In: Hasanuzzaman M, Tanveer M (eds) Salt and drought stress tolerance in plants, signaling and communication in plants. Springer, Cham, pp 337–350

Jalil SU, Ahmad I, Ansari MI (2017) Functional loss of GABA transaminase (GABA-T) expressed early leaf senescence under various stress conditions in *Arabidopsis thaliana*. Curr Plant Biol 9–10:11–22

Jalil SU, Khan MQR, Ansari MI (2019) Role of GABA transaminase in the regulation of development and senescence in *Arabidopsis thaliana*. Curr Plant Biol 19:100119

Ji J, Yue J, Xie T et al (2018) Roles of γ-aminobutyric acid on salinity-responsive genes at transcriptomic level in poplar: involving in abscisic acid and ethylene signalling pathways. Planta 248:675–690. https://doi.org/10.1007/s00425-018-2915-9

Jurkonienė S, Gavelienė V, Mockevičiūtė R, Jankovska-Bortkevič E, Šveikauskas V, Jankauskienė J, Žalnierius T, Kozeko L (2025) GABA and proline application induce drought resistance in oilseed rape. Plants 14:860. https://doi.org/10.3390/plants14060860

Kabała K, Janicka M (2024) Relationship between the GABA pathway and signaling of other regulatory molecules. Int J Mol Sci 25:10749. https://doi.org/10.3390/ijms251910749

Kalhor MS, Aliniaeifard S, Seif M, Asayesh EJ, Bernard F, Hassani B, Li T (2018) Enhanced salt tolerance and photosynthetic performance: implication of γ-aminobutyric acid application in salt-exposed lettuce (*Lactuca sativa* L.) plants. Plant Physiol Biochem 130:157–172

Khan MIR, Jalil SU, Chopra P, Chhillar H, Ferrante A, Khan NA, Ansari MI (2021) Role of GABA in plant growth, development and senescence. Plant Gene 26:100283

Khan Z, Jan R, Asif S, Farooq M, Kim KM (2024) Exogenous GABA enhances copper stress resilience in rice plants via antioxidant defense mechanisms, gene regulation, mineral uptake, and copper homeostasis. Antioxidants 13:700. https://doi.org/10.3390/antiox13060700

Khanna RR, Jahan B, Iqbal N, Khan NA, AlAjmi MF, Rehman MT, Khan MIR (2021) GABA reverses salt-inhibited photosynthetic and growth responses through its influence on NO-mediated nitrogen–sulfur assimilation and antioxidant system in wheat. J Biotechnol 325:73–82. https://doi.org/10.1016/j.jbiotec.2020.11.015

Kinnersley AM, Lin F (2000) Receptor modifiers indicate that 4-aminobutyric acid (GABA) is a potential modulator of ion transport in plants. J Plant Growth Regul 32:65–76

Kinnersley AM, Turano FJ (2000) γ-Aminobutyric acid (GABA) and plant responses to stress. Crit Rev Plant Sci 19:479–509

Lee J, Kim S, Kim S et al (2021) Production of γ-aminobutyric acid and its supplementary role in the TCA cycle in rice (Oryza sativa L.) seedlings. J Plant Growth Regul 40:78–90. https://doi.org/10.1007/s00344-020-10066-8

Li W, Liu J, Ashraf U, Li G, Li Y, Lu W, Gao L, Han F, Hu J (2016) Exogenous-aminobutyric acid (GABA) application improved early growth, net photosynthesis, and associated physio-biochemical events in maize. Front Plant Sci 7:919. https://doi.org/10.3389/fpls.2016.00919

Li Y, Fan Y, Ma Y, Zhang Z, Yue H, Wang L, Li J, Jiao Y (2017) Effects of exogenous γ-aminobutyric acid (GABA) on photosynthesis and antioxidant system in pepper (Capsicum annuum L.) seedlings under low light stress. J Plant Growth Regul 36:436–449

Li X, Xu X, Shen F, Li W, Qiu C, Wu T, Wang Y, Xu X, Han Z, Zhang X (2021) γ-Aminobutyric acid participates in the adult-phase adventitious rooting recalcitrance. J Plant Growth Regul 40:1981–1991. https://doi.org/10.1007/s00344-020-10251-9

Liang J, Guo F, Cao S, Zhao K, Zhao K, Wang H, Shao X, Wei Y, Zhang C, Zheng Y, Xu F (2022) γ-Aminobutyric acid (GABA) alleviated oxidative damage and programmed cell death in fresh-cut pumpkins. Plant Physiol Biochem 180:9–16. https://doi.org/10.1016/j.plaphy.2022.03.029

Liu JP, Xiao FL, Zhu HF, Zhu Y, Hou XL (2016) Effects of γ-aminobutyric acid on the growth and photosynthesis of pakchoi under waterlogging stress. Acta Agric Shanghai

Liu C, Wang H, Zhang X, Ma F, Guo T, Li C (2021) Activation of the ABA signal pathway mediated by GABA improves the drought resistance of apple seedlings. Int J Mol Sci 22:12676. https://doi.org/10.3390/ijms222312676
Ma Y, Wang P, Gu Z, Tao Y, Shen C, Zhou Y, Han Y, Yang R (2019) Ca^{2+} involved in GABA signal transduction for phenolics accumulation in germinated hulless barley under NaCl stress. Food Chem X 2:100023. https://doi.org/10.1016/j.fochx.2019.100023
Mekonnen DW, Flügge U-I, Ludewig F (2016) Gamma-aminobutyric acid depletion affects stomata closure and drought tolerance of Arabidopsis thaliana. Plant Sci 245:25–34. https://doi.org/10.1016/j.plantsci.2016.01.005
Mittler R (2002) Oxidative stress, antioxidants and stress tolerance. Trends Plant Sci 7:405–410
Mittler R, Vanderauwera S, Suzuki N, Miller G, Tognetti VB, Vandepoele K, Gollery M, Shulaev V, Van Breusegem F (2011) ROS signaling: the new wave? Trends Plant Sci 16:300–309
Mohammadi M, Aelaei M, Saidi M (2021) Pre-harvest spray of GABA and spermine delays post-harvest senescence and alleviates chilling injury of gerbera cut flowers during cold storage. Sci Rep 11:14166. https://doi.org/10.1038/s41598-021-93377-4
Navrot N, Rouhier N, Gelhaye E, Jacquot JP (2007) Reactive oxygen species generation and antioxidant systems in plant mitochondria. Physiol Plant 129:185–195
Nejad-Alimoradi F, Nasibi F, Kalantari KM (2019) 24-Epibrassinolide pre-treatment alleviates the salt-induced deleterious effects in medicinal pumpkin (Cucurbita pepo) by enhancement of GABA content and enzymatic antioxidants. S Afr J Bot 124:111–117. https://doi.org/10.1016/j.sajb.2019.04.027
Noctor G, Veljovic-Jovanovic S, Driscoll S, Novitskaya L, Foyer CH (2002) Drought and oxidative load in the leaves of C3 plants: a predominant role for photorespiration? Ann Bot 89:841–850
Oldroyd GED, Engstrom EM, Long SR (2001) Ethylene inhibits the nod factor signal transduction pathway of Medicago truncatula. Plant Cell 13:1835–1849
Partelli FL, Batista S (2001) Characterization of the main lipid components of chloroplast membranes and cold induced changes in *Coffea* spp. Environ Exp Bot 74:194–204
Pei L, Zhao Y, Shi X, Chen R, Yan J et al (2022) The role of γ-aminobutyric acid (GABA) in the occurrence of adventitious roots and somatic embryos in woody plants. Plants 11:3512. https://doi.org/10.3390/plants11243512
Pospíšilová H, Jiskrová E, Vojta P et al (2016) Transgenic barley overexpressing a cytokinin dehydrogenase gene shows greater tolerance to drought stress. Nat Biotechnol 33:692–705. https://doi.org/10.1016/j.nbt.2015.12.005
Pottosin I, Velarde-Buendía AM, Bose J, Zepeda-Jazo I, Shabala S, Dobrovinskaya O (2014) Cross-talk between reactive oxygen species and polyamines in regulation of ion transport across the plasma membrane: implications for plant adaptive responses. J Exp Bot 65:1271–1283
Priya M, Sharma L, Kaur R et al (2019) GABA (γ-aminobutyric acid), as a thermo-protectant, to improve the reproductive function of heat-stressed mungbean plants. Sci Rep 9:7788. https://doi.org/10.1038/s41598-019-44163-w
Ramesh SA, Kamran M, Sullivan W, Chirkova L, Okamoto M, Degryse F, McLaughlin M, Gilliham M, Tyerman SD (2018) Aluminum-activated malate transporters can facilitate GABA transport. Plant Cell 30:1147–1164. https://doi.org/10.1105/tpc.17.00864
Roberts JKM, Callis J, Wemmer D, Walbot V, Jardetzky O (1984) Mechanism of cytoplasmic pH regulation in hypoxic maize root tips and its role in survival under hypoxia. Proc Natl Acad Sci USA 81:3379–3383
Roberts JKM, Hooks MA, Miaullis AP, Edwards S, Webster C (1992) Contribution of malate and amino acid metabolism to cytoplasmic pH regulation in hypoxic maize root tips studied using nuclear magnetic resonance spectroscopy. Plant Physiol 98:480–487
Salah A, Zhan M, Cao C et al (2019) γ-Aminobutyric acid promotes chloroplast ultrastructure, antioxidant capacity, and growth of waterlogged maize seedlings. Sci Rep 9:484. https://doi.org/10.1038/s41598-018-36334-y
Seifikalhor M, Niknam V, Aliniaeifard S et al (2022) The regulatory role of γ-aminobutyric acid in chickpea plants depends on drought tolerance and water scarcity level. Sci Rep 12:7034. https://doi.org/10.1038/s41598-022-10571-8

Shi SQ, Zheng S, Jiang ZP et al (2010) Effects of exogenous GABA on gene expression of *Caragana intermedia* roots under NaCl stress: regulatory roles for H_2O_2 and ethylene production. Plant Cell Environ 33:149–162. https://doi.org/10.1111/j.1365-3040.2009.02065.x

Shi Y, Li Y, Liu T, Guo C, Liang W, Ma F, Li C (2024) γ-Aminobutyric acid enhances salt tolerance by sustaining ion homeostasis in apples. Plant Physiol Biochem 206:108306. https://doi.org/10.1016/j.plaphy.2023.108306

Srivastava V, Mishra S, Chowdhary AA, Lhamo S, Mehrotra S (2021) The γ-aminobutyric acid (GABA) towards abiotic stress tolerance. In: Wani SH, Gangola MP, Ramadoss BR (eds) Compatible solutes engineering for crop plants facing climate change. Springer, Cham. https://doi.org/10.1007/978-3-030-80674-3_7

Tarkowski ŁP, Van De Poel B, Höfte M, Van Den Ende W (2019) Sweet immunity: inulin boosts resistance of lettuce (*Lactuca sativa*) against grey mold (*Botrytis cinerea*) in an ethylene-dependent manner. Int J Mol Sci 20:1052

Ullah A, Ali I, Noor J, Zeng F, Bawazeer S, Eldin SM, Asghar MA, Javed HH, Saleem K, Ullah S et al (2023) Exogenous γ-Aminobutyric acid (GABA) mitigated salinity-induced impairments in mungbean plants by regulating their nitrogen metabolism and antioxidant potential. Front Plant Sci 13:1081188

Wang Y, Gu W, Meng Y, Xie T (2017) γ-Aminobutyric acid imparts partial protection from salt stress injury to maize seedlings by improving photosynthesis and upregulating osmoprotectants and antioxidants. Sci Rep 7:43609

Wang G, Kong J, Cui D, Zhao H, Niu Y, Xu M, Wang W (2019) Resistance against *Ralstonia solanacearum* in tomato depends on the methionine cycle and the γ-aminobutyric acid metabolic pathway. Plant J 97:1032–1047

Wang X, Gao L, Li Y et al (2025) Effects of exogenous γ-aminobutyric acid (GABA) on nitrogen metabolism of oriental melon seedlings under elevated rhizosphere CO_2. Plant Growth Regul 105:553. https://doi.org/10.1007/s10725-025-01299-2

Wu Q, Su N, Huang X, Cui J, Shabala L, Zhou M, Yu M, Shabala S (2021) Hypoxia-induced increase in GABA content is essential for restoration of membrane potential and preventing ROS-induced disturbance to ion homeostasis. Plant Commun 2:100188. https://doi.org/10.1016/j.xplc.2021.100188

Xiang L, Hu L, Xu W, Zhen A, Zhang L, Hu X (2016) Exogenous γ-aminobutyric acid improves the structure and function of photosystem II in muskmelon seedlings exposed to salinity alkalinity stress. PLoS One 11:e0164847. https://doi.org/10.1371/journal.pone.0164847

Xu B, Long Y, Feng X, Zhu X, Sai N, Chirkova L, Betts A, Herrmann J, Edwards EJ, Okamoto M et al (2021a) GABA signalling modulates stomatal opening to enhance plant water use efficiency and drought resilience. Nat Commun 12:1952. https://doi.org/10.1038/s41467-021-21694-3

Xu B, Sai N, Gilliham M (2021b) The emerging role of GABA as a transport regulator and physiological signal. Plant Physiol 187:2005–2016. https://doi.org/10.1093/plphys/kiab347

Yang A, Cao S, Yang Z, Cai Y, Zheng Y (2011) γ-Aminobutyric acid treatment reduces chilling injury and activates the defence response of peach fruit. Food Chem 129:1619–1622

Yu X, Wang X, Zhang P et al (2024) Exogenous γ-aminobutyric acid (GABA) regulates the response of sugar beet seedlings to salt stress through GABA branched metabolism. J Plant Growth Regul 44:2977. https://doi.org/10.1007/s00344-024-11590-7

Yuan D, Wu X, Gong B, Huo R, Zhao L, Li J, Lü G, Gao H (2023) GABA metabolism, transport and their roles and mechanisms in the regulation of abiotic stress (hypoxia, salt, drought) resistance in plants. Meta 13:347. https://doi.org/10.3390/metabo13030347

Zarbakhsh S, Shahsavar AR (2023) Exogenous γ-aminobutyric acid improves the photosynthesis efficiency, soluble sugar contents, and mineral nutrients in pomegranate plants exposed to drought, salinity, and drought-salinity stresses. BMC Plant Biol 23:543. https://doi.org/10.1186/s12870-023-04568-2

Zhang Y, Deng M, Lin B et al (2025a) Physiological and transcriptomic evidence revealed the role of exogenous GABA in enhancing salt tolerance in strawberry seedlings. BMC Genomics 26:196. https://doi.org/10.1186/s12864-025-11368-5

Zhang ZW, Dang TT, Yang XY, Xie LB, Chen YE, Yuan M, Chen GD, Zeng J, Yuan S (2025b) γ-Aminobutyric acid alleviates programmed cell death in two *Brassica* species under cadmium stress. Int J Mol Sci 26:129. https://doi.org/10.3390/ijms26010129

Zhao M, Li J, Shi X, Malik MS, Quan Y, Guo D, Wang L, Wang S (2023) Effects of exogenous plant regulators on growth and development of "Kyoho" grape under salt alkali stress. Front Plant Sci 14:1274684

Zhen A, Zhang Z, Jin X, Liu T, Ren W, Hu X (2018) Exogenous GABA application improves the NO_3^--N absorption and assimilation in $Ca(NO_3)_2$-treated muskmelon seedlings. Sci Hortic 227:117–123. doi:https://doi.org/10.1016/j.scienta.2017.09.025

Influence of Applied GABA on Growth and Yield

18

Abstract

Exogenous gamma-aminobutyric acid (GABA) application impacts plant physiological and biochemical processes that manifest as improved growth and yield. GABA regulates metabolism through the GABA shunt, enhancing energy production and supporting cell proliferation. It maintains ion homeostasis/cellular membrane integrity, modulates root and shoot development, and improves photosynthetic efficiency. Its interaction with plant hormones further enhances resilience and reproductive success. As a result, GABA-treated plants exhibit stress resilience together with high seed germination and biomass accumulation that translate into augmented flowering/fruit set and fruit/grain yield/quality. The mode of GABA applications in agriculture includes foliar sprays, seed priming, and soil/hydroponic incorporation to boost plant performance. The chapter highlights GABA's potential as a sustainable agricultural tool for enhancing productivity and stress resilience.

Keywords

Photosynthetic efficiency · Root growth · Seed germination · Shoot development · Yields

Highlights

- Gamma-aminobutyric acid (GABA) stimulates plant development, enhances the accumulation of biomass, and results in higher yields.
- GABA regulates metabolism through the GABA shunt, improving energy production and stress resilience.
- GABA modulates root and shoot architecture by influencing auxin and cytokinin signalling.

S. U. Jalil et al., *GABA Signaling System and Abiotic Stress Tolerance in Plants*,
https://doi.org/10.1007/978-981-95-0437-4_18

- GABA improves photosynthetic efficiency, increasing chlorophyll content and CO_2 fixation.
- GABA enhances seed germination, flowering, fruit set, and overall crop productivity.
- GABA mitigates abiotic stress by maintaining ion homeostasis, stabilizing membranes, and scavenging ROS.

18.1 Introduction

Plants are constantly exposed to various environmental and physiological challenges that affect their growth and yield. To cope with these challenges, plants have evolved intricate metabolic pathways and signalling mechanisms. One such key molecule involved in plant adaptation is gamma-aminobutyric acid (GABA), which regulates several physiological processes. GABA is synthesized in response to developmental cues and stress conditions, playing a significant role in balancing energy metabolism and enhancing stress resilience. Understanding how exogenous GABA influences plant growth and yield requires further investigation into its underlying mechanisms and practical implementation strategies.

Through its role in the GABA shunt pathway, GABA significantly affects C/N metabolism, contributing to energy balance and maintaining metabolic homeostasis (Ansari et al. 2014). In recent years, the treatment of plants/seeds with GABA has attracted considerable attention as a potential approach for enhancing plant growth and improving crop productivity (Jalil et al. 2019; Du et al. 2020; Jin et al. 2023). Research shows that GABA influences root and shoot development, enhances photosynthesis, and promotes flowering and fruit formation, ultimately increasing plant biomass and productivity (Li et al. 2016). Experiments with poplar plants using different concentrations of GABA, overexpression of the GAD2 gene, and application of a GABA-T enzyme inhibitor (vigabatrin) record the increased endogenous GABA levels and inhibited adventitious root differentiation and growth (Xie et al. 2020). In Arabidopsis, GABA guides the growth of pollen tubes toward the ovary for fertilization. However, reduced activity of the POP2 gene, which encodes the GABA-T enzyme, leads to excessive GABA accumulation, causing abnormal pollen tube growth (Palanivelu et al. 2003). Further studies on duckweed (*Lemna minor*) have shown that higher GABA concentrations in the growing medium directly increase plant dry weight. For instance, supplementing the medium with GABA results in a threefold increase in plant biomass compared to untreated plants (Kinnersley and Turano 2000). In addition to biomass accumulation, GABA affects plant mineral content and overall development. Studies on *Malus hupehensis* seedlings establish that applying GABA helps the plants tolerate alkaline stress by promoting biomass growth, strengthening the antioxidant defence system to ROS, and improving root development (Liu et al. 2021).

GABA modulates hormonal crosstalk by interacting with key plant growth regulators such as Abscisic acid (ABA), auxins, cytokinins, and Gibberellic acid (GA), fine-tuning plant development and stress responses. GABA also performs the role of a signalling molecule under adverse conditions like drought, salinity, heat, and heavy metal toxicity and elicits an efficacious stress mitigation response in plants. Its accumulation triggers defence mechanisms that regulate ion homeostasis, stabilize cellular membranes, and scavenge ROS, thereby enhancing plant resilience and sustaining productivity. Given its broad-spectrum benefits, agricultural applications of GABA—through foliar sprays, seed priming, and soil supplementation—have shown promising results in improving germination rates, strengthening root systems, and enhancing fruit and grain quality. However, further research is required to optimize GABA concentrations, application timings, and crop-specific responses to maximize its agronomic potential. This chapter explores the multifaceted role of applied GABA in promoting plant growth and yield, providing insights into its metabolic functions, physiological impact, and practical applications in agriculture. A deeper understanding of GABA-mediated growth enhancement can help researchers and agronomists harness its potential to develop sustainable strategies for improving crop production and resilience in changing environmental conditions.

18.2 GABA and Plant Growth and Development

Both exogenous and endogenous GABA influences diverse physio-biochemical processes that modify both root and shoot growth, flowering, and fruit set, and ultimately crop economic yield improvement (Fig. 18.1).

18.2.1 Metabolism Regulation

GABA operates through the GABA shunt pathway for its own synthesis and extended role in regulating energy generation and carbon and nitrogen metabolism, especially under stress conditions. GABA shunt serves as an alternative and supplementary pathway to the Tricarboxylic acid (TCA) cycle for uninterrupted energy production and steady production of intermediary metabolites in adverse conditions. The GABA shunt allows plants to convert glutamate into succinate more efficiently, effectively circumventing two steps of the traditional TCA cycle. This metabolic flexibility is crucial for optimizing energy production, especially under conditions where the standard TCA cycle may be less efficient or compromised. By facilitating the rapid conversion of glutamate to succinate, the GABA shunt ensures a steady supply of intermediates that are essential for cellular respiration and energy generation. This enhanced energy production supports critical physiological processes such as cell proliferation and root elongation, which are fundamental to plant growth and development. Furthermore, the increased availability of succinate and other metabolic intermediates contributes to the overall vigour of the plant, promoting healthier and more robust growth. The regulation of metabolism through the GABA shunt also highlights the adaptability of plants in responding to environmental stresses, as GABA accumulation often increases under adverse conditions such as drought, salinity, or pathogen attack. By modulating metabolic pathways, GABA

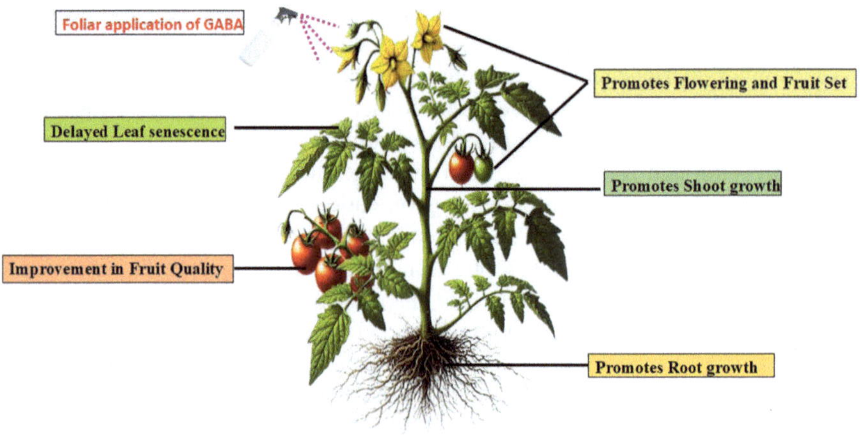

Fig. 18.1 GABA and various physiological and developmental processes in plants. GABA application delays leaf senescence by enhancing chlorophyll content and photosynthetic efficiency, optimizing stomatal conductance, improving CO_2 fixation, and preserving chloroplast structure, leading to sustained leaf function and prolonged photosynthetic activity. It also improves fruit quality by increasing carbohydrate and protein content, enhancing fruit firmness and antioxidant activity, regulating ethylene signalling, delaying ripening, and extending shelf life, thereby improving post-harvest quality and stress resistance. GABA promotes flowering and fruit set by regulating flowering genes, enhancing pollen viability, guiding pollen tube growth, maintaining pollen tube trajectory, modulating calcium signalling, and ensuring fruit set, thereby improving reproductive success. It supports shoot growth by regulating ethylene biosynthesis, increasing shoot biomass, enhancing carbon and nitrogen metabolism, and aiding in shoot elongation, contributing to overall vegetative development. Additionally, GABA enhances root growth by regulating auxin and cytokinin pathways, improving water and nutrient absorption, and increasing root biomass, leading to a stronger root system and better plant resilience

helps plants maintain homeostasis and sustain growth even under challenging circumstances. This intricate interplay between GABA and metabolic regulation underscores its importance not only as a signalling molecule but also as a key player in optimizing plant growth and yield. The ability of GABA to influence these metabolic processes ultimately translates into improved plant performance, making it a critical factor in agricultural productivity and crop resilience.

18.2.2 GABA and Photosynthetic Efficiency

The application of GABA significantly enhances photosynthetic efficiency, a key factor in biomass accumulation and overall crop productivity. GABA influences photosynthesis through multiple mechanisms. First, it increases chlorophyll content in leaves, allowing plants to absorb more light and enhance photosynthetic rates. Second, GABA optimizes stomatal conductance, ensuring a balanced CO_2 uptake and water loss, which helps maintain higher photosynthetic activity even under drought stress. Third, it enhances CO_2 fixation, promoting the conversion of CO_2 into organic compounds essential for growth. By improving photosynthetic efficiency, GABA ultimately boosts crop productivity and yield. Exogenous application of GABA improves photosynthetic performance in several plant species. GABA application preserves chloroplast ultrastructure and reduces chlorophyll degradation, thereby sustaining the functionality of the photosynthetic machinery. Thus, the beneficial impact of GABA on photosynthesis can be partly attributed to its ability to preserve chloroplast integrity and enhance chlorophyll biosynthesis. Maize treated with exogenous GABA improves photosynthetic efficiency, gas exchange, and enzymatic activities, as well as the regulation of stomatal conductance, intercellular CO_2 concentration, and transpiration. The positive physio-biochemical changes manifest as promotion of seedling elongation, biomass accumulation, and chlorophyll content (SPAD values), collectively contributing to overall plant development (Li et al. 2016). In another investigation, GABA treatment significantly improves leaf gas exchange parameters, leading to enhanced photosynthesis in maize seedlings subjected to waterlogging stress (Salah et al. 2019). Similarly, in chickpea plants, GABA application improves photosynthetic performance and decreases oxidative stress by enhancing antioxidant enzyme activities (Seifikalhor et al. 2022).

18.3 GABA and Seed Germination/Seedling Establishment

GABA activates antioxidant enzymes and alleviates oxidative stress, preventing cellular damage and supporting early seedling development. GABA also regulates key metabolic pathways, including carbohydrate metabolism, by modulating enzymes responsible for starch degradation, ensuring a steady energy supply for germinating seeds. Its accumulation during germination has been observed in multiple plant species, including soybean (Matsuyama et al. 2009), oats (Xu et al. 2010), barnyard millet (Sharma et al. 2016), adlay (Xu et al. 2017), rice (Zhao et al. 2017), wheat (Kim et al. 2018; AL-Quraan et al. 2019), barley (Sheng et al. 2018), and Chinese wild rice (Chu et al. 2019), suggesting a conserved mechanism across different crops. Excessive GABA can suppress primary root elongation and inhibit hypocotyl growth in dark conditions (Renault et al. 2011). However, GABA, in a concentration-dependent manner, stimulates the expression of α-amylase genes, facilitating starch breakdown and seed germination (Sheng et al. 2018). In recalcitrant Chinese chestnut seeds, exogenous GABA application disrupts carbon and nitrogen metabolism, maintaining dormancy and preventing both seed germination and primary root growth (Du et al. 2020).

Under stress conditions, GABA contributes to improved seed germination by enhancing osmotic balance, boosting antioxidant activity, and upregulating DREB-related DHNs expression, which aids in drought tolerance (Zhou et al. 2021). In perennial ryegrass, GABA enhances salinity resistance by improving seed germination, seedling growth, carbon and nitrogen homeostasis, and antioxidant defence mechanisms (Tang et al. 2022). Similarly, exogenous GABA application enhances cotton seed germination under salt stress (Dong et al. 2024). Furthermore, priming pepper seeds with GABA activates the GABA shunt, ensuring a stable C:N ratio, regulating metabolism, and optimizing amino acid synthesis. It also strengthens

ROS-scavenging mechanisms, increasing CaDHN3 expression, which helps prevent lipid peroxidation, maintains metabolic stability, protects enzyme functionality, and reduces dehydration, ultimately supporting seed germination under saline conditions (AL-Quraan et al. 2024).

18.4 Root and Shoot Growth Modulation

GABA significantly influences root and shoot growth by modulating hormone (particularly auxins and cytokinins) signalling pathways, which regulate plant architecture and resource allocation.

Exogenous GABA enhances root length and lateral root formation, improving water and nutrient uptake, especially under suboptimal conditions (Michaeli et al. 2011). Application of vigabatrin, a GABA-T inhibitor, reduces root growth in *Brassica napus* (Deleu et al. 2013). It may be reiterated that GABA-T breaks down GABA and produces succinic semialdehyde that avoids excessive GABA accumulation in the cell. It appears that root growth and endogenous GABA level are fine-tuned; excess GABA accumulation inhibits the process of root growth, albeit senses a stress scenario, triggering conservation of biological energy and resources to survive in adverse conditions. In consonant, excessive GABA inhibits primary root elongation in Arabidopsis (Renault et al. 2011) and results in leaf chlorosis and hypocotyl growth suppression (Renault et al. 2013). In Chinese chestnut seeds, exogenous GABA alters C/N metabolism, preventing primary root growth and maintaining seed dormancy (Du et al. 2020). In poplar, increased GABA—through α-ketoglutarate dehydrogenase inhibition or *PagGAD2* overexpression—suppresses adventitious root formation by altering carbon metabolism and phytohormone (IAA, ABA, and ethylene) signalling (Yue et al. 2018; Xie et al. 2020). Evidence from transgenic poplar plants overexpressing *PagGAD2*, along with exogenous GABA treatments and suppression of GABA degradation, further supports that alterations in carbon and nitrogen metabolism and hormonal signalling negatively regulate adventitious root development (Xie et al. 2020). Thus, Pei et al. (2022) infer that GABA-associated metabolic factors critically influence not only adventitious root formation but also plant adaptation.

Beyond root growth, GABA plays a role in plant-microbe interactions. It enhances ectomycorrhizal symbiosis in oak trees by promoting synthesis and reducing mycorrhizal degradation, further supporting plant development (Sebastiana et al. 2021). GABA application also increases shoot biomass, indicating its role in promoting shoot development (Jin et al. 2023). Ethylene, a key plant hormone, is known to regulate shoot growth by inducing the triple response, characterized by reduced stem elongation, hypocotyl swelling, and altered growth direction (Kathiresan et al. 1997; Jin et al. 2023). In most plant tissues, the rate-limiting step in ethylene biosynthesis is the activity of ACC synthase. GABA upregulates ACC synthase transcription to induce ethylene production in sunflower (Kathiresan et al. 1997). ACC inhibitors, such as AVG or AIBA, suppress GABA-induced ethylene production and confirm the role of GABA in ethylene biosynthesis. Furthermore, exogenous GABA treatment strongly induces ethylene release in sunflower tissues, reinforcing its role in ethylene-mediated shoot growth regulation. Ethylene biosynthesis in higher plants involves the enzymatic action of ACC synthase and ACC oxidase, which play a crucial role in regulating growth responses under various conditions (Fatma et al. 2022).

18.5 GABA and Fertilization Success

Sexual reproduction in flowering plants is essential for completing their life cycle. As plants transition from vegetative to reproductive growth, precise molecular signalling ensures successful flowering, fertilization, and fruit set. Among these signals, GABA emerges as a crucial regulator, modulating the expression of flowering-related genes, enhancing pollen viability, and improving fertilization rates. During pollination, pollen grains adhere to the stigma and hydrate, triggering germination and the initiation of polarized pollen tube growth. This tube serves as a conduit for sperm cell delivery to the ovule. The precise guidance of pollen tube elongation is critical for successful fertilization. Researchers have discovered that GABA concentration within floral tissues functions as a chemotropic signal, directing pollen tube growth (Ma 2003; Palanivelu et al. 2003; Yang 2003). A well-defined GABA gradient forms along the pistil, with the highest concentration near the micropyle, ensuring accurate pollen tube targeting and ovule penetration.

In *Arabidopsis*, the *POP2* gene encodes GABA-T, an enzyme responsible for GABA catabolism and gradient maintenance. Mutational studies in *pop2* Arabidopsis plants have revealed that excessive GABA accumulation disrupts pollen tube guidance. Instead of growing toward the ovule, many pollen tubes exhibit aberrant trajectories, leading to fertilization failure (Palanivelu et al. 2003; Renault et al. 2011). Beyond *Arabidopsis*, researchers investigating *Picea wilsonii* observe that GABA influences pollen tube elongation in a concentration-dependent manner. Low GABA concentrations stimulate pollen tube extension, whereas excessive levels suppress elongation (Ling et al. 2013). Further studies reveal that high GABA concentrations disrupt Ca^{2+} influx through Ca^{2+}-permeable channels, impairing intracellular signalling necessary for pollen tube growth. Additionally, GABA decarboxylase (GAD), an enzyme involved in GABA biosynthesis, functions downstream of the Ca^{2+}/calmodulin (CaM) regulatory mechanism, further modulating pollen tube dynamics (Yu et al. 2014). Through these mechanisms, GABA emerges as more than just a metabolic byproduct—it acts as a key signalling molecule orchestrating pollen tube guidance, fertilization, and ultimately, fruit set.

18.6 GABA and Grain and Fruit Quality

GABA plays a significant role in improving the nutritional quality, post-harvest stability, and overall yield of grains and fruits. Studies indicate that GABA enhances carbohydrate and protein content, increases fruit firmness, and boosts antioxidant activity, thereby reducing post-harvest losses. Additionally, GABA actively participates in fruit ripening, influencing ethylene signalling and sugar metabolism to regulate fruit quality and longevity.

GABA accumulation fluctuates throughout fruit development. In tomato (*Solanum lycopersicum*), GABA levels peak during the mature green stage when seeds are immature and rapidly decline as the fruit ripens and seeds mature (Takayama and Ezura 2015). Genetic studies further highlight the regulatory role of GABA in fruit development. Downregulating *SlGAD* genes in tomatoes leads to reduced GABA levels without affecting normal plant growth (Takayama et al. 2015). Conversely, upregulation of *SlGAD2* elevates GABA levels but results in stunted growth, delayed flowering, and reduced fruit yield (Nonaka et al. 2017). Interestingly, overexpression of *SlGAD3* using the fruit-specific *E8* promoter increases GABA levels without causing morphological defects (Takayama et al. 2017). However, in tomato lines with C-terminal-truncated *SlGAD3* overexpression, ripening is disrupted due to delayed ethylene production and decreased ethylene sensitivity (Takayama et al. 2017).

GABA enhances cereal grain yield and quality. In rice (*Oryza sativa*), it increases grain weight, spikelets per plant, and overall yield while regulating yield-related traits under varying nitrogen conditions by improving filled grain percentage and grains per panicle (Li et al. 2019b). Furthermore, GABA affects the accumulation of 2-acetyl-1-pyrroline (2AP), a key aromatic compound in fragrant rice, thereby enhancing its quality (Gao et al. 2020).

Exogenous GABA treatment improves post-harvest quality in fruits. GABA modulates ethylene synthesis and polyamine metabolism and preserves apples (*Malus domestica*) fruit quality (Han et al. 2018; Li et al. 2021a, b, c). GABA application, by enhancing malate levels and suppressing ethylene production, delays ripening and extends the shelf life of fruits (Han et al. 2018). In pears (*Pyrus* spp.), GABA strengthens mitochondrial oxidative defence system, maintains mitochondrial integrity, and finally delays peel browning (Li et al. 2019a). Similarly, GABA improves the quality of 'Cripps Pink' apples by increasing the sugar-to-acid ratio, enhancing flesh firmness and pericarp texture, and reducing acidity (Cheng et al. 2023).

GABA also mitigates chilling injury in olive fruits (*Olea europaea*), as GABA treatment effectively preserves fruit quality under low-temperature storage conditions (Fàn et al. 2022). In soybeans (*Glycine max*), exogenous GABA integrates into the TCA cycle, promotes carbon assimilation, and significantly increases sucrose, fructose, and glucose contents (Yu et al. 2022). Additionally, GABA application before harvest improves fruit sugar content by enhancing sucrose synthesis, partially inhibiting glycolysis, and increasing trehalose 6-phosphate levels, as demonstrated in recent research (Wu et al. 2024).

18.7 GABA and Growth and Yield Enhancement

GABA brings the following positive changes to harness luxuriant growth and post-harvest yield benefits in agricultural crops.

18.7.1 Stress Mitigation

Environmental stress poses a major constraint on crop growth and grain yield, with significant repercussions for economic development and food security. In this context, GABA assumes a prominence as it regulates ion homeostasis, stabilizes cellular membranes, and scavenges ROS (Yuan et al. 2023) and mitigates abiotic stresses such as drought, salinity, and heat (Kaspal et al. 2021; Xu et al. 2021; Balfagón et al. 2022; Dabravolski and Isayenkov 2023). Consequently, GABA brings yield stability and improvement under an adverse environment.

GABA functions as a nontoxic osmolyte, aiding in the maintenance of cellular osmotic balance during stress events such as salinity and drought. GABA seed priming enhances soluble sugars, antioxidants, and enzyme activities, reducing osmotic potential and oxidative stress under drought. It also upregulates dehydrin proteins, DHN-encoding genes, and drought-responsive transcription factors, improving seed germination under water-limited conditions (Zhou et al. 2021). By modulating ion uptake and transport, GABA helps in reducing the accumulation of harmful ions like Na^+ and maintaining a favourable Na^+/K^+ ratio, resulting in improved growth and biomass of mung bean under salinity stress (Ullah et al. 2023). During salinity stress, exogenous GABA improves plant height, biomass, and mitigates leaf yellowing. In cotton seedlings, the application of GABA mitigates salinity stress by enhancing proline and soluble sugar accumulation, improving the K^+/Na^+ balance, stimulating antioxidant enzyme activities, and minimizing oxidative damage (Dong et al. 2024).

18.7.2 Hormonal Crosstalk

GABA emerges as a versatile signalling molecule, a crucial regulator in plant physiology, coordinating with key phytohormones, including ABA, GA, auxins, JA, ethylene, and Salicylic acid (SA). The complex interplay between GABA and these hormones governs physiological, morphological, stress adaptation, and reproductive success in plants.

GA-impaired poplar (*Populus* spp.) mutants (gai and rgl1) exhibit a three-fold increase in GABA accumulation in roots but reduced levels in leaves, linking GA signalling to GABA regulation (Busov et al. 2006). In grapevine (*Vitis* spp.), GA-induced early expression of VvGAD1 coincides with GABA overproduction during full bloom, suggesting its involvement in parthenocarpic fruit set (Jung et al. 2017). GA- and GABA-related genes in *Auricularia fibrillifera* are upregulated during rehydration, highlighting their roles in dormancy breaking and post-drought recovery (Wang et al. 2022). It implies that GABA influences the balance between ABA and GA, two hormones with antagonistic roles in dormancy and germination. ABA primarily promotes dormancy and stress responses, whereas GA facilitates germination and growth. Studies have revealed that GABA modulates this balance, thereby affecting seed germination and stress adaptation (Kabała and Janicka 2024).

GABA application in sweet orange (*Citrus sinensis*) significantly increases auxin levels (IAA and IPA), along with SA, JA, and ABA, through the upregulation of biosynthetic genes, including CitTAA, CitYUC2 (auxin pathway), CitPAL, CitCS, CitICS (SA pathway), and CitLOX, CitAOS, CitAOC (JA pathway), reinforcing its role in stress mitigation (Hijaz et al. 2018). In fact, GABA exerts profound effects on auxin homeostasis and transport, prompting root architecture and AR formation. GABA treatment suppresses auxin-dependent upregulation, delaying AR development by reducing polar auxin transport (Li et al. 2021a, b, c). This suggests that GABA-mediated modulation of auxin distribution is crucial in early rooting processes (Pei et al. 2022).

GABA accumulation in plant tissues often corresponds with an enhanced ethylene biosynthesis, which plays a pivotal role in plant growth and senescence. In excised sunflower tissues, GABA significantly increases ethylene production by upregulating ACC synthase and ACC oxidase gene expression, leading to elevated ACC levels, the key ethylene precursor (Kathiresan et al. 1997). Similarly, in *Stellaria longipes*, high GABA concentrations trigger ACC synthase transcription and ethylene production, resulting in inhibited stem elongation (Kathiresan et al. 1998). Conversely, GABA treatment reduces ethylene biosynthesis in detached soybean leaves (Turano et al. 1997) and apple fruits, where it downregulates MdACS, MdACO, and ethylene signalling genes (MdERF), contributing to prolonged fruit storage and quality maintenance (Han et al. 2018; Li et al. 2021a, b, c).

The interaction between GABA and JA (jasmonoyl-isoleucine, JA-Ile, and its precursor 12-oxophytodienoic acid, OPDA) elucidates tendril development in grapevine, for non-coiled tendrils exhibit higher OPDA, JA, JA-Ile, and ABA levels. But GABA accumulation occurs in both coiled and non-coiled tendrils, indicating its role in tendril morphogenesis. Further, JA or GABA treatments independently induce tendril coiling without touch, suggesting that these two molecules function through separate yet overlapping pathways in tendril movement and mechanical perception (Malabarba et al. 2019). In contrast, GABA and SA exhibit complementary roles in delaying senescence and enhancing stress tolerance. Babarabie et al. (2019) have reported that in cut tuberose (*Polianthes tuberosa*) flowers, exogenous application of GABA and SA extends vase life by improving water uptake, fresh weight, and chlorophyll and protein content. Additionally, these treatments facilitate antioxidant enzyme activity, suggesting that GABA and SA delay senescence through activation of antioxidant defence mechanisms.

18.7.3 Defence Mechanism Activation

Plants constantly encounter biotic and abiotic challenges that threaten their growth, productivity, and survival. To counter these stresses, they have evolved intricate defence mechanisms, many of which GABA regulates. GABA enhances plant resilience by modulating defence responses, secondary metabolism, and antioxidative pathways, ultimately contributing to improved growth and yield.

Upon GABA application, plants exhibit a heightened expression of defence-related genes and synthesis of secondary metabolites such as flavonoids, alkaloids, and polyphenols. These bioactive compounds act as natural protectants, bolstering the plant's ability to withstand pest attacks and pathogen invasions. For instance, GABA treatment upregulates key enzymes involved in the phenylpropanoid pathway, including phenylalanine ammonia lyase, cinnamate-4-hydroxylase, 4-coumarate/coenzyme A ligase, and polyphenol oxidase. This biochemical shift not only strengthens structural defences but also enhances antioxidant activity, thereby mitigating oxidative damage. Notably, GABA-induced modulation of phenylpropanoid and ROS metabolism in blueberries contributes to prolonged shelf life arising from delayed senescence due to elevation of total phenolics and flavonoids (Ge et al. 2018). Beyond its role in secondary metabolism, GABA also influences nitrogen assimilation and flavonoid biosynthesis, essential factors in improving nutritional quality. In edamame, exogenous GABA application leads to a significant enrichment of amino acids, bioactive peptides, and flavonoids, driven by the upregulation of key metabolic enzymes (Yu et al. 2022). This metabolic enhancement not only improves plant growth but also enriches the nutritional profile, enriching sugars and organic acids essential for flavour and health benefits.

The regulatory impact of GABA extends to carbon and nitrogen metabolism, as observed in in vitro blueberry plantlets. Transcriptomic and metabolomic analyses reveal that GABA treatment promotes the accumulation of secondary metabolites, including flavonoids, steroids, and terpenes, which are vital for stress adaptation and plant development (Liu et al. 2023). Furthermore, under moderate salinity stress, GABA application plays a crucial role in optimizing physiological and biochemical stress responses. Ahmadi et al. (2023) have reported that GABA-treated pineapple mint accumulates secondary metabolite accumulation (total polyphenols and flavonoids) with increased antioxidant capacity (DPPH-radical scavenging), improved water retention (relative water content and proline accumulation), and elevated photosynthetic efficiency (chlorophyll content and maximum quantum yield of PSII). Importantly, predictive modelling suggests that a strategic combination of lower NaCl levels and dilute GABA treatment maximizes shoot biomass, essential oil yield, and piperitenone oxide content, underscoring GABA's potential in improving crop performance under saline conditions.

18.7.4 GABA Signalling, Anion Transporters, and Plant Growth

GABA application in tobacco enhances the influx of Ca^{2+} through pollen tube channels. This influx apparently plays a guiding role in directing the pollen tube toward the ovules, potentially improving fertilization efficiency and reproductive success (Yu et al. 2014). Further, GABA interacts with plant-specific anion transporters, which are essential for maintaining ion homeostasis. Anion flux modulation by these transporters maintains ion homeostasis, ensuring optimal nutrient uptake, cell expansion, and overall metabolic stability. By influencing anion flux, GABA plays a crucial role in key developmental processes, including seed germination, root architecture modulation, stomatal dynamics, flowering initiation, and stress adaptation (Ramesh et al. 2015).

Expanding our understanding of GABA's involvement in anion transport, recent studies highlight its interaction with ALMTs. By inhibiting ALMT activity, GABA modulates root growth patterns and enhances plant resilience under acidic and alkaline pH conditions, as well as aluminium stress (Ramesh et al. 2015). The influence of GABA on ALMTs extends to wheat (*Triticum aestivum* L.), where the ALMT1 transporter governs malate exudation to confer aluminium tolerance under acidic soil conditions. Interestingly, this transporter is also activated under alkaline pH, regulating GABA transport and facilitating rhizosphere acidification. Through this mechanism, TaALMT1 enables wheat to adapt to fluctuating soil pH by coordinating the release of both malate and GABA, thereby optimizing nutrient availability and stress resistance (Kamran et al. 2020). These findings underscore the intricate crosstalk between GABA signalling and anion transporters, revealing a sophisticated regulatory network that enhances plant growth, stress resilience, and yield potential.

18.8 Conclusion

The utilization of GABA presents a promising avenue for promoting plant growth, enhancing yield, and boosting resilience against various stresses. As a versatile signalling molecule, GABA modulates several crucial physiological and biochemical processes, including hormonal interactions, carbon and nitrogen metabolism, secondary metabolite production, and antioxidant defence systems. Its significant contribution to alleviating abiotic stresses such as drought, salinity, heat, and heavy metal toxicity highlights its value in strengthening crop performance under adverse environmental conditions. In agricultural practices, GABA has shown remarkable benefits when applied through methods like foliar spraying, seed priming, and soil enrichment, leading to better germination, improved root architecture, enhanced nutrient absorption, and greater overall productivity. However, to fully harness its potential, more research is essential to refine application techniques, understand crop-specific responses, and uncover the detailed molecular mechanisms involved in GABA-mediated growth promotion. Advancing knowledge in these areas will support the integration of GABA strategies into sustainable farming systems, contributing to the development of climate-resilient crops.

References

Ahmadi H, Farhadi H, Morshedloo MR, Maggi F (2023) Modeling and optimizing concentration of exogenous application of g-aminobutyric acid on NaCl stressed pineapple mint (Mentha suaveolens) using response surface methodology: an investigation into secondary metabolites and physiological parameters. BMC Plant Biol 23(1):1–16. https://doi.org/10.1186/s12870-023-04312-w

AL-Quraan NA, AL-Ajlouni ZI, Obedat DI (2019) The GABA shunt pathway in germinating seeds of wheat (Triticum aestivum L.) and barley (Hordeum vulgare L.) under salt stress. Seed Sci Res 29:1–11. https://doi.org/10.1017/S0960258519000230

AL-Quraan NA, Samarah NH, AM AL-F (2024) The physiological effect of GABA priming on pepper (*Capsicum annuum* L.) during seed germination under salt stress. Plant Growth Regul 104:953–974. https://doi.org/10.1007/s10725-024-01207-0

Ansari MI, HasanS, Jalil SU (2014) Leaf Senescence and GABA Shunt. Bioinformation 10:730–732

Babarabie M, Zarei H, Eskandari A (2019) The impact of pre-harvest treatment with gamma-aminobutyric acid (GABA) and SA on vase life and post-harvest traits of tuberose cut flowers. Acta Sci Pol Hortorum Cultus 18:83–92. https://doi.org/10.24326/asphc.2019.4.8

Balfagón D, Gómez-Cadenas A, Rambla JL, Granell A, de Ollas C et al (2022) γ-aminobutyric acid plays a key role in plant acclimation to a combination of high light and heat stress. Plant Physiol 188(4):2026–2038

Busov V, Meilan R, Pearce DW, Rood SB, Ma C, Tschaplinski TJ, Strauss SH (2006) Transgenic modification of Gai or Rgl1 causes dwarfing and alters gibberellins, root growth, and metabolite profiles in Populus. Planta 224:288–299. https://doi.org/10.1007/s00425-005-0213-9

Cheng P, Yue Q, Zhang Y, Zhao S, Khan A, Yang X, He J, Wang S, Shen W, Qian Q, Wanshan D, Ma F, Zhang D, Guan Q (2023) Application of γ-aminobutyric acid (GABA) improves fruit quality and rootstock drought tolerance in apple. J Plant Physiol 280:153890. https://doi.org/10.1016/j.jplph.2022.153890

Chu C, Yan N, Du Y, Liu X, Chu M, Shi J, Zhang H, Liu Y, Zhang Z (2019) iTRAQ-based proteomic analysis reveals the accumula tion of bioactive compounds in Chinese wild rice (Zizania latifolia) during germination. Food Chem 289:635–644. https://doi.org/10.1016/j.foodchem.2019.03.092

Dabravolski SA, Isayenkov SV (2023) The role of the γ-aminobutyric acid (GABA) in plant salt stress tolerance. Horticulturae 9(2):230. https://doi.org/10.3390/horticulturae9020230

Deleu C, Faes P, Niogret MF, Bouchereau A (2013) Effects of the inhibitor of the γ-aminobutyrate-transaminase, vinyl-γ-aminobutyrate, on development and nitrogen metabolism in Brassica napus seedlings. Plant Physiol Biochem 64(5):60–69. https://doi.org/10.1016/j.plaphy.2012.12.007

Dong Z, Huang J, Qi T, Meng A, Fu Q, Fu Y, Xu F (2024) Exogenous γ-aminobutyric acid can improve seed germination and seedling growth of two cotton cultivars under salt stress. Plants 13(1):82. https://doi.org/10.3390/plants13010082

Du C, Chen W, Wu Y, Wang G, Zhao J, Sun J, Ji J, Yan D, Jiang Z, Shi S (2020) Effects of GABA and vigabatrin on the germination of Chinese chestnut recalcitrant seeds and its implications for seed dormancy and storage. Plants (Basel) 9:4

Fan Z, Lin B, Lin H, Lin M, Chen J, Lin Y (2022) Gamma-aminobutyric acid treatment reduces chilling injury and improves quality maintenance of cold-stored Chinese olive fruit. Food Chem X 13:100208

Fatma M, Asgher M, Iqbal N, Rasheed F, Sehar Z et al (2022) Ethylene signaling under stressful environments: analyzing collaborative knowledge. Plants 11(17):2211. https://doi.org/10.3390/plants11172211

Gao Z, Xie W, Ashraf U, Li Y, Ma L, Gui R et al (2020) Exogenous γ-aminobutyric acid (GABA) application at different growth stages regulates 2-acetyl-1-pyrroline, yield, quality and antioxidant attributes in fragrant rice. J Plant Interact 15(1):139–152. https://doi.org/10.1080/17429145.2020.1769210

Ge Y, Duan B, Li C, Tang Q, Li X, Wei M, Chen Y, Li J (2018) γ-aminobutyric acid delays senescence of blueberry fruit by regulation of reactive oxygen species metabolism and phenylpropanoid pathway. Sci Hortic 240:303–309., ISSN 0304-4238. https://doi.org/10.1016/j.scienta.2018.06.044

Han S, Nan Y, Qu W, He Y, Ban Q, Lv Y, Rao J (2018) Exogenous gamma- aminobutyric acid treatment that contributes to regulation of malate metabolism and ethylene synthesis in apple fruit during storage. J Agric Food Chem 66:13473–13482

Hijaz F, Nehela Y, Killiny N (2018) Application of gamma-aminobutyric acid increased the level of phytohormones in Citrus sinensis. Planta 248:909–918. https://doi.org/10.1007/s00425-018-2947-1

Jalil SU, Khan MQR, Ansari MI (2019) Role of GABA transaminase in the regulation of development and senescence in Arabidopsis thaliana. Curr Plant Biol 19:100119. https://doi.org/10.1016/j.cpb.2019.100119

Jin Y, Zhi L, Tang X, Chen Y, Hancock JT, Hu X (2023) The function of GABA in plant cell growth, development and stress response. Phyton-Int J Exp Bot 92(8):2211–2225. https://doi.org/10.32604/phyton.2023.026595

Jung CJ, Hur YY, Moon JS, Jung S-M (2017) Pre-bloom application of gibberellin in 'Tamnara' grape increases γ-aminobutyric acid (GABA) production at full bloom. Hortic Environ Biotechnol 58:568–575. https://doi.org/10.1007/s13580-017-0062-z

Kabała K, Janicka M (2024) Relationship between the GABA pathway and signaling of other regulatory molecules. Int J Mol Sci 25(19):10749. https://doi.org/10.3390/ijms251910749. PMID: 39409078; PMCID: PMC11476557

Kamran M, Ramesh SA, Gilliham M, Tyerman SD, Bose J (2020) Role of TaALMT1 malate-GABA transporter in alkaline pH tolerance of wheat. Plant Cell Environ 43(10):2443–2459. https://doi.org/10.1111/pce.13845

Kaspal M, Kanapaddalagamage MH, Ramesh SA (2021) Emerging roles of γ aminobutyric acid (GABA) gated channels in plant stress tolerance. Plants 10(10):2178. https://doi.org/10.3390/plants10102178

Kathiresan A, Tung P, Chinnappa CC, Reid DM (1997) Gamma-aminobutyric acid stimulates ethylene biosynthesis in sunflower. Plant Physiol 115(1):129–135. https://doi.org/10.1104/pp.115.1.129

Kathiresan A, Miranda J, Chinnappa CC, Reid DM (1998) γ–aminobutyric acid promotes stem elongation in Stellaria longipes: the role of ethylene. Plant Growth Regul 26:131–137. https://doi.org/10.1023/A:1006107815064

Kim MJ, Kwak HS, Kim SS (2018) Effects of germination on protein, γ- aminobutyric acid, phenolic acids, and antioxidant capacity in wheat. Molecules 23(9). https://doi.org/10.3390/molecules23092244

Kinnersley AM, Turano FJ (2000) Gamma aminobutyric acid (GABA) and plant responses to stress. CRC Crit Rev Plant Sci 19:479–509. https://doi.org/10.1080/07352680091139277

Li W, Liu J, Ashraf U, Li G, Li Y, Lu W, Gao L, Han F, Hu J (2016) Exogenous γ-aminobutyric acid (GABA) application improved early growth, net photosynthesis, and associated physio-biochemical events in maize. Front Plant Sci 7:919. https://doi.org/10.3389/fpls.2016.00919

Li J, Zhou X, Wei B, Cheng S, Zhou Q, Ji S (2019a) GABA application improves the mitochondrial antioxidant system and reduces peel browning in 'Nanguo' pears after removal from cold storage. Food Chem 297:124903. https://doi.org/10.1016/j.foodchem.2019.05.177

Li Y, Lai R, Li W, Liu J, Huang M, Tang Y, Tang X, Pan S, Duan M, Tian H, Wu L, Wang S, Mo Z (2019b) γ-aminobutyric acid regulates grain yield formation in different fragrant rice genotypes under different nitrogen levels. J Plant Growth Regul 39:738–750

Li C, Zhu J, Sun L, Cheng Y, Hou J, Fan Y, Ge Y (2021a) Exogenous γ-aminobutyric acid maintains fruit quality of apples through regulation of ethylene anabolism and polyamine metabolism. Plant Physiol Biochem 169:92–101. https://doi.org/10.1016/j.plaphy.2021.11.008. Epub 2021 Nov 10. PMID: 34773806

Li L, Dou N, Zhang H, Chunxia W (2021b) The versatile GABA in plants. Plant Signal Behav 16(3):1862565. https://doi.org/10.1080/15592324.2020.1862565

Li X, Xu X, Shen F, Li W, Qiu C, Wu T, Wang Y, Xu X, Han Z, Zhang X (2021c) γ-Aminobutyric acid participates in the adult-phase adventitious rooting recalcitrance. J Plant Growth Regul 40:1981–1991. https://doi.org/10.1007/s00344-020-10251-9

Ling Y, Chen T, Jing Y, Fan L, Wan Y, Lin J (2013) γ-Aminobutyric acid (GABA) homeostasis regulates pollen germination and polarized growth in Picea wilsonii. Plant 238:831–843

Liu X, Jin Y, Tan K, Zheng J, Gao T, Zhang Z, Zhao Y, Ma F, Li C (2021) MdTyDc Overexpression Improves Alkalinity Tolerance in Malus domestica. Front Plant Sci. 12:625890. https://doi.org/10.3389/fpls.2021.625890. PMID: 33664760; PMCID: PMC7921794

Liu M, Bai M, Yue J, Fei X, Xia X (2023) Integrating transcriptome and metabolome to explore the growth-promoting mechanisms of GABA in blueberry plantlets. Front Plant Sci 14:1319700. https://doi.org/10.3389/fpls.2023.1319700

Ma H (2003) Plant reproduction: GABA gradient, guidance and growth. Curr Biol 13:R834–R836. https://doi.org/10.1016/j.cub.2003.10.015

Malabarba J, Reichelt M, Pasquali G, Mithöfer A (2019) Tendril coiling in grapevine: jasmonates and a new role for GABA? J Plant Growth Regul 38:39–45. https://doi.org/10.1007/s00344-018-9807-x

Matsuyama A, Yoshimura K, Shimizu C, Murano Y, Takeuchi H, Ishimoto M (2009) Characterization of glutamate decarboxylase mediating γ-amino butyric acid increase in the early germination stage of soybean (Glycine max [L.] Merr). J Biosci Bioeng 107:538–543. https://doi.org/10.1016/j.jbiosc.2009.01.012

Michaeli S, Fait A, Lagor K, Nunes-Nesi A, Grillich N et al (2011) A mitochondrial GABA permease connects the GABA shunt and the TCA cycle, and is essential for normal carbon metabolism. Plant J 67(3):485–498. https://doi.org/10.1111/j.1365-313X.2011.04612.x

Nonaka S, Arai C, Takayama M, Matsukura C, Ezura H (2017) Efficient increase of ɣ-aminobutyric acid (GABA) content in tomato fruits by targeted mutagenesis. Sci Rep 7(1):7057. https://doi.org/10.1038/s41598-017-06400-y

Palanivelu R, Brass L, Edlund AF, Preuss D (2003) Pollen tube growth and guidance is regulated by POP2, an Arabidopsis gene that controls GABA levels. Cell 114(1):47–59. https://doi.org/10.1016/S0092-8674(03)00479-3

Pei L, Zhao Y, Shi X, Chen R, Yan J et al (2022) The role of γ-aminobutyric acid (GABA) in the occurrence of adventitious roots and somatic embryos in woody plants. Plants 11(24):3512. https://doi.org/10.3390/plants11243512

Ramesh SA, Tyerman SD, Xu B, Bose J, Kaur S, Conn V, Domingos P, Ullah S, Wege S, Shabala S, Feijó JA, Ryan PR, Gilliham M (2015) GABA signalling modulates plant growth by directly regulating the activity of plant-specific anion transporters. Nat Commun 6:7879. https://doi.org/10.1038/ncomms8879. Erratum in: Nat Commun 2015 Aug 28;6:8293. https://doi.org/10.1038/ncomms9293

Renault H, El Amrani A, Palanivelu R, Updegraff EP, Yu A et al (2011) GABA accumulation causes cell elongation defects and a decrease in expression of genes encoding secreted and cell wall-related proteins in Arabidopsis thaliana. Plant Cell Physiol 52(5):894–908. https://doi.org/10.1093/pcp/pcr041

Renault H, El Amrani A, Berger A, Mouille G, Soubigou-Taconnat L et al (2013) γ-aminobutyric acid transaminase deficiency impairs central carbon metabolism and leads to cell wall defects during salt stress in Arabidopsis roots. Plant Cell Environ 36(5):1009–1018. https://doi.org/10.1111/pce.12033

Salah A, Zhan M, Cao C et al (2019) γ-Aminobutyric acid promotes chloroplast ultrastructure, antioxidant capacity, and growth of waterlogged maize seedlings. Sci Rep 9:484. https://doi.org/10.1038/s41598-018-36334-y

Sebastiana M, Gargallo-Garriga A, Sardans J, Pérez-Trujillo M, Monteiro F et al (2021) Metabolomics and transcriptomics to decipher molecular mechanisms underlying ectomycorrhizal root colonization of an oak tree. Sci Rep 11(1):8576. https://doi.org/10.1038/s41598-021-87886-5

Seifikalhor M, Niknam V, Aliniaeifard S et al (2022) The regulatory role of γ-aminobutyric acid in chickpea plants depends on drought tolerance and water scarcity level. Sci Rep 12:7034. https://doi.org/10.1038/s41598-022-10571-8

Sharma S, Saxena DC, Riar CS (2016) Analysing the effect of germination on phenolics, dietary fibres, minerals and γ-aminobutyric acid contents of barnyard millet (Echinochloa frumentaceae). Food Biosci 13:60–68. https://doi.org/10.1016/j.fbio.2015.12.007

Sheng Y, Xiao H, Guo C, Wu H, Wang X (2018) Effects of exogenous gamma-aminobutyric acid on α-amylase activity in the aleurone of barley seeds. Plant Physiol Biochem 127:39–46

Takayama M, Ezura H (2015) How and why does tomato accumulate a large amount of GABA in the fruit? Front Plant Sci 6:612

Takayama M, Koike S, Kusano M, Matsukura C, Saito K, Ariizumi T, Ezura H (2015) Tomato glutamate decarboxylase genes SlGAD2 and SlGAD3 play key roles in regulating γ-aminobutyric acid levels in tomato (Solanum lycopersicum). Plant Cell Physiol 56(8):1533–1545. https://doi.org/10.1093/pcp/pcv075

Takayama M, Matsukura C, Ariizumi T, Ezura H (2017) Activating glutamate decarboxylase activity by removing the autoinhibitory domain leads to hyper γ-aminobutyric acid (GABA) accumulation in tomato fruit. Plant Cell Rep 36:103–116

Tang J, Li M, Mao P et al (2022) Effects of gamma-aminobutyric acid on seed germination, ion balance, and metabolic activity in perennial ryegrass under salinity stress. J Plant Growth Regul 41:1835–1844. https://doi.org/10.1007/s00344-021-10426-y

Turano FJ, Kramer GF, Wang CY (1997) The effect of methionine, ethylene and polyamine catabolic intermediates on polyamine accumulation in detached soybean leaves. Physiol Plant 101:510–518. https://doi.org/10.1111/j.1399-3054.1997.tb01031.x

Ullah A, Ali I, Noor J, Zeng F, Bawazeer S, Eldin SM, Asghar MA, Javed HH, Saleem K, Ullah S, Ali H (2023) Exogenous g-aminobutyric acid (GABA) mitigated salinity-induced impairments in mungbean plants by regulating their nitrogen metabolism and antioxidant potential. Front Plant Sci 13:1081188. https://doi.org/10.3389/fpls.2022.1081188

Wang Y, Yang Z, Shi L, Yang R, Guo H, Zhang S, Geng G (2022) Transcriptome analysis of Auricularia fibrillifera fruit-body responses to drought stress and rehydration. BMC Genomics 23:58. https://doi.org/10.1186/s12864-021-08284-9

Wu X, Huo R, Yuan D, Zhao L, Kang X, Gong B, Lü G, Gao H (2024) Exogenous GABA improves tomato fruit quality by contributing to regulation of the metabolism of amino acids, organic acids and sugars. Sci Hortic 338:113750. https://doi.org/10.1016/j.scienta.2024.113750

Xie T, Ji J, Chen W, Yue J, Du C et al (2020) GABA negatively regulates adventitious root development in poplar. J Exp Bot 71(4):1459–1474. https://doi.org/10.1093/jxb/erz520

Xu JG, Hu QP, Duan JL, Tian CR (2010) Dynamic changes in gamma-aminobutyric acid and glutamate decarboxylase activity in oats (Avena nuda L.) during steeping and germination. J Agric Food Chem 58(17):9759–9763. https://doi.org/10.1021/jf101268a

Xu L, Chen L, Ali B, Yang N, Chen Y, Wu F, Jin Z, Xu X (2017) Impact of germination on nutritional and physicochemical properties of adlay seed (Coixlachryma-jobi L.). Food Chem 229:312–318

Xu B, Long Y, Feng X, Zhu X, Sai N et al (2021) GABA signalling modulates stomatal opening to enhance plant water use efficiency and drought resilience. Nat Commun 12(1):1952. https://doi.org/10.1038/s41467-021-21694-3

Yang Z (2003) GABA, a new player in the plant mating game. Dev Cell 5:185–186. https://doi.org/10.1016/S1534-5807(03)00236-3

Yu GH, Zou J, Feng J, Peng XB, Wu JY et al (2014) Exogenous γ-aminobutyric acid (GABA) affects pollen tube growth via modulating putative Ca2+-permeable membrane channels and is coupled to negative regulation on glutamate decarboxylase. J Exp Bot 65(12):3235–3248. https://doi.org/10.1093/jxb/eru171

Yu G, Chen F, Wang Y, Chen Q, Liu H, Tian J, Wang M, Ren C, Zhao Q, Yang S, Fengjun Y, Wei J, Zhang Y (2022) Exogenous γ-aminobutyric acid strengthens phenylpropanoid and nitrogen metabolism to enhance the contents of flavonoids, amino acids, and the derivatives in edamame. Food Chem X 16:100511

Yuan D, Wu X, Gong B, Huo R, Zhao L et al (2023) GABA metabolism, transport and their roles and mechanisms in the regulation of abiotic stress (hypoxia, salt, drought) resistance in plants. Meta 13(3):347. https://doi.org/10.3390/metabo13030347

Yue J, Du C, Ji J, Xie T, Chen W et al (2018) Inhibition of α-ketoglutarate dehydrogenase activity affects adventitious root growth in poplar via changes in GABA shunt. Planta 248(4):963–979. https://doi.org/10.1007/s00425-018-2929-3

Zhao GC, Xie MX, Wang YC, Li JY (2017) Molecular mechanisms under lying γ-aminobutyric acid (GABA) accumulation in giant embryo rice seeds. J Agric Food Chem 65:4883–4889. https://doi.org/10.1021/acs.jafc.7b00013

Zhou M, Hassan MJ, Peng Y, Liu L, Liu W, Zhang Y, Li Z (2021) γ-aminobutyric acid (GABA) priming improves seed germination and seedling stress tolerance associated with enhanced antioxidant metabolism, *DREB* expression, and dehydrin accumulation in white clover under water stress. Front Plant Sci 12:776939. https://doi.org/10.3389/fpls.2021.776939. PMID: 34925419; PMCID: PMC8678086

GABA Protection in Photosynthesis

19

Abstract

Gamma-aminobutyric acid (GABA) modulates photosynthesis and maintains a delicate balance between carbon and nitrogen metabolism under both optimal and stress conditions. As a non-proteinogenic amino acid, GABA is actively involved in redox homeostasis and signalling pathways that regulate photosynthetic efficiency. GABA metabolism interplays with key photosynthetic processes, including CO_2 assimilation, ETC dynamics, and ATP synthesis. Thus, understanding the mechanistic basis of GABA function in photosynthetic protection provides insights into improving plant resilience and productivity through metabolic engineering and agronomic interventions.

Keywords

Chlorophyll · CO_2 assimilation · GABA · Photosynthesis · Stomatal conductance

Highlights

- Gamma-aminobutyric acid (GABA) restores redox homeostasis and chloroplast/chlorophyll integrity and protects photosystem function from photoinhibition under drought, salinity, and high light conditions.
- Along with influencing stomatal conductance, GABA constitutes favourable environment for efficient photosynthesis and CO_2 assimilation.
- GABA-mediated strategies provide opportunities to provide crop improvement, enhancing plant productivity.

S. U. Jalil et al., *GABA Signaling System and Abiotic Stress Tolerance in Plants*,
https://doi.org/10.1007/978-981-95-0437-4_19

19.1 Introduction

Plants constantly face environmental challenges that threaten their ability to maintain efficient photosynthesis. Among the diverse regulatory molecules involved in stress adaptation, GABA has emerged as a crucial modulator of plant metabolism, growth regulation, and stress responses. Recent research highlights its significant protective role in sustaining photosynthetic efficiency, particularly under abiotic stresses such as drought, salinity, high light intensity, and waterlogging.

One of the fundamental ways GABA enhances photosynthesis is by promoting chlorophyll biosynthesis and optimizing gas exchange parameters. In maize seedlings, exogenous application of GABA substantially increases chlorophyll content and net photosynthetic rates, ultimately improving plant growth and productivity (Li et al. 2016). This suggests that GABA directly contributes to the stabilization of the photosynthetic machinery, ensuring higher efficiency even under fluctuating environmental conditions. Beyond its influence on chlorophyll production, GABA plays a pivotal role in protecting the chloroplast structure. Under stress conditions such as waterlogging, the ultrastructure of chloroplasts is often compromised, leading to impaired photosynthetic function. However, Salah et al. (2019) have demonstrated that GABA treatment preserves chloroplast integrity, preventing structural deterioration and maintaining the efficiency of the photosynthetic apparatus. This protective function highlights the importance of GABA in sustaining plant metabolism during stress episodes. Stomatal regulation is another critical aspect of photosynthesis that GABA modulates. Stomatal openings control the balance between CO_2 uptake for photosynthesis and water loss through transpiration. During drought stress, maintaining this balance is vital for plant survival. Xu et al. (2021) have reported that GABA fine-tunes stomatal behaviour, optimizing water-use efficiency while ensuring sufficient CO_2 availability for photosynthesis. This regulatory role underscores GABA's ability to enhance photosynthetic stability under water-limited conditions.

Beyond structural and physiological functions, GABA serves as a key player in stress tolerance and photoprotection. High light intensity and heat stress can induce photoinhibition, damaging the photosynthetic machinery and reducing overall efficiency. GABA orchestrates acclimation responses, enabling plants to withstand excessive light exposure and temperature fluctuations (Balfagón et al. 2022). By reinforcing protective mechanisms, GABA contributes to long-term photosynthetic sustainability in challenging environments. Another major stress factor impacting photosynthesis is oxidative stress, which arises from the ROS overproduction under environmental stressors. Excess ROS can damage cellular components, leading to reduced photosynthetic efficiency and overall plant health. However, Seifikalhor et al. (2022) reveal that GABA enhances antioxidant enzyme activity, effectively mitigating oxidative stress. This ROS-scavenging ability ensures the protection of chloroplasts and other cellular structures, safeguarding photosynthesis and promoting plant resilience.

Together, these findings establish GABA as a central regulator of photosynthetic protection and efficiency. By integrating multiple protective strategies—chlorophyll biosynthesis, chloroplast maintenance, stomatal regulation, photoprotection, and oxidative stress mitigation—GABA helps plants navigate and adapt to adverse environmental conditions. Understanding these mechanisms not only expands our knowledge of plant stress responses but also opens new avenues for enhancing crop resilience through targeted metabolic interventions.

19.2 Role of GABA in Photosynthetic Regulation

GABA performs the function as a key regulator of photosynthesis. GABA-induced modulation of various physiological and biochemical processes restores photosynthetic efficiency. From optimizing stomatal conductance to protecting photosynthetic machinery under stress conditions, GABA acts as a vital metabolite linking carbon and nitrogen metabolism, cellular signalling, and energy homeostasis (Fig. 19.1).

19.2.1 Regulation of Stomatal Conductance and CO_2 Assimilation

Stomatal regulation is critical for maintaining an optimal balance between CO_2 uptake and water conservation, thereby influencing photosynthetic efficiency. GABA regulates stomatal aperture, enhancing CO_2 diffusion while reducing transpirational water loss. Studies in *Arabidopsis thaliana* indicate that GABA

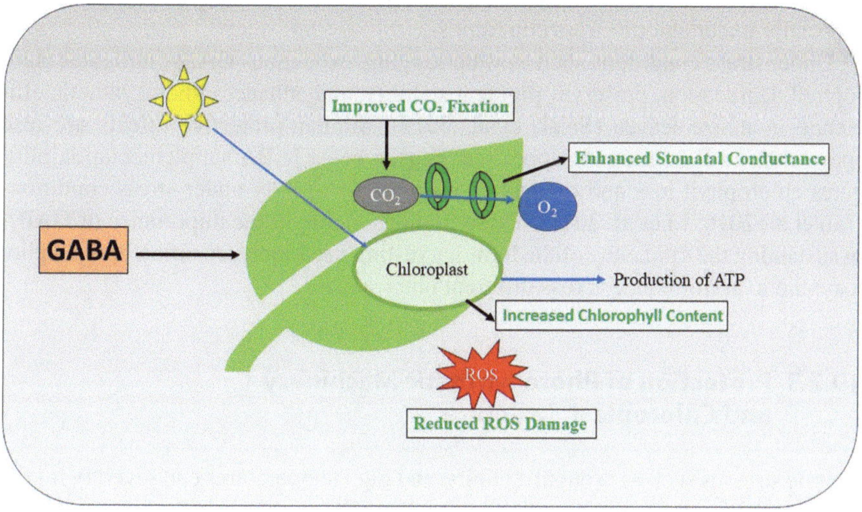

Fig. 19.1 GABA and photosynthesis. GABA enhances photosynthetic efficiency by improving CO_2 fixation and optimizing stomatal conductance, leading to increased ATP production. It also boosts chlorophyll content and reduces ROS damage, thereby preserving chloroplast structure and function. The overall effect contributes to improved plant stress tolerance and photosynthetic efficiency

accumulation in guard cells reduces stomatal opening, thereby improving water-use efficiency and drought tolerance through the inhibition of a tonoplast-localized anion transporter (Xu et al. 2021). This stress-specific accumulation of GABA has been shown to regulate stomatal behaviour in multiple plant species, including both dicots and monocots (Mekonnen et al. 2016). By modulating stomatal movement, GABA enables plants to maintain CO_2 assimilation rates under water-deficient conditions, thus ensuring photosynthetic sustainability.

While GABA effectively regulates stomatal conductance, its role in CO_2 assimilation remains complex. Piechatzek et al. (2024) have demonstrated that GABA does not directly influence stomatal CO_2 signalling in *Arabidopsis*, and previously observed changes in stomatal CO_2 sensitivity in GABA-deficient mutants are attributed to factors beyond GABA's regulatory influence. This suggests that GABA's impact on photosynthetic efficiency primarily operates through water balance optimization rather than direct CO_2 signalling pathways.

19.2.2 Enhancement of Chlorophyll Biosynthesis and Light Harvesting

GABA positively influences chlorophyll biosynthesis, which is essential for efficient light absorption and energy transfer within the photosynthetic apparatus. In maize seedlings, exogenous GABA application significantly increases chlorophyll content, thereby improving photosynthetic performance (Li et al. 2016). By stabilizing chlorophyll pigments and sustaining the structural integrity of light-harvesting complexes, GABA ensures the sustained operation of photosynthetic processes, especially under abiotic stress conditions.

Under stress conditions such as waterlogging, exogenous GABA mitigates chlorophyll degradation, preserves pigment stability, and enhances photosynthetic efficiency in maize leaves (Salah et al. 2019). Similar protective effects are also spectacular in Pakchoi and pepper seedlings, where GABA supplementation mitigates chlorophyll loss and renders photosynthetic activity under stress conditions (Liu et al. 2016; Li et al. 2017). These findings highlight the importance of GABA in sustaining the efficiency of the light-harvesting machinery, ensuring optimal photosynthetic performance across different plant species.

19.2.3 Protection of Photosynthetic Machinery and Chloroplast Structure

Abiotic stresses such as drought, salinity, and high temperatures can severely impair photosynthetic processes by disrupting the ETC and causing photoinhibition. GABA plays a critical role in stabilizing chloroplast structures and reducing damage to photosynthetic machinery. In waterlogged plants, GABA preserves chloroplast ultrastructure, ensuring the sustained functionality of the photosynthetic apparatus despite oxygen deprivation (Salah et al. 2019).

Exogenous GABA alleviates stress-induced damage on the acceptor side of PSII and restores its energy distribution and reaction centre stability. In muskmelon seedlings, GABA supplementation enhances the ability of PSII to repair damaged reaction centres under salinity-alkalinity stress, highlighting its protective role in preserving the structural and functional integrity of the chloroplast (Xiang et al. 2016).

19.2.4 Antioxidant Defence and Reduction of Oxidative Stress

One of GABA's most crucial roles in photosynthesis is its involvement in mitigating oxidative stress. Under high light intensity or stress conditions, ROS accumulate excessively, leading to oxidative damage to lipids, proteins, and pigments within the photosynthetic system. GABA enhances the activity of antioxidant enzymes such as Superoxide dismutase (SOD) and Catalase (CAT), thereby reducing ROS accumulation and preventing oxidative damage (Seifikalhor et al. 2022).

This antioxidative function of GABA is particularly relevant under stress conditions that induce photoinhibition. By reinforcing antioxidant defences, GABA helps maintain photosynthetic efficiency by preventing damage to PSII and sustaining the overall photochemical activity of the chloroplast.

19.2.5 TCA Cycle Interaction and Balance Energy

GABA is an integral component of the GABA shunt, a metabolic pathway that interacts with the TCA cycle to regulate energy metabolism. This pathway links carbon and nitrogen metabolism, thereby maintaining ATP production and providing metabolic intermediates necessary for sustaining photosynthetic efficiency. Under stress conditions, the GABA shunt allows plants to generate energy more efficiently while minimizing damage to cellular structures.

In muskmelon seedlings exposed to hypoxia, GABA supplementation confers PSII photochemical efficiency, underscoring its role in maintaining energy production under stress (Xia et al. 2011). Additionally, GABA contributes to carbon–nitrogen balance through its interaction with the TCA cycle, further supporting plant metabolic stability. By sustaining cell turgor, promoting chlorophyll biosynthesis, and mitigating oxidative stress, GABA plays a pivotal role in enhancing net photosynthesis and overall plant resilience (Wang et al. 2017).

19.3 Photosynthesis Protection Against Abiotic Stresses

GABA has emerged as a critical regulator of photosynthesis under abiotic stress conditions. Its exogenous application boosts photosynthetic performance by modulating key physiological and biochemical pathways. For instance, in maize seedlings exposed to salt stress, GABA treatment significantly promotes net photosynthetic rates and upregulates genes linked to photosynthesis, indicating its role in facilitating better carbon assimilation and energy production under adverse environmental conditions (Wang et al. 2017).

19.3.1 Drought Stress

Drought stress imposes severe limitations on plant growth by reducing CO_2 availability due to stomatal closure, thereby diminishing photosynthetic efficiency. Under such conditions, GABA accumulation serves as an adaptive response, acting as a signalling molecule by binding to ALMT proteins and modulating H^+-ATPase activity through interactions with 14-3-3 proteins. This regulatory mechanism helps optimize stomatal movement, thereby improving water-use efficiency (Mekonnen et al., 2016).

Exogenous application of GABA plays a protective role by bolstering antioxidant defence mechanisms, stabilizing cellular membranes, and maintaining photosynthetic function in chickpea subjected to drought stress (Seifikalhor et al. 2022). In pomegranate cultivars, GABA application sustains transpiration rates, net photosynthesis, intercellular CO_2 concentration, and stomatal conductance, thereby mitigating stress-induced declines and promoting overall plant growth and productivity (Zarbakhsh and Shahsavar 2023).

19.3.2 Salinity Stress

Salt stress disrupts ion homeostasis and impairs photosynthesis and plant growth (Chen et al. 2015). The primary physiological challenges posed by salinity include toxic ion accumulation, osmotic stress, and disruption of the photosynthetic machinery. To cope with such stress, plants adopt various mechanisms, such as ion exclusion, osmotic tolerance, and tissue tolerance (Tyerman et al. 2019).

One of the detrimental effects of salt stress is the restriction of carbon assimilation due to damage to the photosynthetic system (Chen et al. 2015). Exogenous application of GABA alleviates these adverse effects by reducing the accumulation of toxic ions, maintaining cellular integrity, and enhancing the efficiency of photosynthetic machinery. Specifically, in maize seedlings, GABA application improves salt tolerance by sustaining chlorophyll fluorescence parameters and preserving net photosynthesis (Wang et al. 2017). GABA application to salt-stressed strawberry seedlings promotes gene expression of chlorophyll metabolism/biosynthesis and photosynthesis that enhances chlorophyll production and consequent photosynthesis, leading to improved plant health and growth (Zhang et al. 2025).

19.3.3 High Light Intensity and Photoinhibition

Excessive light intensity can trigger photoinhibition, primarily affecting the efficiency of PSII. Under such conditions, GABA safeguards the photosynthetic apparatus by enhancing non-photochemical quenching mechanisms, which dissipate excess energy and prevent oxidative damage (Kaur et al. 2015).

In mung bean plants subjected to heat stress, GABA treatment mitigates damage to cellular membranes and preserves the functionality of key photosynthetic components such as chlorophyll concentration, chlorophyll fluorescence, and Rubisco activity. Moreover, it stimulates carbon assimilation and sucrose synthesis and utilization, thereby ensuring continued growth under stress conditions (Priya et al. 2019). In Arabidopsis, GABA dedicates in regulating the expression of genes associated with light-harvesting complexes, and optimizes energy absorption, minimizing photodamage under high light and heat stress conditions (Balfagón et al. 2022).

19.4 Conclusion

GABA engages in protecting and optimizing photosynthesis, particularly under abiotic stress conditions. It regulates stomatal conductance, balancing CO_2 assimilation and water conservation, while also enhancing chlorophyll biosynthesis and stabilizing photosynthetic pigments to sustain light-harvesting and energy conversion. By protecting the photosynthetic ETC and improving PSII efficiency, GABA mitigates photoinhibition and preserves chloroplast structure. In addition, GABA routinely strengthens antioxidant defence mechanisms, reducing oxidative damage caused by ROS under stress conditions such as drought, salinity, and high light intensity. Through its integration into the GABA shunt, it supports carbon and nitrogen metabolism, maintaining cellular energy homeostasis and ensuring continuous ATP production. Given its multifunctional role in photosynthesis and stress tolerance, GABA holds significant potential as a bio-stimulant for improving crop resilience and productivity. Future research should explore its molecular mechanisms and interactions with other metabolic pathways to enhance its application in sustainable agriculture.

References

Balfagón D, Gómez-Cadenas A, Rambla JL, Granell A, de Ollas C, Bassham DC, Mittler R, Zandalinas SI (2022) γ-Aminobutyric acid plays a key role in plant acclimation to a combination of high light and heat stress. Plant Physiol 188(4):2026–2038. https://doi.org/10.1093/plphys/kiac010

Chen TW, Kahlen K, Stützel H (2015) Disentangling the contributions of osmotic and ionic effects of salinity on stomatal, mesophyll, biochemical and light limitations to photosynthesis. Plant Cell Environ 38(8):1528–1542. https://doi.org/10.1111/pce.12504

Kaur R, Bains TS, Bindumadhava H, Nayyar H (2015) Responses of mungbean (Vigna *radiata* L.) genotypes to heat stress: effects on reproductive biology, leaf function and yield traits. Sci Hortic 197:527–541

Li W, Liu J, Ashraf U, Li G, Li Y, Lu W, Gao L, Han F, Hu J (2016) Exogenous-aminobutyric acid (GABA) application improved early growth, net photosynthesis, and associated physio-biochemical events in maize. Front Plant Sci 7:919. https://doi.org/10.3389/fpls.2016.00919

Li Y, Fan Y, Ma Y, Zhang Z, Yue H (2017) Effects of exogenous γ-aminobutyric acid (gaba) on photosynthesis and antioxidant system in pepper (*Capsicum annuum* L.) seedlings under low light stress. J Plant Growth Regul 36:1–14

Liu JP, Xiao FL, Zhu HF, Zhu Y, Hou XL (2016) Effects of γ-aminobutyric acid on the growth and photosynthesis of pakchoi under waterlogging stress. Acta Agriculture Shanghai 9. https://doi.org/10.1038/s41598-018-36334-y

Mekonnen DW, Flügge U-I, Ludewig F (2016) Gamma-aminobutyric acid depletion affects stomata closure and drought tolerance of Arabidopsis thaliana. Plant Sci 245:25–34. https://doi.org/10.1016/j.plantsci.2016.01.005

Piechatzek A, Feng X, Sai N, Yi C, Hurgobin B, Lewsey M, Herrmann J, Dittrich M, Ache P, Müller T, Kromdijk J, Hedrich R, Xu B, Gilliham M (2024) GABA does not regulate stomatal CO_2 signalling in Arabidopsis. J Exp Bot 75(21):6856–6871. https://doi.org/10.1093/jxb/erae168

Priya M, Sharma L, Kaur R, Bindumadhava H, Nair RM, Siddique KHM, Nayyar H (2019) GABA (γ-aminobutyric acid), as a thermo-protectant, to improve the reproductive function of heat-stressed mungbean plants. Sci Rep 9(1):7788. https://doi.org/10.1038/s41598-019-44163-w

Salah A, Zhan M, Cao C et al (2019) γ-aminobutyric acid promotes chloroplast ultrastructure, antioxidant capacity, and growth of waterlogged maize seedlings. Sci Rep 9:484. https://doi.org/10.1038/s41598-018-36334-y

Seifikalhor M, Niknam V, Aliniaeifard S et al (2022) The regulatory role of γ-aminobutyric acid in chickpea plants depends on drought tolerance and water scarcity level. Sci Rep 12:7034. https://doi.org/10.1038/s41598-022-10571-8

Tyerman SD, Munns R, Fricke W, Arsova B, Barkla BJ et al (2019) Energy costs of salinity tolerance in crop plants. New Phytol 221(1):25–29. https://doi.org/10.1111/nph.15555

Wang Y, Gu W, Meng Y, Xie T (2017) γ-aminobutyric acid imparts partial protection from salt stress injury to maize seedlings by improving photosynthesis and up regulating osmoprotectants and antioxidants. Sci Rep 7:43609. https://doi.org/10.1038/srep43609

Xia QP, Gao HB, Li JR (2011) Effects of gamma-aminobutyric acid on the photosynthesis and chlorophyll fluorescence parameters of muskmelon seedlings under hypoxia stress. J Appl Ecol 22(4):999

Xiang L, Hu L, Xu W, Zhen A, Zhang L, Hu X (2016) Exogenous γ-aminobutyric acid improves the structure and function of photosystem II in muskmelon seedlings exposed to salinity-alkalinity stress. PLoS One 11(10):e0164847. https://doi.org/10.1371/journal.pone.0164847. PMID: 27764179; PMCID: PMC5072714

Xu B, Long Y, Feng X et al (2021) GABA signalling modulates stomatal opening to enhance plant water use efficiency and drought resilience. Nat Commun 12:1952. https://doi.org/10.1038/s41467-021-21694-3

Zarbakhsh S, Shahsavar AR (2023) Exogenous γ-aminobutyric acid improves the photosynthesis efficiency, soluble sugar contents, and mineral nutrients in pomegranate plants exposed to drought, salinity, and drought-salinity stresses. BMC Plant Biol 23(1):543. https://doi.org/10.1186/s12870-023-04568-2

Zhang Y, Deng M, Lin B et al (2025) Physiological and transcriptomic evidence revealed the role of exogenous GABA in enhancing salt tolerance in strawberry seedlings. BMC Genomics 26:196. https://doi.org/10.1186/s12864-025-11368-5

GABA Genetic Engineering/Genome Editing Aspects 20

Abstract

GABA, a crucial signalling molecule, influences stress tolerance, growth, and metabolic regulation in plants. Advances in genetic engineering and genome editing techniques, particularly clustered regularly interspaced short palindromic repeats/CRISPR-associated protein 9 (CRISPR–Cas9), have facilitated precise modifications in GABA biosynthesis and signalling pathways, offering new avenues for crop improvement. GABA biosynthesis pathways in plants engage three enzymes viz., Glutamate decarboxylase (GAD), GABA transaminase (GABA-T), and succinic semialdehyde dehydrogenase (SSADH). Engineering of these pathways via gene overexpression, silencing, or targeted genome editing promotes stress resilience and plant productivity. This advocates for a discussion on the current strategies for genetic manipulation and recent advancements in CRISPR-based approaches for fine-tuning GABA pathways, mitigating stress tolerance, and augmenting crop performance. While genetic modifications of GABA metabolism present promising opportunities, challenges such as regulatory hurdles and potential off-target effects must be addressed. Future research integrating multi-omics approaches and synthetic biology tools would help in accomplishing efficacious GABA bioengineering and the development of climate-resilient crops, ensuring sustainable agricultural productivity and food security.

Keywords

CRISPR–Cas9 · Crop improvement · GABA metabolism · Genetic engineering · Stress tolerance

S. U. Jalil et al., *GABA Signaling System and Abiotic Stress Tolerance in Plants*,
https://doi.org/10.1007/978-981-95-0437-4_20

Highlights

- GABA is essential for regulating plant stress responses, growth, and metabolic processes.
- Genetic engineering and clustered regularly interspaced short palindromic repeats/CRISPR-associated protein 9 (CRISPR–Cas9) enable precise modifications in GABA pathways.
- Overexpression of Glutamate decarboxylase *(GAD)* genes enhances GABA accumulation and stress resilience.
- Genome editing of GABA transaminase (*GABA-T*) and succinic semialdehyde dehydrogenase *(SSADH)* reduces GABA degradation, improving yield.
- GABA bioengineering enhances drought, salinity, and heat stress tolerance in crops.
- Future research should integrate multi-omics and synthetic biology for optimized GABA regulation.

20.1 Introduction

GABA in plants performs metabolic regulation and signalling pathways and confers resilience against various stresses. Its accumulation is tightly regulated by enzymatic activities (Fig. 20.1), primarily through glutamate decarboxylase (GAD), which catalyses the conversion of glutamate to GABA (Bouche and Fromm 2004; Jalil et al. 2017; Ansari et al. 2021). Environmental factors, such as abiotic stresses, including drought, salinity, and temperature fluctuations, can significantly induce GABA accumulation in plants as part of their adaptive mechanism (Islam et al. 2024). Additionally, GABA functions in modulating plant growth, enhancing nitrogen metabolism, and influencing fruit development and ripening (Khan et al. 2021), making it an essential target for genetic engineering to improve both plant resilience and nutritional quality.

Advancements in genome editing technologies, particularly clustered regularly interspaced short palindromic repeats/CRISPR-associated protein 9 (CRISPR–Cas 9) systems, have revolutionized plant biotechnology by enabling precise modifications of genes involved in GABA biosynthesis and metabolism (Jalil and Ansari 2020). Unlike conventional breeding, which is time-consuming and often results in linkage drag, genome editing allows targeted alterations in key genes such as GAD, GABA transaminase (GABA-T), and succinic semialdehyde dehydrogenase (SSADH) (Busch and Fromm 1999) (Fig. 20.2). These genetic modifications have been successfully applied in crops like tomato and rice to enhance GABA accumulation, improve stress tolerance, and promote growth under adverse conditions (Clark et al. 2009; Li et al. 2018; Chen et al. 2024; Sakthivel et al. 2025). The ability to engineer GABA pathways through genome editing presents new opportunities for developing biofortified crops with enhanced nutritional and functional properties.

One of the most notable achievements in GABA genetic engineering is the development of the Sicilian Rouge High GABA tomato, the first gene-edited food

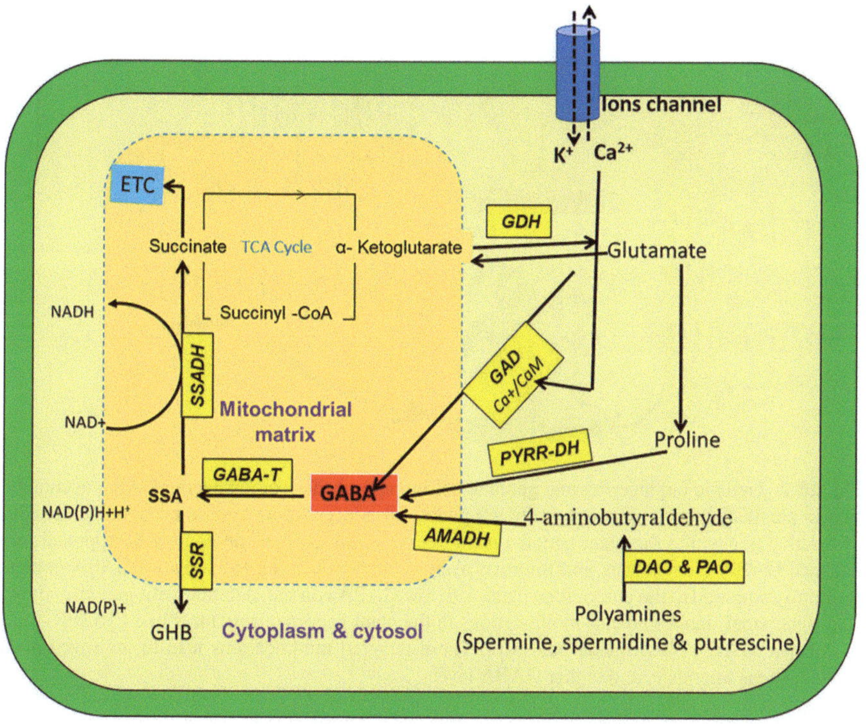

Fig. 20.1 Schematic representation of genes involved in GABA biosynthesis and metabolism in plants. GDH contributes to glutamate metabolism, which feeds into the GABA shunt. Glutamate Decarboxylase. GAD catalyses the conversion of glutamate to GABA, a process regulated by Ca^{2+}/calmodulin (CaM). GABA transaminase (GABA-T) converts GABA into uccinic Semialdehyde (SSA), which is further processed by SSADH into succinate, linking it to the tricarboxylic acid (TCA) cycle. Additionally, diamine oxidases (DAO) and polyamine oxidase (PAO) convert polyamines (PAs) into 4-aminobutyraldehyde that amino methyltransferase dehydrogenase (AMADH) picks up as a substrate for GABA biosynthesis. The pyrroline dehydrogenase (PYRR-DH) enzyme plays a role in GABA biosynthesis from proline. The pathway plays a crucial role in stress responses, energy metabolism, and osmotic balance in plants

approved for commercial sale. By modifying the autoinhibitory domain of the GAD gene using CRISPR–Cas9, researchers significantly increased GABA levels in tomato fruits without compromising their yield or fruit quality (Lee et al. 2018). Similar approaches in rice have demonstrated that overexpressing or knocking out GAD genes can influence GABA accumulation, leading to improved salt stress tolerance (Kanchitanuraka et al. 2023). Such breakthroughs highlight the potential of genome editing not only for enhancing GABA-enriched functional foods but also for improving crop resilience to environmental stresses, thereby contributing to sustainable agriculture.

Despite these advancements, challenges remain in the widespread adoption of GABA genome editing in plants. Regulatory hurdles, public acceptance, and potential off-target effects of CRISPR-based modifications need to be carefully addressed

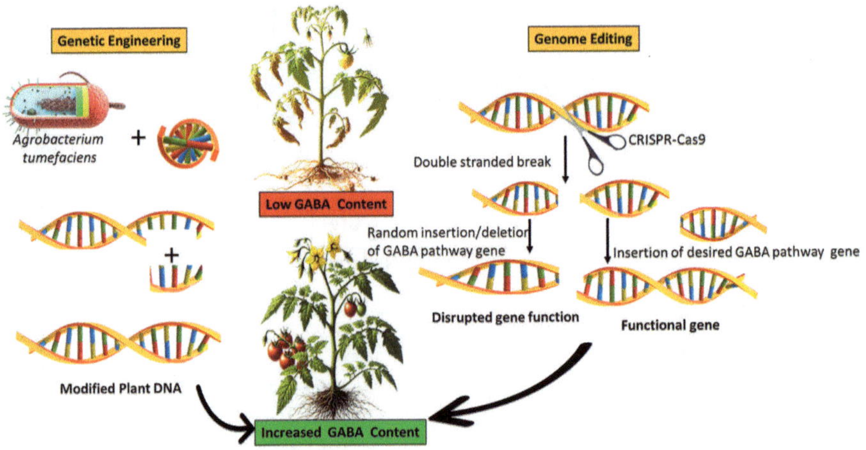

Fig. 20.2 Genetic engineering and genome editing approaches for enhancing GABA accumulation in plants. Genetic engineering (the left panel) introduces foreign genes or modified versions of existing genes into the plant genome using recombinant DNA technology. This approach can enhance GABA biosynthesis and improve plant resilience, leading to increased GABA content (bottom) compared to the unmodified plant with low GABA content. Alternatively, genome editing (the right panel) makes precise modifications in the plant genome using tools like CRISPR–Cas9 to enhance GABA biosynthetic pathways. These targeted modifications lead to improved plant growth, stress tolerance, and higher GABA levels

to ensure the safe deployment of gene-edited crops. Future research should focus on refining gene-editing techniques, identifying novel regulatory elements controlling GABA biosynthesis, and conducting long-term field evaluations to assess the stability and agronomic benefits of GABA-enriched crops. With continued progress, genome editing holds great promise for revolutionizing plant metabolic engineering, paving the way for the development of nutritionally enhanced, stress-tolerant, and high-yielding crop varieties.

20.2 Key Genes Involved in GABA Regulation

GABA regulation is a complex interplay between biosynthesis, degradation, and transcriptional responses to environmental stimuli. Over the years, researchers have identified key genes governing GABA metabolism (Fig. 20.1), shedding light on its significance in plant physiology.

20.2.1 Biosynthesis of GABA in Plants

The primary route of GABA biosynthesis in plants is the decarboxylation of glutamate, a reaction catalysed by GAD (Yogeswara et al. 2020). GAD, encoded by GAD genes, represents the rate-limiting enzyme in GABA production. This irreversible decarboxylation process is influenced by various factors, including glutamate availability, cytosolic pH, and environmental stress signals. Glutamate itself is synthesized via nitrogen assimilation pathways such as the glutamine synthetase/ glutamine:2-oxoglutarate aminotransferase (GS/GOGAT) cycle, operating in both the cytoplasm and plastids (Ramos-Ruiz et al. 2019).

Environmental stresses, like drought, salinity, and hypoxia, induce intracellular Ca^{2+} accumulation, which activates CaM genes (Bose et al. 2011; Behera et al. 2018). The Ca^{2+}/CaM complex interacts with the C-terminal domain of GAD, enhancing its enzymatic activity and driving GABA production, particularly at an optimal pH of 7.0–7.5 (Takayama and Ezura 2015; Shelp et al. 2018). Additionally, stress-induced cytoplasmic acidification further stimulates GAD activity, ensuring rapid GABA accumulation under adverse conditions (Bouché et al. 2004; Fait et al. 2008).

GAD genes in various plant species perform diverse roles linked to GABA metabolism. In *Arabidopsis*, AtGAD1 regulates GABA synthesis in roots (Akçay et al. 2012; Ji et al. 2017), while in rice, OsGAD2 plays a major role in GABA accumulation, exhibiting significantly higher activity in transgenic plants (Liu et al. 2004). In tomatoes, SlGAD1 is linked to fruit growth rather than GABA content, whereas SlGAD2 and SlGAD3 enhance GABA levels in the fruit (Shelp et al. 2012a, b; Takayama and Ezura 2015). However, SlGAD4 in tomatoes and AtGAD4 in *Arabidopsis* do not appear to contribute significantly to GABA accumulation or plant development.

Recent findings in transgenic rice overexpressing the OsCam1-1 CaM gene suggest an enhanced GAD activity during germination, leading to elevated GABA levels. These transgenic lines exhibited significantly higher GABA accumulation than wild-type plants, particularly under salt stress conditions, further reinforcing the role of CaM in GABA synthesis regulation (Kanchitanuraka et al. 2023).

20.2.2 Alternative Pathways of GABA Synthesis

PA degradation pathway provides an alternative route for GABA synthesis in plants (Fig. 20.1) (Li et al. 2021). PAs like putrescine (Put), spermine (Spm), and spermidine (Spd) are broken down to produce GABA. Put is primarily formed from ornithine via ornithine decarboxylase (ODC) or from arginine through arginine decarboxylase (ADC) (Planas-Portell et al. 2013). In a sequence of biochemical events, DAO and PAO perform PA degradation to produce 4-aminobutanal (ABAL) (Tavladoraki et al. 2016), which AMADH uses as a substrate to produce GABA (Shelp et al. 2012b).

Environmental stress conditions, such as anaerobic stress in broad beans and salt stress in soybean roots, can upregulate polyamine synthesis and GABA accumulation by enhancing DAO, PAO, and AMADH activities (Yang et al. 2013, 2016; Al-Quraan et al. 2013). However, despite its role in GABA biosynthesis, the polyamine degradation pathway is less efficient than the GABA shunt, especially in monocot plants (Flores and Filner 1985; Torrigiani et al. 2010).

Another alternative pathway involves the non-enzymatic conversion of proline to GABA. This occurs through hydroxyl radicals attacking proline's nitrogen atom, leading to decarboxylation and formation of Δ^1-pyrroline, which is subsequently converted into GABA by PYRR-DH (Forlani et al. 2019). This pathway becomes more active under stress conditions due to increased reactive oxygen species (ROS) production (Aleksza et al. 2017; Halliwell 2006).

20.3 GABA Catabolism: The GABA Shunt

GABA is primarily degraded via the GABA shunt, a metabolic pathway that links amino acid metabolism to the TCA cycle. This process occurs in the mitochondrial matrix and involves two key enzymes: GABA-T, which converts GABA into SSA. Subsequently, SSADH oxidizes SSA into succinate for entry into the TCA cycle (Jacoby et al. 2011; Michaeli and Fromm 2015; Yuan et al. 2023).

Loss-of-function analyses using RNAi in transgenic tomato plants reveal that SlGABA-T1 is crucial for GABA metabolism. RNAi suppression of SlGABA-T1 leads to severe dwarfism and infertility, with significantly increased GABA levels in mature green and red ripe fruits. Three GABA-T genes have been detected in tomato fruits, yet their precise roles remain under investigation (Koike et al. 2013). Nevertheless, GABA in tomatoes accumulates at high levels in immature fruits but is rapidly converted to succinate during ripening through GABA-T and SSADH activities. The suppression of GABA-T or SSADH genes results in GABA accumulation, highlighting their essential roles in GABA catabolism (Gramazio et al. 2020).

In tomatoes, SlGABA-T3 is primarily expressed in plastids and supports plant growth, while SlSSADH facilitates the final conversion of SSA into succinate, ensuring efficient GABA turnover (Akihiro et al. 2008). AtPOP2 in *Arabidopsis* encodes a functional GABA-T enzyme that regulates GABA gradients essential for pollen tube growth, root development, and shoot formation and senescence (Palanivelu et al. 2003; Al-Quraan and Al-Share 2016; Jalil et al. 2019).

γ-Hydroxybutyrate (GHB) formation is the culmination step of GABA catabolism (Fig. 20.1), a reversible reaction catalysed by succinate semialdehyde reductase (SSR) that converts SSA into GHB (Bouché et al. 2003). In tomatoes, two SSR genes, SlSSR1 and SlSSR2, regulate SSA conversion to GHB and play a critical role in GABA metabolism during fruit ripening (Xing et al. 2007). The process occurs in multiple cellular compartments. Environmental stress, particularly hypoxia, promotes GHB formation as a detoxification mechanism (Yin et al. 2018).

Together, these studies highlight the intricate genetic regulation of GABA metabolism in plants, underscoring its essential roles in development and stress responses. Understanding these pathways provides valuable insights for improving crop resilience through metabolic engineering.

20.4 Genetic Engineering Approaches to Enhance GABA in Plants

GABA plays a crucial role in plant metabolism, particularly in stress responses. Efforts to enhance GABA accumulation in plants through genetic engineering have focused on modulating its biosynthesis and degradation pathways (Fig. 20.2). One of the earliest insights into GABA enhancement emerges from studies on rice, attributing that the transcription factor OsMYB55 could regulate GABA levels by activating the GAD3 enzyme. Overexpression of OsMYB55 significantly increases GABA content, especially under high-temperature stress, leading to improved stress tolerance and yield (El-Kereamy et al. 2012). Expanding on these findings, a groundbreaking approach has helped rice achieve stable GABA accumulation by modifying the GABA shunt pathway. The approach, with the help of RNAi, follows the overexpression of a truncated version of the GAD2 gene and the suppression of GABA-T. The dual strategy has resulted in transgenic rice lines with significantly higher GABA content (75–350 mg/100 g polished rice), along with elevated levels of other amino acids, such as alanine, serine, and valine. Notably, the GABA-enriched rice retained its high GABA levels for up to 6 months after harvest, making it a promising approach for biofortified rice production (Shimajiri et al. 2013). Around the same time, researchers have successfully employed RNAi to suppress the GABA transaminase 1 (OsGABA-T1) gene in the japonica rice variety "Ning Japonica 1," leading to significant GABA accumulation in rice grains (Zhou et al. 2015).

In tomatoes, natural GABA accumulation is high in immature fruits but declines during ripening due to the action of GABA-T and SSADH. To understand the role of GABA-T in fruit ripening, researchers employ RNAi to silence the SlGABA-T1 gene. The resulting transgenic plants exhibit severe dwarfism and infertility, with GABA levels in red ripe fruits increasing by 6.8–9.2 times compared to wild-type plants. The study elucidates that SlGABA-T1 is a critical enzyme responsible for reducing GABA levels during fruit ripening (Koike et al. 2013).

Further studies have explored GABA metabolism under stress conditions. Identification and cloning of the MuGABA-T gene in mulberry reveal that salt stress and hormones, abscisic acid (ABA) and salicylic acid (SA) significantly induce its expression. Overexpression of MuGABA-T in *Arabidopsis* and mulberry plants leads to reduced GABA levels, increased ROS accumulation, and heightened sensitivity to salt stress. However, exogenous GABA application activates antioxidant enzymes, mitigates oxidative damage, and provides protection against salt stress. These findings have underscored the role of MuGABA-T in regulating GABA metabolism and ROS homeostasis during stress responses (Zhang et al. 2022).

In maize, genetic engineering efforts to boost GABA levels by introducing a deregulated AtGAD1 gene successfully enhance GABA accumulation. However, this increase comes at a cost, as plants exhibit chlorosis and stunted growth. In contrast, the suppression of GABA biosynthesis and transport genes has no visible effect, while the knockout of GABA catabolism genes results in significant growth defects, highlighting the delicate balance of GABA homeostasis in plant growth and development (Rajni et al. 2021).

Another recent advancement involves the use of a CaM overexpression strategy in rice. Transgenic "Khao Dawk Mali 105" (KDML105) rice overexpressing the OsCam1-1 CaM gene maintains the enhanced salt and drought stress tolerance. The transgenic lines exhibit a rapid increase in GABA content during germination, particularly under salt stress, with levels peaking after 6–8 days. In fact, a gain in GAD activity is responsible for GABA accumulation in transgenic lines with OsCam1-1 overexpression (Kanchitanuraka et al. 2023).

In addition to genetic modifications, metabolic engineering strategies have been employed to optimize GABA biosynthesis. Advanced techniques such as pathway optimization and enzyme engineering have been successfully applied in microorganisms, offering promising insights for plant-based approaches (Su et al. 2021). Collectively, these studies demonstrate that genetic engineering approaches targeting GABA biosynthesis and catabolism can effectively enhance GABA levels in plants. However, achieving a balance between increased GABA accumulation and maintaining plant growth remains a key challenge for future research.

20.5 Genome Editing of GABA Pathways Using CRISPR–Cas

The advancement of genome editing technologies has revolutionized plant metabolic engineering, with the CRISPR–Cas system emerging as a precise and efficient tool for targeted genetic modifications (Fig. 20.2). Among its applications, enhancing GABA levels in crops has garnered significant attention due to GABA's role in stress tolerance and its health-promoting benefits (Sakthivel et al. 2025). While conventional breeding strategies for increasing GABA content are time-intensive and are often limited by linkage drag, genome editing provides a direct and rapid approach to modifying key metabolic pathways.

One of the earliest breakthroughs in GABA-enriched crops has been achieved in tomatoes using CRISPR–Cas9 (Nonaka et al. 2017). By targeting and deleting the autoinhibitory domains of GAD genes (SlGAD2 and SlGAD3), researchers have successfully increased GABA accumulation in tomato fruits by 7- to 15-fold. Remarkably, this genetic modification does not interfere with fruit ripening, preserving the overall quality and yield of the crop. The study demonstrates the feasibility of precise genome editing to enhance nutritional traits without adverse pleiotropic effects.

Further research has explored the potential of CRISPR-mediated genome editing in metabolic pathway regulation (Li et al. 2018). Using the pYLCRISPR–Cas9 system, scientists have targeted the tomato phytoene desaturase gene, producing albino phenotypes with a 61% success rate in gene editing. Additionally, employing a multiplex CRISPR–Cas9 approach has simultaneously targeted five key genes in the GABA shunt pathway. The strategy has resulted in 53 genome-edited tomato plants, including single to quadruple mutants, with GABA accumulation significantly increasing—quadruple mutants exhibit a 19-fold higher GABA content in leaves compared to wild-type plants. Hence, there exists the immense potential of multiplex genome editing in metabolic engineering and trait improvement.

Expanding on these findings, an integration of genome editing with breeding strategies has led to the development of high-GABA hybrid tomatoes (Lee et al. 2018). Researchers have crossed genome-edited tomato lines with the pure-line cultivar "Aichi First" to assess the stability of GABA enhancement in hybrid offspring. The hybrid lines demonstrate a consistently high GABA accumulation with minimal impact on other fruit traits. The results confirm that high GABA content is a dominant trait, making genome-edited tomatoes valuable parental lines for hybrid cultivar development. The success of this approach highlights genome editing as an efficacious transformative tool for rapid breeding programmes.

Beyond tomatoes, GABA metabolism has been a focal point in stress tolerance studies, particularly in rice. Given rice's susceptibility to salt stress, researchers have explored the role of GABA in enhancing resilience under adverse conditions (Chen et al. 2024). CRISPR-mediated modifications of GAD genes in rice have created both overexpression (gad3-ox1 and gad3-ox2) and knockout (gad1/3-ko) lines. The overexpression lines exhibit increased endogenous GABA levels, leading to enhanced salt resistance. On the other hand, the knockout lines display a reduced GABA content and heightened sensitivity to stress. Elevated GABA levels correlate with improved antioxidant activity/osmotic regulation and reduced membrane lipid peroxidation and Na^{+} accumulation. The findings provide a strong theoretical foundation for engineering salt-tolerant rice varieties through targeted genome editing.

The application of CRISPR–Cas in metabolic enhancement has also led to the commercialization of gene-edited crops. By 2020, Japan had introduced over 300 functional food products containing GABA, surpassing the popularity of indigestible dextrin in the health food market (Minako et al. 2023). The development and release of the Sicilian Rouge High GABA tomato as the world's first gene-edited food approved for commercialization is a landmark achievement. Utilizing CRISPR–Cas9, scientists have modified the autoinhibitory domain of the GAD gene to enable high-density accumulation of GABA. Consumer acceptance studies and in-house monitoring surveys further validate its nutritional benefits and market viability, paving the way for future gene-edited functional foods.

The integration of CRISPR–Cas genome editing with metabolic pathway engineering and speed breeding has unlocked new possibilities in functional food development. By precisely targeting key regulatory genes, researchers have successfully enhanced GABA content in multiple crop species, demonstrating the potential of genome editing to improve both stress resilience and nutritional quality. As genome editing technologies continue to evolve, their application in food biofortification and sustainable agriculture will undoubtedly expand, shaping the future of plant-based nutrition and crop improvement.

20.6 Conclusion

Genome editing has emerged as a powerful tool for enhancing GABA biosynthesis in plants, offering a precise and efficient alternative to traditional breeding and transgenic approaches. By targeting key genes such as GAD using CRISPR–Cas technology, researchers have successfully developed GABA-enriched crops like tomatoes and rice with improved nutritional value and enhanced stress tolerance. Multiplex CRISPR strategies further enable the simultaneous modification of multiple genes, optimizing GABA accumulation while maintaining normal plant growth and development. Despite these advancements, challenges such as off-target effects, regulatory constraints, and consumer acceptance need to be addressed. Future research should focus on refining genome editing techniques, exploring novel regulatory elements, and assessing the long-term impacts of GABA biofortification on plant physiology. With continued innovations, genome editing holds great promise for developing functional foods and stress-resilient crops, contributing to sustainable agriculture, global food security, and human health.

References

Akçay N, Bor M, Karabudak T, Ozdemir F, Türkan I (2012) Contribution of gamma amino butyric acid (GABA) to salt stress responses of Nicotiana sylvestris CMSII mutant and wild type plants. J Plant Physiol 169:452–458

Akihiro T, Koike S, Tani R, Tominaga T, Watanabe S, Iijima Y, Aoki K, Shibata D, Ashihara H, Matsukura C et al (2008) Biochemical mechanism on GABA accumulation during fruit development in tomato. Plant Cell Physiol 49:1378–1389

Aleksza D, Horváth GV, Sándor G, Szabados L (2017) Proline accumulation is regulated by transcription factors associated with phosphate starvation. Plant Physiol 175:555–567

Al-Quraan NA, Al-Share AT (2016) Characterization of the-aminobutyric acid shunt pathway and oxidative damage in Arabidopsis thaliana pop2 mutants under various abiotic stresses. Biol Plantarum 60:132–138

Al-Quraan NA, Sartawe FA, Qaryouti MM (2013) Characterization of-aminobutyric acid metabolism and oxidative damage in wheat (Triticum aestivum L.) seedlings under salt and osmotic stress. J Plant Physiol 170:1003–1009

Ansari MI, Jalil SU, Ansari SA et al (2021) GABA shunt: a key-player in mitigation of ROS during stress. Plant Growth Regul 94:131–149

Behera S, Zhaolong X, Luoni L, Bonza MC, Doccula FG, DeMichelis MI, Morris RJ, Schwarzländer M, Costa A (2018) Cellular Ca2+ signals generate defined pH signatures in plants. Plant Cell 30:2704–2719

Bose J, Pottosin II, Shabala SS, Palmgren MG, Shabala S (2011) Calcium efflux systems in stress signaling and adaptation in plants. Front Plant Sci 2:85

Bouché N, Fromm H (2004) GABA in plants: just a metabolite? Trends Plant Sci 9:110–115

Bouché N, Fait A, Bouchez D, Møller SG, Fromm H (2003) Mitochondrial succinic-semialdehyde dehydrogenase of the aminobutyrate shunt is required to restrict levels of reactive oxygen intermediates in plants. Proc Natl Acad Sci USA 100:6843

Bouché N, Fait A, Zik M, Fromm H (2004) The root-specific glutamate decarboxylase (GAD1) is essential for sustaining GABA levels in Arabidopsis. Plant Mol Biol 55:315–325

Busch KB, Fromm H (1999) Plant succinic semialdehyde dehydrogenase. Cloning, purification, localization in mitochondria, and regulation by adenine nucleotides. Plant Physiol 121:589–597

Chen M, Zhu C, Zhang H, Chen S, Wang X, Gan L (2024) Endogenous γ-aminobutyric acid accumulation enhances salinity tolerance in rice. Plants 13(19):2750. https://doi.org/10.3390/plants13192750

Clark SM, Di Leo R, Van Cauwenberghe OR, Mullen RT, Shelp BJ (2009) Subcellular localization and expression of multiple tomato gamma-aminobutyrate transaminases that utilize both pyruvate and glyoxylate. J Exp Bot 60:3255–3267

El-Kereamy A, Bi YM, Ranathunge K, Beatty PH, Good AG, Rothstein SJ (2012) The rice R2R3-MYB transcription factor OsMYB55 is involved in the tolerance to high temperature and modulates amino acid metabolism. PLoS One 7:e52030. https://doi.org/10.1371/journal.pone.0052030

Fait A, Fromm H, Walter D, Galili G, Fernie AR (2008) Highway or byway: the metabolic role of the GABA shunt in plants. Trends Plant Sci 13:14–19

Flores HE, Filner P (1985) Polyamine catabolism in higher plants: characterization of pyrroline dehydrogenase. Plant Growth Regul 3:277–291

Forlani G, Trovato M, Funck D, Signorelli S (2019) Regulation of proline accumulation and its molecular and physiological functions in stress defence. In: Hossain M, Kumar V, Burritt D, Fujita M, Mäkelä P (eds) Osmoprotectant-mediated abiotic stress tolerance in plants. Springer, Cham, pp 73–97

Gramazio P, Takayama M, Ezura H (2020) Challenges and prospects of new plant breeding techniques for GABA improvement in crops: tomato as an example. Front Plant Sci 11:577980. https://doi.org/10.3389/fpls.2020.577980

Halliwell B (2006) Reactive species and antioxidants. Redox biology is a fundamental theme of aerobic life. Plant Physiol 141:312–322

Islam SNU, Kouser S, Hassan P, Asgher M, Shah AA, Khan NA (2024) Gamma-aminobutyric acid interactions with phytohormones and its role in modulating abiotic and biotic stress in plants. Stress Biol 4(1):36. https://doi.org/10.1007/s44154-024-00180-y

Jacoby RP, Taylor NL, Millar AH (2011) The role of mitochondrial respiration in salinity tolerance. Trends Plant Sci 16:614–623

Jalil SU, Ansari MI (2020) Genome editing for the improvement of Brassicaceae for abiotic stress tolerance. In: The plant family Brassicaceae. Springer Nature, Singapore, pp 473–482

Jalil SU, Ahmad I, Ansari MI (2017) Functional loss of GABA transaminase (GABA-T) expressed early leaf senescence under various stress conditions in *Arabidopsis thaliana*. Curr Plant Biol 9–10:11–22

Jalil SU, Khan MQR, Ansari MI (2019) Role of GABA transaminase in the regulation of development and senescence in *Arabidopsis thaliana*. Curr Plant Biol 19:100119

Ji J, Zheng L, Yue J, Yao X, Chang E, Xie T, Deng N, Chen L, Huang Y, Jiang Z et al (2017) Identification of two CiGADs from Caragana intermedia and their transcriptional responses to abiotic stresses and exogenous abscisic acid. PeerJ 5:e3439

Kanchitanuraka P, Chadchawan S, Buaboocha T (2023) Effects of calmodulin overexpression on gamma-aminobutyric acid (GABA) levels and glutamate decarboxylase activity in rice seedlings. ScienceAsia 49:646–652. https://doi.org/10.2306/scienceasia1513-1874.2023.065

Khan MIR, Jalil SU, Chopra P, Chhillar H, Ferrante A, Khan NA, Ansari MI (2021) Role of GABA in plant growth, development and senescence, plant. Gene 26:100283

Koike S, Matsukura C, Takayama M, Asamizu E, Ezura H (2013) Suppression of γ-aminobutyric acid (GABA) transaminases induces prominent GABA accumulation, dwarfism and infertility in the tomato (Solanum lycopersicum L.). Plant Cell Physiol 54(5):793–807. https://doi.org/10.1093/pcp/pct035

Lee J, Nonaka S, Takayama M, Ezura H (2018) Utilization of a genome-edited tomato (Solanum lycopersicum) with high gamma aminobutyric acid content in hybrid breeding. J Agric Food Chem 66(4):963–971. https://doi.org/10.1021/acs.jafc.7b05171

Li R, Li R, Li X, Fu D, Zhu B, Tian H, Luo Y, Zhu H (2018 Feb) Multiplexed CRISPR/Cas9-mediated metabolic engineering of γ-aminobutyric acid levels in Solanum lycopersicum. Plant Biotechnol J 16(2):415–427. https://doi.org/10.1111/pbi.12781

Li L, Dou N, Zhang H, Wu C (2021) The versatile GABA in plants. Plant Signal Behav 16:1862565

Liu LL, Zhao L, Li Q, Jiang L, Wan JM (2004) Molecular cloning and expression of a novel glutamate decarboxylase gene in rice. Rice Genet 21:39–42
Michaeli S, Fromm H (2015) Closing the loop on the GABA shunt in plants: are GABA metabolism and signaling entwined? Front Plant Sci 6:419
Minako S, Yatabe J, Ezura H, Takayama M, Nagata H, Kawasaki T (2023) PS-C23-11: creation of tomatoes with high gaba accumulation by gene editing and their hypotensive effects. J Hypertens 41(Suppl 1):e407. https://doi.org/10.1097/01.hjh.0000916984.02228.01
Nonaka S, Arai C, Takayama M et al (2017) Efficient increase of ɣ-aminobutyric acid (GABA) content in tomato fruits by targeted mutagenesis. Sci Rep 7:7057. https://doi.org/10.1038/s41598-017-06400-y
Palanivelu R, Brass L, Edlund AF, Preuss D (2003) Pollen tube growth and guidance is regulated by POP2, an Arabidopsis gene that controls GABA levels. Cell 114:47–59
Planas-Portell J, Gallart M, Tiburcio AF, Altabella T (2013) Copper-containing amine oxidases contribute to terminal polyamine oxidation in peroxisomes and apoplast of Arabidopsis thaliana. BMC Plant Biol 13:109
Rajni SM, Bedair MF, Li H, Duff SMG (2021) Phenotypic effects from the expression of a deregulated AtGAD1 transgene and GABA pathway suppression mutants in maize. PLoS One 16(12):e0259365. https://doi.org/10.1371/journal.pone.0259365
Ramos-Ruiz R, Felix M, Gertrude K (2019) The effects of GABA in plants. Cogent Food Agric 5:1670553
Sakthivel K, Balasubramanian R, Sampathrajan V et al (2025) Transforming tomatoes into GABA-rich functional foods through genome editing: a modern biotechnological approach. Funct Integr Genomics 25:27. https://doi.org/10.1007/s10142-025-01538-9
Shelp BJ, Mullen RT, Waller JC (2012a) Compartmentation of GABA metabolism raises intriguing questions. Trends Plant Sci 17:57–59
Shelp BJ, Bozzo GG, Trobacher CP, Zarei A, Deyman KL, Brikis CJ (2012b) Hypothesis/review: contribution of putrescine to 4-aminobutyrate (GABA) production in response to abiotic stress. Plant Sci 193–194:130–135
Shelp BJ, Bown AW, Zarei A (2018) 4-aminobutyrate (GABA): a metabolite and signal with practical significance. Botany 95:11
Shimajiri Y, Oonishi T, Ozaki K, Kainou K, Akama K (2013) Genetic manipulation of the γ-aminobutyric acid (GABA) shunt in rice: overexpression of truncated glutamate decarboxylase (GAD2) and knockdown of γ-aminobutyric acid transaminase (GABA-T) lead to sustained and high levels of GABA accumulation in rice kernels. Plant Biotechnol J 11:594–604
Su A, Yu Q, Luo Y, Yang J, Wang E, Yua H (2021) Metabolic engineering of microorganisms for the production of multifunctional non-protein amino acids: γ-aminobutyric acid and δ-aminolevulinic acid. Microb Biotechnol 14:2279–2290
Takayama M, Ezura H (2015) How and why does tomato accumulate a large amount of GABA in the fruit? Front Plant Sci 6:612
Tavladoraki P, Cona A, Angelini R (2016) Copper-containing amine oxidases and FAD-dependent polyamine oxidases are key players in plant tissue differentiation and organ development. Front Plant Sci 7:824
Torrigiani P, Scoccianti V, Bagni N (2010) Polyamine oxidase activity and polyamine content in maize during seed germination. Physiol Plantarum 74:427–432
Xing SG, Jun YB, Hau ZW, Liang LY (2007) Higher accumulation of gamma-aminobutyric acid induced by salt stress through stimulating the activity of diamine oxidases in Glycine max (L.) Merr. roots. Plant Physiol Biochem 45:560–566
Yang R, Guo Q, Gu Z (2013) GABAshunt and polyamine degradation pathway on-aminobutyric acid accumulation in germinating fava bean (Vicia faba L.) under hypoxia. Food Chem 136:152–159
Yang R, Feng L, Wang S, Yu N, Gu Z (2016) Accumulation of-aminobutyric acid in soybean by hypoxia germination and freeze–thawing incubation. J Sci Food Agric 96:2090–2096

Yin Y, Cheng C, Fang W (2018) Effects of the inhibitor of glutamate decarboxylase on the development and GABA accumulation in germinating fava beans under hypoxia-NaCl stress. RSC Adv 8:20456–20461

Yogeswara IBA, Maneerat S, Haltrich D (2020) Glutamate decarboxylase from lactic acid bacteria – a key enzyme in GABA synthesis. Microorganisms 8:1923

Yuan D, Wu X, Gong B, Huo R, Zhao L, Li J, Lü G, Gao H (2023) GABA metabolism, transport and their roles and mechanisms in the regulation of abiotic stress (hypoxia, salt, drought) resistance in plants. Meta 13(3):347. https://doi.org/10.3390/metabo13030347. PMID: 36984787; PMCID: PMC10051757

Zhang M, Liu Z, Fan Y, Liu C, Wang H, Li Y, Xin Y, Gai Y, Ji X (2022) Characterization of GABA-transaminase gene from mulberry (*Morus multicaulis*) and its role in salt stress tolerance. Genes (Basel) 13(3):501. https://doi.org/10.3390/genes13030501

Zhou L, Shen B, Bai S, Liu X, Jiang L, Zhai H, Wan J (2015) RNA interference of OsGABA-T1 gene expression induced GABA accumulation in rice grain. Acta Agron Sin 41:1305–1312. https://doi.org/10.3724/SP.J.1006.2015.01305

MIX
Papier aus verantwortungsvollen Quellen
Paper from responsible sources
FSC® C105338

If you have any concerns about our products,
you can contact us on
ProductSafety@springernature.com

In case Publisher is established outside the EU,
the EU authorized representative is:
Springer Nature Customer Service Center GmbH
Europaplatz 3, 69115 Heidelberg, Germany

Printed by Libri Plureos GmbH
in Hamburg, Germany